Finite Mathematics
SECOND EDITION

The Smith Business Series

This book is part of the Smith Business Series of textbooks, which includes:

Business Mathematics, Second Edition.
This is an arithmetic-based business mathematics textbook. Published by Wm. C. Brown.

Mathematics for Business Students.
This is an algebra-based business mathematics textbook. Published by Wm. C. Brown.

Finite Mathematics, Second Edition.
This is a standard finite mathematics textbook. Published by Brooks/Cole.

Calculus with Applications.
This is a business calculus textbook for students in management or the life or social sciences. Published by Brooks/Cole.

College Mathematics and Calculus with Applications to Management, Life, and Social Sciences.
This textbook combines the material in the finite mathematics and business calculus books. Published by Brooks/Cole.

Computer Aided Finite Mathematics and Calculus.
This text was written by Chris Avery and Charles Barker to accompany books in this mathematics series. Published by Brooks/Cole.

Other Brooks/Cole Titles by Karl J. Smith

Mathematics: Its Power and Utility, Second Edition

The Nature of Mathematics, Fifth Edition

Essentials of Trigonometry, Second Edition

Trigonometry for College Students, Fourth Edition

Precalculus Mathematics, Fifth Edition

Algebra and Trigonometry

The Smith and Boyle Precalculus Series Published by Brooks/Cole

Beginning Algebra for College Students, Third Edition

Intermediate Algebra for College Students, Third Edition

Study Guide for Intermediate Algebra for College Students, Third Edition

College Algebra, Third Edition

Finite Mathematics

SECOND EDITION

Karl J. Smith

 Brooks/Cole Publishing Company
Pacific Grove, California

Brooks/Cole Publishing Company   A Division of Wadsworth, Inc.

Printed in the United States of America

10   9   8   7   6   5   4   3   2   1

Library of Congress Cataloging-in-Publication Data

Smith, Karl J.
    Finite mathematics/Karl J. Smith—2nd ed.
        p.   cm.

    Includes bibliographical references and index.
    ISBN 0-534-08904-6
    1. Mathematics—1961–   I. Title.
QA39.2.S59   1988   87-19401   CIP

510—dc19

Sponsoring Editor: Jeremy Hayhurst
Editorial Assistant: Maxine Westby
Production Services Coordinator: Joan Marsh
Production: Cece Munson, The Cooper Company
Manuscript Editor: Betty Berenson
Interior Design: Jamie Sue Brooks
Cover Design: Katherine Minerva
Cover Photo: Lee Hocker
Technical Illustration: Carl Brown
Typesetting: Polyglot Pte Ltd, Singapore
Cover Printing: Phoenix Color Corp.
Printing and Binding: R. R. Donnelley & Sons, Harrisonburg, Virginia

This book is dedicated, with love, to my son,
Shannon J. Smith

Preface

Finite Mathematics provides the noncalculus mathematics background necessary for students in business, management, or the life or social sciences. Emphasis throughout is to enhance students' understanding of the modeling process and how mathematics is used in real world applications. The prerequisite for this course is intermediate algebra.

It seems as if every new book claims innovation, state-of-the-art production, supplementary materials, readability, abundant problems, and relevant applications. How, then, is one book chosen over another? More specifically, how does this book differ from other books for this course, and what factors were taken into consideration as it was written?

Content

First, every book must cover the appropriate topics, hopefully in the right order. Finite mathematics has evolved and changed considerably since it was first introduced in the 1960s. This text focuses on matrices to solve linear systems of equations and inequalities (linear programming), sets, combinatorics, and probability, and some supplementary topics such as Markov chains, game theory, and mathematics of finance. New recommendations regarding discrete mathematics are also influencing finite mathematics courses today, so I have included appendices on logic and mathematical induction.

The primary skill addressed in this book is that of problem solving and building a mathematical model. Most books simply pay lip service to building models, but rarely *develop* the skill. In this book, an entire section is devoted to the nature of a linear programming problem and how to formulate an appropriate model (see Section 3.2). In addition, the modeling applications at the end of each chapter illustrate the model-building process in real life situations. These applications are open-ended assignments that require a mathematical model-building approach for their development. An essay written in response to each of these applications is given in its entirety in the *Student's Solutions Manual* to illustrate how model building can be developed. These model-building applications, even if not assigned, demonstrate, in a very real way, how the material developed in the rest of the chapter can be used to answer some nontrivial questions (see, for example, the Modeling Application on Ecology at the end of Chapter 4).

The text is divided into sections of nearly equal size that each take about one class day to develop. Since there are 55 numbered sections (including 17 optional sections), there is ample opportunity to select material that tailors the book to individual classes. The interdependence of the chapters is shown below:

Chapter	Prerequisite Chapters	
1	None	Chapter 1 may be
2	1	assumed if the
3	1	students have had a
4	1, 3	recent intermediate
5	1	algebra course
6	1, 5	
7	1, 5, 6	
8	1, 2, 6	
9	1, 2, 6	
10	1	

Style

An author's writing style also distinguishes one textbook from another. My writing style is informal, and I always write with the student in mind. I offer study hints along the way and let the students know what is important. Frequent and abundant examples are provided so that students can understand each step before proceeding to the next. A second color is used to highlight important steps or particular parts of an equation or formula. The chapter reviews list important terms and provide review problems for the material covered in the chapter. Together they emphasize the important ideas in the course.

Problems

The third, and one of the most important factors in deciding on a textbook, is the number—and quality—of the problems presented. This is where I have spent a great deal of effort in developing this book. Problems should help to develop students' understanding of the material, and not inhibit or thwart that understanding by being obscure. Problems are presented here in matched pairs with an answer for one provided at the back. There are about three times more problems than are needed for assignments, so students have the opportunity to practice additional problems, both for the midterm and for the final. (There are almost 3000 problems in this text.) Each problem set presents drill problems to develop manipulative skills and a large number of applications to show how the material can be used in business, management, and the life and social sciences. The types of problems include:

1. *Drill.* There are a large number of drill problems that provide adequate practice for the student to develop a clear understanding of each topic.
2. *Applications.* These are self-contained problems that provide relevance and practicality for the topic at hand.
3. *Modeling applications.* Each chapter concludes with an optional real life application that allows the material to be applied to a *real* (rather than a textbook) problem, but at a level of difficulty that is manageable for the student.
4. *CPA, CMA, and Actuary exam questions.* Actual questions from Actuary, CPA, and CMA exams are scattered throughout the textbook. These test questions provide a link between textbook and profession.
5. *Historical questions.* Historical notes provide insight into the humanness of

mathematics, but instead of being superfluous commentaries, these are integrated into the problem sets to give students a taste of some of the ways the topics were originally developed as solutions to mathematical problems.

6. *Chapter review problems.*   Each chapter ends with a sample test to make it clear to the student what specific skills from the chapter need to be mastered.

Supplementary Materials

The final factor often used in selecting a textbook is the type and quality of available supplements. In addition to the answers in the back of the book, several supplements are available:

1. *Student's Supplement:*   This provides complete solutions to the odd-numbered problems in the book. It also includes essays to illustrate modeling and lists of objectives that define the necessary skills studied in each chapter.
2. *Instructor's Supplement:*   This includes answers to all of the problems in the book. It also has additional questions to accompany the modeling applications.
3. *Testing program:*   There is a computerized test bank with text-editing capabilities that allows you to create an almost unlimited number of tests or retests of the material. This test bank is also available in printed form for instructors without access to a computer.
4. *Computer Aided Finite Mathematics:*   This is a computer supplement prepared by Chris Avery and Charles Barker. Appendix D lists the programs available in this supplement.

As the author, I am also available to help you create any other set of supplementary materials that you feel are necessary or worthwhile for your course.

Acknowledgments

The production of a textbook is a team effort. I would like to thank Cece Munson, Joan Marsh, Carl Brown, Jamie Sue Brooks, Katherine Minerva, and Jeremy Hayhurst for their extraordinary effort and help in producing this book.

The accuracy of a mathematics textbook is very important, and I am especially grateful for the checking of all of the examples and problems that I received from Terry Shell, Pat Bannantine, Michael Anderson, and Donna Szott. Their meticulous checking, as well as their numerous suggestions, is greatly appreciated.

The reviewers of the manuscript have also offered many valuable suggestions, and I would like to offer each of them my sincere thanks.

Craig Benham
    University of Kentucky
Thomas Covington
    Northwestern State University
Joe S. Evans
    Middle Tennessee State University
Matthew Gould
    Vanderbilt University

Edwin Klein
    University of Wisconsin
Jacqueline Payton
    Virginia State University
Roland Sink
    Pasadena City College
Donald Zalewski
    Northern Michigan University

And last, but not least, my thanks go to my family, Linda, Missy, and Shannon, for their love, support, and continued understanding of my involvement in writing this book.

Karl J. Smith
Sebastopol, California

Contents

    * Optional sections

CHAPTER 1
Linear Models

APPLICATIONS

Management (*Business, Economics, Finance, and Investments*)

Cost analysis (1.1, Problems 43–48; 1.6, Problem 1)
Income taxes (1.1, Problem 51; 1.2, Problem 51)
Depreciation (1.2, Problems 52–54)
Scrap value (1.2, Problem 55)
Demand equation (1.3, Problem 94)
Supply equation (1.3, Problem 95)
Cost equations (1.3, Problems 101–102)
Supply and demand (1.4, Problems 41–46)
Break-even analysis (1.4, Problems 47–53)
Value of a stock portfolio (1.4, Problems 54–55)

Life sciences (*Biology, Ecology, Health, and Medicine*)

Determining temperature (1.1, Problems 39–42)
Cost of a multivitamin (1.1, Problem 49; 1.2, Problem 49)
Mixing chemicals to obtain proper grades of fertilizer (1.4, Problem 57)

Social sciences (*Demography, Political Science, Population, Psychology, Society, and Sociology*)

Predicting populations (1.3, Problems 96–97)
Voter demographics (1.4, Problem 56)
Louvre Tablet (1.4, Problem 58)

General interest

Model for a free-falling object (1.1, Problems 29–38)
Modeling as discussed in the *American Mathematical Monthly* (1.1, Problem 52)
Descartes' discovery of a coordinate system (1.2, Problem 56)
Best price for a car rental (1.3, Problems 98–99; 1.4, Problems 37–40)
Cost of operating a car (1.3, Problem 100)

Modeling application—Gaining a Competitive Edge in Business

CHAPTER
OVERVIEW
Here we learn what is meant by a mathematical model, which forms the foundation for much of the remaining material in the text. Mathematical modeling, in short, is the application of mathematics to real life situations.

PREVIEW
We begin by defining and discussing a mathematical model and the very fundamental unifying idea of a set. Next, the concepts of formulas, coordinate systems, lines, and systems of equations are reviewed, forming the foundation on which this course is built.

PERSPECTIVE
Many of the preliminary ideas needed for this course are introduced in this chapter, such as working with formulas, graphs, lines, and systems of equations. We will be able to apply these ideas to a variety of applied problems as we progress through this book.

1.1   Mathematical Modeling

We are about to begin the study of finite mathematics, specifically organized for students in management, life sciences, or social sciences. This is a recent course in mathematics that was first developed in the 1960s in the now classic book by Kemeney, Snell, and Thompson by that name. Over the years, the course has evolved to include the topics which are now included in this text, but *finite mathematics* is not well defined in the sense that algebra or calculus is well defined. We might say that it is everything that is not included in calculus. But, more specifically, finite mathematics involves *applying* the elementary mathematics of sets, matrices, linear programming, probability, and statistics to real life problems.

A real life situation does not easily lend itself to mathematical analysis because the real world is far too complicated to be precisely and mathematically defined. It is therefore necessary to develop what is known as a **mathematical model**. The model is based on certain assumptions about the real world and is modified by experimentation and accumulation of data. It is then used to predict some future occurrence in the real world. A mathematical model is not static or unchanging but is continually being revised and modified as additional relevant information becomes known.

Some mathematical models are quite accurate, particularly in the physical sciences. For example, the path of a projectile, the distance that an object falls in a vacuum, or the time of sunrise tomorrow all have mathematical models that provide very accurate predictions about future occurrences. On the other hand, in the fields of social science, psychology, and management, models provide much less accurate predictions because they must deal with situations that are often random in character. It is therefore necessary to consider two types of models:

Types of Models

A **deterministic model** predicts the exact outcome of a situation because it is based on certain known laws.

A **probabilistic model** deals with situations that are random in character and can predict the outcome within a certain stated or known degree of accuracy.

How do we construct a model? We need to observe a real world problem and make assumptions about the influencing factors. This is called *abstraction*.

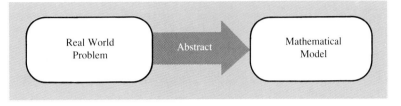

You must know enough about the mechanics of mathematics to *derive results* from the model.

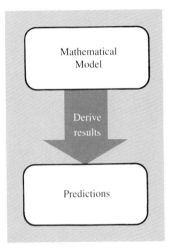

The next step is to gather data. Does the prediction given by the model fit all the known data? If not, we use the data to *modify* the assumptions used to create the model. This is an ongoing process.

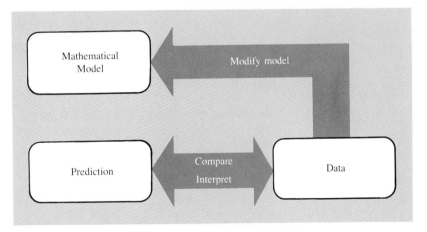

We begin this course by reviewing some mathematics.

Sets and Numbers

One of the most fundamental ideas in mathematics is the idea of a **set**. The objects of a set S are called the *elements*, or *members*, of S. A set with no elements is called the *empty*, or *null*, set and is denoted by the symbol $\varnothing$. The set

$$\{\ldots, -3, -2, -1, 0, 1, 2, 3, \ldots\}$$

is called the **set of integers**. Certain *subsets* of this set are

$$\left.\begin{array}{l} Natural\ numbers \\ Counting\ numbers \\ Positive\ integers \end{array}\right\} \quad \{1, 2, 3, \ldots\}$$

$$Negative\ integers \qquad \{-1, -2, -3, \ldots\}$$

$$\left.\begin{array}{l} Nonnegative\ integers \\ Whole\ numbers \end{array}\right\} \quad \{0, 1, 2, 3, \ldots\}$$

Another important set of numbers is the set of **rational numbers**, Q, consisting of all *quotients* of integers:

$$Q = \left\{ \frac{p}{q} \;\middle|\; p \text{ is an integer and } q \text{ is a nonzero integer} \right\}$$

The above line is read "Q is the set of all $\frac{p}{q}$ such that p is an integer and q is a nonzero integer." The notation used is sometimes called **set-builder notation**. Examples of rational numbers are

$$\tfrac{2}{3}, \quad \tfrac{-1}{2}, \quad \tfrac{15}{4}, \quad 5, \quad -2, \quad 0, \quad \tfrac{36}{11}, \quad \ldots$$

Note that each integer is also a rational number, since, for example, $5 = \frac{5}{1}$, the quotient of two integers.

We also use *decimal representation* for rational numbers. Every rational number can be represented as a terminating decimal or as a repeating decimal:

Terminating decimals:	0.5,	0.75,	0.006333, ...
Repeating decimals:	0.555...,	0.757575...,	0.006333..., ...

A number whose decimal representation does not terminate or repeat is called an **irrational number**. Examples of irrational numbers are π, $\sqrt{2}$, $\sqrt{3}$, $\sqrt{5}$, $\sqrt{6}$, $\sqrt{7}$, $\sqrt{8}$, $\sqrt{10}, \ldots$.

If we put together all the rational numbers and all the irrational numbers, the resulting set is called the set of **real numbers**. The real numbers can most easily be visualized by using a *one-dimensional coordinate system* called the **real number line** (see Figure 1.1).

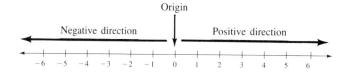

Figure 1.1 A real number line

A *one-to-one correspondence* is established between all real numbers and all points on a real number line:

1. Every point on the line corresponds to precisely one real number.
2. Every real number corresponds to precisely one point.

A point associated with a particular number is called the **graph** of that number.

EXAMPLE 1 Graph the following numbers on a real number line: $4, -3, 2.5, 1.313311333111\ldots$, $\frac{2}{3}, \pi, -\sqrt{2}$.

Solution When graphing, the exact positions of the points are approximated.

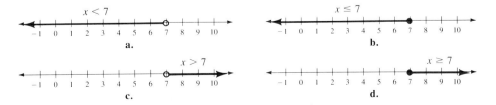

Linear Relationships

There are certain relationships between real numbers with which you need to be familiar:

Less than $a < b$, read "a is less than b," means the graph of a is to the left of the graph of b.

Greater than $a > b$, read "a is greater than b," means the graph of a is to the right of the graph of b.

Equal to $a = b$, read "a is equal to b," means that a and b represent the same point on the number line.

Less than or Equal to $a \le b$, read "a is less than or equal to b," means that either $a < b$ or $a = b$ (but not both).

Greater than or Equal to $a \ge b$, read "a is greater than or equal to b," means that either $a > b$ or $a = b$ (but not both).

Between $b < a < c$, read "a is between b and c," means *both* $b < a$ and $a < c$. Additional "between" relationships are also used:

$b \le a \le c$ means $b \le a$ and $a \le c$.

$b \le a < c$ means $b \le a$ and $a < c$.

$b < a \le c$ means $b < a$ and $a \le c$.

Graphs of inequality statements are also drawn on a one-dimensional coordinate system. For example, $x < 7$ denotes the interval shown in Figure 1.2a. Since $x \ne 7$, this fact is shown by an open circle as the end point of the graph. In Figure 1.2b, compare this with $x \le 7$; the end point $x = 7$ is included, as indicated by the solid dot. To sketch the graph, we darken (or color) the appropriate portion. Graphs for $x > 7$ and $x \ge 7$ are drawn in a similar way (Figures 1.2c and 1.2d).

Figure 1.2 Graphs of inequality statements

Some Models

In the first part of this text we focus on deterministic models for which the abstraction step has already been completed. Models are often stated in terms of formulas. A **formula** is an equation or other relationship given in mathematical symbols that can be applied to some specific situation. For example, the formula

$$d = 16t^2$$

is used to predict the distance an object will fall in a vacuum as a function of time. A **function** is a rule that assigns to each number of one set, called the **domain**, exactly one number from another set, called the **range**. In this example, for each nonnegative value of time (t) there is exactly one value for the distance (d) the object falls. The variable t is called the **independent variable**, and d is called the **dependent variable**. The formula for a free-falling object uses an **exponent**.

Exponential Notation

If b is any real number and n is a positive integer, then

$$b^n = \underbrace{b \cdot b \cdot b \cdot \cdots \cdot b}_{n \text{ factors}}$$

$$b^0 = 1 \qquad b \neq 0$$

$$b^{-n} = \frac{1}{b^n} \qquad b \neq 0$$

EXAMPLE 2   **Model for a Free-Falling Object**   A rock is dropped from the top of a canyon. How far does it fall in 1 second? In 2 seconds? If it hits bottom in 4 seconds, how high is the point from which it was dropped?

Solution   The formula for this model was found through experimentation. We assume that air resistance is negligible and that the formula $d = 16t^2$ is correct, where d is the distance in feet and t is the time in seconds.

For 1 second, let $t = 1$ and *evaluate the formula*:

$$d = 16 \cdot 1^2$$
$$= 16$$

The rock falls 16 feet in 1 second.

For 2 seconds, we let $t = 2$:

$$d = 16 \cdot 2^2$$
$$= 16 \cdot 4$$
$$= 64$$

The rock falls 64 feet in 2 seconds.

For 4 seconds, we let $t = 4$:

$$d = 16 \cdot 4^2$$
$$= 16 \cdot 16$$
$$= 256$$

The rock was dropped from a height of 256 feet.   ∎

EXAMPLE 3   If a ball is dropped from a 100-foot tower, how long will it take to hit the ground (neglecting air resistance)?

Solution   The formula for this model is the same as that given in Example 2:

$$d = 16t^2$$

Let $d = 100$: Then

$$100 = 16t^2$$

This is a **quadratic equation**.

$$16t^2 - 100 = 0$$
$$(4t - 10)(4t + 10) = 0$$
$$t = \frac{10}{4}, -\frac{10}{4}$$

The negative value of time is meaningless for this model, so $t = \frac{10}{4}$ or $2\frac{1}{2}$ seconds.

■

EXAMPLE 4   **Model from Biology**   It has been noticed that the rate at which certain crickets chirp depends on the temperature. Build a model to answer the following questions: What is the temperature when 20 chirps are counted in 15 seconds? When 50 chirps are counted? If the temperature is 20°C, how many times will the cricket chirp in 1 minute?

Solution   To build a model, many observations must be made and much data gathered. We will assume that this work has been completed and that the formula for the temperature in degrees Celsius (C) is given as a function of the number of chirps (n) in 15 seconds:

$$C = 0.6n + 4$$

For 20 chirps, we let $n = 20$ and *evaluate the formula*:

$$C = 0.6(20) + 4 = 16$$

The temperature is 16°C.
    For 50 chirps, we let $n = 50$:

$$C = 0.6(50) + 4 = 34$$

The temperature is 34°C.
    For 20°C, we let $C = 20$:

$$20 = 0.6n + 4$$
$$16 = 0.6n \qquad \text{Divide both sides by 0.6 and simplify}$$
$$\frac{80}{3} = n$$

If $\frac{80}{3}$ chirps are heard in 15 seconds, then there will be $\frac{80}{3}(4) = 106\frac{2}{3}$ chirps in 1 minute. Since we cannot count $\frac{2}{3}$ of a chirp, the answer is 106 chirps in 1 minute at 20°C.

■

Do you see why we did not round to 107 chirps for the last answer in Example 4? In general, *never round off in the middle of the problem*, only when stating the final answer.

Cost Analysis

In business, costs and prices can be analyzed over a long period or a short period. **Short-run analysis** has traditionally referred to a time period over which costs and prices remain constant. Over longer periods, economic factors such as inflation or supply and demand tend to influence costs and prices. The cost of every manufacturing process can be divided into fixed and variable costs. Certain costs, such as rent, taxes, insurance, and utilities, exist even if no product is actually manufactured. These are called **fixed costs** and usually remain constant over the short run. Other costs, such as materials, labor, and distribution, depend directly on the number of items actually produced. These are called **variable costs** and increase as more items are produced. The *total cost* can be given by $C = ax + b$, where a represents the variable costs and b represents the fixed costs over the short-run manufacture of x items.

EXAMPLE 5   If the fixed costs for a certain item total $2,100 and the variable costs total $.80 per item, what is the cost of producing 2000 items? 5000 items? How many items can be produced for $10,000?

Solution   In the formula $C = ax + b$, we let $a = 0.8$ and $b = 2100$ to obtain

$$C = 0.8x + 2100$$

For $x = 2000$:

$$C = 0.8(2000) + 2100 = 3700$$

It costs $3,700 to produce 2000 items.
For $x = 5000$:

$$C = 0.8(5000) + 2100 = 6100$$

It costs $6,100 to produce 5000 items.
If the cost is $10,000, then

$$10{,}000 = 0.8x + 2100$$
$$7900 = 0.8x$$
$$9875 = x$$

The company could produce 9875 items for a cost of $10,000.   ■

EXAMPLE 6   Suppose each item in Example 5 sells for $1.50. The **revenue** (R) is the amount collected and depends only on the price and number of items sold (x), according to the formula

$$R = px$$

If $p = 1.50$, what is the revenue for 2000 items? For 5000 items? How many items must be sold for a revenue of $10,000?

Solution   We use the formula $R = 1.5x$.
For $x = 2000$:

$$R = 1.5(2000) = 3000$$

The revenue is $3,000 for 2000 items.
For $x = 5000$:

$$R = 1.5(5000) = 7,500$$

The revenue is $7,500 for 5000 items.
For $R = 10,000$:

$$10,000 = 1.5x$$
$$6,666\tfrac{2}{3} = x$$

Since the domain permits only positive integers, the company must sell 6667 items to have $10,000 in revenue.   ∎

Problem Set 1.1

Answers to the odd-numbered problems are given on page 403.

1. What is a mathematical model?

2. Why are mathematical models necessary or useful?

3. Graph the following numbers on a real number line: $2, -\frac{3}{2}, 1.923, -0.1212212221\ldots, \pi/2, \sqrt{3}$.

4. Graph the following numbers on a real number line: $-2.5, -\frac{9}{4}, -0.05, -\sqrt{3}/2, -\pi/3, -0.9090909\ldots$.

In Problems 5–16 use the definition of exponent to write each expression in simplest exponent form.

5. xxx

6. $xxxxx$

7. $\dfrac{1}{x}$

8. $\dfrac{1}{xx}$

9. $\dfrac{1}{xxxx}$

10. $\dfrac{1}{xxx}$

11. $2xx$

12. $(2x)(2x)$

13. $-xx$

14. $(-x)(-x)$

15. $-2 \cdot 2$

16. $(-2)(-2)$

Use the definition of exponent to write each expression in Problems 17–28 without exponents.

17. x^2

18. x^4

19. x^3

20. x^0

21. x^{-2}

22. x^{-3}

23. x^{-4}

24. x^7

25. $-x^2$

26. $(-x)^2$

27. $-x^4$

28. $(-x)^4$

APPLICATIONS

Model for a Free-Falling Object *Use the model given in Examples 2 and 3 for Problems 29–38.*

29. How far would an object fall in 3 seconds?

30. How far would an object fall in 10 seconds?

31. How far would an object fall in the second second?

32. How far would an object fall in the third second?

33. If a ball is dropped from a 49-foot tower, how long will it take to hit the bottom?

34. If a rock is dropped into a well 64 feet deep, how long will it take to hit the bottom?

35. If a ball is dropped from a 32-foot tower, how long will it take (to the nearest second) to hit the bottom?

36. If a rock is dropped into a well 80 feet deep, how long (to the nearest second) will it take to hit the bottom?

37. Repeat Problem 35 for a 75-foot tower.

38. Repeat Problem 36 for a well 120 feet deep.

Model from Biology *Use the model given in Example 4 for Problems 39–42.*

39. If 10 chirps are counted in 15 seconds, what is the Celsius temperature (to the nearest degree)?

40. If 10 chirps are counted in 15 seconds, what is the Fahrenheit temperature (to the nearest degree)?

41. If the temperature is 40°C, how many cricket chirps would be heard in 1 minute?

42. If the temperature is 40°F, how many cricket chirps would be heard in 1 minute?

Cost Analysis *Use the model given in Examples 5 and 6 for Problems 43–48.*

43. A manufacturer has variable costs of $8.50 and fixed costs of $3,600. What is the cost of producing 10 items?

44. A manufacturer has variable costs of $8.50 and fixed costs of $3,600. What is the cost of producing 1000 items?

45. A manufacturer has variable costs of $8.50 and fixed costs of $3,600. What is the cost of producing 10,000 items?

46. If the item described in Problems 43–45 is to be sold for $17.50, what is the revenue for 100 items?

47. If the item described in Problems 43–45 is to be sold for $17.50, what is the revenue for 1000 items?

48. If the item described in Problems 43–45 is to be sold for $17.50, what is the revenue for 10,000 items?

Problems 49–51 deal with models taken from several different disciplines.

49. The cost of a multivitamin depends on the weight purchased according to the formula $C = 3w$, where C is the cost in dollars and w is the weight in grams. Give the cost for each of the following weights: $w = 2$; $w = 10$; $w = 30$.

50. The admission price for a performance of *Cats* is $30 for adults and $15 for students. Since the capacity of the theater is 700 seats, the formula relating the number of adults (a) and the number of students (s) in the theater is $a + s = 700$. Find the number of student admissions for the following numbers of adult admissions: $a = 50$; $a = 100$; $a = 210$.

51. The 1986 regulations of the Internal Revenue Service allow taxpayers a $1,080 deduction for each dependent and a 15% standard deduction on the amount of income. For a person with income I and four dependents (including herself), the amount A to be taxed (taxable income) can be found by the formula

$$A = \text{Income} - \overset{\text{dependent}}{\text{deduction}} - \overset{\text{standard}}{\text{deduction}}$$
$$= I - 4(1080) - 0.15I$$
$$= 0.85I - 4320$$

In 1986, if a taxpayer's income was between $26,550 and $32,270, the tax was approximately 16% of taxable income. If T represents the amount of tax, then

$$T = 0.16A = 0.16(0.85I - 4320)$$

Find T for the following amounts of income: $I = \$26,550$; $I = \$30,000$; $I = \$32,270$

52. **Historical Question** The notion of a mathematical model is not new. More than 20 years ago, John Synge wrote an article for the Mathematical Education Notes section of the October 1961 issue of *The American Mathematical Monthly* (Vol. 68, p. 799). He described the modeling process as consisting of three stages:

1. A dive from the world of reality into the world of mathematics.
2. A swim in the world of mathematics.
3. A climb from the world of mathematics back to the world of reality, carrying a prediction in our teeth.

Relate these stages to the process of mathematical modeling described in this section.

1.2 Cartesian Coordinates and Graphs

The mathematical models we just considered relate two variables via a formula. When making predictions, it is extremely beneficial to draw a picture or graph of these relationships. A **two-dimensional coordinate system** can be introduced by considering two perpendicular coordinate lines in a plane. Usually, one of the coordinate lines is horizontal with the positive direction to the right; the other is vertical with the positive direction upward. These coordinate lines are called **coordinate axes**, and the point of intersection is called the **origin**. Note in Figure 1.3

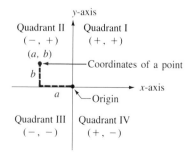

Figure 1.3 Cartesian coordinate system

that the axes divide the plane into four parts called **quadrants I, II, III, IV**. This two-dimensional coordinate system is also called a **Cartesian coordinate system** in honor of René Descartes (1596–1650), who first described a coordinate system in mathematical detail.

Points of a plane are denoted by **ordered pairs**, two real numbers represented by (a, b), where a is called the **first component**, or **coordinate**, and b is the **second component** or **coordinate** (see Figure 1.3). The order in which the components are listed is important since $(a, b) \neq (b, a)$ if $a \neq b$.

Usually the horizontal number line is called the **x-axis** (sometimes called the *axis of abscissas*), and x represents the first component of the ordered pair. The vertical number line is called the **y-axis** (sometimes called the *axis of ordinates*), and y represents the second component of the ordered pair. The plane determined by the x- and y-axes is called a **coordinate plane, Cartesian plane**, or **xy-plane**. When we refer to a point (x, y), we are referring to a point in the coordinate plane whose abscissa is x and whose ordinate is y. To **plot a point** (x, y) means to locate the point with coordinates (x, y) in the plane and represent its location by a dot.

To **graph** a relation means to draw a picture of the ordered pairs that satisfy the equation in a one-to-one fashion. This process of graphing is shown for two of the models of Section 1.1 in Examples 1 and 2 below. (Program 1 on the computer disk accompanying this book plots linear equations as well as giving a table of values to help you with this process.)

EXAMPLE 1 **Model from Biology** The model for determining the Celsius temperature by counting the chirps of a cricket is $C = 0.6n + 4$. Graph this relationship.

Solution We choose values for n and then find the corresponding values for C. Note that the model requires the domain to be $n \geq 0$ and that the ordered pairs are represented as (n, C). Let

$n = 0;$	$C = 0.6(0) + 4 = 4$	The ordered pair is $(0, 4)$
$n = 5;$	$C = 0.6(5) + 4 = 7$	$(5, 7)$
$n = 10;$	$C = 0.6(10) + 4 = 10$	$(10, 10)$
$n = 20;$	$C = 0.6(20) + 4 = 16$	$(20, 16)$

We plot these points as shown in the figure, along with as many others as we wish. The points in the figure seem to lie on a line. If both variables are first degree, the graph is a line.

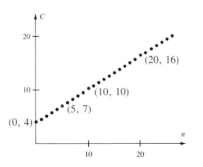

Graph of $C = 0.6n + 4$, where C is the degrees Celsius and n is the number of cricket chirps in 15 seconds

EXAMPLE 2   **Model for a Free-Falling Object**   The model for a free-falling object is given by the equation $d = 16t^2$. Graph this relationship.

Solution   Note that this model requires $t \geq 0$, and the ordered pairs are represented as (t, d). We let

$$t = 0; \quad d = 16 \cdot 0^2 = 0 \qquad \text{The ordered pair is } (0, 0)$$
$$t = 1; \quad d = 16 \cdot 1^2 = 16 \qquad\qquad\qquad\quad (1, 16)$$
$$t = 2; \quad d = 16 \cdot 2^2 = 64 \qquad\qquad\qquad\quad (2, 64)$$
$$t = 3; \quad d = 16 \cdot 3^2 = 144 \qquad\qquad\qquad\; (3, 144)$$
$$t = 4; \quad d = 16 \cdot 4^2 = 256 \qquad\qquad\qquad\; (4, 256)$$

Plot these points. Note that, for convenience, we have chosen different scales in the figure for the axes. These points do not lie on a line. If the points are connected, you see a portion of a curve called a **parabola**.

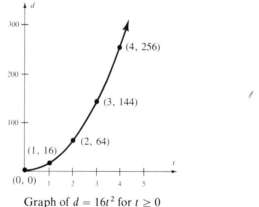

Graph of $d = 16t^2$ for $t \geq 0$

In the remainder of this chapter we focus on equations whose graphs are *lines* or *parts of lines*. These equations have two first-degree variables. Such equations are called **linear equations**.

Plotting points can be used to determine the graph of almost any equation, but as a general method it is too time consuming. Instead, we can employ certain characteristics of the line to simplify our work:

1. A **line** is determined by two points; that is, if you know any two points on a line, you can draw the line by using a straightedge and drawing a line that passes through the two known points.
2. The **x-intercept** is the x-value where a graph crosses the x-axis.
3. The **y-intercept** is the y-value where a graph crosses the y-axis.
4. The **slope** of a line is the steepness of the line. (We discuss the slope of a line in the next section.)

In this section we will make use of the first property of linear equations, namely, using two points on a given line to determine the graph. You can find *any* two points, but frequently a line may be specified in **standard form** and you will want to choose values to simplify the amount of required arithmetic.

Standard Form of the
Equation of a Line

The *standard form equation of a line* is

$$Ax + By + C = 0$$

where (x, y) is any point on the line; A, B, and C are constants (A and B not both zero).

The points on a line that are generally the easiest to find are the intercepts described by properties 2 and 3 above.

To find the y-intercept: Let $x = 0$ and solve for y.
To find the x-intercept: Let $y = 0$ and solve for x.

EXAMPLE 3   Graph $2x + 3y - 6 = 0$.

Solution   We let $x = 0$: $2 \cdot 0 + 3y - 6 = 0$

$$3y = 6$$
$$y = 2 \qquad \text{The point is } (0, 2).$$

Let $y = 0$: $2x + 3 \cdot 0 - 6 = 0$

$$2x = 6$$
$$x = 3 \qquad \text{The point is } (3, 0).$$

Draw the line passing through the plotted intercepts.

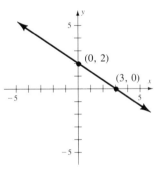

Graph of $2x + 3y - 6 = 0$

EXAMPLE 4   Graph $5x - 2y + 4 = 0$.

Solution   Let $x = 0$: $-2y + 4 = 0$

$$2y = 4$$
$$y = 2 \qquad \text{Plot } (0, 2).$$

Let $y = 0$: $5x + 4 = 0$

$$5x = -4$$
$$x = -\tfrac{4}{5} \qquad \text{Plot } (-\tfrac{4}{5}, 0).$$

If the points are not integers, you can often adjust the scale so that the points are easy to plot, as shown in the figure.

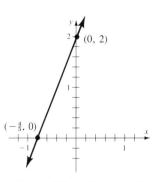

Graph of $5x - 2y + 4 = 0$

EXAMPLE 5   Graph $y - 3 = 0$.

Solution   Note that the given equation is equivalent to $y = 3$. This means that the second component must be 3; plot at least two points with the second component 3, as shown in the figure, and draw the line. The result is a **horizontal line**.

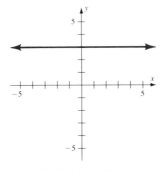

Graph of $y - 3 = 0$ ∎

EXAMPLE 6   Graph $x + 5 = 0$.

Solution   This equation is equivalent to $x = -5$, which means that the first component for all points on this line is -5. Plot at least two points and draw the line: it is a **vertical line**.

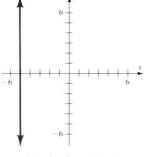

Graph of $x + 5 = 0$ ∎

Examples 5 and 6 illustrate two special cases of linear equations:

Horizontal Line
Vertical Line

A *horizontal line* passing through the point (h, k) has as its equation $y = k$.

A *vertical line* passing through the point (h, k) has as its equation $x = h$.

Depreciation

Many of the costs of doing business are deductible as business expenses, but the Internal Revenue Service does not allow the purchase price (also called the *cost* or *basis*) of an asset to be deducted in a single year. Instead, it has guidelines for determining the **useful life** of an asset, or the length of time it takes to use up an asset. At the end of its life, the asset either has no **salvage value** or it has a **scrap value**. The method by which the value of the asset is spread over its useful life is called

depreciation. There are several methods for calculating depreciation, but the easiest assumes that the original cost is used up equally over the life of the asset. This method is called **straight-line depreciation**.

Straight-Line Depreciation

> The *current value*, *y*, of an asset depreciated according to the *straight-line method* is given by
>
> $$y = \text{basis} - \left(\frac{\text{basis} - \text{scrap value}}{\text{useful life}} \right) x$$
>
> where *x* is the number of years that the asset has been depreciated. (The basis is the amount paid for the item.)

EXAMPLE 7   A delivery truck cost $11,000 and has an estimated life of 3 years with a scrap value of $2,000. Graph the current value.

Solution $y = 11,000 - \left(\dfrac{11,000 - 2000}{3} \right) x$

$= 11,000 - 3000x$

The easiest method of graphing this is to plot points.
If $x = 0$, then $y = 11,000$; and
if $x = 3$, then $y = 2000$,
so the line passes through the points $(0, 11,000)$ and $(3, 2000)$.

Note in the figure that the graph requires that the scales on the *x*- and the *y*-axes not be the same.

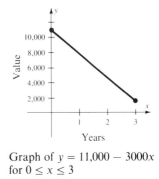

Graph of $y = 11,000 - 3000x$
for $0 \le x \le 3$

There are two other methods of depreciation often used in business: *double-declining balance* and *sum-of-the-years'-digits* methods. However, these methods do not have linear models, so they will not be discussed here.

Problem Set 1.2

Find the x- and y-intercepts for Problems 1–12.

1. $2x - 3y + 6 = 0$ **2.** $2x + 3y + 6 = 0$

3. $3x + 2y - 6 = 0$ **4.** $3x - 2y + 6 = 0$

5. $5x - 2y + 10 = 0$ **6.** $5x + 2y + 10 = 0$

7. $3x + 5y - 15 = 0$ **8.** $3x - 5y - 15 = 0$

9. $5x + 3y + 15 = 0$ **10.** $5x - 3y - 15 = 0$

11. $4x + 3y - 12 = 0$ **12.** $3x - 4y + 12 = 0$

Graph each of the linear equations in Problems 13–36.

13. $2x - 3y + 6 = 0$ **14.** $2x + 3y + 6 = 0$

15. $3x + 2y - 6 = 0$ **16.** $3x - 2y + 6 = 0$

17. $5x - 2y + 10 = 0$ **18.** $5x + 2y + 10 = 0$

19. $3x + 5y - 15 = 0$ **20.** $3x - 5y - 15 = 0$

21. $5x + 3y + 15 = 0$ **22.** $5x - 3y - 15 = 0$

23. $4x + 3y - 12 = 0$ **24.** $3x - 4y + 12 = 0$

25. $2x + 3y + 9 = 0$ **26.** $3x - 2y + 10 = 0$

27. $6x - 5y - 12 = 0$ **28.** $x + 5y + 3 = 0$

29. $2x - y + 3 = 0$ **30.** $5x - y + 6 = 0$

31. $x + 1 = 0$ **32.** $x - 4 = 0$ **33.** $x - 3 = 0$

34. $y - 5 = 0$ **35.** $y + 2 = 0$ **36.** $y + 1 = 0$

Graph each of the equations in Problems 37–48. Identify each as linear or nonlinear.

37. $2x - 3y + 8 = 0$ **38.** $4x - 3y + 5 = 0$

39. $y = \frac{2}{3}x + 5$ **40.** $y = -\frac{1}{2}x - 4$

41. $y = \frac{1}{4}x^2$ **42.** $y = -\frac{1}{3}x^2$

43. $y = 8x^2$

44. $y = 10x^2$

45. $xy = 1$

46. $xy = -1$

47. $y = 1 - \dfrac{1}{x}$

48. $y = 3 + \dfrac{1}{x}$

APPLICATIONS

49. The cost of a multivitamin depends on the weight purchased according to the formula $C = 3w$, where C is the cost in dollars and w is the weight in grams. Graph (w, C) for $0 \le w \le 10$.

50. The admission price for a performance of *Cats* is $30 for adults and $15 for students. Since the capacity of the theater is 700 seats, the formula relating the variables is $a + s = 700$, where a is the number of paid adult admissions and s is the number of paid student admissions. Graph (a, c) for $0 \le a \le 700$.

51. The taxable income for certain taxpayers is given by the formula $T = 0.16(0.85I - 4000)$. Graph (I, T) for $28{,}000 \le I \le 32{,}000$.

52. A heat-sensitive machine cost $32,000 and has a useful life of 6 years. If the scrap value is $2,000, graph the current value for the years 0 through 6, assuming straight-line depreciation.

53. A fourplex has a value of $250,000 with a useful life of 30 years (no scrap value). Graph the current value for years 0 through 30, assuming straight-line depreciation.

54. A new roof cost $8,000 and has a useful life of 8 years (no scrap value). Graph the current value for years 0 through 8, assuming straight-line depreciation.

55. A distributor bought a vending machine for $8,000. The machine has a probable scrap value of $200 at the end of its expected 10-year life. The value V at the end of n years is given by

$$V = 8000 - 780n$$

Find V for the following numbers of years: $n = 1$; $n = 5$; $n = 10$.

56. **Historical Question** In 1637 René Descartes published a book containing an appendix called *La Géométrie*.* This appendix introduced the world to *analytic geometry*. One legend concerning Descartes' discovery of this subject was that it came to him in a dream on November 10, 1619. Although Descartes never said that the dream was about analytic geometry, he did claim that the dream changed his life. He said that the dream clarified his purpose in life and revealed to him "a marvelous science." One of the equations that Descartes derived in La Géométrie was

$$y^2 = ay - bxy + cx - dx^2$$

Graph this curve for $a = 6$, $b = 0$, $c = 8$, and $d = 1$ by plotting points.

1.3 Lines

We have seen that the graph of a line is determined by two points. The standard form of the equation of a line was defined as

$$Ax + By + C = 0$$

In this section we develop a simplified method of graphing a line by algebraically solving the equation for y. If $B \ne 0$, then we can solve for y to put the equation into the form

$$y = mx + b$$

(*Note:* This lowercase b is not the same number as the capital B.) The constants b and m give us important information about the line we want to graph.

Find the y-intercept by letting $x = 0$:

$$y = m \cdot 0 + b$$
$$y = b$$

This means that the y-intercept is b.

---

* The title of the book is *Discours de la Méthode pour Bien Conduire sa Raison et Chercher la Vérité dans les Sciences.*

The second constant, m, which is the coefficient of x when the equation is solved for y, tells us the steepness or **slope** of the line. To define what is meant by the slope of a line we need to understand what is meant by the vertical change (**rise**) in a graph relative to a horizontal change (**run**). These ideas are illustrated in Figure 1.4.

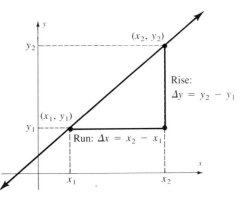

Figure 1.4 Slope of a line

Let Δx represent the horizontal change and Δy represent the vertical change. Note in Figure 1.4 that $\Delta x = x_2 - x_1$ and $\Delta y = y_2 - y_1$.

Slope

> Let (x_1, y_1) and (x_2, y_2) be points on a line such that $x_1 \neq x_2$. Then
>
> $$\text{Slope} = \frac{\text{vertical change}}{\text{horizontal change}} = \frac{\Delta y}{\Delta x} \quad \text{or} \quad \frac{\text{Rise}}{\text{Run}} = \frac{y_2 - y_1}{x_2 - x_1}$$
>
> If $\Delta x = 0$, then the line is vertical and has *no* slope ($\Delta y / 0$ is undefined).
> If $\Delta y = 0$, then the line is horizontal and has *zero* slope ($0/\Delta x = 0$).

To show that m in the equation $y = mx + b$ is the slope, consider the line specified by the equation $y = mx + b$ that passes through (x_1, y_1) and (x_2, y_2) with $x_1 \neq x_2$. This means that $y_1 = mx_1 + b$ and $y_2 = mx_2 + b$ so that

$$
\begin{aligned}
\text{Slope} = \frac{\Delta y}{\Delta x} &= \frac{y_2 - y_1}{x_2 - x_1} \\
&= \frac{(mx_2 + b) - (mx_1 + b)}{x_2 - x_1} \qquad \text{Substitution} \\
&= \frac{mx_2 - mx_1}{x_2 - x_1} \\
&= \frac{m(x_2 - x_1)}{x_2 - x_1} \\
&= m
\end{aligned}
$$

This discussion tells you that you can find the y-intercept and slope for linear equations of the form $y = mx + b$ by inspection, as shown in Example 1.

EXAMPLE 1   Find the slope and y-intercept.

		Slope	y-intercept
a.	$y = \frac{1}{2}x + 3$	$m = \frac{1}{2}$	$b = 3$
b.	$y = x - 3$	$m = 1$	$b = -3$
c.	$y = -\frac{2}{3}x + \frac{5}{2}$	$m = -\frac{2}{3}$	$b = \frac{5}{2}$
d.	$3x + 4y + 8 = 0$		

Solve for y:

$$4y = -3x - 8$$
$$y = -\frac{3}{4}x - \frac{8}{4} \qquad m = -\frac{3}{4} \qquad b = -\frac{8}{4} = -2$$ ∎

Since the slope and y-intercept are easy to find after the linear equation is solved for y, we give a special name to this form of the equation: the **slope-intercept form**.

Slope-Intercept Form of the Equation of a Line

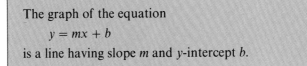

The graph of the equation

$$y = mx + b$$

is a line having slope m and y-intercept b.

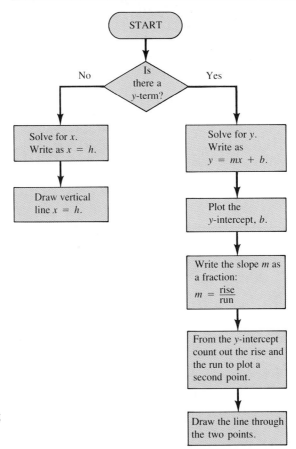

Figure 1.5 Procedure for graphing a line by using the slope-intercept method

This form of the equation of a line can be used for graphing certain lines where it is not convenient to plot points. The procedure is summarized in Figure 1.5. Carefully study this procedure.

EXAMPLE 2 Graph $y = \frac{1}{2}x + 3$.

Solution By inspection, the y-intercept is 3 and the slope is $\frac{1}{2}$; the line is graphed by first plotting the y-intercept $(0, 3)$ and then find a second point by counting out the slope: up 1 and over 2.

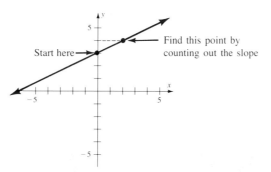

EXAMPLE 3 Graph $2x + 3y - 6 = 0$.

Solution Solve for y:

$$3y = -2x + 6$$
$$y = -\frac{2}{3}x + 2$$

The y-intercept is 2 and the slope is $-\frac{2}{3}$; the line is graphed by first plotting the y-intercept $(0, 2)$ and then finding a second point by counting out the slope: down 2 and over 3.

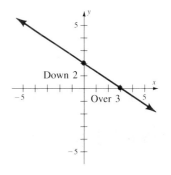

EXAMPLE 4 Graph $4x + 2y - 5 = 0$ for $-1 \le x \le 3$.

Solution Solve for y:

$$2y = -4x + 5$$
$$y = -2x + \frac{5}{2}$$

The y-intercept is $\frac{5}{2}$ and the slope is -2. The graph of this line is shown as a dashed line in the figure. Because of the restriction on the domain in this example, the graph is that part of the line with x values between -1 and 3 (inclusive), as shown by the solid color line segment in the figure.

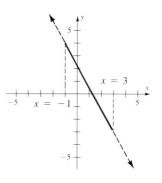

If you are given two points, it is also possible to draw the line and find the slope using $m = \Delta y / \Delta x$.

EXAMPLE 5   Sketch the line passing through the points whose coordinates are given. Then find the slope of each line.

a. $(2, -3)$ and $(-1, 2)$ **b.** $(-4, -1)$ and $(1, 3)$
c. $(-3, 4)$ and $(5, 4)$ **d.** $(-3, 2)$ and $(-3, 4)$

Solution   **a.** $m = \dfrac{2 - (-3)}{-1 - 2} = \dfrac{5}{-3} = -\dfrac{5}{3}$ **b.** $m = \dfrac{3 - (-1)}{1 - (-4)} = \dfrac{4}{5}$

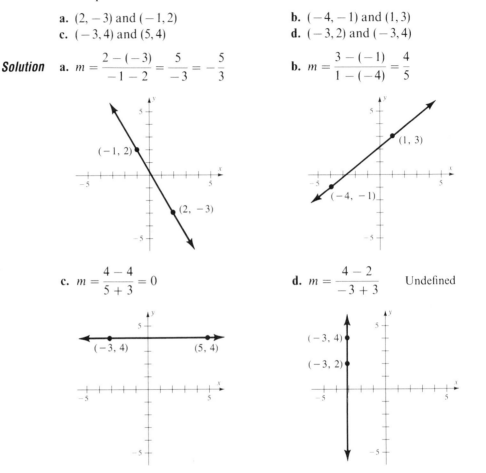

c. $m = \dfrac{4 - 4}{5 + 3} = 0$ **d.** $m = \dfrac{4 - 2}{-3 + 3}$ Undefined

In constructing mathematical models it is often necessary to write a linear equation using available or given information about the line. Example 6 shows how to do this if you know the slope and y-intercept.

EXAMPLE 6   Find the equation of the line with y-intercept 5 and slope $-\frac{2}{3}$.

Solution   Use the equation $y = mx + b$, where $b = 5$ and $m = -\frac{2}{3}$:

$$y = -\tfrac{2}{3}x + 5$$

More often than not, unfortunately, when you need the equation of a line you will not know the y-intercept. You may, however, know a point and the slope, or two points. In these cases it is easier to use another form of the equation of a line called the **point-slope form**.

Point-Slope Form of the Equation of a Line

A nonvertical line having slope m and passing through (x_1, y_1) has the equation

$$y - y_1 = m(x - x_1)$$

EXAMPLE 7   Find the equation of the line with slope 3 passing through $(-2, -5)$.

Solution   Use the equation $y - y_1 = m(x - x_1)$, where $m = 3$, $x_1 = -2$, and $y_1 = -5$:

$$y - (-5) = 3[x - (-2)]$$
$$y + 5 = 3(x + 2)$$ ∎

If you know two points and want the equation, first find the slope and *then* use the point-slope form.

EXAMPLE 8   Find the equation of the line passing through $(-2, 3)$ and $(4, -1)$.

Solution   First find the slope:

$$m = \frac{\Delta y}{\Delta x} = \frac{-1 - 3}{4 - (-2)} = \frac{-4}{6} = -\frac{2}{3}$$

Now use the point-slope form (you can use *either* of the given points):

Using $(-2, 3)$:   Using $(4, -1)$:

$y - 3 = -\frac{2}{3}(x + 2)$ $y + 1 = -\frac{2}{3}(x - 4)$

It is not easy to see that these two equations are the same. For this reason, we often ask that questions of lines be algebraically changed so that they appear in standard form.

Change $y - 3 = -\frac{2}{3}(x + 2)$ to the standard form $Ax + By + C = 0$:

$$3(y - 3) = -2(x + 2) \qquad \text{Multiply both sides by 3}$$
$$3y - 9 = -2x - 4 \qquad \text{Eliminate parentheses}$$
$$2x + 3y - 5 = 0 \qquad \text{Add } 2x \text{ and 4 to both sides}$$

Now change $y + 1 = -\frac{2}{3}(x - 4)$ to standard form:

$$3(y + 1) = -2(x - 4) \qquad \text{Multiply both sides by 3}$$
$$3y + 3 = -2x + 8 \qquad \text{Eliminate parentheses}$$
$$2x + 3y - 5 = 0 \qquad \text{Add } 2x \text{ and subtract 8 on both sides}$$

In standard form the equations are identical.   ∎

When deriving the equation of a line, do not forget the special cases for horizontal and vertical lines.

EXAMPLE 9   Find the equation of the line passing through $(7, -2)$ with no slope.

Solution   Do not confuse *no* slope (vertical line) with *zero* slope (horizontal line). This is a vertical line, so the equation has the form $x = h$ when it passes through (h, k). Thus $x = 7$ is the equation. In standard form, $x - 7 = 0$.   ∎

Price/Demand Equation

A mathematical model is often constructed by assuming that the relationship between two variables is linear and then writing an equation using two known data points, as illustrated in Example 10.

EXAMPLE 10 The demand for a new movie video is linearly related to its price. Market research shows that at $30 about 3 million tapes would be sold, but at $60 only 1 million would be sold. Write the standard form equation to represent this information.

Solution Let x be the price and y be the number of tapes sold (in millions). Then the data points are $(30, 3)$ and $(60, 1)$. First, find m:

$$m = \frac{1 - 3}{60 - 30} = \frac{-2}{30} = \frac{-1}{15}$$

Use the point-slope form to find the equation:

$$y - 3 = \tfrac{-1}{15}(x - 30)$$

Finally, find the standard form:

$$15y - 45 = -x + 30$$
$$x + 15y - 75 = 0$$

Parametric Form of the Equation of a Line

Sometimes it is convenient to define x and y in terms of another variable, say, t. For example, suppose that the location of a point (x, y) is defined in terms of the time, t (in seconds), as follows:

$$x = 1 + t$$
$$y = 2t$$

These equations might be interpreted by saying that the first component has a "1 second head start" but that the rate at which the second component is changing (with respect to time) is twice as fast as the first component. We can tabulate the values of x and y by choosing values for t, where $t \geq 0$ (since time cannot be negative).

t	0	1	2	3	4	5
x	1	2	3	4	5	6
y	0	2	4	6	8	10

Graph of $x = 1 + t,\quad y = 2t$

The variable t is called a **parameter** and the equations $x = 1 + t$ and $y = 2t$ are called *parametric equations*. The graph of the parametric equations looks like a line.

You can prove that it is a line by *eliminating the parameter.* Solve the first equation for t:

$$t = x - 1$$

and substitute into the second equation:

$$y = 2t = 2(x - 1) = 2x - 2$$

which you can recognize as a linear equation.

Parametric Form of the Equation of a Line

> The graph of the parametric equations
>
> $$x = x_1 + at \qquad \text{and} \qquad y = y_1 + bt$$
>
> is a line passing through (x_1, y_1) with slope $m = b/a$.

It is easy to derive this result by solving one of the equations for t and then substituting the result into the other equation.

$$x - x_1 = at$$

$$\frac{x - x_1}{a} = t$$

Now, by substitution,

$$y = y_1 + bt = y_1 + b\left(\frac{x - x_1}{a}\right)$$

$$y - y_1 = \frac{b}{a}(x - x_1)$$

This is the equation of a line passing through (x_1, y_1) with slope b/a.

The various forms for linear equations are summarized in the box:

Forms of a Linear Equation

STANDARD FORM:	$Ax + By + C = 0$	(x, y) is any point on the line; A, B, and C are constants; A and B are not both zero
SLOPE-INTERCEPT FORM:	$y = mx + b$	m is the slope; b is the y-intercept
POINT-SLOPE FORM:	$y - y_1 = m(x - x_1)$	m is the slope; (x_1, y_1) is the known point
HORIZONTAL LINE:	$y = k$	(h, k) is a point on the line
VERTICAL LINE:	$x = h$	(h, k) is a point on the line
PARAMETRIC FORM:	$x = x_1 + at$ $y = y_1 + bt$	(x_1, y_1) is a known point; b/a is the slope

Problem Set 1.3

Find the slope and y-intercept in Problems 1–24.

1. $y = 2x + 4$
2. $y = 5x - 3$
3. $y = 9x + 1$
4. $y = -4x - 1$
5. $y = -3x + 4$
6. $y = -x - 6$
7. $y = -\frac{1}{2}x + 5$
8. $y = \frac{5}{6}x - 5$
9. $y = -\frac{4}{3}x + \frac{2}{3}$
10. $x + y = 3$
11. $x - y = 5$
12. $5x + y = 6$
13. $2x + y - 5 = 0$
14. $5x - y + 3 = 0$
15. $2x - y - 5 = 0$
16. $4x + 3y + 4 = 0$
17. $3x + 2y - 7 = 0$
18. $5x - 4y + 3 = 0$
19. $y + 2 = 0$
20. $x - 4 = 0$
21. $3x + 1 = 0$
22. $100x - 250y + 500 = 0$
23. $2x - 5y - 1200 = 0$
24. $3x - 4y + 900 = 0$

Graph the lines whose equations are given in Problems 25–48 by finding the slope and y-intercept (found in Problems 1–24).

25. $y = 2x + 4$
26. $y = 5x - 3$
27. $y = 9x + 1$
28. $y = -4x - 1$
29. $y = -3x + 4$
30. $y = -x - 6$
31. $y = -\frac{1}{2}x + 5$
32. $y = \frac{5}{6}x - 5$
33. $y = -\frac{4}{3}x + \frac{2}{3}$
34. $x + y = 3$
35. $x - y = 5$
36. $5x + y = 6$
37. $2x + y - 5 = 0$
38. $5x - y + 3 = 0$
39. $2x - y - 5 = 0$
40. $4x + 3y + 4 = 0$
41. $3x + 2y - 7 = 0$
42. $5x - 4y + 3 = 0$
43. $y + 2 = 0$
44. $x - 4 = 0$
45. $3x + 1 = 0$
46. $100x - 250y + 500 = 0$
47. $2x - 5y - 1200 = 0$
48. $3x - 4y + 900 = 0$

Graph the lines in Problems 49–60 from their parametric equations.

49. $x = 2 + 3t$
    $y = 1 + t$
50. $x = 1 + 2t$
    $y = 2 + t$
51. $x = 5t$
    $y = 2 + t$
52. $x = 3t$
    $y = 1 + 2t$
53. $x = 1 - 2t$
    $y = 5 + 3t$
54. $x = 2 - 2t$
    $y = 1 + t$
55. $x = -1 + t$
    $y = 3 - 2t$
56. $x = -2 + t$
    $y = 4 - 5t$
57. $x = -3 + 4t$
    $y = -1 - 3t$
58. $x = -2 - 3t$
    $y = -2 + 5t$
59. $x = 6 - 2t$
    $y = -5 + 3t$
60. $x = 5 - t$
    $y = -3 - 2t$

Graph the line segments in Problems 61–72 by finding a point and the slope. Note that some of the restrictions are on x while other restrictions are on y.

61. $(4, 11)$, $m = -\frac{5}{2}$; $4 \le x \le 8$
62. $(7, -3)$, $m = -1$; $-3 \le y \le 0$
63. $(-9, 3)$, $m = -3$; $-3 \le y \le 0$
64. $(0, -3)$, $m = 0$; $-7 \le x \le 7$
65. $(8, 1)$, $m = 0$; $-6 \le x \le 8$
66. $(4, 5)$, no slope; $0 \le y \le 11$
67. $y + 2 = \frac{1}{12}(x - 1)$; $1 \le x \le 13$
68. $y - 1 = \frac{4}{9}(x + 2)$; $-3 \le y \le 1$
69. $y - 1 = -\frac{2}{5}(x - 8)$; $-7 \le x \le 13$
70. $y + 2 = -\frac{9}{8}(x - 1)$; $-7 \le x \le 1$
71. $y + 6 = \frac{5}{2}(x - 11)$; $-11 \le y \le -1$
72. $y + 11 = \frac{9}{4}(x - 9)$; $5 \le x \le 9$

Sketch the line passing through the points whose coordinates are given and find the slope of each line in Problems 73–81.

73. $(2, 3)$ and $(5, 6)$
74. $(0, 7)$ and $(3, 0)$
75. $(-1, -2)$ and $(4, 11)$
76. $(4, -2)$ and $(7, -3)$
77. $(-6, -4)$ and $(-9, 3)$
78. $(6, 0)$ and $(-3, 0)$
79. $(0, 0)$ and $(0, 3)$
80. $(4, -3)$ and $(4, 1)$
81. $(-1, 2)$ and $(3, 2)$

Find the standard form of the equation of the line satisfying the given conditions in Problems 82–93.

82. y-intercept 6; slope 5
83. y-intercept -3; slope -2
84. y-intercept 0; slope 0
85. y-intercept 5; slope 0
86. Slope 3; passing through $(2, 3)$
87. Slope -1; passing through $(-4, 5)$
88. Slope $\frac{1}{2}$; passing through $(5, 3)$
89. Slope $\frac{2}{5}$; passing through $(5, -2)$
90. Passing through $(-4, -1)$ and $(4, 3)$
91. Passing through $(5, 6)$ and $(1, -2)$
92. Passing through $(4, -2)$ and $(4, 5)$
93. Passing through $(5, 6)$ and $(7, 6)$

APPLICATIONS

Problems 94–102 provide some real world examples of line graphs. One way of finding the equation of the line is to write two data points from the given information and then use those two points to write the equation. Use the given information to write a standard form equation of the line.

94. The demand for a certain product is related to the price of the item. Suppose a new line of stationery is tested at two stores. It is found that 25 boxes are sold within a month at

$1 while 15 boxes are sold at $2 in the same time. Let x be the price and let y be the number of boxes sold.

95. An important factor related to the demand for a product is the supply. The amount of stationery in Problem 94 that can be supplied is also related to the price. At $1 each, 10 boxes can be supplied, and at $2 each, 20 boxes can be supplied. Let x be the price and let y be the number of boxes supplied.

96. The population of Florida in 1970 was roughly 6.8 million, and in 1980 it was 9.7 million. Let x be the year (let 1950 be the base year; that is, $x = 0$ represents 1950 and $x = 10$ represents 1960) and let y be the population. Use this equation to predict the population in 1990.

97. The population of Texas in 1970 was roughly 11.2 million, and in 1980 it was 14.2 million. Let x be the year (let 1950 be the base year; that is, $x = 0$ represents 1950 and $x = 10$ represents 1960) and let y be the population. Use this equation to predict the population in 1990.

98. It costs $90 to rent a car driven 100 miles and $140 for one driven 200 miles. Let x be the number of miles driven and let y be the total cost of the rental.

99. It costs $60 to rent a car driven 50 miles and $60 for one driven 260 miles. Let x be the number of miles driven and let y be the total cost of the rental.

100. Suppose it costs $100 for the maintenance and repairs of a 3-year-old car driven 1000 miles and $650 for maintenance and repairs to drive 6500 miles. Let x be the number of miles and let y be the cost for repairs and maintenance.

101. It costs $65,000 to produce 1000 items and $45,000 to produce 500 items. Let x be the number of items produced and let y be the total cost.

102. It costs $175,000 to produce 100 items and $625,000 to produce 1000 items. Let x be the number of items produced and let y be the total cost.

1.4 Systems of Equations

Many situations involve two variables or unknowns related in some specific fashion, as in Example 1.

EXAMPLE 1 **Cost Analysis** Linda is on a business trip and needs to rent a car for a day. The prices at two agencies are different, and she wants to rent the least expensive car.

> Agency A: $10 per day plus 50¢ per mile
> Agency B: $40 per day plus 25¢ per mile

Which agency should she rent from?

Solution The answer to this question depends on the number of miles Linda intends to drive. Let c be the cost (in dollars) of the rental and let m be the number of miles driven. These variables are related as follows:

> COST = BASIC CHARGE + MILEAGE CHARGE

where the mileage charge is the cost per mile times the number of miles driven. Thus

> Agency A: $c = 10 + 0.5m$
> Agency B: $c = 40 + 0.25m$

We use a Cartesian graph as a model for this problem and graph the equations for agencies A and B. The point of intersection looks like the point $(120, 70)$. What does this mean? It means that if Linda plans on driving less than 120 miles, she should rent from agency A, and if she plans on driving more than 120 miles, she should rent from agency B.

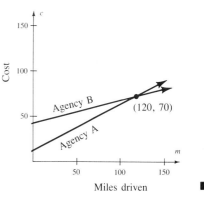

Miles driven

The model in Example 1 is called a **system of equations**. Solving systems of equations is a procedure that is used throughout mathematics. In this text we will solve a variety of systems of equations and inequalities, so we now introduce a notation that will be easy to generalize in later applications. We begin our study of this topic by considering an arbitrary system of two equations with two variables:

$$\begin{cases} a_{11}x_1 + a_{12}x_2 = b_1 \\ a_{21}x_2 + a_{22}x_2 = b_2 \end{cases}$$

The notation may seem strange, but it will prove useful. The variables are x_1 and x_2, and the constants are $a_{11}, a_{12}, a_{21}, a_{22}, b_1,$ and b_2. By a *system of two equations with two variables*, we mean any two equations in those variables. The **simultaneous solution** of a system is the intersection of the solution sets of the individual equations. The brace in front of the equations indicates that this intersection is the desired solution. If all the equations in a system are linear, it is called a **linear system**. We limit our study in this section to linear systems of equations with two unknowns.

Since the graph of each equation in a system of linear equations in two variables is a line, the solution set for the system is the intersection of two lines. In two dimensions, two lines must be related to each other in one of three possible ways:

1. The lines intersect at a single point. They are called **consistent**.
2. The graphs are parallel lines. In this case, the solution set is empty, and the equations are called **inconsistent**.
3. The graphs are the same line. In this case, there are infinitely many points in the solution set, and any solution of one equation is also a solution of the other. The equations of such a system are said to be **dependent**. If the equations of a system are not dependent, then they are called **independent**.

EXAMPLE 2   Relate the system $\begin{cases} 2x - 3y = -8 \\ x + y = 6 \end{cases}$ to the general system $\begin{cases} a_{11}x_1 + a_{12}x_2 = b_1 \\ a_{21}x_1 + a_{22}x_2 = b_2 \end{cases}$ and solve by graphing.

Solution   The variables are $x_1 = x$ and $x_2 = y$. The subscripts in $a_{11}, a_{12}, a_{21},$ and a_{22} are called *double subscripts* and indicate *position*. That is, a_{12} should not be read "a sub twelve," but rather as "a sub one-two." This means that it represents the second constant in the first equation:

$$a_{12}$$

$$\text{Equation number} \underset{}{\overset{}{\text{⌐}}} \text{Coefficient number in equation}$$

This effort in using notation may seem rather complicated when dealing with only two equations and two variables, but we want to develop a notation that can be used with many variables and many equations.

For this example, the constants are

$$a_{11} = 2 \qquad a_{12} = -3 \qquad b_1 = -8$$
$$a_{21} = 1 \qquad a_{22} = 1 \qquad b_2 = 6$$

The solution is $x = 2$, $y = 4$, or $(2, 4)$, as shown in the figure. ∎

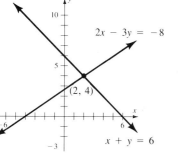

Graph of the system
$$\begin{cases} 2x - 3y = -8 \\ x + y = 6 \end{cases}$$

EXAMPLE 3   Relate the system $\begin{cases} 2x - 3y = -8 \\ 4x - 6y = -2 \end{cases}$ to the general system and solve by graphing.

Solution   The variables are $x_1 = x$, $x_2 = y$, and the constants are

$$a_{11} = 2 \qquad a_{12} = -3 \qquad b_1 = -8$$
$$a_{21} = 4 \qquad a_{22} = -6 \qquad b_2 = -2$$

Since the graphs of the lines are distinct and parallel, as shown in the figure, there is no point of intersection. These equations are inconsistent.

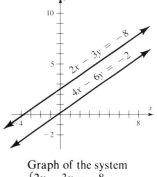

Graph of the system
$$\begin{cases} 2x - 3y = -8 \\ 4x - 6y = -2 \end{cases}$$

$\blacksquare$

EXAMPLE 4   Relate the system $\begin{cases} 2x - 3y = -8 \\ y = \frac{2}{3}x + \frac{8}{3} \end{cases}$ to the general system and solve by graphing.

Solution   To relate to the general system, both equations must be arranged in the usual form; the second equation needs to be rewritten:

$$y = \tfrac{2}{3}x + \tfrac{8}{3}$$
$$3y = 2x + 8 \qquad \text{Multiply by 3}$$
$$-2x + 3y = 8$$

Thus, in standard form, the system is

$$\begin{cases} 2x - 3y = -8 \\ -2x + 3y = 8 \end{cases}$$

This means that the variables are $x_1 = x$, $x_2 = y$, and the constants are

$$a_{11} = 2 \qquad a_{12} = -3 \quad b_1 = -8$$
$$a_{21} = -2 \quad a_{22} = 3 \qquad b_2 = 8$$

The equations represent the same line, as shown in the figure, and are dependent.

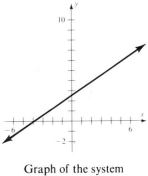

Graph of the system
$$\begin{cases} 2x - 3y = -8 \\ y = \tfrac{2}{3}x + \tfrac{8}{3} \end{cases}$$

$\blacksquare$

The **graphing method** can give solutions only as accurate as the graphs we can draw, and, consequently, it is inadequate for most applications. We therefore need more efficient methods.

In general, given a system, the procedure is to write a simpler equivalent system. Two systems are said to be **equivalent** if they have the same solution set. In this chapter we limit ourselves to finding only real roots. There are several ways to go about writing equivalent systems. The first nongraphical method we consider comes from the substitution property of real numbers and leads to a **substitution method** for solving systems.

Substitution Method for
Solving Systems of
Equations

1. *Solve* one of the equations for one of the variables.
2. *Substitute* the expression obtained into the other equation.
3. *Solve* the resulting equation in a single variable for the value of that variable.
4. *Substitute* that value into either of the original equations to determine the value of the other variable.
5. *State* the solution.

EXAMPLE 5   Solve $\begin{cases} 2p + 3q = 5 \\ \quad q = -2p + 7 \end{cases}$ by substitution.

Solution   Since $q = -2p + 7$, substitute $-2p + 7$ for q in the other equation:

$$2p + 3(-2p + 7) = 5$$
$$2p - 6p + 21 = 5$$
$$-4p = -16$$
$$p = 4$$

Substitute 4 for p in either of the given equations:

$$q = -2p + 7$$
$$= -2(4) + 7$$
$$q = -1$$

The solution is $(p, q) = (4, -1)$.   ∎

A third method for solving systems is called the **linear combination method**. It involves substitution and the idea that if equal quantities are added to equal quantities, the resulting equation is equivalent to the original system. In general, such addition will not simplify matters unless the numerical coefficients of one or more terms are opposites. However, we can often force them to be opposites by multiplying one or both of the given equations by appropriate nonzero constants.

Linear Combination Method
for Solving Systems of
Equations

1. *Multiply* one or both of the equations by a constant or constants, so that the coefficients of one of the variables become opposites.
2. *Add* corresponding members of the equations to obtain a new equation in a single variable.
3. *Solve* the derived equation for that variable.
4. *Substitute* the value of the found variable into either of the original equations and solve for the second variable.
5. *State* the solution.

EXAMPLE 6   Solve: $\begin{cases} 3x + 5y = -2 \\ 2x + 3y = 0 \end{cases}$.

Solution   Multiply both sides of the first equation by 2 and both sides of the second equation by -3. This procedure, shown below, forces the coefficients of x to be opposites:

$$2 \begin{cases} 3x + 5y = -2 \\ -3 \begin{cases} 2x + 3y = 0 \end{cases} \end{cases}$$

This means you should add the equations of the system

$$+ \begin{cases} 6x + 10y = -4 \\ -6x - 9y = 0 \end{cases}$$

$$[6x + (-6x)] + [10y + (-9y)] = -4 + 0 \qquad \text{This step should be done mentally}$$
$$y = -4$$

If $y = -4$, then

$$2x + 3y = 0$$
$$2x + 3(-4) = 0$$
$$x = 6$$

The solution is $(6, -4)$.   ∎

Later we will need to find the corner points for certain regions in the plane and the methods of this section, as illustrated in Example 7, will be helpful.

EXAMPLE 7   Find the corner points labeled A, B, C, and D in the illustrated region.

Solution   Point A is obvious, it is the origin $(0,0)$. Points B and C are relatively easy to find because they are the intercepts of the lines.

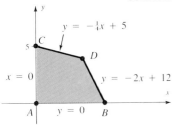

For point B, the x-intercept is found when $y = 0$ in the equation $y = -2x + 12$. Thus

$$0 = -2x + 12$$
$$-12 = -2x$$
$$6 = x$$

Point B is $(6,0)$.

For point C, the y-intercept is found by inspection; it is the point $(0, 5)$.

Point D is the intersection of the lines whose equations are given, namely:

$$\begin{cases} y = -\tfrac{1}{4}x + 5 \\ y = -2x + 12 \end{cases}$$

Solve these by substitution,

$$-2x + 12 = -\tfrac{1}{4}x + 5$$
$$-8x + 48 = -x + 20$$
$$-7x = -28$$
$$x = 4$$

Finally, $y = -2(4) + 12 = 4$, so D is the point $(4, 4)$.   ∎

There are two very important business applications of systems: *supply and demand* and *break-even analysis.*

Supply and Demand

Let p be the price of an item and n the number of items available at price p; that is, consider the ordered pair (p, n). There are two curves relating p and n. The first, called the **supply curve**, expresses the relationship between p and n from a manufacturer's point of view. For every value of p, the supply curve gives the number n that the manufacturer is willing to produce at the price p. The higher the price, the more the manufacturer is willing to supply. Therefore, the supply curve rises when viewed from left to right. The other curve, the **demand curve**, expresses the relationship between p and n from the consumer's point of view. For every value of p, the demand curve gives the number n that the consumer is willing to buy at price p. The higher the price, the less consumers will buy, so demand curves fall when viewed from left to right.

If supply and demand curves are drawn on the same coordinate system, the intersection point (if it exists) is called the **equilibrium point** and represents values for which supply equals demand.

In this section, the supply and demand curves are linear, but in most real life situations they are more general curves, as shown by the illustrations in the margin.

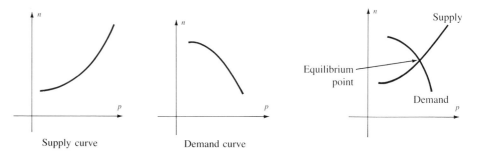

Supply curve   Demand curve

EXAMPLE 8   A product has a supply curve given by the equation $n = 500p - 8100$ and a demand curve given by the equation $n = -100p + 15{,}000$. Find the equilibrium point.

Solution   Find the intersection point by using substitution:

$$-100p + 15{,}000 = 500p - 8100$$
$$600p = 23{,}100$$
$$p = \frac{23{,}100}{600}$$
$$= 38.5$$

Now, if $p = 38.5$, then

$$n = -100(38.5) + 15{,}000$$
$$= 11{,}150$$

Thus equilibrium is reached when the price is $38.50 and the number is 11,150 items.   ∎

Break-Even Analysis

In order to make a profit, revenue must exceed cost. The point at which revenue equals the cost is called the **break-even point**.

EXAMPLE 9   A company producing bicycle reflectors has fixed costs of $2,100 and variable costs of 80¢ per reflector. The selling price of each reflector is $1.50. What is the break-even point?

Solution   Let x be the number of items produced, C be the cost, and R be the revenue. Then

$$C = 0.8x + 2100$$

and

$$R = 1.5x$$

Although the graph in the figure illustrates what is meant by break-even analysis and profit and loss, it is often not accurate enough to find the break-even point by inspection. Therefore we will also solve this system of equations by substitution:

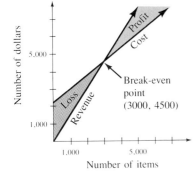

$1.5x = 0.8x + 2100$	Since $R = C$ at the break-even point
$0.7x = 2100$	
$x = 3000$	If $x = 3000$, then $R = 1.5(3000) = 4500$

If 3000 reflectors are produced, the cost and revenue will both be $4,500.   ■

Problem Set 1.4

Solve the systems in Problems 1–6 by graphing.

1. $\begin{cases} y = 3x - 7 \\ y = -2x + 8 \end{cases}$

2. $\begin{cases} x + y = 1 \\ 3x + y = -5 \end{cases}$

3. $\begin{cases} y = \frac{2}{3}x - 7 \\ 2x + 3y = 3 \end{cases}$

4. $\begin{cases} y = \frac{3}{5}x + 2 \\ 3x - 5y = 10 \end{cases}$

5. $\begin{cases} 2x - 3y = 9 \\ y = \frac{2}{3}x - 3 \end{cases}$

6. $\begin{cases} 2x - 3y = 0 \\ y = \frac{2}{3}x - 2 \end{cases}$

Solve the systems in Problems 7–12 by substitution. Relate each system to the general system $\begin{cases} a_{11}x_1 + a_{12}x_2 = b_1 \\ a_{21}x_1 + a_{22}x_2 = b_2 \end{cases}$.

7. $\begin{cases} a = 3b - 7 \\ a = -2b + 8 \end{cases}$

8. $\begin{cases} s + t = 1 \\ 3s + t = -5 \end{cases}$

9. $\begin{cases} m = \frac{2}{3}n - 7 \\ 2n + 3m = 3 \end{cases}$

10. $\begin{cases} v = \frac{3}{5}u + 2 \\ 3u - 5v = 10 \end{cases}$

11. $\begin{cases} 2p - 3q = 9 \\ q = \frac{2}{3}p - 3 \end{cases}$

12. $\begin{cases} 3t_1 + 5t_2 = 1541 \\ t_2 = 2t_1 + 160 \end{cases}$

Solve the systems in Problems 13–18 by linear combinations. Relate each system to the general system $\begin{cases} a_{11}x_1 + a_{12}x_2 = b_1 \\ a_{21}x_1 + a_{22}x_2 = b_2 \end{cases}$.

13. $\begin{cases} c + d = 2 \\ 2c - d = 1 \end{cases}$

14. $\begin{cases} 2s_1 + s_2 = 10 \\ 5s_1 - 2s_2 = 16 \end{cases}$

15. $\begin{cases} 3q_1 - 4q_2 = 3 \\ 5q_1 + 3q_2 = 5 \end{cases}$

16. $\begin{cases} 9x + 3y = 5 \\ 3x + 2y = 2 \end{cases}$

17. $\begin{cases} 7x + y = 5 \\ 14x - 2y = -2 \end{cases}$

18. $\begin{cases} 2x + 3y = 1 \\ 3x - 2y = 0 \end{cases}$

Solve the systems in Problems 19–30 by any method.

19. $\begin{cases} 5x + 4y = 5 \\ 15x - 2y = 8 \end{cases}$

20. $\begin{cases} 3x + 2y = 1 \\ 6x + 4y = 2 \end{cases}$

21. $\begin{cases} 4x - 2y = -28 \\ y = \frac{1}{2}x + 5 \end{cases}$

22. $\begin{cases} 12x - 5y = -39 \\ y = 2x + 9 \end{cases}$

23. $\begin{cases} y = 2x - 1 \\ y = -3x - 9 \end{cases}$

24. $\begin{cases} y = \frac{2}{3}x - 5 \\ y = -\frac{4}{3}x + 7 \end{cases}$

25. $\begin{cases} 2x - 3y - 5 = 0 \\ 3x - 5y + 2 = 0 \end{cases}$

26. $\begin{cases} 2x + 3y = a \\ x - 5y = b \end{cases}$

27. $\begin{cases} ax + by = 1 \\ bx + ay = 0 \end{cases}$

28. $\begin{cases} 0.12x + 0.06y = 210 \\ x + y = 2000 \end{cases}$

29. $\begin{cases} 0.12x + 0.06y = 228 \\ x + y = 2000 \end{cases}$

30. $\begin{cases} 0.12x + 0.06y = 1140 \\ x + y = 10,000 \end{cases}$

Find the coordinates of the corner points for each of the regions in Problems 31–36.

31.

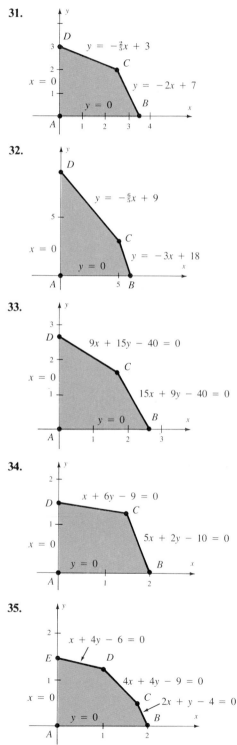

32.

33.

34.

35.

36.

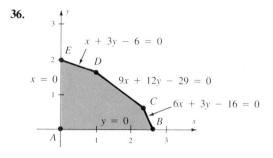

APPLICATIONS

37. Suppose a car rental agency gives the following choices:

> Option A: $30 per day plus 40¢ per mile
>
> Option B: Flat $50 per day with unlimited mileage

At what mileage are both rates the same?

38. Suppose a car rental agency gives the following options:

> Option A: $40 per day plus 50¢ per mile
>
> Option B: Flat $60 per day with unlimited mileage

At what mileage are both rates the same?

39. A paint sprayer rents for $4 per hour or $24 per day. At how many hours are the rates the same?

40. Two car rental agencies have the following rates:

> Agency A: $15 per day plus 20¢ per mile
>
> Agency B: $25 per day plus 15¢ per mile

At what mileage are both rates the same?

41. The supply curve for a certain commodity is $n = 250p - 2500$, and the demand curve for the same commodity is $n = 3500 - 150p$. What should the price of the commodity be in order for the market to be stable?

42. The supply curve for a new product is $n = 2000p + 6000$, and the demand curve for the same product is $n = 13,000 - 1500p$. What is the equilibrium point for this product?

43. The supply curve for a new software product is $n = 2.5p - 500$, and the demand curve for the same product is $n = 200 - 0.5p$.
   a. At $250 for the product, how many items would be supplied? How many demanded?
   b. At what price would no items be supplied?
   c. At what price would no items be demanded?
   d. What is the equilibrium price for this product?
   e. How many units will be produced at the equilibrium price?

44. The supply curve for a new commodity is $n = 2500p - 500$, and the demand curve for the same product is $n = 31,500 - 1500p$.
  a. At \$15 per commodity, how many items would be supplied? How many demanded?
  b. At what price would no items be supplied?
  c. At what price would no items be demanded?
  d. What is the equilibrium price for this product?
  e. How many units will be produced at the equilibrium price?

45. Suppose you want to sell T-shirts on your campus. You are trying to decide on a price between \$1 and \$7, so you conduct a market-research survey and find that 800 students will buy the shirt for \$1, but none will buy it for \$7. You also find a supplier who can supply 200 shirts if they are sold for \$1 and 600 if they are sold for \$7. At what price will supply equal demand, and how many could you expect to sell if both supply and demand are linear?

46. A new greeting card is introduced in a test market. Within the month, 3000 cards are sold priced at 50¢ and 1500 are sold priced at \$2. On the other hand, 2500 cards could be supplied at a cost of \$2 but only 1500 could be supplied at a cost of \$1. At what price does the supply equal the demand, assuming that both are linear?

47. A firm producing sunglasses has fixed costs of \$1,200 and variable costs of 80¢ per pair. Find the break-even point if the sunglasses sell for \$2 per pair.

48. Repeat Problem 47 if the sunglasses sell for \$3 per pair.

49. Microdrop Corporation produces floppy computer disks at a cost of \$1.20 per disk with total fixed costs of \$18,000. If the disks sell for \$3 each, what is the break-even point?

50. Repeat Problem 49 if the disks sell for \$2.50 each.

51. Hallmark introduces a new greeting card that has fixed costs of \$2,800 and variable costs of 75¢ per card. If the cards sell for \$1.50 each, what is the break-even point?

52. Repeat Problem 51 if the cards sell for \$1.25 each.

53. Repeat Problem 51 if the cards sell for \$2 each.

54. An investor owns Xerox and Standard Oil stock. The closing prices of each stock for the week are given in the table. If the value of the portfolio was \$19,500 on Monday and \$20,300 on Friday, how many shares of each stock does the investor own?

	Standard Oil	Xerox
Monday	45	60
Tuesday	$45\frac{1}{2}$	$62\frac{1}{8}$
Wednesday	46	63
Thursday	$47\frac{1}{4}$	$61\frac{1}{2}$
Friday	47	62

55. Another investor owns the same stocks as the investor described in Problem 54. This investor's portfolio was valued at \$34,500 on Monday and at \$35,600 on Wednesday. How many shares of each stock are owned by this investor?

56. The total number of registered Democrats and Republicans in a certain community is 100,005. Voter turnout in a recent election showed that 50% of the Democrats voted, 60% of the Republicans voted, and a total of 55,200 votes were cast, which also included Independents and write-in votes. If the registered Democrats outnumber the registered Republicans by 2 to 1, how many registered Democrats and Republicans are there in the community and how many independents or write-in votes were there in the election?

57. Two chemicals are combined to form two grades of fertilizer. One unit of grade A fertilizer requires 10 pounds of potassium and 3 pounds of calcium, while a unit of grade B fertilizer requires 9 pounds of potassium and 4 pounds of calcium. If 780 pounds of potassium and 260 pounds of calcium are available, how many units of each grade of fertilizer can be mixed?

58. Historical Question The Louvre Tablet from the Babylonian civilization is dated at about 1500 B.C. It shows a system equivalent to

$$\begin{cases} xy = 1 \\ x + y = a \end{cases}$$

Even though this is not a linear system, you can solve it by substitution if you let $a = 2$. Find x and y.

1.5 Summary and Review

IMPORTANT TERMS

Break-even analysis [1.4]
Cartesian coordinate system [1.2]
Components [1.2]
Consistent [1.4]
Coordinate axes [1.2]
Cost analysis [1.1]
Dependent [1.4]
Dependent variable [1.1]

Depreciation [1.2]	Plot a point [1.2]
Deterministic model [1.1]	Point-slope form [1.3]
Domain [1.1]	Probabilistic model [1.1]
Equilibrium point [1.4]	Quadrant [1.2]
Equivalent systems [1.4]	Quadratic equation [1.1]
Exponent [1.1]	Range [1.1]
Fixed costs [1.1]	Rational numbers [1.1]
Formula [1.1]	Real number line [1.1]
Function [1.1]	Real numbers [1.1]
Graph [1,1, 1.2]	Revenue [1.1]
Graphing method [1.4]	Rise [1.3]
Horizontal line [1.2]	Run [1.3]
Inconsistent [1.4]	Set [1.1]
Independent [1.4]	Short-run analysis [1.1]
Independent variable [1.1]	Simultaneous solution [1.4]
Integer [1.1]	Slope [1.2, 1.3]
Irrational numbers [1.1]	Slope-intercept form [1.3]
Line [1.2]	Standard form [1.2]
Linear combination method [1.4]	Straight-line depreciation [1.2]
Linear equation [1.2, 1.3]	Substitution model [1.4]
Linear system [1.4]	Supply and demand [1.4]
Mathematical model [1.1]	System of equations [1.4]
Ordered pair [1.2]	Two-dimensional coordinate
Origin [1.2]	system [1.2]
Parabola [1.2]	Variable costs [1.1]
Parameter [1.3]	Vertical line [1.2]
Parametric form [1.3]	x-intercept [1.2]
Plane [1.2]	y-intercept [1.2]

SAMPLE TEST *For additional practice there are a large number of review problems categorized by objective in the Student Solutions Manual. The following sample test (40 minutes) is intended to review the main ideas of the chapter.*

1. A model used to find total cost is $C = v + f$, where v represents the variable costs and f is the fixed costs. Futhermore, $v = px$ where p is the price of each item and x is the number of items. If each item costs \$125 to produce and there are fixed costs of \$55,000, how much will it cost to produce 5000 items? What is the average cost per item if 5000 items are produced?

Graph the lines or curves in Problems 2–8.

2. $xy - 2 = 0$ 3. $y + 2 = 0$ 4. $5x = 10$

5. $y = -\frac{4}{5}x + 2$ 6. $5x + 3y - 10 = 0$ 7. $x = 3t, \quad y = -2t$

8. $4x + y + 800 = 0$ for $0 \le x \le 100$

9. Find the slope of the line passing through $(0, 0)$ and $(-3, 5)$.

10. Find the equation of the line described in Problem 9.

11. Find the equation of the line with slope $-\frac{3}{5}$ passing through $(-1, 3)$.

12. Find the equation of the horizontal line passing through $(1, 5)$.

13. Find the equation of the line with slope -3 and intercept $\frac{1}{2}$.

Solve the systems in Problems 14–16. Use a different Method for each.

14. $\begin{cases} 5x - 3y = 6 \\ 3x + 3y = 10 \end{cases}$

15. $\begin{cases} y = 4x - 1 \\ y = -5x + 8 \end{cases}$

16. $\begin{cases} x + y = 18 \\ 4x - 3y = 58 \end{cases}$

17. Find the corner points for the given region.

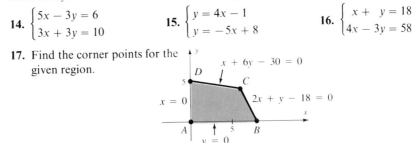

18. The tax (T) for certain taxpayers with income I is given by $T = 0.18I - 3000$. Graph (I, T) for $22,000 \leq I \leq 24,000$.

19. Two car rental agencies have the following rate schedule:

    Agency A: $40 per day plus 25¢ per mile
    Agency B: $35 per day plus 30¢ per mile

At what mileage are both rates the same?

20. A company is producing an item that costs $1 each with fixed costs of $3,500. If the item sells for $3, what is the break-even point?

Modeling Application 1

Gaining a Competitive Edge in Business

Solartex manufactures solar collector panels. During the first year of operation, rent, insurance, utilities, and other fixed costs averaged $8,500 per month. Each panel sold for $320 and cost the company $95 in materials and $55 in labor. Since company resources are limited, Solartex cannot spend more than $20,000 in any one month. Sunenergy, another company selling similar panels, competes directly with Solartex. Last month, Sunenergy manufactured 85 panels at a total cost of $20,475, but the previous month produced only 60 panels at a total cost of $17,100.

Mathematical modeling involves creating mathematical equations and procedures to make predictions about the real world. Typical textbook problems focus on limited, specific skills, but when confronted with a real life example you are often faced with a myriad of "facts" without specific clues on how to fit them together to make predictions. In this book, you will be given a modeling application and asked to write a paper using the given information. You will need to do some research to have adequate data. There are no "right answers" for these papers. In the *Student's Solution Manual*, a paper is presented for each modeling application, but your paper could certainly take a quite different direction.

For this first application, the following questions might help you get started: What is the cost equation for Solartex? How many panels can be manufactured by Solartex given the available capital? What is the minimum number of panels that should be produced? In order to make a profit, revenue must exceed cost; what is Solartex's break-even point? How would you compare Solartex and Sunenergy?

Feel free to supply information that is not given above. For example, a study of market demand could be useful. Suppose you find that at $75 per panel, Solartex (or a real company you have studied) could sell 200 panels per month, but at $450, they could sell only 20 panels per month. On the other hand, if Solartex had to sell the panels at $225 each, they could only afford material that would limit them to 30 panels per month, but at $450 each, they could afford sufficient material to supply 100 panels per month. What is the equilibrium point for this information?

Write a paper based on this modeling application.

CHAPTER 2
Matrix Theory

APPLICATIONS

Management (*Business, Economics, Finance, and Investments*)

Inventory control (2.1, Problem 5)
Delivery routes (2.1, Problem 64)
Airline routes (2.1, Problem 65)
Construction costs (2.1, Problem 66)
Mixing an alloy (2.2, Problem 50)
Manufacturing a candy (2.2, Problem 51)
Leontief models (2.4, Problems 7–17)
Study of the Israeli economy (2.4, Problems 18–20)

Life sciences (*Biology, Ecology, Health, and Medicine*)

Preparing cattle feed (2.2, Problem 48; 2.3, Problems 55–57)
Preparing a pesticide (2.2, Problem 49)
Preparing a balanced diet (2.3, Problems 58–60)

Social sciences (*Demography, Political Science, Population, Psychology, Society, and Sociology*)

Diplomatic communication channels (2.1, Problem 67)
Voter demographics (2.2, Problem 52)

General interest

Matrix of car rental costs (2.1, Problem 6)
Development of matrix theory (2.1, Problems 68–73)
Best-value ice cream cone (2.2, Problem 53)

Modeling application—Leontief Historical Economic Model

CHAPTER OVERVIEW   Here we learn an extremely powerful concept in mathematical modeling—the *matrix*.

PREVIEW   We first learn what matrices are, then how to manipulate them, and, finally, how to use them to solve systems of equations. The modeling application is a study of the U.S. economy. Wassily Leontief (1906–) received the Nobel Prize in Economics in 1973 for describing the economy in terms of an input–output model. We look at three extended applications of the Leontief model: a *closed model* in which we try to find the relative income of each participant within a system, an *open model* in which we try to find the amount of production needed to achieve a forecast demand, and a *historical economic model* used by Leontief.

PERSPECTIVE The important idea of this chapter—one that must be mastered in order to succeed with the material that follows later in the book—is that of *Gauss–Jordan elimination*. It is used over and over again in this course, so work hard to understand the procedure.

2.1 Introduction to Matrices

One of the biggest problems in applying mathematics to the real world is developing a means of systematizing and handling the great number of variables and large amounts of data that are inherent in real life situations. One step in handling large amounts of data is the creation of a mathematical model involving what are called **matrices**.

Definition of Matrix

> A *matrix* is a rectangular array of numbers. (The plural is *matrices*.)

You are already familiar with matrices from everyday experiences. For example, Table 2.1 shows a car rental chart in the form of a matrix.

TABLE 2.1
Example of a matrix: Costs of weekly car rentals in Europe

Country	Fiat Panda	Ford Fiesta	Opel Kadett	Renault R14	Volkswagen Microbus
Austria	US $149	US $179	US $219	US $269	US $289
Belgium	130	143	222	273	425
Denmark	164	212	269	408	480
France	189	214	257	326	386
Great Britain	160	174	206	215	243
Holland	156	183	213	259	305
Ireland	176	185	198	222	233
Italy	179	213	259	353	408
Luxembourg	130	143	222	273	425
Spain	156	188	206	247	306
Sweden	178	205	246	281	315

In mathematics we enclose the array of numbers in brackets and denote the entire array by using a capital letter. The size of the matrix is called the **order** and is specified by naming the number of **rows** (rows are horizontal) first and then the number of **columns** (columns are vertical). A matrix with order $m \times n$ (read "m by n") means that the matrix has m rows and n columns. The order of the matrix in Table 3.1 is 11×5 (eleven by five). The price of the Renault in Denmark is found in row 3, column 4.

EXAMPLE 1 Name the order of each given matrix.

$$A = \begin{bmatrix} 1 & 6 & 3 \\ 4 & 8 & -2 \end{bmatrix} \qquad B = \begin{bmatrix} 6 & 1 \\ 4 & -3 \\ 2 & 0 \\ -1 & 4 \end{bmatrix} \qquad C = \begin{bmatrix} 5 \\ 8 \\ 1 \end{bmatrix}$$

$$D = \begin{bmatrix} 4 & 8 & 6 & 2 \end{bmatrix} \qquad E = \begin{bmatrix} 6 & 8 & -2 \\ 5 & 4 & -1 \\ 0 & -5 & 7 \end{bmatrix}$$

Solution The order of A is 2×3; B is 4×2; C is 3×1; D is 1×4; and E is 3×3. ■

If the matrix has only one column (for example, matrix C), then it is called a **column matrix**; if it has only one row (matrix D, for example), then it is called a **row matrix**; and if the number of rows and columns are the same, it is called a **square matrix**. The numbers in a matrix are called the **entries**, or **components**, of a matrix. If all the entries of a matrix are zero, it is called a **zero matrix** and is represented by **0**.

Two matrices are said to be **equal** if they have the same order and if the corresponding entries are equal. (Corresponding entries are numbers in the same position—that is, same row and column location.)

EXAMPLE 2 Find x and y, if possible, so that the pairs of matrices are equal.

a. $\begin{bmatrix} 2 & x \\ y & -3 \end{bmatrix} = \begin{bmatrix} 2 & 5 \\ -1 & -3 \end{bmatrix}$ **b.** $\begin{bmatrix} 6 & 4 \\ x & y \end{bmatrix} = \begin{bmatrix} 6 & 0 \\ 4 & -7 \end{bmatrix}$

By inspection, $x = 5$ and $y = -1$.   No values of x and y will make these matrices equal since $4 \neq 0$ in the upper right entries.

c. $\begin{bmatrix} x & 8 \\ y & -4 \end{bmatrix} = \begin{bmatrix} 2 & 8 & -1 \\ 3 & -4 & 0 \end{bmatrix}$ No values of x and y will make these matrices equal since the matrices must have the same order to be equal. ■

EXAMPLE 3 **Inventory Control** Suppose a company produces two models of wireless telephones and manufactures them at four factories. The number of units produced at each factory can be summarized by the following matrix:

	Factory			
	A	B	C	D
Standard model	16	25	15	8
Deluxe model	10	0	14	3

If the entries represent thousands of units, answer the following questions by reading the matrix:

a. How many standard models are produced at factory D?
b. How many standard models are produced at factory A?
c. How many deluxe models are produced at factory C?
d. At which factory are no deluxe models produced?

Solution   **a.** 8000   **b.** 16,000   **c.** 14,000   **d.** Factory B   ■

Sometimes entries 1 and 0 are used to summarize information in matrix form, as illustrated by Example 4.

EXAMPLE 4   **Communication Theory**   Use matrix notation to summarize the following:

The United States has diplomatic relations with the Soviet Union and Mexico, but not with Cuba.

Mexico has diplomatic relations with the United States and the Soviet Union, but not with Cuba.

The Soviet Union has diplomatic relations with the United States, Mexico, and Cuba.

Cuba has diplomatic relations with the Soviet Union, but not with the United States and Mexico.

Solution   Use 1 if the row entry and the corresponding column entry have diplomatic relations and 0 if they do not. Then:

$$
\begin{array}{c c}
 & \begin{array}{cccc} \text{U.S.} & \text{U.S.S.R.} & \text{Cuba} & \text{Mexico} \end{array} \\
\begin{array}{c} \text{U.S.} \\ \text{U.S.S.R.} \\ \text{Cuba} \\ \text{Mexico} \end{array} &
\left[\begin{array}{cccc}
0 & 1 & 0 & 1 \\
1 & 0 & 1 & 1 \\
0 & 1 & 0 & 0 \\
1 & 1 & 0 & 0
\end{array}\right]
\end{array}
\qquad \blacksquare
$$

The solution to Example 4 is called a **communication matrix** or *incidence matrix*.

If two matrices have the same order, we define the **sum of the matrices** as the matrix whose components result from the sum of the corresponding components of the given matrices. We also say the original matrices are **conformable**. If the matrices do not have the same order, then we say the matrices are **not conformable** for addition.

EXAMPLE 5   Let

$$
\mathbf{0} = \begin{bmatrix} 0 & 0 & 0 \end{bmatrix} \qquad A = \begin{bmatrix} 4 & 2 & 6 \end{bmatrix}
$$

$$
B = \begin{bmatrix} 6 \\ 1 \end{bmatrix} \qquad C = \begin{bmatrix} 1 & 7 & 2 \end{bmatrix} \qquad D = \begin{bmatrix} 1 \\ 2 \end{bmatrix}
$$

$$
E = \begin{bmatrix} 2 & 1 & 0 \\ 4 & 7 & 3 \\ -2 & 0 & 1 \end{bmatrix} \qquad F = \begin{bmatrix} 6 & 1 & 2 \\ 3 & -10 & 4 \\ 1 & 3 & -2 \end{bmatrix}
$$

Find (wherever possible):

a. $A + C$ **b.** $B + D$ **c.** $A + \mathbf{0}$ **d.** $A + B$ **e.** $E + F$

Solution   **a.** $A + C = \begin{bmatrix} 4 + 1 & 2 + 7 & 6 + 2 \end{bmatrix} = \begin{bmatrix} 5 & 9 & 8 \end{bmatrix}$

b. $B + D = \begin{bmatrix} 6 + 1 \\ 1 + 2 \end{bmatrix} = \begin{bmatrix} 7 \\ 3 \end{bmatrix}$

c. $A + \mathbf{0} = \begin{bmatrix} 4 + 0 & 2 + 0 & 6 + 0 \end{bmatrix} = \begin{bmatrix} 4 & 2 & 6 \end{bmatrix}$

d. Not conformable

e. $E + F = \begin{bmatrix} 2+6 & 1+1 & 0+2 \\ 4+3 & 7+(-10) & 3+4 \\ -2+1 & 0+3 & 1+(-2) \end{bmatrix} = \begin{bmatrix} 8 & 2 & 2 \\ 7 & -3 & 7 \\ -1 & 3 & -1 \end{bmatrix}$ ∎

EXAMPLE 6 Let S_{1987} and S_{1988} represent the sales (in thousands of dollars) for a company in 1987 and 1988, respectively:

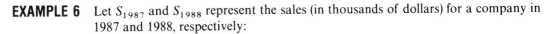

$$S_{1987} = \begin{bmatrix} 150 & 200 & 350 \\ 100 & 150 & 50 \end{bmatrix} \begin{matrix} \text{Wholesale} \\ \text{Retail} \end{matrix} \qquad S_{1988} = \begin{bmatrix} 175 & 300 & 400 \\ 110 & 100 & 100 \end{bmatrix} \begin{matrix} \text{Wholesale} \\ \text{Retail} \end{matrix}$$

What is the combined sales (in thousands of dollars) for the company for the given years?

Solution The answer is found by finding the sum of the matrices:

$$S_{1987} + S_{1988} = \begin{bmatrix} 150+175 & 200+300 & 350+400 \\ 100+110 & 150+100 & 50+100 \end{bmatrix}$$

$$= \begin{bmatrix} 325 & 500 & 750 \\ 210 & 250 & 150 \end{bmatrix}$$ ∎

For the company in Example 6, we might be interested in finding the increase (or decrease) in sales between 1987 and 1988. We see that the difference in wholesale sales in Chicago is found by subtracting:

$$175 - 150 = 25$$

The difference for the other components is found similarly. The **difference of matrices** A and B of the same order, $A - B$, is found by subtracting the entries of B from the corresponding entries of A.

EXAMPLE 7 Find the change in sales between 1987 and 1988 for the company with sales S_{1987} and S_{1988} in Example 6.

Solution The answer is found by finding the difference of the matrices:

$$S_{1988} - S_{1987} = \begin{bmatrix} 175-150 & 300-200 & 400-350 \\ 110-100 & 100-150 & 100-50 \end{bmatrix}$$

$$= \begin{bmatrix} 25 & 100 & 50 \\ 10 & -50 & 50 \end{bmatrix}$$

The negative entry indicates a decrease in sales. ∎

Continuing with the same example, suppose we want to find the profit generated by the sales in each category and have found that 3% of the gross sales for a given year is actual profit. Then, we find the 1988 profit from wholesale sales in Chicago by multiplying 0.03 (3%) by 175 (the wholesale sales in thousands of dollars in Chicago in 1988):

$$0.03 \cdot 175 = 5.25$$

To find the profit in each category, we multiply the matrix

$$S_{1985} = \begin{bmatrix} 175 & 300 & 400 \\ 110 & 100 & 100 \end{bmatrix}$$

by 0.03. This operation of multiplying each entry of a matrix A by a number c is denoted by cA and is called **scalar multiplication**.

EXAMPLE 8 Find the profit (in thousands of dollars) for S_{1988}.

Solution Calculate $(0.03)S_{1988}$.

$$(0.03)S_{1988} = \begin{bmatrix} (0.03)(175) & (0.03)(300) & (0.03)(400) \\ (0.03)(110) & (0.03)(100) & (0.03)(100) \end{bmatrix}$$

$$= \begin{bmatrix} 5.25 & 9 & 12 \\ 3.3 & 3 & 3 \end{bmatrix}$$ ∎

You now know how to represent information in matrix form, how to add and subtract matrices, and how to multiply a matrix and a number (scalar multiplication). We now turn our attention to multiplying two matrices.

It might seem natural to multiply two matrices by multiplying corresponding entries. However, such a definition has not proved to be useful, and it turns out that the definition that is useful is rather unusual, so we will consider several examples before formally defining matrix multiplication.

Example 8 illustrates scalar multiplication by considering a 3% profit in three sales locations, as given by

$$(0.03)S_{1988} = (0.3)\begin{bmatrix} 175 & 300 & 400 \\ 110 & 100 & 100 \end{bmatrix} \begin{matrix} \text{Wholesale} \\ \text{Retail} \end{matrix}$$

with columns labeled Chicago, Los Angeles, New York.

Now, however, suppose an inventory tax of 3% is imposed on wholesale sales and a 6% tax is imposed on retail sales. What is the total tax (in thousands of dollars) at each location? First, the necessary calculations to answer this question must be summarized:

	Wholesale tax	+	Retail tax		Total tax
Chicago tax:	$0.03(175)$	+	$0.06(110)$	$= 5.25 + 6.6$	$= 11.85$
Los Angeles tax:	$0.03(300)$	+	$0.06(100)$	$= 9 + 6$	$= 15$
New York tax:	$0.03(400)$	+	$0.06(100)$	$= 12 + 6$	$= 18$

These calculations summarize the operation of matrix multiplication.

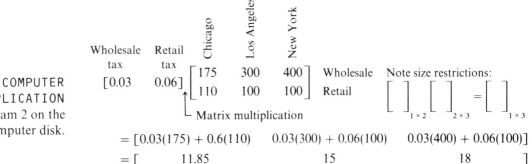

COMPUTER
APPLICATION
See Program 2 on the
computer disk.

$$= [0.03(175) + 0.6(110) \quad 0.03(300) + 0.06(100) \quad 0.03(400) + 0.06(100)]$$

$$= [\quad 11.85 \quad\quad\quad 15 \quad\quad\quad 18 \quad\quad]$$

The operation of matrix multiplication has widespread applications, but it is difficult to understand, so we will proceed slowly through several examples, each progressing from the previous one.

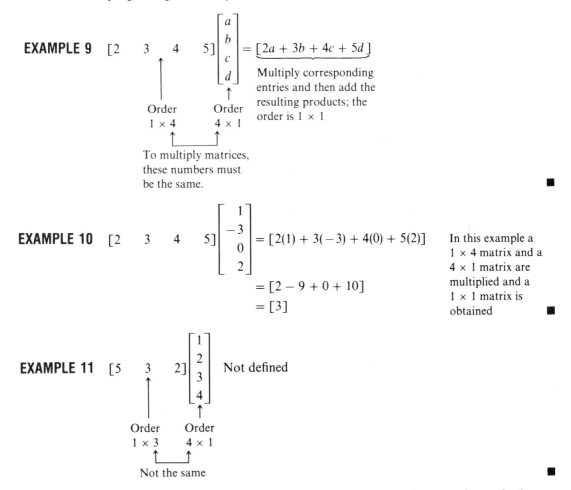

EXAMPLE 9

$$[2 \quad 3 \quad 4 \quad 5]\begin{bmatrix} a \\ b \\ c \\ d \end{bmatrix} = [\underline{2a + 3b + 4c + 5d}]$$

Multiply corresponding entries and then add the resulting products; the order is 1×1

Order
1×4

Order
4×1

To multiply matrices, these numbers must be the same.

EXAMPLE 10

$$[2 \quad 3 \quad 4 \quad 5]\begin{bmatrix} 1 \\ -3 \\ 0 \\ 2 \end{bmatrix} = [2(1) + 3(-3) + 4(0) + 5(2)]$$

$$= [2 - 9 + 0 + 10]$$

$$= [3]$$

In this example a 1×4 matrix and a 4×1 matrix are multiplied and a 1×1 matrix is obtained

EXAMPLE 11

$$[5 \quad 3 \quad 2]\begin{bmatrix} 1 \\ 2 \\ 3 \\ 4 \end{bmatrix} \quad \text{Not defined}$$

Order
1×3

Order
4×1

Not the same

The process illustrated in Examples 9–11 is repeated over and over for larger-order matrices. If two matrices cannot be multiplied because of their order (as in Example 11), then they are said to be **nonconformable** for multiplication.

You can multiply matrices that are not both row or column matrices. For example,

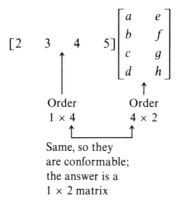

$$[2 \quad 3 \quad 4 \quad 5]\begin{bmatrix} a & e \\ b & f \\ c & g \\ d & h \end{bmatrix}$$

Order
1 × 4

Order
4 × 2

Same, so they
are conformable;
the answer is a
1 × 2 matrix

If there are two columns, then the product is found in two steps:

$$[2 \quad 3 \quad 4 \quad 5]\begin{bmatrix} a & e \\ b & f \\ c & g \\ d & h \end{bmatrix} = [2a + 3b + 4c + 5d \qquad]$$

Row 1, column 1

Answer for position 1, 1

$$[2 \quad 3 \quad 4 \quad 5]\begin{bmatrix} a & e \\ b & f \\ c & g \\ d & h \end{bmatrix} = [2a + 3b + 4c + 5d \qquad 2e + 3f + 4g + 5h]$$

Row 1, column 2

Answer for position 1, 2

EXAMPLE 12
$$[2 \quad 3 \quad 4 \quad 5]\begin{bmatrix} 1 & 2 \\ -3 & 1 \\ 0 & -2 \\ 2 & 3 \end{bmatrix} = [2(1) + 3(-3) + 4(0) + 5(2) \qquad]$$

$$[2 \quad 3 \quad 4 \quad 5]\begin{bmatrix} 1 & 2 \\ -3 & 1 \\ 0 & -2 \\ 2 & 3 \end{bmatrix} = [\mathbf{3} \qquad]$$

$$[2 \quad 3 \quad 4 \quad 5]\begin{bmatrix} 1 & 2 \\ -3 & 1 \\ 0 & -2 \\ 2 & 3 \end{bmatrix} = [3 \qquad 2(2) + 3(1) + 4(-2) + 5(3)]$$

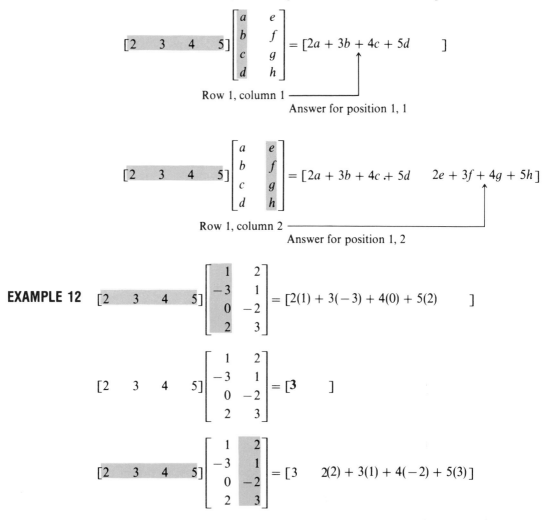

$$[2 \quad 3 \quad 4 \quad 5] \begin{bmatrix} 1 & 2 \\ -3 & 1 \\ 0 & -2 \\ 2 & 3 \end{bmatrix} = [3 \quad 14] \qquad \blacksquare$$

EXAMPLE 13
$$[2 \quad -1 \quad 2] \begin{bmatrix} 4 & 2 & -1 \\ 1 & 0 & 2 \\ 3 & -1 & 3 \end{bmatrix} = [2(4) + (-1)1 + 2(3) \qquad \quad]$$

$$[2 \quad -1 \quad 2] \begin{bmatrix} 4 & 2 & -1 \\ 1 & 0 & 2 \\ 3 & -1 & 3 \end{bmatrix} = [13 \qquad 2(2) + (-1)(0) + 2(-1) \qquad]$$

$$[2 \quad -1 \quad 2] \begin{bmatrix} 4 & 2 & -1 \\ 1 & 0 & 2 \\ 3 & -1 & 3 \end{bmatrix} = [13 \qquad 2 \qquad 2(-1) + (-1)(2) + 2(3)]$$

$$= [13 \quad 2 \quad 2] \qquad \blacksquare$$

Example 13 is written out to clearly demonstrate what is happening. Your work, however, would probably look like this:

Be sure you know where these numbers came from; this is the way work will be shown in subsequent examples

$$[2 \quad -1 \quad 2] \begin{bmatrix} 4 & 2 & -1 \\ 1 & 0 & 2 \\ 3 & -1 & 3 \end{bmatrix} = [8 - 1 + 6 \qquad 4 + 0 - 2 - 2 - 2 + 6]$$

$$= [13 \quad 2 \quad 2]$$

EXAMPLE 14
$$\begin{bmatrix} 1 & 3 & 2 \\ -1 & 4 & 3 \end{bmatrix} \begin{bmatrix} 4 & 2 \\ -1 & 3 \\ 2 & -3 \end{bmatrix} = \begin{bmatrix} 4-3+4 & 2+9-6 \\ -4-4+6 & -2+12-9 \end{bmatrix} = \begin{bmatrix} 5 & 5 \\ -2 & 1 \end{bmatrix} \qquad \blacksquare$$

Now, to define matrix multiplication, let the symbol

$$A = [a_{ij}]_{m,n}$$

mean A is a matrix of order m by n with $1 \le i \le m$ and $1 \le j \le n$. That is, a_{23} is the entry in the second row, third column, or, more generally, a_{ij} is an element of A in the ith row, jth column. Now let $A = [a_{ij}]_{m,r}$, $B = [b_{ij}]_{p,n}$, and $C = [c_{ij}]_{m,n}$, and write $AB = C$, where C is called the **product** of A and B. If $p = r$, then matrix multiplication is possible, so we can define multiplication for

These numbers tell the size of the product matrix

$$A = [a_{ij}]_{m,r} \qquad \text{and} \qquad B = [b_{ij}]_{r,n}$$

Note that these must be the same for multiplication to be defined

Matrix Multiplication

$AB = C$, where

$$[c_{ij}]_{m,n} = \begin{bmatrix} a_{i1} & a_{i2} & \cdots & a_{ir} \end{bmatrix} \begin{bmatrix} b_{1j} \\ b_{2j} \\ \vdots \\ b_{rj} \end{bmatrix}$$

$$= [a_{i1}b_{1j} + a_{i2}b_{2j} + \cdots + a_{ir}b_{rj}]$$

EXAMPLE 15 Give the *order* of the following products and give the *position* of the entry found by the highlighted product. If multiplication is not possible, say so.

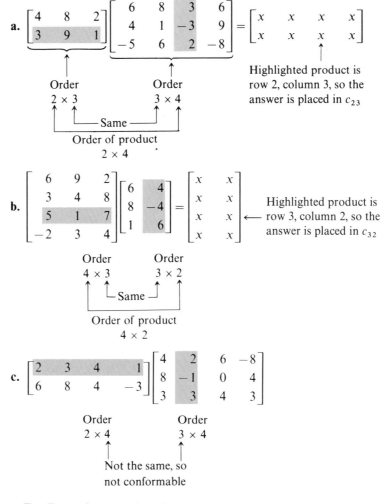

For Examples 16 and 17, let

$$A = \begin{bmatrix} 1 & 2 \\ 3 & 4 \end{bmatrix} \quad \text{and} \quad B = \begin{bmatrix} -1 & 2 \\ 1 & 3 \end{bmatrix}$$

EXAMPLE 16 $AB = \begin{bmatrix} 1 & 2 \\ 3 & 4 \end{bmatrix} \begin{bmatrix} -1 & 2 \\ 1 & 3 \end{bmatrix} = \begin{bmatrix} -1+2 & 2+6 \\ -3+4 & 6+12 \end{bmatrix} = \begin{bmatrix} 1 & 8 \\ 1 & 18 \end{bmatrix}$ ∎

EXAMPLE 17 $BA = \begin{bmatrix} -1 & 2 \\ 1 & 3 \end{bmatrix} \begin{bmatrix} 1 & 2 \\ 3 & 4 \end{bmatrix} = \begin{bmatrix} -1+6 & -2+8 \\ 1+9 & 2+12 \end{bmatrix} = \begin{bmatrix} 5 & 6 \\ 10 & 14 \end{bmatrix}$ ∎

Note from Examples 16 and 17 that $AB \neq BA$. That is, matrix multiplication is *not commutative*. This means you must be careful not to switch the order of the matrix factors when working with matrix products.

This section concludes with some models that use matrix multiplication.

EXAMPLE 18 A developer builds a housing complex featuring two-, three-, and four-bedroom units. Each unit comes in two different floor plans. The matrix P (for production) tells the number of each type of unit for this development.

$$P = \begin{matrix} & \text{Plan I} & \text{Plan II} \\ \begin{bmatrix} 10 & 5 \\ 25 & 10 \\ 15 & 10 \end{bmatrix} & \begin{matrix} \text{2 bedrooms} \\ \text{3 bedrooms} \\ \text{4 bedrooms} \end{matrix} \end{matrix}$$

Many materials are used in building these homes, but this model will be simplified to include only lumber, concrete, fixtures, and labor. The matrix M (for materials) gives the amounts of these materials used (in appropriate units of each).*

$$M = \begin{matrix} \text{Lumber} & \text{Concrete} & \text{Fixtures} & \text{Labor} \\ \begin{bmatrix} 7 & 8 & 9 & 20 \\ 8 & 9 & 9 & 22 \end{bmatrix} & \begin{matrix} \text{Plan I} \\ \text{Plan II} \end{matrix} \end{matrix}$$

The matrix product PM gives the amount of material needed for the development.

$$PM = \begin{matrix} & \text{I} & \text{II} \\ 2 \\ 3 \\ 4 \end{matrix} \begin{bmatrix} 10 & 5 \\ 25 & 10 \\ 15 & 10 \end{bmatrix} \begin{matrix} \text{Lumber} & \text{Concrete} & \text{Fixtures} & \text{Labor} \\ \begin{bmatrix} 7 & 8 & 9 & 20 \\ 8 & 9 & 9 & 22 \end{bmatrix} \end{matrix}$$

$$= \begin{matrix} \text{Lumber} & \text{Concrete} & \text{Fixtures} & \text{Labor} \\ \begin{bmatrix} 70+40 & 80+45 & 90+45 & 200+110 \\ 175+80 & 200+90 & 225+90 & 500+220 \\ 105+80 & 120+90 & 135+90 & 300+220 \end{bmatrix} & \begin{matrix} \text{2 bedrooms} \\ \text{3 bedrooms} \\ \text{4 bedrooms} \end{matrix} \end{matrix}$$

$$= \begin{matrix} \text{Lumber} & \text{Concrete} & \text{Fixtures} & \text{Labor} \\ \begin{bmatrix} 110 & 125 & 135 & 310 \\ 255 & 290 & 315 & 720 \\ 185 & 210 & 225 & 520 \end{bmatrix} & \begin{matrix} \text{2 bedrooms} \\ \text{3 bedrooms} \\ \text{4 bedrooms} \end{matrix} \end{matrix}$$

---

* For example, lumber in 1000 board feet, concrete in cubic yards, etc.

If the cost of each unit of material is given by the matrix C (for cost), then the total cost of each model for this development is $(PM)C$.

Cost per unit

$$C = \begin{bmatrix} 800 \\ 90 \\ 1000 \\ 1000 \end{bmatrix} \begin{array}{l} \text{Lumber} \\ \text{Concrete} \\ \text{Fixtures} \\ \text{Labor} \end{array}$$

$$(PM)C = \begin{bmatrix} 110 & 125 & 135 & 310 \\ 255 & 290 & 315 & 720 \\ 185 & 210 & 225 & 520 \end{bmatrix} \begin{bmatrix} 800 \\ 90 \\ 1000 \\ 1000 \end{bmatrix}$$

$$= \begin{bmatrix} 544,250 \\ 1,265,100 \\ 911,900 \end{bmatrix} \begin{array}{l} \text{2 bedrooms} \\ \text{3 bedrooms} \\ \text{4 bedrooms} \end{array}$$

Since 15 of the units have two bedrooms, the average cost per unit is $36,283 ($544,250 ÷ 15). For three-bedroom units the average cost per unit is $36,146, and for four-bedroom units the average cost is $36,476.   ■

EXAMPLE 19   In Example 4 a matrix was developed to represent the diplomatic relations of the United States, the Soviet Union, Cuba, and Mexico. Let A represent the communication matrix from that example:

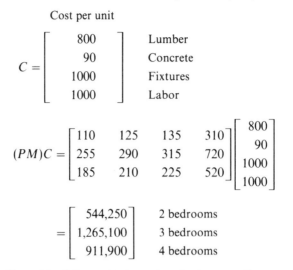

$$\begin{array}{cccc} \text{U.S.} & \text{U.S.S.R.} & \text{Cuba} & \text{Mexico} \end{array}$$

$$A = \begin{bmatrix} 0 & 1 & 0 & 1 \\ 1 & 0 & 1 & 1 \\ 0 & 1 & 0 & 0 \\ 1 & 1 & 0 & 0 \end{bmatrix} \begin{array}{l} \text{U.S.} \\ \text{U.S.S.R.} \\ \text{Cuba} \\ \text{Mexico} \end{array}$$

How many channels of communication are open to the various countries if they are willing to speak through an intermediary?

Solution   Consider a particular example, say, Mexico sending a message to Cuba through a single intermediary. Mexico can send a message to the United States, but the United States cannot pass the message along to Cuba; Mexico can send a message to the Soviet Union and then the Soviet Union can forward it to Cuba; Mexico to Cuba direct does not qualify as sending the message through an intermediary. Therefore, there is only one way that Mexico can send a message to Cuba via an intermediary. How about the United States sending a message to the United States through an intermediary? (Perhaps to test for leaks in the communications network!) The United States can send a message to the Soviet Union and then back to the United States, as well as to Mexico and then back. Thus there are two ways that the United States can send a message to itself through an intermediary. Do you see that there are three ways that the Soviet Union can send a round-trip message through one intermediary? All of this information can be summarized in matrix form:

	U.S.	U.S.S.R.	Cuba	Mexico
U.S.	2	1	1	1
U.S.S.R.	1	3	0	1
Cuba	1	0	1	1
Mexico	1	1	1	2

This matrix is the same as $AA = A^2$:

$$A^2 = \begin{bmatrix} 0 & 1 & 0 & 1 \\ 1 & 0 & 1 & 1 \\ 0 & 1 & 0 & 0 \\ 1 & 1 & 0 & 0 \end{bmatrix} \begin{bmatrix} 0 & 1 & 0 & 1 \\ 1 & 0 & 1 & 1 \\ 0 & 1 & 0 & 0 \\ 1 & 1 & 0 & 0 \end{bmatrix}$$

$$= \begin{bmatrix} 0+1+0+1 & 0+0+0+1 & 0+1+0+0 & 0+1+0+0 \\ 0+0+0+1 & 1+0+1+1 & 0+0+0+0 & 1+0+0+0 \\ 0+1+0+0 & 0+0+0+0 & 0+1+0+0 & 0+1+0+0 \\ 0+1+0+0 & 1+0+0+0 & 0+1+0+0 & 1+1+0+0 \end{bmatrix}$$

	U.S.	U.S.S.R.	Cuba	Mexico	
$=$	2	1	1	1	U.S.
	1	3	0	1	U.S.S.R.
	1	0	1	1	Cuba
	1	1	1	2	Mexico

■

In general, given a communication matrix A:

One intermediary message—one multiplication: $AA = A^2$

Two intermediary messages—two multiplications: $AAA = A^3$

Three intermediary messages—three multiplications: $AAAA = A^4$

And so on.

Another common application of communication matrices is the network of airline routes in the United States. For example, there are several different ways to fly from Kansas City to San Francisco: direct, through one intermediate city, two intermediate cities, and so on. Problem 65 in Problem Set 2.1 involves this type of application.

Problem Set 2.1

State the order of each matrix in Problems 1–2 and if it is a square matrix, a column matrix, or a row matrix.

1. a. $A = \begin{bmatrix} 6 & 1 & 4 \\ 7 & 9 & 2 \\ 1 & 5 & 3 \end{bmatrix}$ **b.** $B = \begin{bmatrix} 2 \\ 1 \\ 5 \end{bmatrix}$

c. $C = \begin{bmatrix} 4 & 9 \\ 1 & 6 \\ 7 & 5 \end{bmatrix}$ **d.** $D = \begin{bmatrix} 4 \\ 0 \\ 1 \\ 3 \end{bmatrix}$

e. $E = \begin{bmatrix} 4 & 1 & 7 \\ 9 & 6 & 5 \end{bmatrix}$

f. $F = \begin{bmatrix} 5 & 0 & 1 & 2 \end{bmatrix}$

g. $G = \begin{bmatrix} 6 & 9 \end{bmatrix}$

2. a. $H = \begin{bmatrix} 6 & 5 \\ 9 & 2 \end{bmatrix}$ **b.** $I = \begin{bmatrix} 1 & 0 & 0 \\ 0 & 1 & 0 \\ 0 & 0 & 1 \end{bmatrix}$

c. $J = \begin{bmatrix} 0 & 1 \\ 1 & 0 \end{bmatrix}$ **d.** $K = \begin{bmatrix} 1 & 0 \\ 0 & 1 \\ 0 & 0 \\ 0 & 0 \\ 0 & 0 \end{bmatrix}$

e. $L = \begin{bmatrix} 1 & 0 & 0 & 0 \\ 0 & 1 & 0 & 0 \end{bmatrix}$

f. $M = [5]$

In Problems 3–4 find replacements of the variables that make the matrices equal. If it is not possible to make the matrices equal, state that.

3. a. $\begin{bmatrix} 4 & 8 \\ -1 & x \end{bmatrix} = \begin{bmatrix} y & 8 \\ -1 & 2 \end{bmatrix}$

b. $\begin{bmatrix} 6 \\ x \\ y \end{bmatrix} = \begin{bmatrix} z \\ 4 \\ -9 \end{bmatrix}$

c. $\begin{bmatrix} 1 & 0 \\ 0 & 1 \end{bmatrix} = \begin{bmatrix} a & b \\ c & d \end{bmatrix}$

4. a. $\begin{bmatrix} 0 & x \\ 0 & 1 \end{bmatrix} = \begin{bmatrix} 1 & 2 \\ 0 & y \end{bmatrix}$

b. $\begin{bmatrix} 2 & 4 & 6 \\ 5 & 9 & 3 \end{bmatrix} = \begin{bmatrix} 2 & x & y \\ z & 9 & 3 \end{bmatrix}$

c. $\begin{bmatrix} -1 & 2 & 3 \\ 4 & x & 0 \end{bmatrix} = \begin{bmatrix} -1 & 2 & y \\ z & 2 & 1 \end{bmatrix}$

5. A company supplies four parts for General Motors. The parts are manufactured at four factories. Summarize the following production information using a matrix: Factory A produces 25 units of part one, 42 units of part two, and 193 units of part three; factory B produces 16 units of part one, 39 units of part two, and 150 units of part three; factory C produces 50 units of each part; and factory D produces 320 units of part four only.

6. a. Write Table 2.1 in matrix form.
   b. What is the entry in row 2, column 3?

c. What is the entry in row 3, column 2?
d. In which row and column is the price of a Volkswagen Microbus in Sweden?
e. In which row and column is the price of an Opel Kadett in Holland?
f. Would the prices of a Volkswagen Microbus in various countries be represented as a row matrix or a column matrix?

In Problems 7–8 give the order of the product and tell the position of the entry found by the highlighted product. If multiplication is not possible, state that.

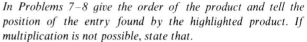

7. a. $\begin{bmatrix} \boxed{1} & \boxed{2} & \boxed{3} \\ 8 & 4 & 5 \\ 6 & -1 & 3 \end{bmatrix} \begin{bmatrix} 4 & \boxed{6} & 1 \\ 2 & \boxed{5} & 1 \\ 4 & \boxed{9} & -3 \end{bmatrix}$

b. $\begin{bmatrix} 1 & 2 & 3 \\ 8 & 4 & 5 \\ \boxed{6} & \boxed{-1} & \boxed{3} \end{bmatrix} \begin{bmatrix} 4 & \boxed{6} & 1 \\ 2 & \boxed{5} & 1 \\ 4 & \boxed{9} & -3 \end{bmatrix}$

c. $\begin{bmatrix} 1 & 2 & 3 \\ \boxed{8} & \boxed{4} & \boxed{5} \\ 6 & -1 & 3 \end{bmatrix} \begin{bmatrix} \boxed{4} & 6 & 1 \\ \boxed{2} & 5 & 1 \\ \boxed{4} & 9 & -3 \end{bmatrix}$

d. $\begin{bmatrix} \boxed{6} & \boxed{-3} & \boxed{2} & \boxed{-4} & \boxed{5} \\ 3 & 5 & 1 & 2 & 3 \end{bmatrix} \begin{bmatrix} 6 & 5 \\ 1 & 1 \\ 2 & 7 \\ 3 & -2 \\ 4 & -1 \end{bmatrix}$

e. $\begin{bmatrix} 6 & -3 & 2 & -4 & 5 \\ \boxed{3} & \boxed{5} & \boxed{1} & \boxed{2} & \boxed{3} \end{bmatrix} \begin{bmatrix} 6 & 5 \\ 1 & 1 \\ 2 & 7 \\ 3 & -2 \\ 4 & -1 \end{bmatrix}$

f. $\begin{bmatrix} \boxed{6} & -3 & 2 & -4 & 5 \\ 3 & 5 & 1 & 2 & 3 \end{bmatrix} \begin{bmatrix} 6 & \boxed{5} \\ 1 & \boxed{1} \\ 2 & \boxed{7} \\ 3 & \boxed{-2} \\ 4 & \boxed{-1} \end{bmatrix}$

8. a. $\begin{bmatrix} 6 & \boxed{8} & 1 \\ \boxed{-2} & \boxed{3} & \boxed{5} \end{bmatrix} \begin{bmatrix} 5 & \boxed{6} & -2 & -9 & -5 \\ 2 & \boxed{5} & 1 & 8 & 4 \\ 1 & \boxed{-3} & 7 & 3 & 12 \end{bmatrix}$

b. $\begin{bmatrix} \boxed{6} & \boxed{8} & \boxed{1} \\ -2 & 3 & 5 \end{bmatrix} \begin{bmatrix} 5 & 6 & -2 & -9 & \boxed{-5} \\ 2 & 5 & 1 & 8 & \boxed{4} \\ 1 & -3 & 7 & 3 & \boxed{12} \end{bmatrix}$

c. $\begin{bmatrix} 9 & -8 & 1 & 4 \\ 5 & 2 & 3 & 6 \\ -7 & 6 & -5 & -8 \end{bmatrix} \begin{bmatrix} 1 & 5 & 9 & 13 & 17 \\ 2 & 6 & 10 & 14 & 18 \\ 3 & 7 & 11 & 15 & 19 \\ 4 & 8 & 12 & 16 & 20 \end{bmatrix}$

d. $\begin{bmatrix} 9 & -8 & 1 & 4 \\ 5 & 2 & 3 & 6 \\ -7 & 6 & -5 & -8 \end{bmatrix} \begin{bmatrix} 1 & 5 & 9 & 13 & 17 \\ 2 & 6 & 10 & 14 & 18 \\ 3 & 7 & 11 & 15 & 19 \\ 4 & 8 & 12 & 16 & 20 \end{bmatrix}$

e. $\begin{bmatrix} 6 & 4 & 6 & 3 \\ 9 & -1 & 3 & 4 \end{bmatrix} \begin{bmatrix} 1 & -2 & 4 & 6 \\ 6 & 5 & 8 & -2 \\ 3 & -3 & 7 & 5 \end{bmatrix}$

f. $\begin{bmatrix} 6 & 4 & 6 & 3 \\ 9 & -1 & 3 & 4 \end{bmatrix} \begin{bmatrix} 1 & -2 & 4 & 6 \\ 6 & 5 & 8 & -2 \\ 3 & -3 & 7 & 5 \end{bmatrix}$

Find the indicated matrices in Problems 9–11, if possible. Give the order of your answer.

9. a. $\begin{bmatrix} 3 & 5 & 8 & 9 \end{bmatrix} \begin{bmatrix} w \\ x \\ y \\ z \end{bmatrix}$

b. $\begin{bmatrix} 2 & -3 & 4 & 5 \end{bmatrix} \begin{bmatrix} w \\ x \\ y \\ z \end{bmatrix}$

c. $\begin{bmatrix} 2 & 3 & 5 \end{bmatrix} \begin{bmatrix} a & d \\ b & e \\ c & f \end{bmatrix}$

d. $\begin{bmatrix} 3 & -2 & 1 \end{bmatrix} \begin{bmatrix} a & b \\ c & d \\ e & f \end{bmatrix}$

10. a. $\begin{bmatrix} 1 & -2 & 2 \end{bmatrix} \begin{bmatrix} 1 & 2 & 3 \\ 4 & -1 & 3 \\ -3 & 2 & 1 \end{bmatrix}$

b. $\begin{bmatrix} 3 & 2 & -4 \end{bmatrix} \begin{bmatrix} 1 & 2 & -3 \\ -4 & 3 & 2 \\ 2 & 3 & 1 \end{bmatrix}$

c. $\begin{bmatrix} 1 & 3 & 2 \\ -1 & 1 & 2 \end{bmatrix} \begin{bmatrix} 6 & 3 & 2 & -3 \\ 1 & 1 & -4 & -1 \\ 2 & 1 & 1 & 2 \end{bmatrix}$

d. $\begin{bmatrix} 6 & 1 & 3 \\ 2 & -3 & 5 \end{bmatrix} \begin{bmatrix} 1 & 0 & 2 & 1 & 1 \\ 2 & 0 & 1 & 1 & 0 \\ 3 & 1 & 4 & 0 & 0 \end{bmatrix}$

11. a. $\begin{bmatrix} w \\ x \\ y \\ z \end{bmatrix} \begin{bmatrix} 8 & -1 & 3 & -2 \end{bmatrix}$

b. $\begin{bmatrix} w \\ x \\ y \\ z \end{bmatrix} \begin{bmatrix} 6 & -1 & 3 & 2 \end{bmatrix}$

c. $\begin{bmatrix} 6 \\ 3 \\ -1 \end{bmatrix} \begin{bmatrix} 2 & -4 & 3 \end{bmatrix}$

d. $\begin{bmatrix} 5 \\ -2 \\ -1 \end{bmatrix} \begin{bmatrix} 2 & 6 & 4 \end{bmatrix}$

Find the indicated matrices in Problems 12–51. Let

$$A = \begin{bmatrix} 1 & 0 & 2 \\ 3 & -1 & 2 \\ 4 & 1 & 0 \end{bmatrix} \qquad B = \begin{bmatrix} 1 & 4 & 0 \\ 3 & -1 & 2 \\ -2 & 1 & 5 \end{bmatrix}$$

$$C = \begin{bmatrix} 8 & 1 & 6 \\ 3 & 5 & 7 \\ 4 & 9 & 2 \end{bmatrix} \qquad D = \begin{bmatrix} -2 \\ 1 \\ 3 \end{bmatrix} \qquad E = \begin{bmatrix} 4 \\ 1 \\ 6 \end{bmatrix}$$

$$F = \begin{bmatrix} 2 & 4 & 7 \end{bmatrix} \qquad G = \begin{bmatrix} 6 & -1 & 0 \end{bmatrix}$$

12. $A + B$ **13.** $B + A$ **14.** $A - B$

15. $B - A$ **16.** $C + D$ **17.** $D + C$

18. $C - D$ **19.** $D - C$ **20.** $E + F$

21. $F + E$ **22.** $E - F$ **23.** $F - E$

24. $2A + B$ **25.** $3A - 4B$ **26.** $2C + 3D$

27. $C - 2D$ **28.** $3E - 2D$ **29.** $3E - 2F$

30. $E + 3F$ **31.** $3B + 2C$ **32.** $A + (B + C)$

33. $(A + B) + C$ **34.** $(B + C) + D$ **35.** $B + (C + D)$

36. AB **37.** BA **38.** AC

39. CA **40.** BC **41.** CB

42. $A(BC)$ **43.** $(AB)C$ **44.** $B(AC)$

45. $(BA)C$ **46.** $C(AB)$ **47.** $(CA)B$

48. $B + C$ **49.** $A(B + C)$ **50.** $(B + C)A$

51. $AB + AC$

Find the indicated matrices in Problems 52–63. Let

$$A = \begin{bmatrix} 1 & 2 \\ 4 & 0 \\ -1 & 3 \\ 2 & 1 \end{bmatrix}$$

$$B = \begin{bmatrix} 4 & 2 \\ -1 & 3 \end{bmatrix}$$

$$C = \begin{bmatrix} 1 & 0 & 0 & 0 \\ 0 & 1 & 0 & 0 \\ 0 & 0 & 1 & 0 \\ 0 & 0 & 0 & 1 \end{bmatrix}$$

$$D = \begin{bmatrix} 4 & 1 & 3 & 6 \\ -1 & 0 & -2 & 3 \end{bmatrix}$$

52. AB **53.** BA **54.** B^2

55. CA **56.** BD **57.** DB

58. $(B + C)A$ **59.** $BA + CA$ **60.** CD

61. A^2 **62.** C^3 **63.** B^3

APPLICATIONS

64. A company with five offices in San Francisco operates its own delivery service for office mail. Arrows represent the direction of communication in the figure. Note that A can send mail directly to offices B, D, and E but not to C.

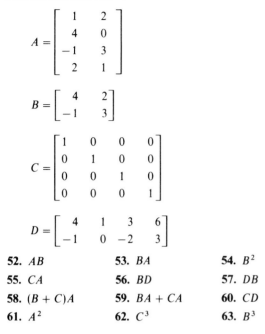

a. Express this delivery network with a communication matrix.
b. Develop a matrix showing the possible two-stop deliveries.
c. Develop a matrix showing the possible three-stop deliveries.

65. Consider the map of airline routes shown in the margin.

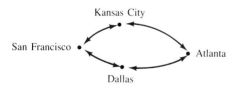

a. Fill in the blanks in the following communication matrix representing the airline routes:

$$T = \begin{array}{c c} & \begin{array}{cccc} \text{SF} & \text{D} & \text{A} & \text{KC} \end{array} \\ \begin{array}{c} \text{SF} \\ \text{D} \\ \text{A} \\ \text{KC} \end{array} & \begin{bmatrix} & & & \\ & & & \\ & & & \\ & & & \end{bmatrix} \end{array}$$

b. In how many ways can you travel from Kansas City to San Francisco making exactly one stop?
c. In how many ways can you travel from San Francisco to Kansas City making two stops?
d. Write the communication matrix showing the number of routes among these cities if you make exactly one stop.
e. Write the communication matrix showing the number of routes among these cities if you make exactly two stops.

66. In Example 18, add 40 units of overhead for the plan I homes and 45 units of overhead for the plan II models. If the cost for overhead is 1000 per unit, find the total cost for the housing complex as well as the average cost for two-, three-, and four-bedroom homes assuming all the other information in Example 18 remains unchanged.

67. If the countries in Example 19 are willing to speak through two intermediaries, find the communication matrix and tell how many different ways the United States can communicate with each of the other countries in this fashion.

Historical Question The English mathematician Arthur Cayley (1821–1895) was responsible for matrix theory. Cayley began his career as a lawyer in 1849, but he was always a mathematician at heart, publishing more than 200 papers in mathematics while practicing law. In 1863 he gave up law to accept a chair in mathematics at Cambridge. He originated the notion of matrices in 1858 and worked on matrix theory over the next several years. His work was entirely theoretical and was not used for any practical purpose until 1925 when matrices were used in quantum mechanics. Since 1950, matrices have played an important role in the social sciences and business. Much of the early work with matrices focused upon their properties. In Problems 68–73 we look at some of these properties. Let A, B, and C be square matrices of order 3 and let a and b be real numbers. Give an example to show that each of the following is false or else explain why you think the property is true.

68. $A + B = B + A$ **69.** $A - B = B - A$

70. $(A - B) - C = A - (B - C)$

71. $(A + B) + C = A + (B + C)$

72. $a(B + C) = aB + aC$

73. $(a + b)C = aC + bC$

2.2 Gauss–Jordan Elimination

In this section we develop a procedure to solve more general linear systems. Consider a system of m equations and n variables:

$$\begin{cases} a_{11}x_1 + a_{12}x_2 + \cdots + a_{1n}x_n = b_1 \\ a_{21}x_1 + a_{22}x_2 + \cdots + a_{2n}x_n = b_2 \\ \quad\vdots \qquad \vdots \qquad \vdots \qquad \vdots \\ a_{m1}x_1 + a_{m2}x_2 + \cdots + a_{mn}x_n = b_m \end{cases}$$

Now consider what is called the **augmented matrix** for this system:

$$\begin{bmatrix} a_{11} & a_{12} & \cdots & a_{1n} & b_1 \\ a_{21} & a_{22} & \cdots & a_{2n} & b_2 \\ \vdots & \vdots & & \vdots & \vdots \\ a_{m1} & a_{m2} & \cdots & a_{mn} & b_m \end{bmatrix}$$

EXAMPLE 1   Write the given systems in augmented matrix form.

a. $\begin{cases} 2x + 3y = 8 \\ 3x + 2y = 7 \end{cases}$

b. $\begin{cases} 2x + y = 3 \\ 3x - y = 2 \\ 4x + 3y = 7 \end{cases}$

c. $\begin{cases} 3x - 2y + z = -2 \\ 4x - 5y + 3z = -9 \\ 2x - y + 5z = -5 \end{cases}$

d. $\begin{cases} 5x - 3y + z = -3 \\ 2x + 5z = 14 \end{cases}$

e. $\begin{cases} x_1 - 3x_3 + x_5 = -3 \\ x_2 + x_4 = -1 \\ x_3 + x_5 = 7 \\ x_1 + x_2 - x_3 + 4x_4 = -8 \\ x_1 + x_2 + x_3 + x_4 + x_5 = 8 \end{cases}$

Solution   Note that some coefficients are negative and some are zeros.

a. $\begin{bmatrix} 2 & 3 & 8 \\ 3 & 2 & 7 \end{bmatrix}$

b. $\begin{bmatrix} 2 & 1 & 3 \\ 3 & -1 & 2 \\ 4 & 3 & 7 \end{bmatrix}$

c. $\begin{bmatrix} 3 & -2 & 1 & -2 \\ 4 & -5 & 3 & -9 \\ 2 & -1 & 5 & -5 \end{bmatrix}$

d. $\begin{bmatrix} 5 & -3 & 1 & -3 \\ 2 & 0 & 5 & 14 \end{bmatrix}$

e. $\begin{bmatrix} 1 & 0 & -3 & 0 & 1 & -3 \\ 0 & 1 & 0 & 1 & 0 & -1 \\ 0 & 0 & 1 & 0 & 1 & 7 \\ 1 & 1 & -1 & 4 & 0 & -8 \\ 1 & 1 & 1 & 1 & 1 & 8 \end{bmatrix}$ ∎

EXAMPLE 2   Write a system of equations (use $x_1, x_2, \ldots, x_n$ for the variables) that has the given augmented matrix.

$$
\textbf{a.} \begin{bmatrix} 2 & 1 & -1 & -3 \\ 3 & -2 & 1 & 9 \\ 1 & -4 & 3 & 17 \end{bmatrix}
\qquad
\textbf{b.} \begin{bmatrix} 1 & 0 & 0 & 3 \\ 0 & 1 & 0 & -2 \\ 0 & 0 & 1 & -5 \end{bmatrix}
\qquad
\textbf{c.} \begin{bmatrix} 1 & 0 & 4 \\ 0 & 1 & 3 \end{bmatrix}
$$

$$
\textbf{d.} \begin{bmatrix} 1 & 0 & -5 \\ 0 & 1 & 2 \\ 2 & 1 & -3 \end{bmatrix}
\qquad
\textbf{e.} \begin{bmatrix} 1 & 0 & 0 & 3 \\ 0 & 1 & 0 & -3 \\ 0 & 0 & 1 & 6 \\ 0 & 0 & 0 & 1 \end{bmatrix}
\qquad
\textbf{f.} \begin{bmatrix} 1 & 0 & 0 & 1 \\ 0 & 1 & 0 & 5 \\ 0 & 0 & 1 & -2 \\ 0 & 0 & 0 & 0 \end{bmatrix}
$$

Solution   **a.** $\begin{cases} 2x_1 + x_2 - x_3 = -3 \\ 3x_1 - 2x_2 + x_3 = 9 \\ x_1 - 4x_2 + 3x_3 = 17 \end{cases}$ **b.** $\begin{cases} x_1 = 3 \\ x_2 = -2 \\ x_3 = -5 \end{cases}$ **c.** $\begin{cases} x_1 = 4 \\ x_2 = 3 \end{cases}$

d. $\begin{cases} x_1 = -5 \\ x_2 = 2 \\ 2x_1 + x_2 = -3 \end{cases}$ **e.** $\begin{cases} x_1 = 3 \\ x_2 = -3 \\ x_3 = 6 \\ 0 = 1 \end{cases}$ **f.** $\begin{cases} x_1 = 1 \\ x_2 = 5 \\ x_3 = -2 \\ 0 = 0 \end{cases}$ ∎

The goal of this section is to solve a system of m equations with n unknowns. We have already looked at systems of two equations with two unknowns. In high school you may have solved three equations with three unknowns. Now, however, we want to be able to solve problems with two equations and five unknowns, or three equations and two unknowns, or any mixture of linear equations or unknowns. The procedure of this section—**Gauss–Jordan elimination**—is a general method for solving all of these types of systems. We write the system in augmented matrix form (as in Example 1), then carry out a process that transforms the matrix until the solution is obvious. Look back at Example 2—the solutions to parts b and c are obvious. Part e shows $0 = 1$ in the last equation, so this system has no solution (0 cannot equal 1), and part f shows $0 = 0$ (which is true for all replacements of the variable), which means that the solution is found by looking at the other equations (namely, $x_1 = 1$, $x_2 = 5$, and $x_3 = -2$). The terms with nonzero coefficients in these examples are arranged on a diagonal, and such a system is said to be in *diagonal form*.

But what process will allow us to transform a matrix into diagonal form? We begin with some steps called **elementary row operations**. Elementary row operations change the *form* of a matrix, but the new form represents an equivalent system. Matrices which represent equivalent systems are called **equivalent matrices** and we introduce these elementary row operations in order to write equivalent matrices. Let us work with a system with three equations and three unknowns (any size will work the same way).

$$
\begin{array}{cc}
\textit{System format} & \textit{Matrix format} \\[4pt]
\begin{cases} 3x - 2y + 4z = 11 \\ x - y - 2z = -7 \\ 2x - 3y + z = -1 \end{cases}
&
\begin{bmatrix} 3 & -2 & 4 & 11 \\ 1 & -1 & -2 & -7 \\ 2 & -3 & 1 & -1 \end{bmatrix}
\end{array}
$$

Interchanging two equations is equivalent to interchanging two rows in matrix format, and, certainly, if we do this, the solution to the system will be the same:

System format

$$\begin{cases} x - y - 2z = -7 \\ 3x - 2y + 4z = 11 \\ 2x - 3y + z = -1 \end{cases}$$

Matrix format

$$\left[\begin{array}{ccc|c} 1 & -1 & -2 & -7 \\ 3 & -2 & 4 & 11 \\ 2 & -3 & 1 & -1 \end{array}\right]$$

The first and second equations (rows) are interchanged

Elementary Row Operation 1: Interchange any two rows. Since multiplying or dividing both sides of any equation by any nonzero number does not change the solution, then, in matrix format, the solution will not be changed if any row is multiplied or divided by a nonzero constant. For example, multiply both sides of the first equation by -3:

System format

$$\begin{cases} -3x + 3y + 6z = 21 \\ 3x - 2y + 4z = 11 \\ 2x - 3y + z = -1 \end{cases}$$

Matrix format

$$\left[\begin{array}{ccc|c} -3 & 3 & 6 & 21 \\ 3 & -2 & 4 & 11 \\ 2 & -3 & 1 & -1 \end{array}\right]$$

Both sides of the first equation are multiplied by -3; the first row is multiplied by -3

Because it is easier to program software by dividing multiplication and division into two separate row operations, we have done the same.

Elementary Row Operation 2: Multiply all the elements of a row by the same nonzero real number.

Elementary Row Operation 3: Divide all the elements of a row by the same nonzero real number.

The last property we need to carry out Gauss–Jordan elimination rests on the property we used with the linear combination method, namely, adding equations to eliminate a variable. In terms of the system format this means that one equation can be replaced by its sum with another equation in the system. For example, if we add the first equation to the second equation we have:

System format

$$\begin{cases} -3x + 3y + 6z = 21 \\ y + 10z = 32 \\ 2x - 3y + z = -1 \end{cases}$$

Matrix format

$$\left[\begin{array}{ccc|c} -3 & 3 & 6 & 21 \\ 0 & 1 & 10 & 32 \\ 2 & --3 & 1 & -1 \end{array}\right]$$

In terms of matrix format, any row can be replaced by its sum with some other row.

This process is usually used in conjunction with Elementary Row Operation 2. That is, an equation is changed by adding to it a nonzero multiple of another equation in the system. Go back to the original system:

System format

$$\begin{cases} 3x - 2y + 4z = 11 \\ x - y - 2z = -7 \\ 2x - 3y + z = -1 \end{cases}$$

Matrix format

$$\left[\begin{array}{ccc|c} 3 & -2 & 4 & 11 \\ 1 & -1 & -2 & -7 \\ 2 & -3 & 1 & -1 \end{array}\right]$$

We can change this system by multiplying the second equation by -3 and adding the result to the first equation. In matrix terminology we would say multiply the second row by -3 and add it to the first row:

System format	*Matrix format*

$$\begin{cases} \quad\ \ y + 10z = 32 \\ x - \ y - \ 2z = -7 \\ 2x - 3y + \quad z = -1 \end{cases} \qquad \begin{bmatrix} 0 & 1 & 10 & \vdots & 32 \\ 1 & -1 & -2 & \vdots & -7 \\ 2 & -3 & 1 & \vdots & -1 \end{bmatrix}$$

Once again, multiply the second row, this time by -2, and add it to the third row. Wait! Why -2? Where did that come from? The idea is the same one we used in the linear combination method—we use a number that will give a zero coefficient to the x in the third equation.

System format	*Matrix format*

$$\begin{aligned} y + 10z &= 32 \\ x - \ y - \ 2z &= -7 \\ -y + \ 5z &= 13 \end{aligned} \qquad \begin{bmatrix} 0 & 1 & 10 & \vdots & 32 \\ 1 & -1 & -2 & \vdots & -7 \\ 0 & -1 & 5 & \vdots & 13 \end{bmatrix}$$

Note that the multiplied row is not changed, that, instead, the changed row is the one to which the multiplied row is added. We call the original row the **pivot row** and the changed row the **target row**.

Elementary Row Operation 4: Multiply all the entries of a row (the pivot row) by a nonzero real number and add each resulting product to the corresponding entry of another specified row (the target row). Note that this operation changes only the target row.

There you have it! You should carry out these four elementary row operations until you have a system for which the solution is obvious, as illustrated in Example 3.

EXAMPLE 3   Solve $\begin{cases} 2x - 5y = 5 \\ \ x - 2y = 1 \end{cases}$.

Solution

System notation	*Matrix notation*

$$\begin{cases} 2x - 5y = 5 \\ \ x - 2y = 1 \end{cases} \qquad \begin{bmatrix} 2 & -5 & \vdots & 5 \\ 1 & -2 & \vdots & 1 \end{bmatrix}$$

Elementary Row Operation 1   Interchange the first and second equations

Interchange the first and second rows

$$\begin{cases} \ x - 2y = 1 \\ 2x - 5y = 5 \end{cases} \qquad \begin{bmatrix} 1 & -2 & \vdots & 1 \\ 2 & -5 & \vdots & 5 \end{bmatrix} \begin{matrix} \longleftarrow \text{Pivot row} \\ \longleftarrow \text{Target row} \end{matrix}$$

Elementary Row Operation 4   Add -2 times the first equation to the second

Add -2 times the first row to the second row

$$\begin{cases} \ x - 2y = 1 \\ \quad\ - \ y = 3 \end{cases} \qquad \begin{bmatrix} 1 & -2 & \vdots & 1 \\ 0 & -1 & \vdots & 3 \end{bmatrix} \begin{matrix} \longleftarrow \text{Pivot row remains unchanged} \\ \longleftarrow \text{Target row changes} \end{matrix}$$

Elementary Row Operation 2   Multiply both sides of the second equation by -1

Multiply row 2 by -1

$$\begin{cases} \ x - 2y = 1 \\ \quad\quad\ y = -3 \end{cases} \qquad \begin{bmatrix} 1 & -2 & \vdots & 1 \\ 0 & 1 & \vdots & -3 \end{bmatrix} \begin{matrix} \\ \longleftarrow \text{New pivot row} \end{matrix}$$

Elementary Row   Add 2 times the second   Add 2 times the second
Operation 4   equation to the first   row to the first row

$$\begin{cases} x \quad\;\; = -5 \\ \quad y = -3 \end{cases} \qquad \left[\begin{array}{cc|c} 1 & 0 & -5 \\ 0 & 1 & -3 \end{array}\right]$$

This is called the **reduced row-echelon form**

The solution, $(-5, -3)$, is now obvious.   ∎

As you study Example 3, first look at how the elementary row operations led to a system equivalent to the first—but one for which the solution is obvious. Next, try to decide *why* a particular row operation was chosen when it was. Most students quickly learn the elementary row operations, but then use a series of (almost random) steps until the obvious solution results. This often works, but is not very efficient. The steps chosen in Example 3 illustrate a very efficient method of using the elementary row operations to determine a system whose solution is obvious. The method was discovered independently by two mathematicians, Karl Friedrich Gauss (1777–1855) and Camille Jordan (1838–1922), so today the process is known as the Gauss–Jordan method. (Since the process was only recently attributed to Jordan, many books refer to the method simply as *Gaussian elimination*.) Before stating the process, however, we will consider a procedure called **pivoting**.

Pivoting

1. Divide all entries in the row in which the pivot appears (called the *pivot row*) by the pivot element so that the pivot entry becomes a 1. This uses Elementary Row Operation 3.
2. Obtain zeros above and below the pivot element by using Elementary Row Operation 4.

EXAMPLE 4   Pivot the given matrix about the circled element.

Solution
$$\left[\begin{array}{cc|c} 15 & ⑤ & 35 \\ 5 & 2 & -3 \end{array}\right] \xrightarrow{R1 \div 5} \left[\begin{array}{cc|c} 3 & 1 & 7 \\ 5 & 2 & -3 \end{array}\right] \xrightarrow{-2R1 + R2} \left[\begin{array}{cc|c} 3 & 1 & 7 \\ -1 & 0 & -17 \end{array}\right]$$

This shorthand notation   Multiply row 1 (pivot row)
shows what we did: by -2 and add it to
divide row 1 by 5   row 2 (target row)

Consider another matrix.

$$\left[\begin{array}{ccc|c} 2 & 3 & 4 & 1 \\ -1 & ② & 3 & -2 \\ 0 & 1 & -1 & 3 \end{array}\right] \xrightarrow{\frac{1}{2}R2} \left[\begin{array}{ccc|c} 2 & 3 & 4 & 1 \\ -\frac{1}{2} & 1 & \frac{3}{2} & -1 \\ 0 & 1 & -1 & 3 \end{array}\right]$$

$$\xrightarrow{-3R2 + R1} \left[\begin{array}{ccc|c} \frac{7}{2} & 0 & -\frac{1}{2} & 4 \\ -\frac{1}{2} & 1 & \frac{3}{2} & -1 \\ 0 & 1 & -1 & 3 \end{array}\right] \xrightarrow{-1R2 + R3} \left[\begin{array}{ccc|c} \frac{7}{2} & 0 & -\frac{1}{2} & 4 \\ -\frac{1}{2} & 1 & \frac{3}{2} & -1 \\ \frac{1}{2} & 0 & -\frac{5}{2} & 4 \end{array}\right]$$ ∎

The four elementary row operations are listed below for easy reference.

Elementary Row Operations

There are *four elementary row operations* for producing equivalent matrices:

1. Interchange any two rows.
2. Multiply all the elements of a row by the same nonzero real number.
3. Divide all the elements of a row by the same nonzero real number.
4. Multiply all the entries of a row (*pivot row*) by a nonzero real number and add each resulting product to the corresponding entry of another specified row (*target row*). (Note that this operation changes only the target row.)

COMPUTER APPLICATION　The disk accompanying this book has a program called Row Reduction. It lists the four elementary row operations as options. The program allows the operator input by asking what to do and then carrying out the arithmetic (in decimal form), or else it automatically carries out the pivoting process.

You are now ready to see the method worked out by Gauss and Jordan. It efficiently uses the elementary row operations to diagonalize the matrix. That is, the first pivot is the first entry in the first row, first column; the second is the entry in the second row, second column; and so on until the solution is obvious.

Gauss–Jordan Elimination

1. Select the element in the first row, first column, as a pivot.
2. Pivot.
3. Select the element in the second row, second column, as a pivot.
4. Pivot.
5. Repeat the process until you arrive at the last row, or until the pivot element is a zero. If it is a zero and you can interchange that row with a row below it, so that the pivot element is no longer a zero, do so and continue. If it is a zero and you cannot interchange rows so that it is not a zero, the process is complete.

EXAMPLE 5　Solve $\begin{cases} 3x - 2y + z = -2 \\ 4x - 5y + 3z = -9. \\ 2x - y + 5z = -5 \end{cases}$

Solution　We will solve this system by choosing the steps according to the Gauss–Jordan method.

$$\left[\begin{array}{ccc|c} ③ & -2 & 1 & -2 \\ 4 & -5 & 3 & -9 \\ 2 & -1 & 5 & -5 \end{array}\right] \xrightarrow{R1 \div 3} \left[\begin{array}{ccc|c} 1 & -\frac{2}{3} & \frac{1}{3} & -\frac{2}{3} \\ 4 & -5 & 3 & -9 \\ 2 & -1 & 5 & -5 \end{array}\right] \xrightarrow{-4R1 + R2} \left[\begin{array}{ccc|c} 1 & -\frac{2}{3} & \frac{1}{3} & -\frac{2}{3} \\ 0 & -\frac{7}{3} & \frac{5}{3} & -\frac{19}{3} \\ 2 & -1 & 5 & -5 \end{array}\right]$$

3 is the pivot　　　　Obtain a 1 in the pivot position　　　　Pivot row is 1; target row is 2; -4 is the opposite of the corresponding element in the target row so that a zero is obtained

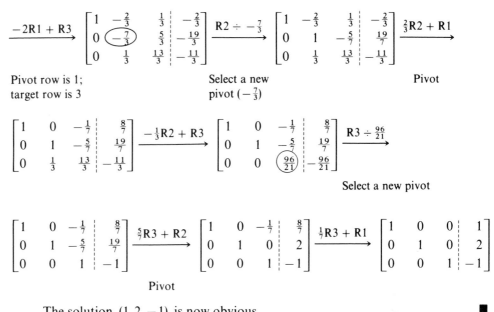

$$\xrightarrow{-2R1+R3} \begin{bmatrix} 1 & -\frac{2}{3} & \frac{1}{3} & -\frac{2}{3} \\ 0 & \boxed{-\frac{7}{3}} & \frac{5}{3} & -\frac{19}{3} \\ 0 & \frac{1}{3} & \frac{13}{3} & -\frac{11}{3} \end{bmatrix} \xrightarrow{R2 \div -\frac{7}{3}} \begin{bmatrix} 1 & -\frac{2}{3} & \frac{1}{3} & -\frac{2}{3} \\ 0 & 1 & -\frac{5}{7} & \frac{19}{7} \\ 0 & \frac{1}{3} & \frac{13}{3} & -\frac{11}{3} \end{bmatrix} \xrightarrow{\frac{2}{3}R2+R1}$$

Pivot row is 1;   Select a new   Pivot
target row is 3   pivot $\left(-\frac{7}{3}\right)$

$$\begin{bmatrix} 1 & 0 & -\frac{1}{7} & \frac{8}{7} \\ 0 & 1 & -\frac{5}{7} & \frac{19}{7} \\ 0 & \frac{1}{3} & \frac{13}{3} & -\frac{11}{3} \end{bmatrix} \xrightarrow{-\frac{1}{3}R2+R3} \begin{bmatrix} 1 & 0 & -\frac{1}{7} & \frac{8}{7} \\ 0 & 1 & -\frac{5}{7} & \frac{19}{7} \\ 0 & 0 & \boxed{\frac{96}{21}} & -\frac{96}{21} \end{bmatrix} \xrightarrow{R3 \div \frac{96}{21}}$$

Select a new pivot

$$\begin{bmatrix} 1 & 0 & -\frac{1}{7} & \frac{8}{7} \\ 0 & 1 & -\frac{5}{7} & \frac{19}{7} \\ 0 & 0 & 1 & -1 \end{bmatrix} \xrightarrow{\frac{5}{7}R3+R2} \begin{bmatrix} 1 & 0 & -\frac{1}{7} & \frac{8}{7} \\ 0 & 1 & 0 & 2 \\ 0 & 0 & 1 & -1 \end{bmatrix} \xrightarrow{\frac{1}{7}R3+R1} \begin{bmatrix} 1 & 0 & 0 & 1 \\ 0 & 1 & 0 & 2 \\ 0 & 0 & 1 & -1 \end{bmatrix}$$

Pivot

The solution, $(1, 2, -1)$, is now obvious.   ■

Note that the Gauss–Jordan method usually introduces (often ugly) fractions. For this reason, many people are turning to readily available computer programs to carry out the drudgery of this method. However, if you do not have access to a computer, you can often reduce the amount of arithmetic by forcing the pivot elements to be 1 by using elementary row operations other than division. Also, you can combine the pivoting steps. In this example we could have forced the first pivot to be 1 by multiplying row 3 by -1 and adding it to the first row instead of dividing by 3. Let us look at the arithmetic in this simplified version:

$$\begin{bmatrix} 3 & -2 & 1 & -2 \\ 4 & -5 & 3 & -9 \\ 2 & -1 & 5 & -5 \end{bmatrix} \longrightarrow \begin{bmatrix} \boxed{1} & -1 & -4 & 3 \\ 4 & -5 & 3 & -9 \\ 2 & -1 & 5 & -5 \end{bmatrix} \longrightarrow \begin{bmatrix} 1 & -1 & -4 & 3 \\ 0 & -1 & 19 & -21 \\ 0 & 1 & 13 & -11 \end{bmatrix} \longrightarrow$$

       $-1R3+R1$ Pivot; combine two steps in one:
                         $-4R1+R2; -2R1+R3$

$$\begin{bmatrix} 1 & -1 & -4 & 3 \\ 0 & \boxed{1} & -19 & 21 \\ 0 & 1 & 13 & -11 \end{bmatrix} \longrightarrow \begin{bmatrix} 1 & 0 & -23 & 24 \\ 0 & 1 & -19 & 21 \\ 0 & 0 & \boxed{32} & -32 \end{bmatrix} \longrightarrow \begin{bmatrix} 1 & 0 & -23 & 24 \\ 0 & 1 & -19 & 21 \\ 0 & 0 & 1 & -1 \end{bmatrix} \longrightarrow \begin{bmatrix} 1 & 0 & 0 & 1 \\ 0 & 1 & 0 & 2 \\ 0 & 0 & 1 & -1 \end{bmatrix}$$

New pivot is   Pivot second row   New pivot is   Pivot third row:
$R2 \div (-1)$ (combine two steps): $R3/32$ $19R3+R2;$
               $R2+R1; -1R2+R3$ $23R3+R1$

The real beauty of Gauss–Jordan elimination is that it works with all sizes of linear systems. Consider the following example consisting of five equations and five unknowns.

EXAMPLE 6 Solve $\begin{cases} x_1 - 3x_3 + x_5 = -3 \\ x_2 + x_4 = -1 \\ x_3 + x_5 = 7 \\ x_1 + x_2 - x_3 + 4x_4 = -8 \\ x_1 + x_2 + x_3 + x_4 + x_5 = 8 \end{cases}$

Solution

$$\left[\begin{array}{ccccc|c} ① & 0 & -3 & 0 & 1 & -3 \\ 0 & 1 & 0 & 1 & 0 & -1 \\ 0 & 0 & 1 & 0 & 1 & 7 \\ 1 & 1 & -1 & 4 & 0 & -8 \\ 1 & 1 & 1 & 1 & 1 & 8 \end{array}\right] \xrightarrow[\substack{-R1 + R5}]{-R1 + R4} \left[\begin{array}{ccccc|c} 1 & 0 & -3 & 0 & 1 & -3 \\ 0 & ① & 0 & 1 & 0 & -1 \\ 0 & 0 & 1 & 0 & 1 & 7 \\ 0 & 1 & 2 & 4 & -1 & -5 \\ 0 & 1 & 4 & 1 & 0 & 11 \end{array}\right]$$

$$\xrightarrow[\substack{-R2 + R5}]{-R2 + R4} \left[\begin{array}{ccccc|c} 1 & 0 & -3 & 0 & 1 & -3 \\ 0 & 1 & 0 & 1 & 0 & -1 \\ 0 & 0 & ① & 0 & 1 & 7 \\ 0 & 0 & 2 & 3 & -1 & -4 \\ 0 & 0 & 4 & 0 & 0 & 12 \end{array}\right] \xrightarrow[\substack{-2R3 + R4 \\ -4R3 + R5}]{3R3 + R1} \left[\begin{array}{ccccc|c} 1 & 0 & 0 & 0 & 4 & 18 \\ 0 & 1 & 0 & 1 & 0 & -1 \\ 0 & 0 & 1 & 0 & 1 & 7 \\ 0 & 0 & 0 & ③ & -3 & -18 \\ 0 & 0 & 0 & 0 & -4 & -16 \end{array}\right]$$

$$\xrightarrow[R4 \div 3]{} \left[\begin{array}{ccccc|c} 1 & 0 & 0 & 0 & 4 & 18 \\ 0 & 1 & 0 & 1 & 0 & -1 \\ 0 & 0 & 1 & 0 & 1 & 7 \\ 0 & 0 & 0 & ① & -1 & -6 \\ 0 & 0 & 0 & 0 & -4 & -16 \end{array}\right] \xrightarrow[(-1)R4 + R2]{} \left[\begin{array}{ccccc|c} 1 & 0 & 0 & 0 & 4 & 18 \\ 0 & 1 & 0 & 0 & 1 & 5 \\ 0 & 0 & 1 & 0 & 1 & 7 \\ 0 & 0 & 0 & 1 & -1 & -6 \\ 0 & 0 & 0 & 0 & ㋔ & -16 \end{array}\right]$$

$$\xrightarrow[R5 \div (-4)]{} \left[\begin{array}{ccccc|c} 1 & 0 & 0 & 0 & 4 & 18 \\ 0 & 1 & 0 & 0 & 1 & 5 \\ 0 & 0 & 1 & 0 & 1 & 7 \\ 0 & 0 & 0 & 1 & -1 & -6 \\ 0 & 0 & 0 & 0 & 1 & 4 \end{array}\right]$$

$$\xrightarrow[\substack{R5 + R4 \\ -R5 + R3 \\ -R5 + R2 \\ -4R5 + R1}]{} \left[\begin{array}{ccccc|c} 1 & 0 & 0 & 0 & 0 & 2 \\ 0 & 1 & 0 & 0 & 0 & 1 \\ 0 & 0 & 1 & 0 & 0 & 3 \\ 0 & 0 & 0 & 1 & 0 & -2 \\ 0 & 0 & 0 & 0 & 1 & 4 \end{array}\right]$$

The solution is $(x_1, x_2, x_3, x_4, x_5) = (2, 1, 3, -2, 4)$. ■

Example 7 is an example of a system of three equations with two unknowns that has a solution. Example 8 shows what Gauss–Jordan elimination looks like when there are three equations with two unknowns and no solution. (Remember, if there is no solution, the system is inconsistent.)

EXAMPLE 7 Solve $\begin{cases} 2x + y = 3 \\ 3x - y = 2. \\ 4x + 3y = 7 \end{cases}$

Solution
$\begin{bmatrix} 2 & 1 & | & 3 \\ 3 & -1 & | & 2 \\ 4 & 3 & | & 7 \end{bmatrix} \xrightarrow{-1R1 + R2} \begin{bmatrix} 2 & 1 & | & 3 \\ 1 & -2 & | & -1 \\ 4 & 3 & | & 7 \end{bmatrix} \begin{array}{c} \text{Interchange} \\ \xrightarrow{\text{rows 1 and 2}} \\ R_1 \leftrightarrow R_2 \end{array} \begin{bmatrix} 1 & -2 & | & -1 \\ 2 & 1 & | & 3 \\ 4 & 3 & | & 7 \end{bmatrix}$

$\xrightarrow[-4R1 + R3]{-2R1 + R2} \begin{bmatrix} ① & -2 & | & -1 \\ 0 & 5 & | & 5 \\ 0 & 11 & | & 11 \end{bmatrix} \xrightarrow{R2 \div 5} \begin{bmatrix} 1 & -2 & | & -1 \\ 0 & ① & | & 1 \\ 0 & 11 & | & 11 \end{bmatrix}$

$\xrightarrow[-11R2 + R3]{2R2 + R1} \begin{bmatrix} 1 & 0 & | & 1 \\ 0 & 1 & | & 1 \\ 0 & 0 & | & 0 \end{bmatrix}$

This final matrix is equivalent to the system: $\begin{cases} x = 1 \\ y = 1. \\ 0 = 0 \end{cases}$

The solution is $(1, 1)$.

$\quad$ *Check:* $\quad 2x + y = 2(1) + 1 = 3 \quad \checkmark$
$\quad\quad\quad\quad\quad 3x - y = 3(1) - 1 = 2 \quad \checkmark$
$\quad\quad\quad\quad\quad 4x + 3y = 4(1) + 3(1) = 7 \checkmark$ ∎

EXAMPLE 8 Solve $\begin{cases} 2x + y = 3 \\ 3x - y = 2. \\ x - 2y = 4 \end{cases}$

Solution
$\begin{bmatrix} 2 & 1 & | & 3 \\ 3 & -1 & | & 2 \\ 1 & -2 & | & 4 \end{bmatrix} \xrightarrow{R1 \leftrightarrow R3} \begin{bmatrix} 1 & -2 & | & 4 \\ 3 & -1 & | & 2 \\ 2 & 1 & | & 3 \end{bmatrix} \xrightarrow[-2R1 + R3]{-3R1 + R2} \begin{bmatrix} 1 & -2 & | & 4 \\ 0 & 5 & | & -10 \\ 0 & 5 & | & -5 \end{bmatrix}$

$\xrightarrow{R2 \div 5} \begin{bmatrix} 1 & -2 & | & 4 \\ 0 & 1 & | & -2 \\ 0 & 5 & | & -5 \end{bmatrix} \xrightarrow[-5R2 + R3]{2R2 + R1} \begin{bmatrix} 1 & 0 & | & 0 \\ 0 & 1 & | & -2 \\ 0 & 0 & | & 5 \end{bmatrix}$

This is equivalent to $\begin{cases} x = 0 \\ y = -2. \\ 0 = 5 \end{cases}$ But, since $0 \neq 5$ regardless of the values of x and y,

this is an inconsistent system. ∎

$\quad$ A dependent system has infinitely many solutions, but just because a system has infinitely many solutions does not mean that it is satisfied by any set of values. For example, $x + y = 5$ has infinitely many solutions, but not just *any* replacements for x and y will satisfy that equation. In fact, we might say if we choose any value, say,

t, for x, then y is fixed to be $5 - t$. In the last chapter we called t a *parameter*, and we could say that any ordered pair of the form $(t, 5 - t)$ will satisfy the equation $x + y = 5$. Example 9 illustrates a dependent system and the use of a parameter in specifying the solution.

EXAMPLE 9   Solve $\begin{cases} 3x - 2y + 4z = 8 \\ x - y - 2z = 5 \\ 4x - 3y + 2z = 13 \end{cases}$.

Solution
$$\begin{bmatrix} 3 & -2 & 4 & \vdots & 8 \\ 1 & -1 & -2 & \vdots & 5 \\ 4 & -3 & 2 & \vdots & 13 \end{bmatrix} \xrightarrow{R_1 \leftrightarrow R2} \begin{bmatrix} 1 & -1 & -2 & \vdots & 5 \\ 3 & -2 & 4 & \vdots & 8 \\ 4 & -3 & 2 & \vdots & 13 \end{bmatrix}$$

$$\xrightarrow[-4R1 + R3]{-3R1 + R2} \begin{bmatrix} 1 & -1 & -2 & \vdots & 5 \\ 0 & 1 & 10 & \vdots & -7 \\ 0 & 1 & 10 & \vdots & -7 \end{bmatrix} \xrightarrow[-R2 + R3]{R2 + R1} \begin{bmatrix} 1 & 0 & 8 & \vdots & -2 \\ 0 & 1 & 10 & \vdots & -7 \\ 0 & 0 & 0 & \vdots & 0 \end{bmatrix}$$

This is equivalent to the system $\begin{cases} x + 8z = -2 \\ y + 10z = -7 \\ 0 = 0 \end{cases}$. If we pick any value for z, say, t, then x and y are, in turn, determined: $x = -2 - 8t$ and $y = -7 - 10t$. This is called a *parametric solution*: $(-2 - 8t, -7 - 10t, t)$. Such a solution means that there are infinitely many ordered triplets satisfying the system (a dependent system), but just *not any* ordered triplet. Let us list a few of the solutions:

if $t = 0$: $(-2, -7, 0)$;

if $t = 1$: $(-10, -17, 1)$;

if $t = \frac{1}{2}$: $(-6, -12, \frac{1}{2})$; and so on.   ∎

You may also need a parameter when there are more unknowns than equations, as illustrated by Example 10.

EXAMPLE 10   Solve $\begin{cases} 5x - 3y + z = -3 \\ 2x + 5z = 14 \end{cases}$.

Solution
$$\begin{bmatrix} 5 & -3 & 1 & \vdots & -3 \\ 2 & 0 & 5 & \vdots & 14 \end{bmatrix} \xrightarrow{-2R2 + R1} \begin{bmatrix} 1 & -3 & -9 & \vdots & -31 \\ 2 & 0 & 5 & \vdots & 14 \end{bmatrix}$$

$$\xrightarrow{-2R1 + R2} \begin{bmatrix} 1 & -3 & -9 & \vdots & -31 \\ 0 & 6 & 23 & \vdots & 76 \end{bmatrix} \xrightarrow{R2 \div 6} \begin{bmatrix} 1 & -3 & -9 & \vdots & -31 \\ 0 & 1 & \frac{23}{6} & \vdots & \frac{38}{3} \end{bmatrix}$$

$$\xrightarrow{3R2 + R1} \begin{bmatrix} 1 & 0 & \frac{5}{2} & \vdots & 7 \\ 0 & 1 & \frac{23}{6} & \vdots & \frac{38}{3} \end{bmatrix}$$

This is equivalent to the system $\begin{cases} x + \frac{5}{2}z = 7 \\ y + \frac{23}{6}z = \frac{38}{3} \end{cases}$.

Here, again, we can choose any z, but then x and y are determined. We could call our choice t (as in the previous examples), but common practice is to let $z = kt$, where k is the least common multiple of the denominators of the fractions. By making this choice, many fractions can be avoided (for those who are not particularly fond of fractions). In this example, we can let $z = 6t$, then

$$x = 7 - \tfrac{5}{2}(6t) = 7 - 15t$$
$$y = \tfrac{38}{3} - \tfrac{23}{6}(6t) = \tfrac{38}{3} - 23t$$

This gives a parametric solution: $(7 - 15t, \tfrac{38}{3} - 23t, 6t)$. Remember, *every* different choice of t gives another solution. ∎

It is possible that you will need more than one parameter for a problem. Suppose you have 2 equations with 4 unknowns, then you will need at least $4 - 2 = 2$ parameters, as illustrated by Example 11.

EXAMPLE 11 Solve $\begin{cases} -w + 2x - 3y + z = 3 \\ 2w - 3x + y - 2z = 4 \end{cases}$.

Solution
$$\begin{bmatrix} -1 & 2 & -3 & 1 & \vdots & 3 \\ 2 & -3 & 1 & -2 & \vdots & 4 \end{bmatrix} \longrightarrow \begin{bmatrix} 1 & -2 & 3 & -1 & \vdots & -3 \\ 2 & -3 & 1 & -2 & \vdots & 4 \end{bmatrix}$$

$$\longrightarrow \begin{bmatrix} 1 & -2 & 3 & -1 & \vdots & -3 \\ 0 & 1 & -5 & 0 & \vdots & 10 \end{bmatrix}$$

$$\longrightarrow \begin{bmatrix} 1 & 0 & -7 & -1 & \vdots & 17 \\ 0 & 1 & -5 & 0 & \vdots & 10 \end{bmatrix}$$

This is equivalent to the system

$$\begin{cases} w - 7y - z = 17 \\ x - 5y \phantom{{}- z} = 10 \end{cases} \quad \text{or} \quad \begin{matrix} w = 17 + 7y + z \\ x = 10 + 5y \end{matrix}$$

You can choose any values for both y and z, and then w and x will be determined; for example, if we use s and t as the parameters so that $y = s$ and $z = t$, then

$$w = 17 + 7s + t$$
$$x = 10 + 5s$$

and the solution is $(17 + 7s + t, 10 + 5s, s, t)$. For example,

if $s = 0$ and $t = 0$, then $(17, 10, 0, 0)$ is a solution;
if $s = 0$ and $t = 1$, then $(18, 10, 0, 1)$ is a solution;
if $s = 1$ and $t = 0$, then $(24, 15, 1, 0)$ is a solution;
if $s = 1$ and $t = 1$, then $(25, 15, 1, 1)$ is a solution; and so on.

Each of these solutions can be checked in the original system to verify that each, in turn, satisfies the system. ∎

EXAMPLE 12 A rancher has to mix three types of feed for her cattle. The analysis shown in the table gives the amounts per bag (100 lb) of grain.

Grain	Protein	Carbohydrates	Sodium
A	7 lb	88 lb	1 lb
B	6 lb	90 lb	1 lb
C	10 lb	70 lb	2 lb

How many bags of each type of grain should the rancher mix to provide 71 pounds of protein, 854 pounds of carbohydrates, and 12 pounds of sodium?

Solution Let a, b, and c be the number of bags of grains A, B, and C, respectively, needed for the mixture. Then:

Grain	Protein	Carbohydrates	Sodium
A	$7a$	$88a$	a
B	$6b$	$90b$	b
C	$10c$	$70c$	$2c$
Total	71	854	12

Thus

$$\begin{cases} 7a + 6b + 10c = 71 \\ 88a + 90b + 70c = 854 \\ a + b + 2c = 12 \end{cases}$$

$$\left[\begin{array}{ccc|c} 7 & 6 & 10 & 71 \\ 88 & 90 & 70 & 854 \\ 1 & 1 & 2 & 12 \end{array}\right] \longrightarrow \left[\begin{array}{ccc|c} 1 & 1 & 2 & 12 \\ 88 & 90 & 70 & 854 \\ 7 & 6 & 10 & 71 \end{array}\right]$$

$$\longrightarrow \left[\begin{array}{ccc|c} 1 & 1 & 2 & 12 \\ 0 & 2 & -106 & -202 \\ 0 & -1 & -4 & -13 \end{array}\right] \longrightarrow \left[\begin{array}{ccc|c} 1 & 1 & 2 & 12 \\ 0 & 1 & -53 & -101 \\ 0 & -1 & -4 & -13 \end{array}\right]$$

$$\longrightarrow \left[\begin{array}{ccc|c} 1 & 0 & 55 & 113 \\ 0 & 1 & -53 & -101 \\ 0 & 0 & -57 & -114 \end{array}\right] \longrightarrow \left[\begin{array}{ccc|c} 1 & 0 & 55 & 113 \\ 0 & 1 & -53 & -101 \\ 0 & 0 & 1 & 2 \end{array}\right]$$

$$\longrightarrow \left[\begin{array}{ccc|c} 1 & 0 & 0 & 3 \\ 0 & 1 & 0 & 5 \\ 0 & 0 & 1 & 2 \end{array}\right]$$

The rancher should mix three bags of grain A, five bags of grain B, and two bags of grain C. ∎

Problem Set 2.2

Write the systems in Problems 1–2 in augmented matrix form.

1. a. $\begin{cases} 4x + 5y = -16 \\ 3x + 2y = 5 \end{cases}$

b. $\begin{cases} x + y + z = 4 \\ 3x + 2y + z = 7 \\ x - 3y + 2z = 0 \end{cases}$

c. $\begin{cases} x_1 + 3x_2 + x_3 + x_4 = 3 \\ x_1 - 2x_3 + 2x_4 = 0 \\ x_3 + 5x_4 = -14 \\ x_2 - 3x_3 - x_4 = 2 \end{cases}$

2. a. $\begin{cases} 2x + y = 0 \\ 3x - 2y = -7 \\ x - 3y = -1 \end{cases}$

b. $\begin{cases} 2x - y + z = 2 \\ 2x - 4z = 32 \end{cases}$

c. $\begin{cases} x_1 - 2x_2 - x_3 - x_4 = 4 \\ 5x_1 + 2x_2 - x_3 + 2x_4 = 23 \\ 3x_1 + 4x_2 - 3x_3 + x_4 = 2 \\ 2x_1 - 2x_2 + x_3 - x_4 = 10 \end{cases}$

Write a system of equations (use $x_1, x_2, x_3, \ldots, x_n$ for the variables) that has the given augmented matrix in Problems 3–4.

3. a. $\begin{bmatrix} 2 & 1 & 4 & 3 \\ 6 & 2 & -1 & -4 \\ -3 & -1 & 0 & 1 \end{bmatrix}$

b. $\begin{bmatrix} 1 & 0 & 0 & 5 \\ 0 & 1 & 0 & -3 \\ 0 & 0 & 1 & 4 \end{bmatrix}$

c. $\begin{bmatrix} 1 & 0 & 0 & 3 \\ 0 & 1 & 0 & 2 \\ 0 & 0 & 1 & -8 \\ 0 & 0 & 0 & 1 \end{bmatrix}$

4. a. $\begin{bmatrix} 1 & 0 & 0 & 5 \\ 0 & 1 & 2 & 4 \end{bmatrix}$

b. $\begin{bmatrix} 1 & 0 & 0 & 0 & 3 \\ 0 & 1 & 0 & 0 & -6 \\ 0 & 0 & 1 & 5 & 4 \end{bmatrix}$

c. $\begin{bmatrix} 1 & 0 & 0 & 0 & 6 \\ 0 & 1 & 0 & 0 & -3 \\ 0 & 0 & 1 & 0 & 5 \\ 0 & 0 & 0 & 1 & 4 \end{bmatrix}$

5. Let $A = \begin{bmatrix} -2 & 2 & 4 & 8 \\ -1 & 3 & -2 & 3 \\ 1 & 5 & 3 & -2 \end{bmatrix}$

Work each part separately: use this matrix for each part and not the results from another part.

a. Interchange two rows to obtain a 1 in the first position of the first row.
b. Multiply each member of row 2 by 2.
c. Multiply each member of row 1 by $\frac{1}{2}$.
d. Multiply row 3 by 2 and add the result to row 1.
e. Multiply row 3 by -1 and add the result to row 2.

6. Let $B = \begin{bmatrix} 2 & -1 & 4 & 8 \\ 3 & -2 & 1 & 4 \\ -5 & 1 & 2 & -7 \end{bmatrix}$

Work each part separately: use this matrix for each part and not the results from another part.

a. Interchange two rows to obtain a 1 in the second position of the second row.
b. Multiply each member of row 1 by $\frac{1}{2}$.
c. Multiply each member of row 3 by -2.
d. Multiply row 1 by 3 and add the result to row 3.
e. Multiply row 2 by 2 and add the result to row 3.

Solve the systems in Problems 7–47 by using Gauss–Jordan elimination.

7. $\begin{cases} x + y = 3 \\ 2x + 3y = 8 \end{cases}$

8. $\begin{cases} x - y = 2 \\ 3x + 2y = 11 \end{cases}$

9. $\begin{cases} 3x - 4y = -2 \\ 2x + 5y = -32 \end{cases}$

10. $\begin{cases} 5x + 2y = 4 \\ 3x + y = -1 \end{cases}$

11. $\begin{cases} 3x - 2y = 10 \\ 6x - 4y = 20 \end{cases}$

12. $\begin{cases} 4x + 5y = 9 \\ 7x + 7y = 14 \end{cases}$

13. $\begin{cases} 2x + y = 0 \\ 3x - 2y = -7 \\ x - 3y = -1 \end{cases}$

14. $\begin{cases} 3x - 5y = 9 \\ 2x - 4y = 8 \\ x + 3y = -11 \end{cases}$

15. $\begin{cases} 2x - 3y = -8 \\ 9x - 7y = -10 \\ 7x + 5y = -6 \end{cases}$

16. $\begin{cases} 3x - 2y = 13 \\ 4x + 5y = 2 \\ -2x - 3y = 0 \\ x + y = 1 \end{cases}$

17. $\begin{cases} 4x + 3y = -19 \\ -x - 5y = 24 \\ 2x + 3y = -17 \\ 5x - 2y = 5 \end{cases}$

18. $\begin{cases} 3x - 2y = 6 \\ 5x + 4y = 2 \\ -2x + 3y = 1 \\ x - 4y = -3 \end{cases}$

19. $\begin{cases} x + y + z = 6 \\ 2x - y + z = 3 \\ x - 2y - 3z = -12 \end{cases}$

20. $\begin{cases} 2x - y + z = 3 \\ x - 3y + 2z = 7 \\ x - y - z = -1 \end{cases}$

21. $\begin{cases} x + y + z = 4 \\ x + 3y + 2z = 4 \\ x - 2y + z = 7 \end{cases}$

22. $\begin{cases} x + 2z = 7 \\ x + y = 11 \\ -2y + 9z = -3 \end{cases}$

23. $\begin{cases} x + 2z = 13 \\ 2x + y = 8 \\ -2y + 9z = 41 \end{cases}$

24. $\begin{cases} 6x + y + 20z = 27 \\ x - y = 0 \\ y + z = 2 \end{cases}$

25. $\begin{cases} 4x + y + 2z = 7 \\ x + 2y = 0 \\ 3x - y - z = 7 \end{cases}$

26. $\begin{cases} 2x - y + 4z = 13 \\ 3x + 6y = 0 \\ 2y - 3z = 3 + 3x \end{cases}$

27. $\begin{cases} 3x - 2y + z = 5 \\ 5x - 3y = 24 \\ 2y + z = -5 \end{cases}$

28. $\begin{cases} x + y + z = 4 \\ x - 2y - z = 1 \\ 3x + y - 2z = -1 \end{cases}$

29. $\begin{cases} x + y + z = 4 \\ 3x + 2y + z = 7 \\ x - 3y + 2z = 2 \end{cases}$

30. $\begin{cases} x + y + z = 6 \\ x - 2y - z = 2 \\ 3x - y - 2z = 1 \end{cases}$

31. $\begin{cases} 2x - y + 3z = 7 \\ -x + 3y - 2z = -13 \\ 3x - 4y + 5z = 20 \end{cases}$

32. $\begin{cases} 3x - 2y + z = 13 \\ x - 5y + 2z = 24 \\ 2x + 3y - z = -11 \end{cases}$

33. $\begin{cases} 6x - y - 2z = 7 \\ 5x - 4y - 5z = 5 \\ x + 3y + 3z = 4 \end{cases}$

34. $\begin{cases} 2x - y + z = 2 \\ 2x - 4z = 32 \end{cases}$

35. $\begin{cases} x - 2y + z = -3 \\ 3y - 7z = -6 \end{cases}$

36. $\begin{cases} 2x - 3y = 1 \\ 4y + 6z = -8 \end{cases}$

37. $\begin{cases} w + x - 2y + 3z = 5 \\ 3w - 2x + y - 5z = 8 \end{cases}$

38. $\begin{cases} -2w + 3x - y - z = 6 \\ 3w - 2x + y + 2z = 5 \end{cases}$

39. $\begin{cases} 2w - 3x + z = 5 \\ 3x + 4y - z = 6 \end{cases}$

40. $\begin{cases} x_1 + 2x_2 + x_4 = 3 \\ 3x_1 + x_2 - 2x_4 = -1 \\ x_1 + 3x_3 - x_4 = 2 \end{cases}$

41. $\begin{cases} 6x_1 - 3x_2 + x_4 = -12 \\ 2x_2 + 4x_3 - x_4 = 1 \\ 3x_1 + 2x_2 + 2x_3 = -3 \end{cases}$

42. $\begin{cases} 3x_2 - x_3 - 4x_4 = 1 \\ 2x_1 + x_3 + 3x_4 = -1 \\ 5x_1 - 3x_2 - 5x_3 = -4 \end{cases}$

43. $\begin{cases} x_1 + 3x_2 + x_3 + x_4 = -1 \\ x_1 - 2x_3 + 2x_4 = -4 \\ x_3 + 5x_4 = -14 \\ x_2 - 3x_3 - x_4 = -1 \end{cases}$

44. $\begin{cases} x_1 - 2x_2 - x_3 - x_4 = -1 \\ 5x_1 + 2x_2 - x_3 + 2x_4 = 10 \\ 3x_1 + 4x_2 - 3x_3 + x_4 = -13 \\ 2x_1 - 2x_2 + x_3 - x_4 = 9 \end{cases}$

45. $\begin{cases} x_1 - 3x_2 + x_3 - x_4 = 3 \\ -x_1 + 2x_2 - 2x_3 + x_4 = -3 \\ 3x_1 - 5x_2 - 6x_3 + 4x_4 = 10 \\ 2x_1 + 3x_2 + 4x_3 + 2x_4 = 7 \end{cases}$

46. $\begin{cases} 2x_1 + 3x_2 + 2x_3 - x_4 = 10 \\ x_1 + 2x_2 - 4x_3 + 3x_4 = -3 \\ x_1 + 5x_2 - 2x_3 + 4x_4 = 1 \\ 3x_1 + 3x_2 + 2x_3 - 3x_4 = 14 \\ 5x_1 - 6x_2 - x_3 + 4x_4 = -1 \end{cases}$

47. $\begin{cases} x_1 - x_2 + 2x_3 - x_4 + 2x_5 = 2 \\ 2x_1 + x_2 + x_3 + 2x_4 - 2x_5 = 0 \\ -x_1 - x_2 - x_3 - 3x_4 - x_5 = 3 \\ 3x_1 + 2x_2 - x_3 - x_4 + x_5 = 7 \end{cases}$

APPLICATIONS

48. A farmer must mix three types of cattle feed. The table gives the amounts per bag (100 lb) of grain.

Grain	Protein	Carbohydrates	Sodium
A	9 lb	75 lb	2 lb
B	5 lb	90 lb	1 lb
C	8 lb	80 lb	1 lb

How many bags of each type of grain should be mixed to provide 58 pounds of protein, 655 pounds of carbohydrates, and 11 pounds of sodium?

49. In order to control a certain type of disease, it is necessary to use 23 liters of pesticide A and 34 liters of pesticide B. The dealer can order commercial spray I, each container of which holds 5 liters of pesticide A and 2 liters of pesticide B, and commercial spray II, each container of which holds 2 liters of pesticide A and 7 liters of pesticide B. How many containers of each type of commercial spray should be used to attain exactly the right proportion of pesticides needed?

50. In order to manufacture a certain alloy it is necessary to use 33 kg of metal A and 56 kg of metal B. The manufacturer mixes alloy I, each bar of which contains 3 kg of metal A and 5 kg of metal B, with alloy II, each bar of which contains 4 kg of metal A and 7 kg of metal B, to make the desired alloy. How much of the two alloys should be used in order to produce the alloy desired?

51. A candy maker mixes chocolate, milk, and coconut to produce three kinds of candy (I, II, and III) with the following proportions:

Candy I: 7 lb chocolate, 5 gal milk, and 1 oz coconut

Candy II: 3 lb chocolate, 2 gal milk, and 2 oz coconut

Candy III: 4 lb chocolate, 3 gal milk, and 3 oz coconut

If 67 pounds of chocolate, 48 gallons of milk, and 32 ounces of coconut are available, how much of each kind of candy can be produced?

52. The total number of registered Democrats, Republicans, and Independents in a certain community is 100,000. Voter turnout in a recent election was tabulated as follows:

50% of the Democrats voted
60% of the Republicans voted
70% of the Independents voted
55,200 votes were cast (assume that there were no write-in votes; everyone voted Democratic, Republican, or Independent)

The ratio of registered Democrats to registered Independents is 9 to 1. How many registered Democrats, Republicans, and Independents are there in the community?

53. Baskin-Robbins stores recently began offering three sizes of ice cream cones:

Small scoop: 2.5 oz for $.70
Medium scoop: 4 oz for $.95
Large scoop: 6 oz for $1.40

Are these sizes consistently priced? If not, which size is the best bargain? [Hint: you must consider not only the price per ounce for ice cream, but also the price of the cone.]

2.3 Inverse Matrices

In the set of real numbers there are two properties that we now extend to matrices.

Identity Elements   ADDITION: $a + 0 = 0 + a = a$ 0 is the identity for addition in the set of real numbers.

MULTIPLICATION: $a \cdot 1 = 1 \cdot a = a$ 1 is the identity for multiplication in the set of real numbers.

Inverse Elements   ADDITION: $a + (-a) = (-a) + a = 0$ For each number a there exists an opposite (*additive inverse*) so that the sum of a and its opposite is 0.

MULTIPLICATION: $a \cdot a^{-1} = a^{-1} \cdot a = 1$ For each nonzero number a there exists a reciprocal (*multiplicative inverse*) so that the product of a and its reciprocal is 1.

For matrices, the identity for addition is the zero matrix **0** so that $A + 0 = 0 + A = A$ for conformable matrices **0** and A. For example,

$$\begin{bmatrix} 1 & 2 & 3 \\ 4 & 5 & 6 \\ 7 & 8 & 9 \end{bmatrix} + \begin{bmatrix} 0 & 0 & 0 \\ 0 & 0 & 0 \\ 0 & 0 & 0 \end{bmatrix} = \begin{bmatrix} 1 & 2 & 3 \\ 4 & 5 & 6 \\ 7 & 8 & 9 \end{bmatrix}$$

Identity matrix
for 3 × 3 matrices
and addition

Identical

For matrix multiplication, the square matrix I of order $n \times n$ consisting of 1s on the main diagonal and zeros elsewhere is called the **identity matrix** of order n, since $IA = AI = A$ for every conformable matrix A. For example,

$$\begin{bmatrix} 1 & 2 & 3 \\ 4 & 5 & 6 \\ 7 & 8 & 9 \end{bmatrix} \begin{bmatrix} 1 & 0 & 0 \\ 0 & 1 & 0 \\ 0 & 0 & 1 \end{bmatrix} = \begin{bmatrix} 1 & 2 & 3 \\ 4 & 5 & 6 \\ 7 & 8 & 9 \end{bmatrix}$$

Identity matrix
for 3 × 3 matrices
and multiplication

Identical

The inverse matrix for addition is simply the matrix whose entries are opposites of the corresponding entries of the original matrix. However, it is the inverse for multiplication that is of particular interest to us.

Inverse of a Matrix

If A is a square matrix and if there exists a matrix A^{-1} such that
$$A^{-1}A = AA^{-1} = I$$
then A^{-1} is called the *inverse of A* for multiplication.

Usually, in the context of matrices, when we talk simply of the inverse of A we mean the inverse of A for multiplication.

EXAMPLE 1   Verify that the inverse of $A = \begin{bmatrix} 2 & 1 \\ 3 & 2 \end{bmatrix}$ is $B = \begin{bmatrix} 2 & -1 \\ -3 & 2 \end{bmatrix}$

Solution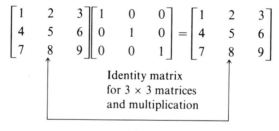
$$AB = \begin{bmatrix} 2 & 1 \\ 3 & 2 \end{bmatrix} \begin{bmatrix} 2 & -1 \\ -3 & 2 \end{bmatrix} = \begin{bmatrix} 4-3 & -2+2 \\ 6-6 & -3+4 \end{bmatrix}$$

$$= \begin{bmatrix} 1 & 0 \\ 0 & 1 \end{bmatrix}$$

$$= I$$

$$BA = \begin{bmatrix} 2 & -1 \\ -3 & 2 \end{bmatrix}\begin{bmatrix} 2 & 1 \\ 3 & 2 \end{bmatrix} = \begin{bmatrix} 4-3 & 2-2 \\ -6+6 & -3+4 \end{bmatrix}$$

$$= \begin{bmatrix} 1 & 0 \\ 0 & 1 \end{bmatrix}$$

$$= I$$

Thus $B = A^{-1}$. ■

EXAMPLE 2   Show that A and B are inverses of each other, where

$$A = \begin{bmatrix} 0 & 1 & 2 \\ -1 & 1 & 2 \\ 1 & -2 & -5 \end{bmatrix} \quad \text{and} \quad B = \begin{bmatrix} 1 & -1 & 0 \\ 3 & 2 & 2 \\ -1 & -1 & -1 \end{bmatrix}$$

Solution $AB = \begin{bmatrix} 0 & 1 & 2 \\ -1 & 1 & 2 \\ 1 & -2 & -5 \end{bmatrix}\begin{bmatrix} 1 & -1 & 0 \\ 3 & 2 & 2 \\ -1 & -1 & -1 \end{bmatrix}$

$$= \begin{bmatrix} 0+3-2 & 0+2-2 & 0+2-2 \\ -1+3-2 & 1+2-2 & 0+2-2 \\ 1-6+5 & -1-4+5 & 0-4+5 \end{bmatrix}$$

$$= \begin{bmatrix} 1 & 0 & 0 \\ 0 & 1 & 0 \\ 0 & 0 & 1 \end{bmatrix}$$

$$= I$$

$$BA = \begin{bmatrix} 1 & -1 & 0 \\ 3 & 2 & 2 \\ -1 & -1 & -1 \end{bmatrix}\begin{bmatrix} 0 & 1 & 2 \\ -1 & 1 & 2 \\ 1 & -2 & -5 \end{bmatrix}$$

$$= \begin{bmatrix} 0+1+0 & 1-1+0 & 2-2+0 \\ 0-2+2 & 3+2-4 & 6+4-10 \\ 0+1-1 & -1-1+2 & -2-2+5 \end{bmatrix}$$

$$= \begin{bmatrix} 1 & 0 & 0 \\ 0 & 1 & 0 \\ 0 & 0 & 1 \end{bmatrix}$$

$$= I$$

Since $AB = I = BA$, then $B = A^{-1}$. ■

If a given matrix has an inverse, we say that it is **nonsingular**. The unanswered question, however, is how to *find* an inverse matrix. Consider the following two examples.

EXAMPLE 3 Find the inverse of $A = \begin{bmatrix} 1 & 2 \\ 1 & 4 \end{bmatrix}$

Solution Find a matrix B (if it exists) so that $AB = I$; since we do not know B, let its entries be variables. That is, let

$$B = \begin{bmatrix} x_1 & x_2 \\ y_1 & y_2 \end{bmatrix}$$

Then

$$AB = \begin{bmatrix} 1 & 2 \\ 1 & 4 \end{bmatrix}\begin{bmatrix} x_1 & x_2 \\ y_1 & y_2 \end{bmatrix} = \begin{bmatrix} x_1 + 2y_1 & x_2 + 2y_2 \\ x_1 + 4y_1 & x_2 + 4y_2 \end{bmatrix} = \begin{bmatrix} 1 & 0 \\ 0 & 1 \end{bmatrix}$$

By the definition of the equality of matrices, we see that

$$\begin{cases} x_1 + 2y_1 = 1 \\ x_1 + 4y_1 = 0 \end{cases} \qquad\qquad \begin{cases} x_2 + 2y_2 = 0 \\ x_2 + 4y_2 = 1 \end{cases}$$

Solve these systems by using Gauss–Jordan elimination:

$$\begin{bmatrix} 1 & 2 & | & 1 \\ 1 & 4 & | & 0 \end{bmatrix} \longrightarrow \begin{bmatrix} 1 & 2 & | & 1 \\ 0 & 2 & | & -1 \end{bmatrix} \qquad\qquad \begin{bmatrix} 1 & 2 & | & 0 \\ 1 & 4 & | & 1 \end{bmatrix} \longrightarrow \begin{bmatrix} 1 & 2 & | & 0 \\ 0 & 2 & | & 1 \end{bmatrix}$$

$$\longrightarrow \begin{bmatrix} 1 & 2 & | & 1 \\ 0 & 1 & | & -\frac{1}{2} \end{bmatrix} \qquad\qquad \longrightarrow \begin{bmatrix} 1 & 2 & | & 0 \\ 0 & 1 & | & \frac{1}{2} \end{bmatrix}$$

$$\longrightarrow \begin{bmatrix} 1 & 0 & | & 2 \\ 0 & 1 & | & -\frac{1}{2} \end{bmatrix} \qquad\qquad \longrightarrow \begin{bmatrix} 1 & 0 & | & -1 \\ 0 & 1 & | & \frac{1}{2} \end{bmatrix}$$

$$\begin{cases} x_1 = 2 \\ y_1 = -\frac{1}{2} \end{cases} \qquad\qquad\qquad \begin{cases} x_2 = -1 \\ y_2 = \frac{1}{2} \end{cases}$$

Therefore the inverse is

$$B = \begin{bmatrix} x_1 & x_2 \\ y_1 & y_2 \end{bmatrix} = \begin{bmatrix} 2 & -1 \\ -\frac{1}{2} & \frac{1}{2} \end{bmatrix}$$ ■

The solution of the two systems is shown side by side so you can easily see that the steps are identical since the two rows on the left are the same; the third columns are

$$\begin{bmatrix} 1 \\ 0 \end{bmatrix}\begin{bmatrix} 0 \\ 1 \end{bmatrix}$$

respectively. It would seem that we might combine steps, but, before we do, consider a three-by-three example.

EXAMPLE 4 Find the inverse for the matrix

$$\begin{bmatrix} 1 & -1 & 0 \\ 3 & 2 & 2 \\ -1 & -1 & -1 \end{bmatrix}$$

(This is matrix B from Example 2.)

Solution We need to find a matrix

$$\begin{bmatrix} x_1 & x_2 & x_3 \\ y_1 & y_2 & x_3 \\ z_1 & z_2 & z_3 \end{bmatrix}$$

so that

$$\begin{bmatrix} 1 & -1 & 0 \\ 3 & 2 & 2 \\ -1 & -1 & -1 \end{bmatrix} \begin{bmatrix} x_1 & x_2 & x_3 \\ y_1 & y_2 & y_3 \\ z_1 & z_2 & z_3 \end{bmatrix} = \begin{bmatrix} 1 & 0 & 0 \\ 0 & 1 & 0 \\ 0 & 0 & 1 \end{bmatrix}$$

$$\begin{cases} x_1 - y_1 + 0z_1 = 1 \\ 3x_1 + 2y_1 + 2z_1 = 0 \\ -x_1 - y_1 - z_1 = 0 \end{cases} \quad \begin{cases} x_2 - y_2 + 0z_2 = 0 \\ 3x_2 + 2y_2 + 2z_2 = 1 \\ -x_2 - y_2 - z_2 = 0 \end{cases} \quad \begin{cases} x_3 - y_3 + 0z_3 = 0 \\ 3x_3 + 2y_3 + 2z_3 = 0 \\ -x_3 - y_3 - z_3 = 1 \end{cases}$$

We could solve these as three separate systems using Gauss–Jordan elimination; however, all the steps would be identical since the variables on the left of the equal signs are the same in each system. Therefore suppose we augment the matrix of the coefficients by the *three* rows and do all three at once.

$$\left[\begin{array}{ccc|ccc} 1 & -1 & 0 & 1 & 0 & 0 \\ 3 & 2 & 2 & 0 & 1 & 0 \\ -1 & -1 & -1 & 0 & 0 & 1 \end{array}\right] \longrightarrow \left[\begin{array}{ccc|ccc} 1 & -1 & 0 & 1 & 0 & 0 \\ 0 & 5 & 2 & -3 & 1 & 0 \\ 0 & -2 & -1 & 1 & 0 & 1 \end{array}\right] \xrightarrow{2R3 + R2}$$

$$\left[\begin{array}{ccc|ccc} 1 & -1 & 0 & 1 & 0 & 0 \\ 0 & 1 & 0 & -1 & 1 & 2 \\ 0 & -2 & -1 & 1 & 0 & 1 \end{array}\right] \longrightarrow \left[\begin{array}{ccc|ccc} 1 & 0 & 0 & 0 & 1 & 2 \\ 0 & 1 & 0 & -1 & 1 & 2 \\ 0 & 0 & -1 & -1 & 2 & 5 \end{array}\right] \longrightarrow$$

$$\left[\begin{array}{ccc|ccc} 1 & 0 & 0 & 0 & 1 & 2 \\ 0 & 1 & 0 & -1 & 1 & 2 \\ 0 & 0 & 1 & 1 & -2 & -5 \end{array}\right]$$

Now, if we relate this back to the original three systems, we see that the inverse is

$$\begin{bmatrix} 0 & 1 & 2 \\ -1 & 1 & 2 \\ 1 & -2 & -5 \end{bmatrix}$$ ∎

By studying Examples 3 and 4, we are led to a procedure for finding the inverse of a nonsingular matrix:

Procedure for Finding the Inverse of a Matrix

1. Augment the given matrix with I; that is, write $[A \vdots I]$, where I is the identity matrix of the same order as the given square matrix A.
2. Perform elementary row operations using Gauss–Jordan elimination in order to change the matrix A into the identity matrix (if possible).
3. If steps 1 and 2 can be performed, the result in the augmented part is the inverse of A.

> **COMPUTER APPLICATION** Most inverses involve (ugly) fractions, and it is often not only convenient, but necessary, to have numerical assistance. Computer programs for finding the inverse of a matrix are fairly easy to obtain and use. Program 4 on the computer disk accompanying this book is helpful if you have access to a computer. There are two options: manual and automatic reduction. With manual reduction, the computer asks what you want to do and then simply carries out the arithmetic. With automatic reduction, the computer not only does the calculations but also makes the decisions as to what step comes next. This program will also generate additional problems so that you can practice the procedure. You should use the automatic reduction option only after you understand the process.

EXAMPLE 5 Find the inverse of the matrix $A = \begin{bmatrix} 1 & 2 \\ 0 & 0 \end{bmatrix}$.

Solution Write the augmented matrix $[A \mid I]$:

$$\left[\begin{array}{cc|cc} 1 & 2 & 1 & 0 \\ 0 & 0 & 0 & 1 \end{array}\right]$$

We want to make the left-hand side look like the corresponding identity matrix. This is impossible since there are no elementary row operations that will put it into the required form. Thus there is no inverse. ∎

EXAMPLE 6 Find the inverse of the matrix $A = \begin{bmatrix} 0 & 1 & 2 \\ 2 & -1 & 1 \\ -1 & 1 & 0 \end{bmatrix}$.

Solution Write the augmented matrix $[A \mid I]$ and make the left-hand side look like the corresponding identity matrix (if possible):

$$\left[\begin{array}{ccc|ccc} 0 & 1 & 2 & 1 & 0 & 0 \\ 2 & -1 & 1 & 0 & 1 & 0 \\ -1 & 1 & 0 & 0 & 0 & 1 \end{array}\right] \xrightarrow{R1 \leftrightarrow R3} \left[\begin{array}{ccc|ccc} -1 & 1 & 0 & 0 & 0 & 1 \\ 2 & -1 & 1 & 0 & 1 & 0 \\ 0 & 1 & 2 & 1 & 0 & 0 \end{array}\right]$$

$$\longrightarrow \left[\begin{array}{ccc|ccc} 1 & -1 & 0 & 0 & 0 & -1 \\ 2 & -1 & 1 & 0 & 1 & 0 \\ 0 & 1 & 2 & 1 & 0 & 0 \end{array}\right] \longrightarrow \left[\begin{array}{ccc|ccc} 1 & -1 & 0 & 0 & 0 & -1 \\ 0 & 1 & 1 & 0 & 1 & 2 \\ 0 & 1 & 2 & 1 & 0 & 0 \end{array}\right]$$

$$\longrightarrow \left[\begin{array}{ccc|ccc} 1 & 0 & 1 & 0 & 1 & 1 \\ 0 & 1 & 1 & 0 & 1 & 2 \\ 0 & 0 & 1 & 1 & -1 & -2 \end{array}\right] \longrightarrow \left[\begin{array}{ccc|ccc} 1 & 0 & 0 & -1 & 2 & 3 \\ 0 & 1 & 0 & -1 & 2 & 4 \\ 0 & 0 & 1 & 1 & -1 & -2 \end{array}\right]$$

Thus $A^{-1} = \begin{bmatrix} -1 & 2 & 3 \\ -1 & 2 & 4 \\ 1 & -1 & -2 \end{bmatrix}$. ∎

Inverses, when they are known, can be used to solve matrix equations. For example, suppose

$$A = \begin{bmatrix} 0 & 1 & 2 \\ 2 & -1 & 1 \\ -1 & 1 & 0 \end{bmatrix} \qquad X = \begin{bmatrix} x \\ y \\ z \end{bmatrix} \qquad B = \begin{bmatrix} 0 \\ -1 \\ 1 \end{bmatrix}$$

Then

$$AX = \begin{bmatrix} 0 & 1 & 2 \\ 2 & -1 & 1 \\ -1 & 1 & 0 \end{bmatrix}\begin{bmatrix} x \\ y \\ z \end{bmatrix} = \begin{bmatrix} y + 2z \\ 2x - y + z \\ -x + y \end{bmatrix}$$

and $AX = B$ means

$$\begin{cases} y + 2z = 0 \\ 2x - y + z = -1 \\ -x + y = 1 \end{cases}$$

since matrices are equal if and only if corresponding entries are equal. If you can write a system in matrix form, then you can solve the system $AX = B$ provided A^{-1} exists:

$$
\begin{array}{ll}
AX = B & \text{Given system} \\
(A^{-1})AX = A^{-1}B & \text{Multiply both sides by } A^{-1} \text{ on the left} \\
(A^{-1}A)X = A^{-1}B & \text{Associative property} \\
IX = A^{-1}B & \text{Inverse property} \\
X = A^{-1}B & \text{Identity property}
\end{array}
$$

This means that to solve a system, all we have to do is multiply A^{-1} and B to find X. Since

$$A = \begin{bmatrix} 0 & 1 & 2 \\ 2 & -1 & 1 \\ -1 & 1 & 0 \end{bmatrix}$$

we see from Example 6,

$$A^{-1} = \begin{bmatrix} -1 & 2 & 3 \\ -1 & 2 & 4 \\ 1 & -1 & -2 \end{bmatrix}$$

so we can solve for x as follows:

$$X = A^{-1}B = \begin{bmatrix} -1 & 2 & 3 \\ -1 & 2 & 4 \\ 1 & -1 & -2 \end{bmatrix}\begin{bmatrix} 0 \\ -1 \\ 1 \end{bmatrix} = \begin{bmatrix} 0 - 2 + 3 \\ 0 - 2 + 4 \\ 0 + 1 - 2 \end{bmatrix} = \begin{bmatrix} 1 \\ 2 \\ -1 \end{bmatrix}$$

Thus $x = 1$, $y = 2$, and $z = -1$.

The method of solving a system by using the inverse matrix is very efficient if you know the inverse. Unfortunately, *finding* the inverse for only one system is usually more work than using another method to solve the system. However, there are certain applications that yield the same coefficient matrix over and over. In these cases the inverse method is worthwhile. And, finally, computers can often find approximations for inverse matrices quite easily.

> COMPUTER APPLICATION Program 5 on the accompanying computer disk is designed to solve systems by using inverses. If the given matrix is singular, you can also use this software to decide if the system has no solutions or an infinite number of solutions.

Problem Set 2.3

Use multiplication in Problems 1–12 to determine whether the matrices are inverses.

1. $\begin{bmatrix} 1 & 2 \\ 2 & 3 \end{bmatrix}$, $\begin{bmatrix} -3 & 2 \\ 2 & -1 \end{bmatrix}$

2. $\begin{bmatrix} 2 & -5 \\ -1 & 2 \end{bmatrix}$, $\begin{bmatrix} -2 & -5 \\ -1 & -2 \end{bmatrix}$

3. $\begin{bmatrix} 3 & 5 \\ 4 & 7 \end{bmatrix}$, $\begin{bmatrix} 7 & -5 \\ -4 & 3 \end{bmatrix}$

4. $\begin{bmatrix} 4 & 7 \\ 5 & 9 \end{bmatrix}$, $\begin{bmatrix} 9 & -7 \\ -5 & 4 \end{bmatrix}$

5. $\begin{bmatrix} 4 & 3 \\ 2 & 2 \end{bmatrix}$, $\begin{bmatrix} 1 & -\frac{3}{2} \\ -1 & 2 \end{bmatrix}$

6. $\begin{bmatrix} 2 & 3 \\ 2 & 1 \end{bmatrix}$, $\begin{bmatrix} -\frac{1}{4} & \frac{3}{4} \\ \frac{1}{2} & -\frac{1}{2} \end{bmatrix}$

7. $\begin{bmatrix} 0 & 1 & 0 \\ 1 & -1 & 0 \\ -1 & 2 & 1 \end{bmatrix}$, $\begin{bmatrix} -1 & -1 & 0 \\ -1 & 0 & 0 \\ 1 & -1 & -1 \end{bmatrix}$

8. $\begin{bmatrix} 1 & 0 & 0 \\ 0 & 1 & 1 \\ 2 & 0 & 1 \end{bmatrix}$, $\begin{bmatrix} 1 & 0 & 0 \\ 2 & 1 & -1 \\ -2 & 0 & 1 \end{bmatrix}$

9. $\begin{bmatrix} 3 & -2 & 4 \\ 2 & 1 & 2 \\ 5 & 3 & 5 \end{bmatrix}$, $\begin{bmatrix} -1 & 22 & -8 \\ 0 & -5 & 2 \\ 1 & -19 & 7 \end{bmatrix}$

10. $\begin{bmatrix} 6 & -1 & -5 \\ -7 & 1 & 5 \\ -10 & 2 & 11 \end{bmatrix}$, $\begin{bmatrix} -1 & -1 & 0 \\ -27 & -16 & -5 \\ 4 & 2 & 1 \end{bmatrix}$

11. $\begin{bmatrix} 1 & 1 & 0 & 1 \\ 2 & 0 & -5 & 8 \\ 0 & 1 & 3 & 2 \\ 0 & 0 & 3 & 31 \end{bmatrix}$, $\begin{bmatrix} 1 & 95 & -1 & -29 \\ 0 & -92 & 1 & 28 \\ 0 & 31 & 0 & -10 \\ 0 & -3 & 0 & 1 \end{bmatrix}$

12. $\begin{bmatrix} 1 & 0 & 0 & 2 \\ 3 & 1 & -1 & 0 \\ 0 & 2 & 1 & 3 \\ 1 & 0 & 0 & 1 \end{bmatrix}$, $\begin{bmatrix} -2 & 0 & 0 & 2 \\ 0 & \frac{1}{3} & \frac{1}{3} & -1 \\ -3 & -\frac{2}{3} & \frac{1}{3} & 5 \\ 1 & 0 & 0 & -1 \end{bmatrix}$

Find the inverses of the matrices in Problems 13–30, if possible.

13. $\begin{bmatrix} 4 & -7 \\ -1 & 2 \end{bmatrix}$

14. $\begin{bmatrix} 4 & 0 \\ 0 & 5 \end{bmatrix}$

15. $\begin{bmatrix} 3 & 5 \\ 1 & 2 \end{bmatrix}$

16. $\begin{bmatrix} 3 & -1 \\ -4 & 2 \end{bmatrix}$

17. $\begin{bmatrix} 1 & 3 \\ 2 & 0 \end{bmatrix}$

18. $\begin{bmatrix} 2 & 3 \\ 1 & -6 \end{bmatrix}$

19. $\begin{bmatrix} 8 & 6 \\ -2 & 4 \end{bmatrix}$

20. $\begin{bmatrix} 2 & 1 \\ 4 & 3 \end{bmatrix}$

21. $\begin{bmatrix} 1 & -\frac{3}{2} \\ -1 & 2 \end{bmatrix}$

22. $\begin{bmatrix} 1 & 0 & 2 \\ 2 & 1 & 0 \\ 0 & -2 & 9 \end{bmatrix}$

23. $\begin{bmatrix} 6 & 1 & 20 \\ 1 & -1 & 0 \\ 0 & 1 & 3 \end{bmatrix}$

24. $\begin{bmatrix} 4 & 1 & 0 \\ 2 & -1 & 4 \\ -3 & 2 & 1 \end{bmatrix}$

25. $\begin{bmatrix} 1 & -1 & 1 \\ 0 & 2 & -1 \\ 2 & 3 & 0 \end{bmatrix}$

26. $\begin{bmatrix} 15 & 4 & -5 \\ -12 & -3 & 4 \\ -4 & -1 & 1 \end{bmatrix}$

27. $\begin{bmatrix} 1 & 0 & 2 \\ 3 & -1 & 2 \\ 4 & 1 & 0 \end{bmatrix}$

28. $\begin{bmatrix} 1 & 0 & 0 & 1 \\ 0 & 2 & 0 & 0 \\ 0 & 0 & 0 & 1 \\ 2 & 0 & 1 & 0 \end{bmatrix}$

29. $\begin{bmatrix} 0 & 1 & 2 & 0 \\ 0 & 0 & 0 & 1 \\ 1 & 1 & 3 & 0 \\ 2 & 4 & 0 & 0 \end{bmatrix}$

30. $\begin{bmatrix} 1 & 2 & 0 & 0 \\ 0 & 0 & 1 & 0 \\ 1 & 3 & 0 & 1 \\ 2 & 4 & 0 & 0 \end{bmatrix}$

Solve the systems in Problems 31–54 by solving the corresponding matrix equation with an inverse, if possible.

Problems 31–33 use the inverse found in Problem 13.

31. $\begin{cases} 4x - 7y = -2 \\ -x + 2y = 1 \end{cases}$ **32.** $\begin{cases} 4x - 7y = -65 \\ -x + 2y = 18 \end{cases}$

33. $\begin{cases} 4x - 7y = 48 \\ -x + 2y = -13 \end{cases}$

Problems 34–36 use the inverse found in Problem 19.

34. $\begin{cases} 8x + 6y = 12 \\ -2x + 4y = -14 \end{cases}$ **35.** $\begin{cases} 8x + 6y = 16 \\ -2x + 4y = 18 \end{cases}$

36. $\begin{cases} 8x + 6y = -6 \\ -2x + 4y = -26 \end{cases}$

Problems 37–39 use the inverse found in Problem 18.

37. $\begin{cases} 2x + 3y = 9 \\ x - 6y = -3 \end{cases}$ **38.** $\begin{cases} 2x + 3y = 2 \\ x - 6y = 16 \end{cases}$

39. $\begin{cases} 2x + 3y = 2 \\ x - 6y = -14 \end{cases}$

Problems 40–42 use the inverse found in Problem 20.

40. $\begin{cases} 2x + y = 5 \\ 4x + 3y = 9 \end{cases}$ **41.** $\begin{cases} 2x + y = 16 \\ 4x + 3y = 2 \end{cases}$

42. $\begin{cases} 2x + y = -3 \\ 4x + 3y = 1 \end{cases}$

Problems 43–45 use the inverse found in Problem 22.

43. $\begin{cases} x + 2z = 7 \\ 2x + y = 16 \\ -2y + 9z = -3 \end{cases}$ **44.** $\begin{cases} x + 2z = 4 \\ 2x + y = 0 \\ -2y + 9z = 19 \end{cases}$

45. $\begin{cases} x + 2z = 7 \\ 2x + y = 0 \\ -2y + 9z = 31 \end{cases}$

Problems 46–48 use the inverse found in Problem 23.

46. $\begin{cases} 6x + y + 20z = 27 \\ x - y = 0 \\ y + 3z = 4 \end{cases}$ **47.** $\begin{cases} 6x + y + 20z = 14 \\ x - y = 1 \\ y + 3z = 1 \end{cases}$

48. $\begin{cases} 6x + y + 20z = 11 \\ x - y = 5 \\ y + 3z = -3 \end{cases}$

Problems 49–51 use the inverse found in Problem 24.

49. $\begin{cases} 4x + y = 6 \\ 2x - y + 4z = 12 \\ -3x + 2y + z = 4 \end{cases}$ **50.** $\begin{cases} 4x + y = 7 \\ 2x - y + 4z = -11 \\ -3x + 2y + z = -12 \end{cases}$

51. $\begin{cases} 4x + y = -10 \\ 2x - y + 4z = 20 \\ -3x + 2y + z = 20 \end{cases}$

Problems 52–54 use the inverse found in Problem 29.

52. $\begin{cases} x + 2y = 5 \\ z = 3 \\ w + x + 3y = 9 \\ 2w + 4x = 8 \end{cases}$ **53.** $\begin{cases} x + 2y = 0 \\ z = -4 \\ w + x + 3y = 4 \\ 2w + 4x = -2 \end{cases}$

54. $\begin{cases} x + 2y = 7 \\ z = -7 \\ w + x + 3y = 16 \\ 2w + 4x = -4 \end{cases}$

APPLICATIONS

55. A rancher has to mix three types of feed. The table gives the amounts per 100-pound bag of grain.

Grain	Protein	Carbohydrates	Sodium
I	10 lb	75 lb	2 lb
II	20 lb	70 lb	3 lb
III	15 lb	80 lb	1 lb

How many bags of each type of grain should be mixed to provide 135 pounds of protein, 740 pounds of carbohydrates, and 22 pounds of sodium?

56. Repeat Problem 55 to provide 350 pounds of protein, 1885 pounds of carbohydrates, and 48 pounds of sodium.

57. Repeat Problem 55 to provide 545 pounds of protein, 3090 pounds of carbohydrates, and 79 pounds of sodium.

58. A dietitian is to arrange a diet of three basic foods. The diet is to include 1800 units of vitamin A, 9800 units of vitamin C, and 1420 units of calcium. The number of units per ounce of each of the foods is given in the table.

Food	Vitamin A	Vitamin C	Calcium
I	50 units	300 units	20 units
II	30 units	200 units	40 units
III	40 units	100 units	30 units

How many ounces of each food should be supplied?

59. Repeat Problem 58 to supply 1200 units of vitamin A, 6000 units of vitamin C, and 900 units of calcium.

60. Repeat Problem 58 to supply 1140 units of vitamin A, 7100 units of vitamin C, and 1030 units of calcium.

2.4 Leontief Models*

Wassily Leontief received the Nobel Prize in Economics in 1973 for describing the economy in terms of an **input–output model**. We now look at two extended applications of the Leontief model: a **closed model** in which we try to find the relative income of each participant within a system, and an **open model** in which we try to find the amount of production needed to achieve a forecast demand.

A Closed Model

Suppose a simplified economy consists of three individuals: a farmer, a tailor, and a builder, who provide the essentials of food, clothing, and shelter for each other. Of the food produced by the farmer, 50% is used by the farmer, 20% by the tailor, and 30% by the builder. The builder's production is utilized 40% by the builder, 40% by the farmer, and 20% by the tailor. The tailor's production is divided among them as 45% for the builder, 20% for the farmer, and 35% for the tailor. Suppose that each person's wages are to be about $1,000 (it could be shells, sheep, or any other convenient unit used as money). How much should each person pay the others?

We begin by writing the given information in matrix form. This is called an **input–output matrix**.

$$
\begin{array}{cccc}
 & \text{Farmer} & \overset{\text{PRODUCER}}{\text{Builder}} & \text{Tailor} \\
A = \begin{bmatrix} 0.5 & 0.4 & 0.2 \\ 0.3 & 0.4 & 0.45 \\ 0.2 & 0.2 & 0.35 \end{bmatrix} & \begin{array}{l} \text{Farmer} \\ \text{Builder} \\ \text{Tailor} \end{array} & \text{USER}
\end{array}
$$

In general, an input–output matrix for a closed model consists of n rows and n columns where the sum of each column is 1 and each entry a_{ij} is a fraction between 0 and 1, inclusive: $0 \le a_{ij} \le 1$. The a_{ij} entry is the fraction used by i and produced by j.

Let X be the column matrix representing the price of each output in the system. That is, let

$$
X = \begin{bmatrix} x_1 \\ x_2 \\ x_3 \end{bmatrix}
$$

where

$$x_1 = \text{Farmer's wages}$$
$$x_2 = \text{Tailor's wages}$$
$$x_3 = \text{Builder's wages}$$

The wages include those wages paid to themselves as part of the economic system. For the wages to come out even, the total amount paid out by each must equal the

---

* This section is optional.

total amount received by each. That is,

$$X = AX$$

$$\begin{bmatrix} x_1 \\ x_2 \\ x_3 \end{bmatrix} = \begin{bmatrix} 0.5 & 0.4 & 0.2 \\ 0.3 & 0.4 & 0.45 \\ 0.2 & 0.2 & 0.35 \end{bmatrix} \begin{bmatrix} x_1 \\ x_2 \\ x_3 \end{bmatrix}$$

This can be written as the system

$$\begin{bmatrix} x_1 \\ x_2 \\ x_3 \end{bmatrix} = \begin{bmatrix} 0.5x_1 + 0.4x_2 + 0.2x_3 \\ 0.3x_1 + 0.4x_2 + 0.45x_3 \\ 0.2x_1 + 0.2x_2 + 0.35x_3 \end{bmatrix} \quad \text{or} \quad \begin{cases} x_1 = 0.5x_1 + 0.4x_2 + 0.2x_3 \\ x_2 = 0.3x_1 + 0.4x_2 + 0.45x_3 \\ x_3 = 0.2x_1 + 0.2x_2 + 0.35x_3 \end{cases}$$

By subtracting to obtain zeros on the right, we have

$$\begin{cases} 0.5x_1 - 0.4x_2 - 0.2x_3 = 0 \\ -0.3x_1 + 0.6x_2 - 0.45x_3 = 0 \\ -0.2x_1 - 0.2x_2 + 0.65x_3 = 0 \end{cases}$$

A system of equations in which the right-hand side is zero is called a **homogeneous system** of equations.

There is, of course, a trivial solution for which $x_1 = x_2 = x_3 = 0$, but that is not the only solution. For a closed model, the system will have a one-parameter solution. This parameter will serve as a scaling factor.

The solution using Gauss–Jordan elimination is shown below:

$$\begin{bmatrix} 0.5 & -0.4 & -0.2 & \vdots & 0 \\ -0.3 & 0.6 & -0.45 & \vdots & 0 \\ -0.2 & -0.2 & 0.65 & \vdots & 0 \end{bmatrix} \longrightarrow \begin{bmatrix} 5 & -4 & -2 & \vdots & 0 \\ -2 & 4 & -3 & \vdots & 0 \\ -4 & -4 & 13 & \vdots & 0 \end{bmatrix}$$

$$\longrightarrow \begin{bmatrix} 1 & -8 & 11 & \vdots & 0 \\ -2 & 4 & -3 & \vdots & 0 \\ -4 & -4 & 13 & \vdots & 0 \end{bmatrix} \longrightarrow \begin{bmatrix} 1 & -8 & 11 & \vdots & 0 \\ 0 & -12 & 19 & \vdots & 0 \\ 0 & -36 & 57 & \vdots & 0 \end{bmatrix}$$

$$\longrightarrow \begin{bmatrix} 1 & -8 & 11 & \vdots & 0 \\ 0 & 1 & -\frac{19}{12} & \vdots & 0 \\ 0 & 0 & 0 & \vdots & 0 \end{bmatrix} \longrightarrow \begin{bmatrix} 1 & 0 & -\frac{5}{3} & \vdots & 0 \\ 0 & 1 & -\frac{19}{12} & \vdots & 0 \\ 0 & 0 & 0 & \vdots & 0 \end{bmatrix}$$

Thus $x_1 = \frac{5}{3}x_3$ and $x_2 = \frac{19}{12}x_3$. Since the choice of x_3 is arbitrary, choose x_3 so that it is about \$1,000 and also so that it is divisible by 12 (to give an integer solution). Let $x_3 = 1008$:

$$x_1 = \tfrac{5}{3}(1008) = 1680$$
$$x_2 = \tfrac{19}{12}(1008) = 1596$$
$$x_3 = 1008$$

The wages to be paid are \$1,680 to the farmer, \$1,596 to the builder, and \$1,008 to the tailor. Note that if you let $x_3 = 636$, then the farmer's wages are \$1,060, the builder's wages \$1,007, and the tailor's wages \$636. No matter what choice is made for x_3, each other participant's share is determined.

An Open Model

Suppose a simplified economy consists of three industries: farming, clothing, and construction, which provide the essentials of food, clothing, and shelter not only for each other, but also for the consumer. Suppose we take $100 to be the total value of the goods produced by the farming industry. The clothing and construction industries require $10 and $15, respectively, of those goods, and the farmers consume $25 worth of their own output. The consumer requires the remaining $50 worth of the goods produced by the farming industry. The construction industry consumes $5 of its own production, the farming industry consumes $15, and the clothing industry consumes $10. The construction industry produces $110 of goods, so the remaining $80 is used by the consumer. The third industry, the clothing industry, uses $5 worth of its own production, while the farming segment uses $10 and the construction industry uses $20. The consumer requires $85 of the total clothing industry's output of $120. Construct a model to analyze how the output of any one of the industries is affected by changes in the other two and how production can be changed to anticipate changes in consumer demand.

Once again, begin by writing the input–output matrix:

$$
A = \begin{bmatrix} \frac{25}{100} & \frac{15}{110} & \frac{10}{120} \\ \frac{15}{100} & \frac{5}{110} & \frac{10}{120} \\ \frac{10}{100} & \frac{20}{110} & \frac{5}{120} \end{bmatrix}
\begin{matrix} \text{Farming} \\ \text{Construction} \\ \text{Clothing} \end{matrix}
$$

PRODUCER: Farming   Construction   Clothing

USER (numerators are the amounts used by each of these industries)

Consumer

$25 + 15 + 10 + 50 = 100$
$15 + 5 + 10 + 80 = 110$
$10 + 20 + 5 + 85 = 120$

denominators are the production of each industry

$$
A \approx \begin{bmatrix} 0.25 & 0.14 & 0.08 \\ 0.15 & 0.05 & 0.08 \\ 0.10 & 0.18 & 0.04 \end{bmatrix}
$$

This is the input–output matrix correct to the nearest hundredth

The matrix A for an open model consists of n rows and n columns where the entries a_{ij} represent the output of industry j required for one unit of output of industry i. This concept is shown by arrows on the upper matrix. Note that the sum of the columns in an open model are each less than or equal to 1.

Let D_i be the column matrix representing the consumer demand in the year i. That is, $i = 0$ represents the *present* consumer demand:

Consumer

$$
D_0 = \begin{bmatrix} 50 \\ 80 \\ 85 \end{bmatrix}
\begin{matrix} \text{Farming} \\ \text{Construction} \\ \text{Clothing} \end{matrix}
$$

Now, to anticipate future demand, let

x_i = Output of the farming industry in year i

y_i = Output of the construction industry in year i

z_i = Output of the clothing industry in year i

so that

$$X_i = \begin{bmatrix} x_i \\ y_i \\ z_i \end{bmatrix}$$

In particular,

$$X_0 = \begin{bmatrix} 100 \\ 110 \\ 120 \end{bmatrix}$$

The product AX_0 represents the amounts required for internal consumption, and $AX_0 + D_0$ represents the total output. That is,

INTERNAL CONSUMPTION + DEMAND = TOTAL OUTPUT

Suppose marketing forecasts predict a 5-year plan setting demand at

$$D_5 = \begin{bmatrix} 110 \\ 200 \\ 200 \end{bmatrix}$$

We then need to solve

$$AX_5 + D_5 = X_5$$

The solution of this type of forecasting problem is called an *open Leontief model.* For the closed model we found the solution for the particular matrix equation, but with the open model we can anticipate finding $X_1, X_2, X_3, \ldots$ for any number of years. For this reason we solve this system using an inverse matrix (if it exists). First, we solve for X_5:

$$AX_5 + D_5 = X_5$$
$$X_5 - AX_5 = D_5$$
$$IX_5 - AX_5 = D_5$$
$$(I - A)X_5 = D_5$$
$$X_5 = (I - A)^{-1}D_5:$$

Now, find the difference

$$I - A = \begin{bmatrix} 1 & 0 & 0 \\ 0 & 1 & 0 \\ 0 & 0 & 1 \end{bmatrix} - \begin{bmatrix} 0.25 & 0.14 & 0.08 \\ 0.15 & 0.05 & 0.08 \\ 0.10 & 0.18 & 0.04 \end{bmatrix}$$

$$= \begin{bmatrix} 0.75 & -0.14 & -0.08 \\ -0.15 & 0.95 & -0.08 \\ -0.10 & -0.18 & 0.96 \end{bmatrix}$$

and form the augmented matrix to find the inverse $(I - A)^{-1}$:

$$\left[\begin{array}{ccc|ccc} 0.75 & -0.14 & -0.08 & 1 & 0 & 0 \\ -0.15 & 0.95 & -0.08 & 0 & 1 & 0 \\ -0.10 & -0.18 & 0.96 & 0 & 0 & 1 \end{array}\right] \longrightarrow \left[\begin{array}{ccc|ccc} 1 & -0.19 & -0.11 & 1.33 & 0 & 0 \\ 0 & 0.92 & -0.10 & 0.20 & 1 & 0 \\ 0 & -0.20 & 0.95 & 0.13 & 0 & 1 \end{array}\right]$$

$$\longrightarrow \left[\begin{array}{ccc|ccc} 1 & -0.19 & -0.11 & 1.33 & 0 & 0 \\ 0 & 1 & -0.10 & 0.22 & 1.09 & 0 \\ 0 & 0 & 0.93 & 0.18 & 0.22 & 1 \end{array}\right] \longrightarrow \left[\begin{array}{ccc|ccc} 1 & -0.19 & -0.11 & 1.33 & 0 & 0 \\ 0 & 1 & -0.10 & 0.22 & 1.09 & 0 \\ 0 & 0 & 1 & 0.19 & 0.23 & 1.08 \end{array}\right]$$

$$\longrightarrow \left[\begin{array}{ccc|ccc} 1 & -.19 & 0 & 1.35 & 0.03 & 0.12 \\ 0 & 1 & 0 & 0.24 & 1.11 & 0.11 \\ 0 & 0 & 1 & 0.19 & 0.23 & 1.08 \end{array}\right] \longrightarrow \left[\begin{array}{ccc|ccc} 1 & 0 & 0 & 1.40 & 0.23 & 0.14 \\ 0 & 1 & 0 & 0.24 & 1.11 & 0.11 \\ 0 & 0 & 1 & 0.19 & 0.23 & 1.08 \end{array}\right]$$

Substitute this inverse matrix $(I - A)^{-1}$ and the given matrix D_5 into the equation $X_5 = (I - A)^{-1}D_5$:

$$X_5 = \begin{bmatrix} 1.40 & 0.23 & 0.14 \\ 0.24 & 1.11 & 0.11 \\ 0.19 & 0.23 & 1.08 \end{bmatrix}\begin{bmatrix} 110 \\ 200 \\ 200 \end{bmatrix} = \begin{bmatrix} 154.00 + 46.00 + 28.00 \\ 26.40 + 222.00 + 22.00 \\ 20.90 + 46.00 + 216.00 \end{bmatrix} = \begin{bmatrix} 228.00 \\ 270.40 \\ 282.90 \end{bmatrix}$$

The results shown above include a considerable amount of round-off error and are provided so you can follow the process. If you use the computer disk and the given fractions, the last column matrix will be 227.27, 270.00, and 283.64. Thus the required output in 5 years is about $230 for farming, $270 for construction, and $280 for clothing.

Problem Set 2.4

Solve the homogeneous systems in Problems 1–6.

1. $\begin{cases} x + 3y - 5z = 0 \\ 3x + y - 3z = 0 \\ 4x + 6y - 11z = 0 \end{cases}$

2. $\begin{cases} 2x - y + 3z = 0 \\ x + 3y - 4z = 0 \\ 8x + 3y + z = 0 \end{cases}$

3. $\begin{cases} 3x + 6y + 2z = 0 \\ 2x - 3y + 6z = 0 \\ 2x + 3y + 2z = 0 \end{cases}$

4. $\begin{cases} 0.4x_1 - 0.5x_2 - 0.2x_3 = 0 \\ -0.1x_1 + 0.6x_2 - 0.3x_3 = 0 \\ -0.3x_1 - 0.1x_2 + 0.5x_3 = 0 \end{cases}$

5. $\begin{cases} 0.3x_1 - 0.3x_2 - 0.65x_3 = 0 \\ -0.25x_1 + 0.5x_2 - 0.2x_3 = 0 \\ -0.5x_1 - 0.2x_2 + 0.85x_3 = 0 \end{cases}$

6. $\begin{cases} 0.7x_1 - 0.5x_2 - 0.45x_3 = 0 \\ -0.4x_1 + 0.5x_2 - 0.35x_3 = 0 \\ -0.3x_1 - 0.25x_2 + 0.8x_3 = 0 \end{cases}$

APPLICATIONS

7. Suppose each person's wages for the closed model discussed in this section should be around $30,000. Give two possible answers to the question if all other information in the problem stays the same.

8. In the closed model discussed in this section, suppose the farmer uses 45% of his own production and gives 25% to the tailor. Answer the question if the other information stays the same.

9. Suppose a merchant joins the closed model and allocates her resources as shown by the following input–output matrix:

	% of production			
Farmer	Builder	Tailor	Merchant	
0.4	0.3	0.1	0.1	Farmer
0.2	0.3	0.3	0.3	Builder
0.2	0.2	0.4	0.3	Tailor
0.2	0.2	0.2	0.3	Merchant

How much should each person pay the others if the wages are to be about $1,000?

Apply the given demand in Problems 10–12 to the open model discussed in this section to find the required output to meet the demand.

10. $D_1 = \begin{bmatrix} 50 \\ 80 \\ 90 \end{bmatrix}$ **11.** $D_3 = \begin{bmatrix} 70 \\ 100 \\ 110 \end{bmatrix}$ **12.** $D_4 = \begin{bmatrix} 100 \\ 180 \\ 170 \end{bmatrix}$

13. Solve the open model discussed in this section for the following input–output matrix, with the other information remaining the same:

$$A = \begin{bmatrix} 0.3 & 0.2 & 0.1 \\ 0.2 & 0.4 & 0.3 \\ 0.1 & 0.1 & 0.2 \end{bmatrix}$$

14. Consider a simple economy with three industries: manufacturing (M), agriculture (A), and transportation (T). Find A and $I - A$ for this system given the information in the table.

Use	Producer M	A	T	External demand	Total output
M	50	20	100	30	200
A	30	80	100	90	300
T	20	10	50	420	500

15. Find $(I - A)^{-1}$ for Problem 14.

16. Find the output in Problem 14 to meet the given demand.

17. Find the output in Problem 14 to meet the external demand if it changes to 200 for M, 90 for A, and 150 for T.

18. **Historical Question** In 1966 Leontief studied a simplified model of the 1958 Israeli economy by dividing it into three sectors—agriculture, manufacturing, and energy—as shown in the table:*

	Agriculture	Manufacturing	Energy
Agriculture	0.293	0	0
Manufacturing	0.014	0.207	0.017
Energy	0.044	0.010	0.216

The exports (in thousands of Israeli pounds) are shown in the following table:

Agriculture	138,213
Manufacturing	17,597
Energy	1786

Find A and $I - A$ for this system.

19. Find $(I - A)^{-1}$ for Problem 18.

20. Find the output in Problem 18 to meet the external demand.

2.5 Summary and Review

IMPORTANT TERMS

Augmented matrix [2.2]
Closed model [2.4]*
Column matrix [2.1]
Communication matrix [2.1]
Component [2.1]
Conformable matrices [2.1]
Difference of matrices [2.1]
Elementary row operations [2.2]
Entry [2.1]
Equal matrices [2.1]
Equivalent matrices [2.2]
Gauss–Jordan elimination [2.2]
Homogeneous system [2.4]*
Identity matrix [2.3]
Input–output model [2.4]*
Inverse matrix [2.3]

Leontief model [2.4]*
Matrix [2.1]
Nonconformable matrices [2.1]
Nonsingular matrix [2.3]
Open model [2.4]*
Order [2.1]
Pivit row [2.2]
Pivoting [2.2]
Product of matrices [2.2]
Reduced row-echelon form [2.2]
Row matrix [2.1]
Scalar multiplication [2.1]
Square matrix [2.1]
Sum of matrices [2.1]
Target row [2.2]
Zero matrix [2.1]

* Wassily Leontief, *Input–Output Economics* (New York: Oxford University Press, 1966), pp. 54–57.

SAMPLE TEST *For additional practice there are a large number of review problems categorized by objective in the Student Solutions Manual. The following sample test (40 minutes) is intended to review the main ideas of the chapter.*

1. Solve the systems using each of the following methods exactly once: graphing, adding, and substitution.

   a. $\begin{cases} 2x + 3y = 2 \\ 5x - 3y = -27 \end{cases}$
   b. $\begin{cases} y = 2x - 4 \\ y = \frac{1}{3}x + 1 \end{cases}$
   c. $\begin{cases} x + y = 13 \\ 3x - 4y = -17 \end{cases}$

Perform the indicated operations in Problems 2–5, if possible.

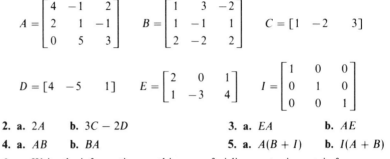

$$A = \begin{bmatrix} 4 & -1 & 2 \\ 2 & 1 & -1 \\ 0 & 5 & 3 \end{bmatrix} \quad B = \begin{bmatrix} 1 & 3 & -2 \\ 1 & -1 & 1 \\ 2 & -2 & 2 \end{bmatrix} \quad C = \begin{bmatrix} 1 & -2 & 3 \end{bmatrix}$$

$$D = \begin{bmatrix} 4 & -5 & 1 \end{bmatrix} \quad E = \begin{bmatrix} 2 & 0 & 1 \\ 1 & -3 & 4 \end{bmatrix} \quad I = \begin{bmatrix} 1 & 0 & 0 \\ 0 & 1 & 0 \\ 0 & 0 & 1 \end{bmatrix}$$

2. a. $2A$ b. $3C - 2D$
3. a. EA b. AE
4. a. AB b. BA
5. a. $A(B + I)$ b. $I(A + B)$
6. a. Write the information on this map of airline routes in matrix form.

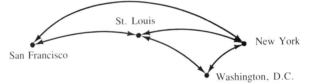

   b. Use a matrix to show the number of ways a person could fly making one stop.
   c. In how many ways can a person fly from San Francisco to New York making two stops?

Solve the systems in Problems 7–8 by Gauss–Jordan elimination.

7. a. $\begin{cases} 4x - 3y = 18 \\ 5x + 2y = 11 \end{cases}$
   b. $\begin{cases} 2x - y = 9 \\ x + 5y = -23 \\ 3x - 4y = 26 \end{cases}$
   c. $\begin{cases} 3x - y + z = 9 \\ 2x + y - 3z = -2 \end{cases}$

8. $\begin{cases} x - 2y + z = 6 \\ y - 3x + w = -2 \\ w - x - y = 2 \\ 2y - 3w + 2z = -2 \end{cases}$

9. a. Find the inverse of $A = \begin{bmatrix} 3 & 3 & -1 \\ -2 & -2 & 1 \\ -4 & -5 & 2 \end{bmatrix}$.

   b. Use the inverse from part a to solve

   $$\begin{cases} 3x + 3y - z = 8 \\ -2x - 2y + z = -1 \\ -4x - 5y + 2z = 3 \end{cases}$$

10. A manufacturer produces two products, I and II. The products require three ingredients: A, B, and C, to be used as follows:

	A	B	C
Product I	6	3	5
Product II	5	4	7

   If the manufacturer has a supply of 3400 units of items A and C and 2000 units of item B, how many of each product can be manufactured?

Modeling Application 2

Leontief Historical Economic Model

In the April 1965 issue of *Scientific American*, Wassily Leontief (see optional Section 2.4) used his model to analyze the 1958 U.S. economy. (A photograph of this model is shown below.) In this model Leontief divided the U.S. economy into 81 sectors. Write a paper based on this model, but for simplicity group these 81 sectors into six related families: FN (final nonmetal: furniture, processed food), FM (final metal: household appliances, motor vehicles), BM (basic metal: machine-shop products, mining), BN (basic nonmetal: agriculture, printing), E (energy: petroleum, coal), and S (services, amusements, real estate). Use the input–out matrix in the table in your paper.

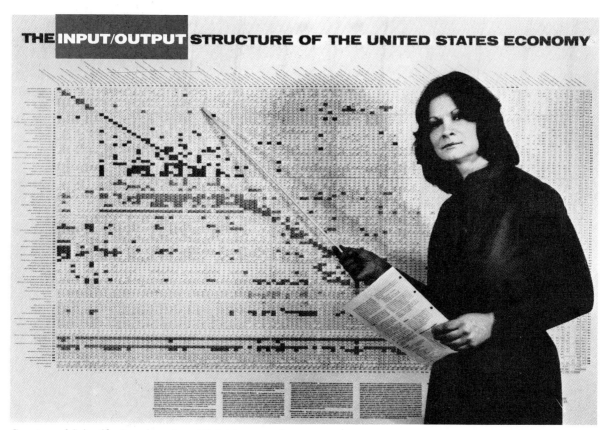

Courtesy of Scientific American, Inc. All rights reserved.

**Internal demands in the 1985
U.S. economy (in millions of
dollars)**

	FN	FM	BM	BN	E	S
FN	0.170	0.004	0.000	0.029	0.000	0.008
FM	0.003	0.295	0.018	0.002	0.004	0.016
BM	0.025	0.173	0.460	0.007	0.001	0.007
BN	0.348	0.037	0.021	0.403	0.011	0.048
E	0.007	0.001	0.039	0.025	0.358	0.025
S	0.120	0.074	0.104	0.125	0.173	0.234

For general guidelines about writing this essay, see the commentary for Modeling Application 1 on page 36.

CHAPTER 3

Linear Programming—The Graphical Method

APPLICATIONS

Management (*Business, Economics, Finance, and Investments*)

Allocation of Resources in Production:
    A farmer planting two crops (3.2, Problem 1; 3.3, Problems 37–39)
    A farmer planting three crops (3.2, Problem 2)
    A manufacturer producing two products (3.2, Problem 13; 3.3, Problem 41)
    Production of tires (3.2, Problem 9)
    Oil production (3.2, Problem 16)
    Manufacturing of tables and shelves (3.2, Problem 17)
    Maximizing profit (3.3, Problems 40–41; 3.4, Problem 10)

Allocation of Resources in Manufacturing:
    Two industrial products (3.2, Problem 3)
    Manufacturing chairs, sofas, and hidabeds (3.2, Problem 4)
    Manufacturing three models of athletic shoes (3.2, Problem 18)
    Wadsworth Widget Company (3.3, Problem 40)

Maximizing yield from an investment (3.2, Problem 7; 3.3, Problems 44–45)
Maximizing profit in baking (3.2, Problem 8)
Office management; storage capacity of computer system (3.2, Problem 10)
Maximizing profits in a motel and restaurant (3.2, Problem 11)
Operating costs; number of days to operate two factories (3.2, Problem 12)
Staff utilization (3.2, Problems 20 and 21; 3.4, Problem 9)
Transportation problem; minimize shipping costs (3.2, Problems 22–24)
Maximizing profit in manufacturing (3.4, Problems 4–8)
CPA examination questions (3.4, Problems 4–10)

Life sciences (*Biology, Ecology, Health, and Medicine*)

Maximum number of animals (2 species) on an island (3.2, Problem 14)
Maximum number of animals (3 species) and limited resources (3.2, Problem 15)
Minimizing exercise time to achieve certain goals (3.2, Problem 19)
Diet problem:
    Nutritional comparison between Kellogg's Corn Flakes and Post Honeycombs (3.2, Problem 5; 3.3, Problem 43)
    Nutritional comparison between Kellogg's Corn Flakes and Kellogg's Raisin Bran (3.2, Problem 6)
    Meeting minimum daily nutritional requirements (3.3, Problem 42)
Modeling application—Air Pollution Control

CHAPTER OVERVIEW   This chapter introduces the topic of linear programming. Linear programming is applied when we are interested in maximizing or minimizing a linear function—for example, maximizing profits, minimizing costs, finding the most efficient shipping schedules, minimizing waste in the production of a product, securing the proper mixes of ingredients in the production of a product, controlling inventories, or finding the most efficient assignment of personnel.

PREVIEW   Graphing systems of linear inequalities is reviewed in the first section and then formulating linear programming problems is discussed. These two ideas are then tied together in the third section to solve linear programming problems by graphing. The extended application concerns a mathematical model that uses the techniques of this chapter.

PERSPECTIVE   This chapter gives us a basic understanding of one of the most important topics in this course—linear programming. However, most real life applications are not limited to two variables, and since graphing is limited to two dimensions, we also need an algebraic method. The next chapter introduces such an algebraic method. This chapter is thus a stepping-stone to the next, which combines matrices, pivoting, and Gauss–Jordan elimination with the foundations of linear programming presented in this chapter.

3.1 Systems of Linear Inequalities

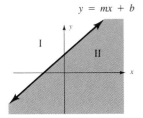

$y = mx + b$

Figure 3.1 Half-planes

A **linear inequality** in two variables can be written in one of the following forms:

$$ax + by < c \qquad ax + by > c$$
$$ax + by \leq c \qquad ax + by \geq c$$

where a, b, and c are constants. The graphing of a linear inequality is similar to the graphing of a linear equation. Any line divides the plane into three regions, as shown in Figure 3.1. Regions I and II are called **half-planes**, so we can say the three sets determined by a line are the two half-planes and the set of points on the line itself. The line is called the **boundary** of the half-planes. If the boundary is included in the half-plane, the half-plane is said to be **closed**; if the boundary is not included, it is said to be **open**. Half-planes are drawn by showing the boundary as a dashed line, as in Figure 3.2c.

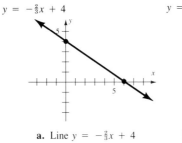

$y = -\tfrac{2}{3}x + 4$

a. Line $y = -\tfrac{2}{3}x + 4$

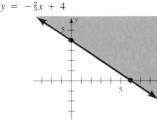

$y = -\tfrac{2}{3}x + 4$

b. Closed half-plane $y \geq -\tfrac{2}{3}x + 4$
   Draw boundary as a solid line

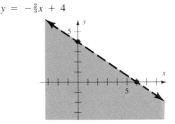

$y = -\tfrac{2}{3}x + 4$

c. Open half-plane $y < -\tfrac{2}{3}x + 4$
   Draw boundary as a dashed line

Figure 3.2 Graphing half-planes

It is apparent in Figure 3.2 that every linear inequality has an associated equation that is the boundary of that half-plane. For example, $2x + 3y < 12$ has the boundary line $2x + 3y = 12$. This line is graphed by solving for y: $y = -\frac{2}{3}x + 4$. After you have drawn the boundary (either solid or dashed), the remaining question is Which half-plane satisfies the inequality? The easiest way to decide is to choose *any* test point not on the boundary line. If it satisfies the inequality, then the solution is the half-plane containing the test point. If the test point does not satisfy the inequality, then the solution is the half-plane that doesn't contain the test point.

EXAMPLE 1   Graph $5x + 2y - 10 \le 0$.

Solution   *Step 1   Graph the boundary line.*   The boundary is included if the given inequality is $\ge$ or $\le$, so draw $5x + 2y - 10 = 0$ as a solid line. The easiest way to sketch this line is to put it into slope-intercept form: $y = -\frac{5}{2}x + 5$.

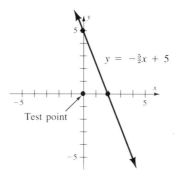

Step 2   Choose a test point.   The point $(0, 0)$ is usually the best choice because of the simplified arithmetic:

$$5x + 2y - 10 \le 0$$

Test $(0, 0)$:

$$5(0) + 2(0) - 10 \le 0$$
$$-10 \le 0 \qquad \text{True}$$

Step 3   Shade the appropriate half-plane.   If the test point satisfies the inequality (it is true, as shown in step 2), shade in the half-plane containing the test point. Otherwise, shade in the other half-plane.

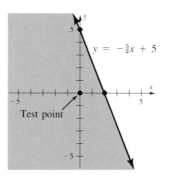

■

EXAMPLE 2 Graph $y > -20x + 110$.

Solution *Step 1* The boundary is $y = -20x + 110$; $m = -\frac{20}{1}$ and $b = 110$. Draw the boundary as a dashed line because it is not included in the half-plane ($>$ symbol).

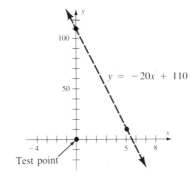

Step 2

$$y > -20x + 110$$

Test point $(0,0)$:

$$0 > -20(0) + 110 \qquad \text{False}$$

Step 3 Graph the half-plane that does not contain the test point.

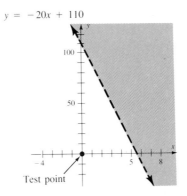

There are many applications in which inequalities play an important role. In Chapter 2 we found the *break-even point*—that is, the place where cost and revenue are the same. Sometimes, breaking even for a business takes the form of a linear equation. Suppose a company knows that it will break even if 50 items are sold at $3 each or if 10 items are sold at $5 each.* This gives a *break-even linear*

---

* In practice, this is more likely to be 50,000 and 10,000 items. We use 50 and 10 here to keep the numbers easy to handle. To find the equation of the line where *x* is the price and *y* is the number of items, we use the points (3, 50) and (5, 10). The slope is

$$m = \frac{10 - 50}{5 - 3} = -20$$

and $y = mx + b$, so $50 = -20(3) + b$ or $b = 110$. Thus $y = -20x + 110$.

equation:

$$y = -20x + 110$$

where x is the price and y is the number of items. Suppose that in order to make a profit it is necessary for

$$y > -20x + 110$$

(This was graphed in Example 2.) However, in this application, it is also necessary that x and y both be positive:

$$x \geq 0$$
$$y \geq 0$$

Furthermore, market research shows that the price must be $8 or less:

$$x \leq 8$$

This information taken together gives a **system of linear inequalities** similar to a system of equations:

$$\begin{cases} y > -20x + 110 \\ x \geq 0 \\ y \geq 0 \\ x \leq 8 \end{cases}$$

The region determined by the system of inequalities is the solution of the system and is similar to the point of intersection for a system of equations because it is the intersection of the individual inequalities in the system.

EXAMPLE 3   Graph the system $\begin{cases} y > -20x + 110 \\ x \geq 0 \\ y \geq 0 \\ x \leq 8 \end{cases}$

Solution   Graph each half-plane, but instead of shading, mark the appropriate regions with arrows. For this example,

$$x \geq 0$$
$$y \geq 0$$

gives the first quadrant, which is marked with arrows as shown.
Next graph $y = -20x + 110$ and use a test point, say, $(0,0)$, to determine the half-plane. Now draw $x = 8$ and mark it with arrows to show that $x \leq 8$. Finally shade the region of the plane that represents the intersection:

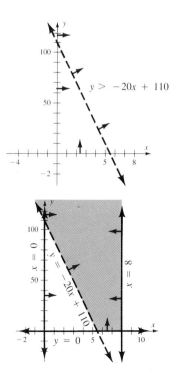

EXAMPLE 4   Graph the solution of the system $\begin{cases} 2x + y \le 3 \\ x - y > 5 \\ x \ge 0 \\ y \ge -10 \end{cases}$

Solution   The graphs of the individual inequalities and their intersections are shown in the figure. Note the use of arrows to show the solutions of the individual inequalities. This device replaces the use of a lot of shading, which can be confusing if there are many inequalities in the system.

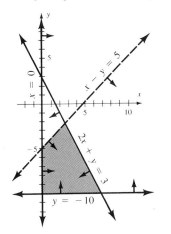

Problem Set 3.1

Graph the solution for each of the linear inequalities in Problems 1–24.

1. $y \ge 2x - 3$ **2.** $y \ge 3x + 2$ **3.** $y \ge 4x + 5$

4. $y \le \frac{1}{2}x + 2$ **5.** $y \le \frac{2}{3}x - 4$ **6.** $y \le \frac{4}{3}x - 3$

7. $x - y > 3$ **8.** $2x + y < -3$ **9.** $3x + y < 4$

10. $x + 2y > 4$ **11.** $x - 3y > 12$ **12.** $3x - 4y > 8$

13. $y \le 6$ **14.** $x \ge 2$ **15.** $y < -2$

16. $y > 0$ **17.** $x < 0$ **18.** $x < 8$

19. $260x - 1040y > 11{,}250$ **20.** $150x - 450y < 7200$

21. $0.06x - 0.05y < 10{,}000$ **22.** $0.03x - 0.05y > 10{,}000$

23. $0.08x + 0.10y \ge 500$ **24.** $0.10x - 0.08y \le 1500$

Graph the solution of each system in Problems 25–51.

25. $\begin{cases} x \ge 0 \\ y \le 0 \end{cases}$ **26.** $\begin{cases} x \ge 0 \\ y \ge 0 \end{cases}$ **27.** $\begin{cases} x \le 0 \\ y \le 0 \end{cases}$

28. $\begin{cases} x \ge 0 \\ y \ge 0 \\ x \le 5 \\ y \le 6 \end{cases}$ **29.** $\begin{cases} x \ge 0 \\ y \ge 0 \\ x < 8 \\ y < 5 \end{cases}$ **30.** $\begin{cases} x \ge 0 \\ y \ge 0 \\ x < 500 \\ y < 1000 \end{cases}$

31. $\begin{cases} y \le 3x - 4 \\ y \ge -2x + 5 \end{cases}$ **32.** $\begin{cases} 2x + y > 3 \\ 3x - y < 2 \end{cases}$ **33.** $\begin{cases} 3x - 2y \ge 6 \\ 2x + 3y \le 6 \end{cases}$

34. $\begin{cases} x + 2y \le 18 \\ x + y \ge 4 \end{cases}$ **35.** $\begin{cases} x + y > 4 \\ x - y > -2 \end{cases}$ **36.** $\begin{cases} 2x + 3y \ge 3 \\ 2x + 3y \le 9 \end{cases}$

37. $\begin{cases} x - 10 \le 0 \\ x \ge 0 \end{cases}$ **38.** $\begin{cases} y - 5 \le 0 \\ y \ge 0 \end{cases}$ **39.** $\begin{cases} y - 25 \le 0 \\ y \ge 0 \end{cases}$

40. $\begin{cases} -5 < x \\ 3 \ge x \\ 5 > y \\ -2 \le y \end{cases}$ **41.** $\begin{cases} -10 \le x \\ x \le 6 \\ 3 < y \\ y < 8 \end{cases}$ **42.** $\begin{cases} -5 < x \\ x \le 2 \\ -4 \le y \\ y < 9 \end{cases}$

43. $\begin{cases} x \ge 0 \\ y \ge 0 \\ x + y \le 8 \\ y \le 4 \\ x \le 6 \end{cases}$ **44.** $\begin{cases} x \ge 0 \\ y \ge 0 \\ x + y \le 9 \\ 2x - 3y \ge -6 \\ x - y \le 3 \end{cases}$

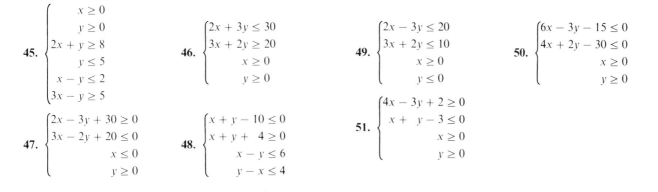

45. $\begin{cases} x \ge 0 \\ y \ge 0 \\ 2x + y \ge 8 \\ y \le 5 \\ x - y \le 2 \\ 3x - y \ge 5 \end{cases}$

46. $\begin{cases} 2x + 3y \le 30 \\ 3x + 2y \ge 20 \\ x \ge 0 \\ y \ge 0 \end{cases}$

49. $\begin{cases} 2x - 3y \le 20 \\ 3x + 2y \le 10 \\ x \ge 0 \\ y \le 0 \end{cases}$

50. $\begin{cases} 6x - 3y - 15 \le 0 \\ 4x + 2y - 30 \le 0 \\ x \ge 0 \\ y \ge 0 \end{cases}$

47. $\begin{cases} 2x - 3y + 30 \ge 0 \\ 3x - 2y + 20 \le 0 \\ x \le 0 \\ y \ge 0 \end{cases}$

48. $\begin{cases} x + y - 10 \le 0 \\ x + y + 4 \ge 0 \\ x - y \le 6 \\ y - x \le 4 \end{cases}$

51. $\begin{cases} 4x - 3y + 2 \ge 0 \\ x + y - 3 \le 0 \\ x \ge 0 \\ y \ge 0 \end{cases}$

3.2 Formulating Linear Programming Models

One of the most difficult tasks in solving a real world problem is the construction of a mathematical model that simulates the situation. In the 1940s a new mathematical model called **linear programming** was found to be applicable to a wide range of situations in which the maximum or minimum value of some variables was needed. In this section, you will learn how to formulate or build a model for a variety of applications (see the examples listed in the box), and in the next section you will learn how to solve these models using graphical techniques. (In the next chapter, an algebraic method, the *simplex method*, for solving linear programming problems is introduced.)

Examples of linear programming models
Allocation of Resources Models
Maximize profit from a manufacturing process
Maximize revenue from a manufacturing process
Minimize costs from a manufacturing process
Maximize production
Find the optimal use of land
Utilization of a sales force
Utilization of office staff
Nutritional (or Diet) Models
Minimize cost of a diet with certain nutritional requirements
Minimize calories given certain nutritional requirements
Manufacture a processed food to meet government regulations
Investment Strategies
Maximize the return on an investment
Allocate funds over several possible investments
Minimize expenses
Transportation Model
Minimize shipping costs

EXAMPLE 1   **Allocation of Resources in Production**   A farmer has 100 acres on which to plant two crops: corn and wheat. To produce these crops, there are certain expenses, as shown in the table.

Item	Cost per acre
Corn	
Seed	$ 12
Fertilizer	58
Planting/care/harvesting	50
Total	$120
Wheat	
Seed	$ 40
Fertilizer	80
Planting/care/harvesting	90
Total	$210

After the harvest, the farmer must usually store the crops while awaiting favorable market conditions. Each acre yields an average of 110 bushels of corn or 30 bushels of wheat. The limitations of resources are

Available capital: $15,000
Available storage facilities: 4000 bushels

If the net profit (the profit *after* all expenses have been subtracted) per bushel of corn is $1.30 and for wheat is $2.00, how should the farmer plant the 100 acres to maximize profits?

Solution   First, you might try to solve this problem by using your intuition. If you plant all 100 acres with wheat, the production is $30 \times 100 = 3000$ bushels, for a net profit of $3000 \times 2 = \$6,000$. But to plant 100 acres with wheat would cost $\$210 \times 100 = \$21,000$, and only $15,000 is available. On the other hand, if 100 acres of corn are planted, the total cost is $\$120 \times 100 = \$12,000$ and the net profit is $110 \times 100 \times \$1.30 = \$14,300$. However, the yield of 11,000 bushels (110×100) cannot be stored since there are facilities to store only 4000 bushels.

To formulate a mathematical model, begin by letting

x = Number of acres to be planted in corn

y = Number of acres to be planted in wheat

There are certain limitations, or **constraints**. The number of acres planted cannot be negative, so

$x \geq 0$ These constraints apply in almost every model, even though they are not
$y \geq 0$ explicitly stated as part of the given problem. Be sure, however, to state them when listing the constraints in the solution

The amount of available land is 100 acres, so

$x + y \leq 100$

Why not $x + y = 100$? It might be more profitable for the farmer to leave some land out of production. That is, it is not *necessary* to plant all the land.

We also know that

$120x$ = Expenses for planting corn

$210y$ = Expenses for planting wheat

The total expenses cannot exceed $15,000; this is the *available capital*:

$120x + 210y \leq 15,000$

The yields are

$110x$ = Yield of acreage planted in corn

$30y$ = Yield of acreage planted in wheat

The total yield cannot exceed the storage capacity of 4000 bushels, so

$110x + 30y \leq 4000$

We summarize the constraints (in boldface above) in the following system:

$$\begin{cases} x \geq 0 \\ y \geq 0 \\ x + y \leq 100 \\ 120x + 210y \leq 15,000 \\ 110x + 30y \leq 4000 \end{cases}$$

Now let P represent the total profit. The farmer wants to maximize this profit. A function that is to be maximized or minimized is called the **objective function**.

$$\text{PROFIT FROM CORN} = \text{VALUE} \cdot \text{AMOUNT}$$
$$= 1.30 \cdot 110x$$
$$= 143x$$

$$\text{PROFIT FROM WHEAT} = \text{VALUE} \cdot \text{AMOUNT}$$
$$= 2.00 \cdot 30y$$
$$= 60y$$

$$P = \text{PROFIT FROM CORN} + \text{PROFIT FROM WHEAT}$$
$$= 143x + 60y$$

The linear programming model is stated as follows:

Maximize: $P = 143x + 60y$

Subject to: $\begin{cases} x \geq 0 \\ y \geq 0 \\ x + y \leq 100 \\ 120x + 210y \leq 15,000 \\ 110x + 30y \leq 4000 \end{cases}$

$\blacksquare$

This linear programming model will be solved in the next section.

EXAMPLE 2 **Allocation of Resources in Manufacturing** The Wadsworth Widget Company manufactures two types of widgets: regular and deluxe. Each widget is produced at a station consisting of a machine and a person who finishes the widgets by hand. The regular widget requires 3 hours of machine time and 2 hours of finishing time. The deluxe widget requires 2 hours of machine time and 4 hours of finishing time. The profit on the regular widget is $25, and on the deluxe widget the profit is $30. If the workday is 8 hours, how many of each type of widget should be produced at each station per day in order to maximize the profit?

Solution We want to maximize the profit, P. There are two types of items, regular and deluxe. Let

x = Number of regular widgets produced

y = Number of deluxe widgets produced

Then

Profit from regular widgets = $25x$

Profit from deluxe widgets = $30y$

P = PROFIT FROM REGULAR WIDGETS + PROFIT FROM DELUXE WIDGETS

= $25x + 30y$

Two constraints are

$x \geq 0$ The number of widgets must be nonnegative

$y \geq 0$

Production time is summarized in the table:

	Machine time (hr)	Manual time (hr)
Regular widget, x	3 each; $3x$ total	2 each; $2x$ total
Deluxe widget, y	2 each; $2y$ total	4 each; $4y$ total
Total workday	$3x + 2y$	$2x + 4y$

Thus

$3x + 2y \leq 8$ Machine workday

$2x + 4y \leq 8$ Manual workday

The linear programming model is:

Maximize: $P = 25x + 30y$

Subject to: $\begin{cases} x \geq 0 \\ y \geq 0 \\ 3x + 2y \leq 8 \\ 2x + 4y \leq 8 \end{cases}$ ∎

EXAMPLE 3 **Diet Problem** A convalescent hospital wishes to provide, at a minimum cost, a diet that has a minimum of 200 grams of carbohydrates, 100 grams of protein, and

120 grams of fats per day. These requirements can be met with two foods:

Food	Carbohydrates	Protein	Fats
A	10 g	2 g	3 g
B	5 g	5 g	4 g

If food A costs 29¢ per ounce and food B costs 15¢ per ounce, how many ounces of each food should be purchased for each patient per day in order to meet the minimum requirements at the lowest cost?

Solution Let

$$x = \text{Number of ounces of food A}$$
$$y = \text{Number of ounces of food B}$$

The *minimum cost, C,* is found by:

Cost of food A $= 0.29x$
Cost of food B $= 0.15y$
$C = 0.29x + 0.15y$

The constraints are

$x \geq 0$ The amounts of food must be nonnegative
$y \geq 0$

The table summarizes the nutrients provided:

Food	Amount (in ounces)	Total consumption (in grams)		
		Carbohydrates	Protein	Fats
A	x	$10x$	$2x$	$3x$
B	y	$5y$	$5y$	$4y$
Total		$10x + 5y$	$2x + 5y$	$3x + 4y$

Daily requirements are:

$$10x + 5y \geq 200$$
$$2x + 5y \geq 100$$
$$3x + 4y \geq 120$$

The linear programming model is:

Minimize: $C = 0.29x + 0.15y$

Subject to:
$$\begin{cases} x \geq 0 \\ y \geq 0 \\ 10x + 5y \geq 200 \\ 2x + 5y \geq 100 \\ 3x + 4y \geq 120 \end{cases}$$

∎

EXAMPLE 4   **Investment**   Brown Bros., Inc., is an investment company analyzing a pension fund for a certain company. A maximum of $10 million is available to invest in two places. No more than $8 million can be invested in stocks yielding 12%, and at least $2 million can be invested in long-term bonds yielding 8%. The stock-to-bond investment ratio cannot be more than 1 to 3. How should Brown Bros. advise their client so that the pension fund will receive the maximum yearly return on investment?

Solution   To build this model you need to use the *simple interest formula*:

$$I = Prt$$

where

I = Interest	The amount paid for the use of another's money
P = Principal	The amount invested
r = Interest rate	Write this as a decimal. It is assumed to be an annual interest rate, unless otherwise stated
t = Time	In years, unless otherwise stated

Let

x = Amount invested (in millions) in stocks (12% yield)

y = Amount invested (in millions) in bonds (8% yield)

Then

Stocks: $I = 0.12x$ $P = x, r = 0.12$, and $t = 1$

Bonds: $I = 0.08y$ $P = y, r = 0.08$, and $t = 1$

The return on investment, R (in millions), is found by

$$R = 0.12x + 0.08y$$

The constraints are:

$x \geq 0$	Investments are nonnegative
$y \geq 0$	
$x + y \leq 10$	Maximum investment is $10 million; note that the unit chosen for this problem is in millions of dollars
$x \leq 8$	No more than $8 million in stocks
$y \geq 2$	No less than $2 million in bonds
$3x \leq y$	Must invest $3 million in bonds for every $1 million invested in stocks for a stock-to-bond ratio of 1 to 3. That is

$$\frac{\text{stock}}{\text{bond}} \leq \frac{1}{3}$$

$$3 \text{ stock} \leq \text{bond}$$

$$3x \leq y$$

Thus the linear programming model for this problem is:

Maximize: $R = 0.12x + 0.08y$

Subject to: $\begin{cases} x \geq 0 \\ y \geq 0 \\ x \leq 8 \\ y \geq 2 \\ x + y \leq 10 \\ 3x \leq y \end{cases}$

$\blacksquare$

EXAMPLE 5 **Transportation Problem** Sears ships a certain air-conditioning unit from factories in Portland, Oregon, and Flint, Michigan, to distribution centers in Los Angeles, California, and Atlanta, Georgia. The shipping costs are summarized in the table:

Source	Destination	Shipping cost
Portland	Los Angeles	$30
	Atlanta	$40
Flint	Los Angeles	$60
	Atlanta	$50

The supply and demand, in number of units, is shown below:

Supply	Demand
Portland, 200	Los Angeles, 300
Flint, 600	Atlanta, 400

How should shipments be made from Portland and Flint to minimize the shipping cost?

Solution The information can be summarized by the following "map."

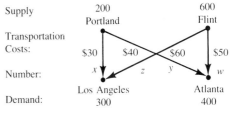

Suppose the following number of units is shipped:

Source	Destination	Number	Shipping cost
Portland	Los Angeles	x	$30x$
	Atlanta	y	$40y$
Flint	Los Angeles	z	$60z$
	Atlanta	w	$50w$

The linear programming problem is then:

Minimize: $C = 30x + 40y + 60z + 50w$

Subject to:
$$\begin{cases} x + y \le 200 \\ z + w \le 600 \\ x + z \ge 300 \\ y + w \ge 400 \\ x \ge 0, \quad y \ge 0, \quad z \ge 0, \quad w \ge 0 \end{cases}$$

Note that as we progressed from Example 1 to Example 5, we were able to formulate the linear programming problem with fewer and fewer words so that the information in the problem was translated almost directly into a linear programming model. Also, notice in Example 5 that we stated the nonnegative constraints together for convenience. In the next section we develop graphical techniques for solving these types of problems.

Problem Set 3.2

Write a linear programming model for Problems 1–24. Do not attempt to solve the problems, but be sure to define all your variables.

APPLICATIONS

1. A farmer has 500 acres on which to plant two crops: corn and wheat. It costs $120 per acre to produce corn and $60 per acre to produce wheat, and there is $24,000 available to pay for this year's production. If the yield per acre is 100 bushels of corn or 40 bushels of wheat and the farmer has contracted to store at least 20,000 bushels, how much should the farmer plant to maximize the profits when the net profit is $1.20 per bushel for corn and $2.50 per bushel for wheat?

2. A farmer has 1500 acres on which to plant three crops: corn, wheat, and soybeans. It costs $150 per acre to produce corn, $80 per acre to produce wheat, and $60 per acre to produce soybeans, and there is $36,000 available to pay for this year's production. If the yield per acre is 100 bushels of corn, or 40 bushels of wheat, or 60 bushels of soybeans, and the farmer has contracted to store at least 20,000 bushels, how much should the farmer plant to maximize the profits when the net profit is $1.00 per bushel for corn, $2 per bushel for wheat, and $1.50 per bushel for soybeans?

3. The Thompson Company manufactures two industrial products: a standard product ($45 profit per item) and an economy product ($30 profit per item). These products are built using machine time and manual labor. The standard product requires 3 hours machine time and 2 hours manual labor. The economy model requires 3 hours machine time and no manual labor. If the week's supply of manual labor is limited to 800 hours and machine time to

1500 hours, how much of each type of product should be produced each week in order to maximize the profit?

4. A furniture manufacturer makes chairs, sofas, and sofabeds. The production process is divided into carpentry, finishing, and upholstery. Manufacture of a chair requires 3 hours of carpentry, 1 hour of finishing, and 2 hours of upholstery. Manufacture of a sofa requires 5 hours of carpentry, 2 hours of finishing, and 4 hours of upholstery. Manufacture of a sofabed requires 8 hours of carpentry, 3 hours of finishing, and 4 hours of upholstery. The manufacturer has available each day 120 hours of manual labor for carpentry, 24 hours of manual labor for finishing, and 96 hours of manual labor for upholstery. The net profit is $50 per chair, $75 per sofa, and $100 per sofabed. How many of each item should be produced each day in order to maximize profits?

5. The nutritional information in the table is found on the sides of the cereal boxes listed (for 1 ounce of cereal with $\frac{1}{2}$ cup of whole milk). What is the minimum cost in order to receive at least 322 grams of starch (and related carbohydrates) and 119 grams of sucrose (and other sugars) by consuming these two cereals if corn flakes cost 7¢ per ounce and Honeycombs cost 19¢ per ounce?

Cereal	Starch and related carbohydrates	Sucrose and other sugars
Kellogg's Corn Flakes	23 g	7 g
Post Honeycombs	14 g	17 g

6. The nutritional information in the table is found on the sides of the cereal boxes listed (for 1 ounce of cereal with $\frac{1}{2}$ cup of whole milk). What is the most cereal a person could consume and not receive more than 322 grams of starch (and related carbohydrates) or more than 126 grams of sucrose (and other sugars)?

Cereal	Starch and related carbohydrates	Sucrose and other sugars
Kellogg's Corn Flakes	23 g	7 g
Kellogg's Raisin Bran	14 g	18 g

7. Your broker tells you about two investments she thinks worthwhile. She recommends a new issue of Pertec stock, which should yield 20% over the next year, and then to balance your account she recommends Campbell Municipal Bonds with a 10% annual yield. The stock-to-bond ratio should be no less than 1 to 3. If you have no more than $100,000 to invest and do not want to invest more than $70,000 in Pertec or less than $20,000 in bonds, how much should you invest in each to maximize your return?

8. Mama's Home Bakery produces two specialties: German chocolate cake and apple strudel. Daily business requires that at least 50 cakes and 100 strudels be baked every day. The capacity of the ovens limits the total number of items that can be baked each day to no more than 250. How many of each item should be baked in order to maximize profits if the net profit on each cake is $1.50 and on each strudel is $1.20?

9. Karlin Manufacturing has discontinued production of a line of tires that has not been profitable. This has created some excess production time, and management is considering devoting this excess to their Glassbelt or Rainbelt tire lines. The amount of excess production available is 200 hours per month on the composition machine and 400 hours per month on production line # 3. The number of hours required for each tire on each of these production lines is given in the table. How many of each tire should Karlin produce to maximize the return if the unit return on each Glassbelt is $52 and on each Rainbelt is $36?

Machine	Glassbelt tire	Rainbelt tire
Composition machine	3 hr	1 hr
Production line # 3	2 hr	2 hr

10. An office manager decides to purchase some personal-size computers. Apple computers cost $900 each, will provide 250 K storage capability, and will take up 5 square feet of desk space. IBM/PC computers cost $1,200 each, will provide 612 K storage, and will take up only 3 square feet of desk space. Proper utilization of space requires that a total of not more than 165 square feet be used for this new equipment. If there is $36,000 available to spend on computers, how many of each type of computer should be purchased if the office manager wishes to maximize storage capacity?

11. Foley's Motel has 200 rooms and a restaurant that seats 50 people. Experience shows that 40% of the commercial guests and 20% of the other guests (i.e., noncommercial) eat in the restaurant. There is a net profit of $4.50 per day from each commercial guest and $3.50 per day from each other guest. Find the number of commercial and other guests needed in order to maximize the net profits. Assume one guest per room and that the capacity of the restaurant will not be exceeded.

12. Karlin Enterprises manufactures two games. Standing orders require that at least 24,000 space-battle games and 5000 football games be produced. The company has two factories: The Gainesville plant can produce 600 space-battle games and 100 football games per day; the Sacramento plant can produce 300 space-battle games and 100 football games per day. If the Gainesville plant costs $20,000 per day to operate and the Sacramento plant costs $15,000 per day, find the number of days per month that each factory should operate to minimize the cost. (Assume that each month has 30 days.)

13. Alco Company manufactures two products: Alpha and Beta. Each product must pass through two processing operations, and all materials are introduced at the first operation. Alco may produce either one product exclusively or various combinations of both products subject to the constraints given in the table. A shortage of technical labor has limited Alpha production to no more

Product	Hours required to produce one unit		Profit per unit
	First process	Second process	
Alpha	1 hr	1 hr	$5
Beta	3 hr	2 hr	$8
Total capacity per day	1200 hr	1000 hr	

than 700 units per day. There are no constraints on the production of Beta other than the hour constraints shown in the table. How many of each product should be manufactured in order to maximize profits?

14. An island is inhabited by two species of animals, A1 and A2, which feed on three types of food, F1, F2, and F3. The first species requires 12 units of food F1, 6 units of F2, and 16 units of F3 to survive. The second species requires 15, 20, and 5 units of foods F1, F2, and F3, respectively. The island can normally supply 1200 units of F1, 1200 units of F2, and 800 units of F3. What is the maximum number of animals that this island can support?

15. An isolated geographic region is inhabited by three species of animals, A1, A2, and A3, which feed on three types of food, F1, F2, and F3. The first species requires 12 units of food F1, 6 units of F2, and 16 units of F3 to survive. The second species requires 15, 20, and 5 units of food F1, F2, and F3, respectively, and the third species requires 20, 10, and 10 of food F1, F2, and F3, respectively. The region can supply 8000 units of F1, 10,000 units of F2, and 6000 units of F3. What is the maximum number of animals that this region can support?

16. The Cosmopolitan Oil Company requires at least 8000 barrels of low-grade oil per month; it also needs at least 12,000 barrels of medium-grade oil per month and at least 2000 barrels of high-grade oil per month. Oil is produced at one of two refineries. The daily production of these refineries is summarized in the table. If it costs $17,000 per day to operate refinery I and $15,000 per day to operate refinery II, how many days per month should each refinery be operated to satisfy the requirements and at the same time minimize the costs? (Assume that each month has 30 days.)

	Oil production (barrels)		
Refinery	**Low grade**	**Medium grade**	**High grade**
I	200	400	100
II	200	300	100

17. Pell, Inc., manufactures two products: tables and shelves. Each product must be processed in each of three departments: machining, assembling, and finishing. The hours needed to produce one unit of product per department and the maximum possible hours per department are shown in the table. Standing orders require that Pell manufacture at least 30 tables and 25 shelves. Pell's net profit is $4 per table and $2 per shelf. How many tables should be manufactured to maximize the net profit?

	Production time per unit (hr)		Maximum capacity
Department	**Tables**	**Shelves**	**(hr)**
Machining	2	1	200
Assembling	2	2	240
Finishing	2	3	300

18. A shoe company produces three models of athletic shoes. Shoe parts are first produced in the manufacturing shop and then put together in the assembly shop. The number of hours of labor required per pair in each shop is given in the table. The company can sell as many pairs of shoes as it can produce. During the next month, no more than 1000 hours can be expended in the manufacturing shop and no more than 800 hours can be expended in the assembly shop. The expected profit from the sale of each pair of shoes is:

Model 1: $12
Model 2: $7
Model 3: $5

The company wants to determine how many pairs of each shoe model to produce to maximize total profits over the next month.*

Shop	**Model 1**	**Model 2**	**Model 3**
Manufacturing	8 hr	5 hr	3 hr
Assembly	5 hr	1 hr	3 hr

19. A woman wants to design a weekly exercise schedule that involves jogging, handball, and aerobic dance. She decides to jog at least 3 hours per week, play handball at least 2 hours per week, and dance at least 5 hours per week. She also wants to devote at least as much time to jogging as to handball because she does not like handball as much as she likes the other exercises. She also knows that jogging consumes 900 calories per hour, handball 600 calories per hour, and dance 800 calories per hour.

---

* This problem is taken from a recent examination administered by the Society of Actuaries. Reprinted by permission of the Society of Actuaries.

Her physician told her that she must burn up a total of at least 9000 calories per week in this exercise program. How many hours should she devote to each exercise if she wishes to minimize her exercise time?

20. To utilize costly equipment efficiently, an assembly line with a capacity of 50 units per shift must operate 24 hours per day. This requires scheduling three shifts with varying labor costs as follows:

Day shift:	Labor cost $100 per unit
Swing shift:	Labor cost $150 per unit
Graveyard shift:	Labor cost $180 per unit

Each unit produced uses 60 linear feet of sheeting material, and there is a total of 1800 feet available for the day shift, 1500 feet for the swing shift, and 3000 feet for the graveyard shift. The day shift must produce at least 20 units and the other shifts must produce at least 50 units together. How many units should be produced on each shift to maximize the revenue if the units are sold for $850 each?

21. Repeat Problem 20 except minimize labor costs rather than maximize revenue.

22. Suppose a computer dealer has stores in Hillsborough and Palo Alto and warehouses in San Jose and Burlingame. The cost of shipping a computer from one location to another as well as the supply and demand at each location are shown on the following "map":

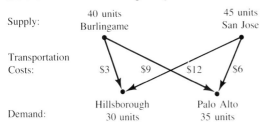

Supply: 40 units Burlingame   45 units San Jose

Transportation Costs: $3 $9 $12 $6

Demand: Hillsborough 30 units   Palo Alto 35 units

How should the dealer ship the sets to minimize the shipping costs?

23. Camstop, Inc., an importer of computer chips, can ship case lots of components to stores in Dallas and Chicago from ports in either Los Angeles or Seattle. The costs and demand are given in the table. How many cases should be shipped from each port to minimize shipping costs?

	Cost to ship (per case)		
Store	Los Angeles port	Seattle port	Store demand
Chicago	$9	$7	80 cases
Dallas	$7	$8	110 cases
Available inventory at each port	90 cases	130 cases	

24. A Texas appliance dealer has stores in Fort Worth and in Houston and warehouses in Dallas and San Antonio. The cost of shipping a refrigerator from Dallas to Fort Worth is $10, from Dallas to Houston it is $15, from San Antonio to Fort Worth it is $14, and from San Antonio to Houston it is $18. Suppose that the Fort Worth store has orders for 15 refrigerators and the Houston store has orders for 35. Also suppose that there are 25 refrigerators in stock at Dallas and 45 at San Antonio. What is the most economical way to supply the requested refrigerators to the two stores?

3.3 Graphical Solution of Linear Programming Problems

In the previous section we looked at some models called *linear programming problems*. In each case the model had a function called an **objective function**, which was to be maximized or minimized while satisfying several conditions, or **constraints**. If there are only two variables, we use a graphical method of solution. We begin with the set of constraints and consider them as a system of inequalities. The solution of this system of inequalities is a set of points, S. Each point of the set S is called a **feasible solution**. The objective function can be evaluated for different feasible solutions to obtain maximum or minimum values.

EXAMPLE 1   Maximize: $K = 4x + 5y$

Subject to: $\begin{cases} 2x + 5y \le 25 \\ 6x + 5y \le 45 \\ x \ge 0, \quad y \ge 0 \end{cases}$

Solution   The constraints give a set of feasible solutions as graphed in the figure. (You may want to review Section 3.1 on graphing systems of inequalities.)

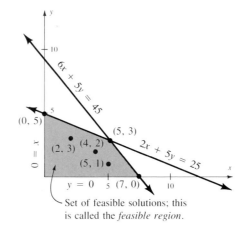

Set of feasible solutions; this is called the *feasible region*.

To solve the linear programming problem, we must now find the feasible solution that makes the objective function as large as possible. Some possible solutions are:

Feasible solution (a point in the solution set of the system)	*Objective function* $K = 4x + 5y$
$(2, 3)$	$4(2) + 5(3) = 8 + 15 = 23$
$(4, 2)$	$4(4) + 5(2) = 16 + 10 = 26$
$(5, 1)$	$4(5) + 5(1) = 20 + 5 = 25$
$(7, 0)$	$4(7) + 5(0) = 28 + 0 = 28$
$(0, 5)$	$4(0) + 5(5) = 0 + 25 = 25$

In this list, the point that makes the objective function the largest is $(7, 0)$. But is this the largest for all feasible solutions? How about $(6, 1)$? Or $(5, 3)$? It turns out that $(5, 3)$ provides the maximum value:

$4(5) + 5(3) = 20 + 15 = 35$ ∎

In Example 1, how did we know that $(5, 3)$ provides the maximum value for K? Obviously, it cannot be done by trial and error as shown in the example. Let us take a closer look. We want to find the maximum value of K. Suppose we try the value of 40. We can graph the line (see Figure 3.3)

$4x + 5y = 40$

or, in slope-intercept form,

$y = -\frac{4}{5}x + 8$

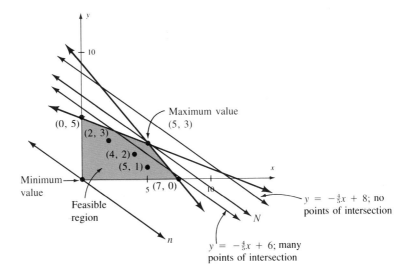

Figure 3.3

Notice that the slope of this line is $-\frac{4}{5}$ and the y-intercept is 8; also notice that it lies above the feasible set—that is, it does not intersect the feasible region, so a value of 40 is clearly unrealistic. Try again; try for the value of K to be 30. Then

$$4x + 5y = 30 \quad \text{or} \quad y = -\tfrac{4}{5}x + 6$$

(Again, see Figure 3.3.) Now there are many points of intersection between our line and the feasible region. Also, note that both of the lines are parallel since they have the same slope.

Now we begin to formulate a procedure. Think of a "family" of lines, all with slope $-\frac{4}{5}$; some of these lines will intersect the feasible region. If we are maximizing K, we want the point that gives the largest value of K and still has an intersection point (see line N); if we are minimizing K, we want the point that gives the lowest value of K and still has an intersection point (see line n).

EXAMPLE 2 Draw (or imagine) lines parallel to the given objective function in the following figures to find the points in the given feasible region that maximize and minimize that function. If there is no such point, state that.

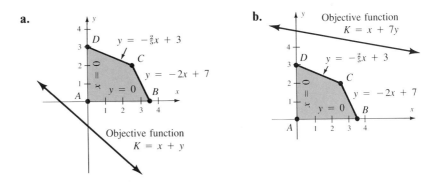

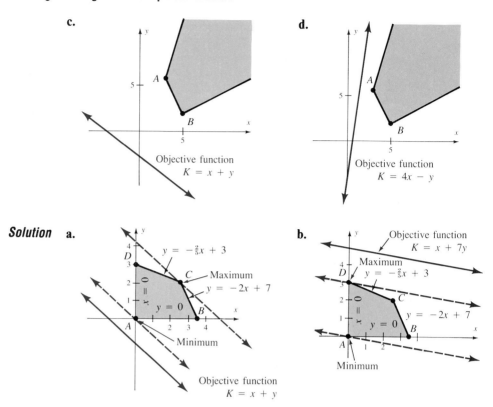

c.

Objective function
$K = x + y$

d.

Objective function
$K = 4x - y$

Solution **a.**

$y = -\frac{2}{5}x + 3$

C — Maximum
$y = -2x + 7$

$x = 0$

$y = 0$

Minimum

Objective function
$K = x + y$

b.

Objective function
$K = x + 7y$

Maximum
$y = -\frac{2}{5}x + 3$

C

$x = 0$

$y = 0$

$y = -2x + 7$

Minimum

Draw lines parallel to the objective function; the maximum value from the feasible region (shaded) is point C and the minimum value is point A.

The maximum value from the feasible region is point D and the minimum value is point A.

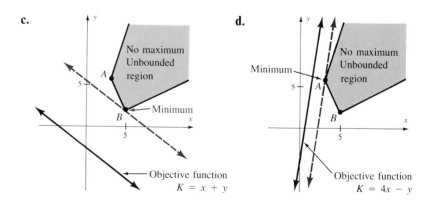

c.

No maximum
Unbounded
region

A

B — Minimum

Objective function
$K = x + y$

d.

Minimum

No maximum
Unbounded
region

A

B

Objective function
$K = 4x - y$

There's no maximum value since the region is unbounded. The minimum value is point B.

There's no maximum value since the region is unbounded; the minimum value is point A.

Note in Example 2 that all of the optimum values occur at **corner points** for the set of feasible solutions. That is, if you consider a family of parallel lines, one member of the family will either coincide with one of the sides of the feasible region or it will intersect it at one point. This follows from the fact that two lines in a plane will either coincide or will intersect at a single point. This fact leads to the **fundamental theorem of linear programming**:

Fundamental Theorem of Linear Programming

A linear expression in two variables,

$$c_1 x + c_2 y$$

defined over a convex set S^* whose sides are line segments, takes on its **maximum value** at a corner point of S and its **minimum value** at a corner point of S. If S is unbounded, there may or may not be an optimum value, but, if there is, then it must occur at a corner point.

With this theorem, we can now summarize the procedure for solving a linear programming problem by graphing:

1. Find the objective expression (the quantity to be maximized or minimized).
2. Find and graph the constraints defined by a system of linear inequalities; the simultaneous solution is called the set S.
3. Find the corner points of S; this may require the solution of a system of two equations with two unknowns, one for each corner point.
4. Find the value of the objective expression for the coordinates of each corner point to find the optimum solutions.

EXAMPLE 3   Solve Example 1 by using the fundamental theorem of linear programming.

Solution   The graph from Example 1 is repeated below with the corner points labeled.
Some corner points can usually be found by inspection. In this case we can see $D = (0, 0)$ and $C = (0, 5)$.

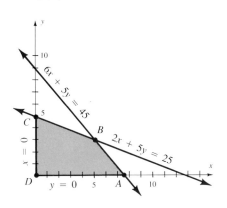

---

* A *convex set* is a set that contains the line segment joining any two of its points.

Some corner points may require some work with the boundary *lines* (use equations of boundaries, not the inequalities giving the regions).

Point *A*   System: $\begin{cases} y = 0 \\ 6x + 5y = 45 \end{cases}$

Solve by substitution: $6x + 5(0) = 45$

$$x = \tfrac{45}{6}$$
$$= \tfrac{15}{2}$$

Point *B*   System: $\begin{cases} 2x + 5y = 25 \\ 6x + 5y = 45 \end{cases}$

Solve by adding: $\begin{cases} -2x - 5y = -25 \\ 6x + 5y = 45 \end{cases}$

$$4x = 20$$
$$x = 5$$

If $x = 5$, then

$$2(5) + 5y = 25$$
$$5y = 15$$
$$y = 3$$

The corner points are thus $(0, 0)$, $(0, 5)$, $(\tfrac{15}{2}, 0)$, and $(5, 3)$. Check these values.

	Objective function
Corner points	$C = 4x + 5y$
$(0, 0)$	$4(0) + 5(0) = 0$
$(0, 5)$	$4(0) + 5(5) = 25$
$(\tfrac{15}{2}, 0)$	$4(\tfrac{15}{2}) + 5(0) = 30$
$(5, 3)$	$4(5) + 5(3) = 35$

Look for largest value of C in this list

The maximum value for C is 35 at $(5, 3)$.   ■

EXAMPLE 4   Maximize $P = x + 5y$ subject to the constraints of Example 1.

Solution   The problem has the same corner points as Example 3.

	Objective function
Corner points	$P = x + 5y$
$(0, 0)$	$0 + 5(0) = 0$
$(0, 5)$	$0 + 5(5) = 25$
$(\tfrac{15}{2}, 0)$	$\tfrac{15}{2} + 5(0) = \tfrac{15}{2}$
$(5, 3)$	$5 + 5(3) = 20$

The maximum value for P is 25 at $(0, 5)$.   ■

EXAMPLE 5   **Allocation of Resources in Production**   A farmer has 100 acres on which to plant two crops: corn and wheat. To produce these crops, there are certain expenses, as shown in the table.

Item	Cost per acre
Corn	
Seed	$ 12
Fertilizer	58
Planting/care/harvesting	50
Total	$120
Wheat	
Seed	$ 40
Fertilizer	80
Planting/care/harvesting	90
Total	$210

After the harvest, the farmer must usually store the crops while awaiting favorable market conditions. Each acre yields an average of 110 bushels of corn or 30 bushels of wheat. The limitations of resources are

Available capital: $15,000
Available storage facilities: 4000 bushels

If the net profit per bushel of corn (after all expenses have been subtracted) is $1.30 and for wheat is $2.00, how should the farmer plant the 100 acres to maximize profits?

Solution   This linear programming problem was formulated in Section 3.2 (Example 1):

Maximize: $P = 143x + 60y$ Where x is the number of acres
$$x \geq 0$$ planted in corn and y is the
$$y \geq 0$$ number of acres planted in wheat
Subject to:
$$x + y \leq 100$$
$$120x + 210y \leq 15{,}000$$
$$110x + 30y \leq 4000$$

First graph the set of feasible solutions by graphing the system of inequalities, as shown in the figure.

Next find the corner points. By inspection, $A = (0,0)$.

Point B: $\begin{cases} 120x + 210y = 15{,}000 \\ x = 0 \end{cases}$

$$120(0) + 210y = 15{,}000$$
$$y = \tfrac{15{,}000}{210}$$
$$= \tfrac{500}{7}$$

Point B: $(0, \tfrac{500}{7})$

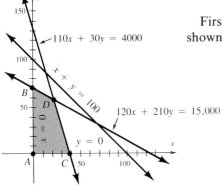

$110x + 30y = 4000$

$120x + 210y = 15{,}000$

$y = 0$

$$\text{Point } C: \quad \begin{cases} 110x + 30y = 4000 \\ \quad\quad\quad\; y = 0 \end{cases}$$

$$110x + 30(0) = 4000$$
$$110x = 4000$$
$$x = \tfrac{400}{11}$$

Point C: $(\tfrac{400}{11}, 0)$

$$\text{Point } D: \quad -7\begin{cases} 110x + \;\;30y = 4000 \\ 120x + 210y = 15,000 \end{cases} \qquad \begin{cases} -770x - 210y = -28,000 \\ \;\;120x + 210y = \;\;\;\;15,000 \end{cases}$$

$$-650x = -13,000$$
$$x = 20$$
$$110(20) + 30y = 4000$$
$$30y = 1800$$
$$y = 60$$

Point D: $(20, 60)$

Use the linear programming theorem and check the corner points:

	Objective function
Corner points	$P = 143x + 60y$
$(0, 0)$	$143(0) \;\; + 60(0) \;\; = 0$
$(0, \tfrac{500}{7})$	$143(0) \;\; + 60(\tfrac{500}{7}) \approx 4286$
$(\tfrac{400}{11}, 0)$	$143(\tfrac{400}{11}) + 60(0) \;\; = 5200$
$(20, 60)$	$143(20) \; + 60(60) = 6460$

The maximum value of P is 6460 at $(20, 60)$. This means that to maximize profits, the farmer should plant 20 acres in corn, plant 60 acres in wheat, and leave 20 acres unplanted. ■

Note from the graph in Example 5 that some of the constraints could be eliminated and everything else would remain unchanged. For example, the boundary $x + y = 100$ was not necessary in finding the maximum value of P. Such a condition is said to be a **superfluous constraint**. It is not uncommon to have superfluous constraints in a linear programming problem. Suppose, however, that the farmer in Example 5 contracted to have the grain stored at a neighboring farm and now the contract calls for *at least* 4000 bushels to be stored. This change from $110x + 30y \leq 4000$ to $110x + 30y \geq 4000$ now makes the condition $x + y \leq 100$ important to the solution of the problem. You must be careful about superfluous constraints even though they do not affect the solution at the present time.

It is also possible to have constraints leading to an empty intersection. For example, consider the constraints

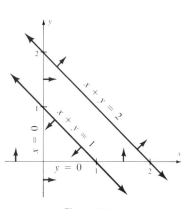

Figure 3.4

$$\begin{cases} x + y \leq 1 \\ x + y \geq 2 \\ x \geq 0, \quad y \geq 0 \end{cases}$$

The graph in Figure 3.4 shows such an empty intersection. We say that this problem has no feasible solution.

EXAMPLE 6   Solve the following linear programming problem.

$$\text{Minimize:} \quad K = 60x + 30y$$

$$\text{Subject to:} \quad \begin{cases} 2x + 3y \geq 120 \\ 2x + y \geq 80 \\ x \geq 0, \quad y \geq 0 \end{cases}$$

Solution   Graph:

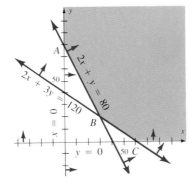

The corner points $A = (0, 80)$ and $C = (60, 0)$ are found by inspection.

$$\text{Point } B: \quad \begin{cases} 2x + 3y = 120 \\ 2x + y = 80 \end{cases}$$

$$2y = 40$$

$$y = 20$$

$$\text{If } y = 20, \text{ then}$$

$$2x + 20 = 80$$

$$2x = 60$$

$$x = 30$$

Point B: $(30, 20)$
Extreme values:

Corner points	Objective function $C = 60x + 30y$
$(0, 80)$	$60(0) + 30(80) = 2400$
$(30, 20)$	$60(30) + 30(20) = 2400$
$(60, 0)$	$60(60) + 30(0) = 3600$

From the list above, there are two minimum values for the objective function: $A = (0, 80)$ and $B = (30, 20)$. In this situation, the objective function will have the same minimum value (2400) at all points along the boundary line segment joining A and B.   ∎

Problem Set 3.3

Find the points in the given feasible region that (a) maximize the given objective function and (b) minimize the given objective function. If, in either case, there is none, state that.

1.

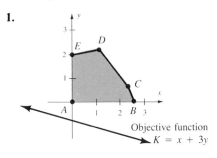

Objective function
$K = x + 3y$

2.

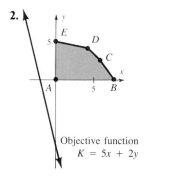

Objective function
$K = 5x + 2y$

3.

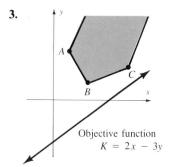

Objective function
$K = 2x - 3y$

4.

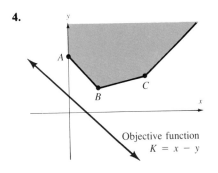

Objective function
$K = x - y$

Find the corner points for each set of feasible solutions in Problems 5–16.

5. $\begin{cases} 2x + y \le 12 \\ x + 2y \le 9 \\ x \ge 0, \quad y \ge 0 \end{cases}$

6. $\begin{cases} 2x + 5y \le 20 \\ 2x + y \le 12 \\ x \ge 0, \quad y \ge 0 \end{cases}$

7. $\begin{cases} 3x + 2y \le 12 \\ x + 2y \le 8 \\ x \ge 0, \quad y \ge 0 \end{cases}$

8. $\begin{cases} x \le 10 \\ y \le 8 \\ 3x + 2y \ge 12 \\ x \ge 0, \quad y \ge 0 \end{cases}$

9. $\begin{cases} x + y \le 8 \\ y \le 4 \\ x \le 6 \\ x \ge 0, \quad y \ge 0 \end{cases}$

10. $\begin{cases} x + y \ge 6 \\ -2x + y \ge -16 \\ y \le 9 \\ x \ge 0, \quad y \ge 0 \end{cases}$

11. $\begin{cases} 3x + 2y \le 8 \\ x + 5y \ge 8 \\ x \ge 0, \quad y \ge 0 \end{cases}$

12. $\begin{cases} x \le 8 \\ y \ge 2 \\ x + y \le 10 \\ x \le 3y \\ x \ge 0, \quad y \ge 0 \end{cases}$

13. $\begin{cases} 10x + 5y \ge 200 \\ 2x + 5y \ge 100 \\ 3x + 4y \ge 120 \\ x \ge 0, \quad y \ge 0 \end{cases}$

14. $\begin{cases} x + y \le 9 \\ 2x - 3y \ge -6 \\ x - y \le 3 \\ x \ge 0, \quad y \ge 0 \end{cases}$

15. $\begin{cases} 2x + y \ge 8 \\ y \le 5 \\ x - y \le 2 \\ 3x - 2y \ge 5 \\ x \ge 0, \quad y \ge 0 \end{cases}$

16. $\begin{cases} 2x + y \ge 8 \\ x - 2y \le 7 \\ x - y \ge -3 \\ x \le 9 \\ x \ge 0, \quad y \ge 0 \end{cases}$

Find the optimum value for each objective function in Problems 17–36.

17. Maximize $W = 30x + 20y$ subject to the constraints of Problem 5.

18. Maximize $P = 40x + 10y$ subject to the constraints of Problem 5.

19. Maximize $T = 100x + 10y$ subject to the constraints of Problem 6.

20. Maximize $V = 20x + 30y$ subject to the constraints of Problem 6.

21. Maximize $P = 100x + 100y$ subject to the constraints of Problem 7.

22. Maximize $T = 30x + 10y$ subject to the constraints of Problem 7.

23. Minimize $C = 24x + 12y$ subject to the constraints of Problem 8.

24. Minimize $K = 6x + 18y$ subject to the constraints of Problem 8.

25. Maximize $F = 2x - 3y$ subject to the constraints of Problem 9.

26. Maximize $P = 5x + y$ subject to the constraints of Problem 9.

27. Minimize $I = 90x + 20y$ subject to the constraints of Problem 10.

28. Minimize $T = 400x + 100y$ subject to the constraints of Problem 10.

29. Maximize $P = 23x + 46y$ subject to the constraints of Problem 10.

30. Minimize $C = 12x + 15y$ subject to the constraints of Problem 10.

31. Maximize $P = 6x + 3y$ subject to the constraints of Problem 14.

32. Maximize $T = x + 6y$ subject to the constraints of Problem 14.

33. Minimize $X = 5x + 3y$ subject to the constraints of Problem 15.

34. Minimize $A = 2x - 3y$ subject to the constraints of Problem 15.

35. Minimize $K = 140x + 250y$ subject to the constraints of Problem 16.

36. Minimize $C = 640x - 130y$ subject to the constraints of Problem 16.

Solve the linear programming problems in Problems 37–45.

APPLICATIONS

37. Suppose the net profit per bushel of corn in Example 5 increases to $2.00 and the net profit for wheat drops to $1.50 per bushel. Maximize the profit if the other conditions of the example remain the same.

38. Suppose the farmer in Example 5 contracted to have the grain stored at a neighboring farm and the contract calls for at least 4000 bushels to be stored. How many acres should be planted in corn and how many in wheat to maximize profits if the other conditions of the example remain the same?

39. A farmer has 500 acres on which to plant two crops: corn and wheat. It costs $120 per acre to produce corn and $60 per acre to produce wheat, and there is $24,000 available to pay for this year's production. If the yield per acre is 100 bushels of corn or 40 bushels of wheat and the farmer has contracted to store at least 18,000 bushels, how much should the farmer plant to maximize profits when the profit is $1.20 per bushel for corn and $2.50 per bushel for wheat?

40. The Wadsworth Widget Company manufactures two types of widgets: regular and deluxe. Each widget is produced at a station consisting of a machine and a person who finishes the widgets by hand. The regular widget requires 2 hours of machine time and 1 hour of finishing time. The deluxe widget requires 3 hours of machine time and 5 hours of finishing time. The profit on the regular widget is $25, and on the deluxe widget it is $30. If the workday is 8 hours, how many of each type of widget should be produced at each station per day in order to maximize the profit?

41. The Thompson Company manufactures two industrial products: standard ($45 profit per item) and economy ($30 profit per item). These items are built using machine time and manual labor. The standard product requires 3 hours of machine time and 2 hours of manual labor. The economy model requires 3 hours of machine time and no manual labor. If the week's supply of manual labor is limited to 800 hours and machine time to 1500 hours, how much of each type of product should be produced each week in order to maximize the profit?

42. A convalescent hospital wishes to provide, at a minimum cost, a diet that has a minimum of 200 grams of carbohydrates, 100 grams of protein, and 120 grams of fats per day. These requirements can be met with two foods:

Food	Carbohydrates	Protein	Fats
A	10 g	2 g	3 g
B	5 g	5 g	4 g

If food A costs 29¢ per ounce and food B costs 15¢ per ounce, how many ounces of each food should be purchased for each patient per day in order to meet the minimum requirements at the lowest cost?

43. The nutritional information in the table below is found on the sides of the cereal boxes listed (for 1 ounce of cereal with $\frac{1}{2}$ cup of whole milk). What is the minimum cost in order to receive at least 322 grams of starch (and related carbohydrates) and 119 grams of sucrose (and other sugars) by consuming these two cereals if corn flakes cost 7¢ per ounce and Honeycombs cost 19¢ per ounce?

Cereal	Starch and related carbohydrates	Sucrose and other sugars
Kellogg's Corn Flakes	23 g	7 g
Post Honeycombs	14 g	17 g

44. Brown Bros., Inc., is an investment company analyzing a pension fund for a certain company. A maximum of $10 million is available to invest in two places. No more than $8 million can be invested in stocks yielding 12%, and at least $2 million must be invested in long-term bonds yielding 8%. The stock-to-bond investment ratio cannot be more than 1 to 3. How should Brown Bros. advise their client so that the pension fund will receive the maximum yearly return on investment?

45. Your broker tells you of two investments she thinks are worthwhile. She recommends a new issue of Pertec stock, which should yield 20% over the next year, and then to balance your account she recommends Campbell Municipal Bonds with a 10% annual yield. The stock-to-bond ratio should be no less than 1 to 3. If you have no more than $100,000 to invest and do not want to invest more than $70,000 in Pertec or less than $20,000 in bonds, how much should you invest in each to maximize your return?

3.4 Summary and Review

IMPORTANT TERMS

Boundary of a half-plane   [3.1]
Closed half-plane   [3.1]
Constraints   [3.2, 3.3]
Corner point   [3.3]
Feasible solution   [3.3]
Fundamental theorem of linear programming   [3.3]
Half-plane   [3.1]
Linear inequality   [3.1]

Linear programming   [3.2]
Linear programming models   [3.2]
Maximum value   [3.3]
Minimum value   [3.3]
Objective function   [3.2, 3.3]
Open half-plane   [3.1]
Superfluous constraint   [3.3]
System of linear inequalities   [3.1]

SAMPLE TEST

For additional practice there are a large number of review problems categorized by objective in the Student Solutions Manual. The following sample test (40 minutes) is intended to review the main ideas of the chapter.

1. Graph $10x - 35y < 210$.

2. Graph the system $\begin{cases} x - y - 5 \le 0 \\ 3x + y - 6 > 0 \\ x \ge 0, \quad y \ge 0 \end{cases}$

3. Maximize: $M = 3x - 4y$

Subject to: $\begin{cases} 2x + \ y \le 420 \\ 2x + 2y \le 500 \\ 2x + 3y \le 600 \\ x \ge 0, \quad y \ge 0 \end{cases}$

Problems 4–10 are reprinted with permission of the American Institute of Certified Public Accountants, Inc. These questions are multiple-choice questions taken from recent CPA examinations. Choose the best, or most appropriate, response for each question.

**CPA Exam
November 1978**

Milligan Company manufactures two models, small and large. Each model is processed as follows:

	Machining	Polishing
Small (X)	2 hr	1 hr
Large (Y)	4 hr	3 hr

The available time for processing the two models is 100 hours a week in the machining department and 90 hours a week in the polishing department. The contribution margin expected is $5 for the small model and $7 for the large model.

4. How would the objective function (maximization of total contribution margin) be expressed?
  A. $5X + 7Y$
  B. $5X + 7Y \leq 190$
  C. $5X(3) + 7Y(7) \leq 190$
  D. $12X + 10Y$

5. How would the restriction (constraint) for the machining department be expressed?
  A. $2(5X) + 4(7Y) \leq 100$
  B. $2x + 4Y$
  C. $2X + 4Y \leq 100$
  D. $5X + 7Y \leq 100$

**CPA Exam
May 1975**

The Random Company manufactures two products, Zeta and Beta. Each product must pass through two processing operations. All materials are introduced at the start of process 1. There are *no* work-in-process inventories. Random may produce either one product exclusively or various combinations of both products subject to the following constraints:

	Hours required to produce 1 unit		Contribution margin per unit
	Process 1	Process 2	
Zeta	1 hr	1 hr	$4.00
Beta	2 hr	3 hr	$5.25
Total capacity in hours per day	1000 hr	1275 hr	

A shortage of technical labor has limited Beta production to 400 units per day. There are *no* constraints on the production of Zeta other than the hour constraints in the above schedule. Assume that all relationships between capacity and production are linear and that all of the above data and relationships are deterministic rather than probabilistic.

6. Given the objective to maximize total contribution margin, what is the production constraint for process 1?
  A. Zeta + Beta $\leq$ 1000
  B. Zeta + 2 Beta $\leq$ 1000
  C. Zeta + Beta $\geq$ 1000
  D. Zeta + 2 Beta $\geq$ 1000

7. Given the objective to maximize total contribution margin, what is the labor constraint for production of Beta?
  A. Beta $\leq$ 400
  B. Beta $\geq$ 400
  C. Beta $\leq$ 425
  D. Beta $\geq$ 425

8. What is the objective function of the data presented?
  A. Zeta + 2 Beta = $9.25
  B. $4.00 Zeta + 3($5.25) Beta = total contribution margin
  C. $4.00 Zeta + $5.25 Beta = total contribution margin
  D. 2($4.00) Zeta + 3($5.25) Beta = total contribution margin

**CPA Exam
November 1977**

9. The Sanch Company plans to expand its sales force by opening several new branch offices. Sanch has $5,200,000 in capital available for new branch offices. Sanch will consider opening only two types of branches: 10-person branches (type *A*) and 5-person branches (type *B*). Expected initial cash outlays are $650,000 for a type *A* branch and $335,000 for a type *B* branch. Expected annual cash inflow, net of income taxes, is $46,000 for a type *A* branch and $18,000 for a type *B* branch. Sanch will hire no more than 100

employees for the new branch offices and will not open more than 10 branch offices. Linear programming will be used to help decide how many branch offices should be opened.

In a system of equations for a linear programming model, which of the following equations would **not** represent a constraint (restriction)?

A. $A + B \leq 10$ **B.** $10A + 5B \leq 100$

C. $\$46{,}000A + \$18{,}000B \leq \$64{,}000$ **D.** $\$650{,}000A + \$335{,}000B \leq \$5{,}200{,}000$

CPA Exam
November 1977

10. The Hale Company manufactures products A and B, each of which requires two processes, polishing and grinding. The contribution margin is $3 for product A and $4 for product B. The graph below shows the maximum number of units of each product that may be processed in the two departments. Considering the constraints (restrictions) on processing, which combination of products A and B maximizes the total contribution margin?

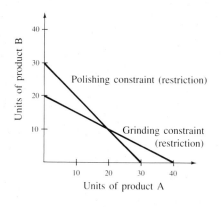

A. 0 units of A and 20 units of B   **B.** 20 units of A and 10 units of B

C. 30 units of A and 0 units of B   **D.** 40 units of A and 0 units of B

Modeling Application 3

Air Pollution Control

Alco Cement Company produces cement. The Environmental Protection Agency has ordered Alco to reduce the amount of emissions released into the atmosphere during production. The company wants to comply, but it also wants to do so at the least possible cost. Present production is 2.5 million barrels of cement, and 2 pounds of dust are emitted for every barrel of cement produced. The cement is produced in kilns that are presently equipped with mechanical collectors. However, in order to reduce the emissions to the required level, the mechanical collectors must be replaced by four-field electrostatic precipitators, which would reduce emissions to 0.5 pound of dust per barrel of cement, or by five-field precipitators, which would reduce emission to 0.2 pound per barrel. The capital and operating costs for the four-field precipitator are 14¢ per barrel of cement produced, and for the five-field precipitator costs are 18¢ per barrel.*

Write a paper that shows the minimum cost for the control methods that Alco should use in order to reduce particulate emissions by 4.2 million pounds. For general guidelines about writing this essay, see the commentary for Modeling Application 1 on page 36.

---

* This application is adapted from R. E. Kohn, "A Mathematical Programming Model for Air Pollution Control," *School Science and Mathematics*, June 1969, pp. 487–499.

CHAPTER 4
Linear Programming—The Simplex Method

APPLICATIONS

Management (*Business, Economics, Finance, and Investments*)

Maximizing the profit in a production process (4.2, Problems 29–30)

Maximizing profits in shoe manufacturing and assembly process (4.2, Problem 31)

Society of Actuaries examination (4.2, Problem 31)

Minimizing the costs in apple production (4.3, Problem 18)

Maximizing profits in machine, assembly, and finishing steps in a production process (4.3, Problems 19–21; 4.5, Problem 10, 4.6, Problem 14)

Minimizing the cost as a function of the days of operation for a plant (4.4, Problem 23)

Management of an investment fund (4.4, Problem 26)

Staff utilization in minimizing labor costs (4.4, Problem 27)

Transportation problem; deciding on the best shipping routes (4.4, Problems 28–31)

Manufacturing process to maximize the profits (4.5, Problems 5–8)

CPA examination questions (4.5, Problems 5–10; 4.6, Problems 16–17)

Deciding how many branch offices to open (4.5, Problem 9)

CMA examination questions (4.6, Problems 18–20)

Life sciences (*Biology, Ecology, Health, and Medicine*)

Determining the proper diet for horses, given certain nutritional requirements (4.3, Problem 17)

Allocation of oil reserves (low, medium, and high grade), (4.4, Problem 24)

Nutritional analysis (4.6, Problem 12)

Maximum number of animals that can be sustained in a limited geographical location (4.6, Problem 13)

Essential amino acids (4.6, Problem 15)

General interest

Allocation of time in an exercise program to achieve certain results (4.4, Problem 25)

Modeling application—Ecology: Maintaining a Profitable Ecological Balance

CHAPTER OVERVIEW   This chapter introduces the *simplex method,* an algebraic method of solving systems of inequalities as well as linear programming problems. This chapter also completes the first part of the course, the study of systems of equations and inequalities.

PREVIEW   We begin by considering slack variables and then review the pivoting process used in Gauss–Jordan elimination. These ideas lead to an algebraic process—the simplex method—for solving linear programming problems. We first apply the method to standard problems. Then we deal with variations of standard problems. The concluding section, Duality, ties together, in a very nice and surprising way, the maximum and minimum type problems.

PERSPECTIVE   This chapter is a continuation of the topic of linear programming. The simplex method is not only an efficient algebraic method for solving linear programming problems, it is also a step-by-step method that can easily be adapted to computer solution. Computer solution is absolutely essential for many real world applications, which may use hundreds of variables and constraints. So, if you have a computer available, you might wish to use Program 6 on the computer disk accompanying this book.

4.1 Slack Variables and the Pivot

In Sections 3.1–3.3 we solved linear programming problems using a graphical procedure. However, since graphical methods are inappropriate for more than two variables, we need an algebraic process leading to a solution of the linear programming problem. In 1946 George Dantzig developed an algebraic process called the **simplex method** for solving linear programming problems. Because of the far-reaching applications of this method, some have called it the most important development in applied mathematics of this century. The process itself is easy, but it takes a considerable amount of discussion to set up the proper terminology and notation. In this section we will:

1. Write the linear programming problem in *standard form.*
2. Introduce *slack variables.*
3. Write the linear programming problem in matrix form, called the *initial simplex tableau.*
4. Introduce a *pivoting process.*

In the next section we solve linear programming problems using the simplex method.

   Throughout this section the objective function is represented by the variable z, and the linear programming variables are represented by $x_1, x_2, x_3, \ldots$. Consider the following linear programming problems using this notation:

Minimize: $z = 0.08x_1 + 0.18x_2$

Subject to:
$$\begin{cases} 23x_1 + 14x_2 \le 322 \\ 7x_1 + 18x_2 \le 111 \\ x_1 \ge 0, \quad x_2 \ge 0 \end{cases}$$

Maximize: $z = 35x_1 + 18x_2$

Subject to:
$$\begin{cases} 0.9x_1 + 0.06x_2 \le 1000 \\ 320x_1 + 118x_2 \le 10{,}000 \end{cases}$$

Maximize: $z = 1.2x_1 + 2.5x_2$

Maximize: $z = 45x_1 + 30x_2$

Subject to: $\begin{cases} 120x_1 + 200x_2 \leq 24{,}000 \\ 100x_1 + 40x_2 \geq 20{,}000 \\ x_1 \geq 0, \quad x_2 \geq 0 \end{cases}$

Subject to: $\begin{cases} 2x_1 \leq 800 \\ 2x_1 + 3x_2 \leq 1500 \\ x_1 \geq 0, \quad x_2 \geq 0 \end{cases}$

Standard Linear Programming Problem

A linear programming problem is said to be a *standard linear programming problem* or, more simply, in **standard form**, if the following three conditions are met:

Condition 1. The objective function is to be maximized.
Condition 2. All variables are nonnegative.
Condition 3. All constraints (except those in condition 2) are less than or equal to ($\leq$) a nonnegative constant.

EXAMPLE 1 Determine whether each of the following linear programming problems is in standard form.

a. Minimize: $z = 0.08x_1 + 0.18x_2$

Subject to: $\begin{cases} 23x_1 + 14x_2 \leq 322 \\ 7x_1 + 18x_2 \leq 111 \\ x_1 \geq 0, \quad x_2 \geq 0 \end{cases}$

This is not a standard form problem since it is a minimization problem and thus fails to meet condition 1.

b. Maximize: $z = 35x_1 + 18x_2$

Subject to: $\begin{cases} 0.9x_1 + 0.06x_2 \leq 1000 \\ 320x_1 + 118x_2 \leq 10{,}000 \end{cases}$

This is not a standard form problem since x_1 and x_2 are not necessarily nonnegative and thus condition 2 has not been met.

c. Maximize: $z = 1.2x_1 + 2.5x_2$

Subject to: $\begin{cases} 120x_1 + 200x_2 \leq 24{,}000 \\ 100x_1 + 40x_2 \geq 20{,}000 \\ x_1 \geq 0, \quad x_2 \geq 0 \end{cases}$

This is not a standard form problem since the second constraint is not less than or equal to a nonzero constant and thus condition 3 has not been met.

d. Maximize: $z = 45x_1 + 30x_2$

Subject to: $\begin{cases} 2x_1 \leq 800 \\ 2x_1 + 3x_2 \leq 1500 \\ x_1 \geq 0, \quad x_2 \geq 0 \end{cases}$

This is a standard linear programming problem. ∎

The goal of the simplex method is to solve a standard linear programming problem by transforming it into a system of linear *equations* that can then be solved using the simplex method. To make this transformation, we introduce new variables to the problems. These variables "take up the slack" created by the inequalities and are consequently called **slack variables**. For example, when we say

$x \leq 5$

we are saying that there exists a nonnegative number y, called a *slack variable*, such that

$x + y = 5$

That is, slack variables allow us to transform linear programming problems from systems of inequalities to corresponding systems of equations.

Slack Variable

> For every nonnegative constant b and each inequality of the form
> $$x \le b$$
> there exists a nonnegative number y, called a *slack variable*, such that
> $$x + y = b$$

In this book we denote slack variables by $y_1, y_2, y_3, \ldots$ to distinguish them from the linear programming variables $x_1, x_2, x_3, \ldots$. Also, a statement such as

$$z = 45x_1 + 30x_2$$

is rewritten as

$$-45x_1 - 30x_2 + z = 0$$

In Examples 2 and 3 we rewrite the given standard linear programming problems using slack variables.

EXAMPLE 2   Maximize: $z = 45x_1 + 30x_2$

Subject to: $\begin{cases} 2x_1 \le 800 \\ 2x_1 + 3x_2 \le 1500 \\ x \ge 0, \quad x_2 \ge 0 \end{cases}$

Solution   Using slack variables,

$$\begin{cases} -45x_1 - 30x_2 + z = 0 \\ 2x_1 + y_1 = 800 \\ 2x_1 + 3x_2 + y_2 = 1500 \\ x_1 \ge 0, \quad x_2 \ge 0 \\ y_1 \ge 0, \quad y_2 \ge 0 \end{cases}$$

Note that the nonnegative requirements are not rewritten using slack variables.   ∎

EXAMPLE 3   Maximize: $z = 3x_1 + 9x_2 + 12x_3 - 5x_4$

Subject to: $\begin{cases} x_1 + x_2 \le 40 \\ x_3 + x_4 \le 45 \\ x_1 + x_3 \le 30 \\ x_2 + x_4 \le 35 \\ x_1 \ge 0, \quad x_2 \ge 0, \quad x_3 \ge 0, \quad x_4 \ge 0 \end{cases}$

Solution   Using slack variables,

$$\begin{cases} -3x_1 - 9x_2 - 12x_3 + 5x_4 + z = 0 \\ x_1 + x_2 + y_1 = 40 \\ x_3 + x_4 + y_2 = 45 \\ x_1 + x_3 + y_3 = 30 \\ x_2 + x_4 + y_4 = 35 \\ x_1 \ge 0, \quad x_2 \ge 0, \quad x_3 \ge 0, \quad x_4 \ge 0 \\ y_1 \ge 0, \quad y_2 \ge 0, \quad y_3 \ge 0, \quad y_4 \ge 0 \end{cases}$$

   ∎

Note that we write the nonnegative conditions $(x_1 \geq 0, x_2 \geq 0, \ldots)$ separately on the bottom lines because the variables are nonnegative for *all* standard linear programming problems. Since we know this, there is no need to include the non-negative variables in the matrix notation. We now write a matrix called the **initial simplex tableau**. When writing this matrix, it is important that you write the variables in the same order so that the proper coefficients are aligned. The standard form maximization statement at the top is generally written at the bottom in the initial simplex tableau, as shown in Examples 4 and 5.

EXAMPLE 4   Maximize: $z = 45x_1 + 30x_2$

Subject to: $\begin{cases} 2x_1 \leq 800 \\ 2x_1 + 3x_2 \leq 1500 \\ x_1 \geq 0, \quad x_2 \geq 0 \end{cases}$

Solution   Align the slack variables for tableau form:

$$\begin{cases} 2x_1 \qquad\; + y_1 \qquad\qquad = 800 \\ 2x_1 + 3x_2 \qquad + y_2 \qquad = 1500 \\ -45x_1 - 30x_2 \qquad\qquad + z = 0 \end{cases}$$

The tableau form is:

$$\begin{array}{ccccc}
x_1 & x_2 & y_1 & y_2 & z \\
\end{array}$$

$$\left[\begin{array}{ccccc|c}
2 & 0 & 1 & 0 & 0 & 800 \\
2 & 3 & 0 & 1 & 0 & 1500 \\
\hline
-45 & -30 & 0 & 0 & 1 & 0
\end{array} \right]$$

These are for reference only and are not part of the tableau form

EXAMPLE 5   Maximize: $z = 3x_1 + 9x_2 + 12x_3 - 5x_4$

Subject to: $\begin{cases} x_1 + x_2 \leq 40 \\ x_3 + x_4 \leq 45 \\ x_1 + x_3 \leq 30 \\ x_2 + x_4 \leq 35 \\ x_1 \geq 0, \quad x_2 \geq 0, \quad x_3 \geq 0, \quad x_4 \geq 0 \end{cases}$

Solution   Align the slack variables for tableau form:

$$\begin{cases} x_1 + x_2 \qquad\qquad\quad + y_1 \qquad\qquad\qquad\qquad = 40 \\ \qquad\quad x_3 + x_4 \qquad\quad + y_2 \qquad\qquad\quad = 45 \\ x_1 \qquad + x_3 \qquad\qquad\qquad + y_3 \qquad\quad = 30 \\ \qquad x_2 \qquad + x_4 \qquad\qquad\qquad + y_4 = 35 \\ -3x_1 - 9x_2 - 12x_3 + 5x_4 \qquad\qquad\qquad\qquad + z = 0 \end{cases}$$

The tableau form is:

$$\begin{array}{ccccccccc}
x_1 & x_2 & x_3 & x_4 & y_1 & y_2 & y_3 & y_4 & z \\
\end{array}$$

$$\left[\begin{array}{ccccccccc|c}
1 & 1 & 0 & 0 & 1 & 0 & 0 & 0 & 0 & 40 \\
0 & 0 & 1 & 1 & 0 & 1 & 0 & 0 & 0 & 45 \\
1 & 0 & 1 & 0 & 0 & 0 & 1 & 0 & 0 & 30 \\
0 & 1 & 0 & 1 & 0 & 0 & 0 & 1 & 0 & 35 \\
\hline
-3 & -9 & -12 & 5 & 0 & 0 & 0 & 0 & 1 & 0
\end{array} \right]$$

Note that the initial simplex tableau looks very much like the augmented matrices we used in Chapter 2 to solve systems of equations. The procedure at that time was to follow a method called *Gauss–Jordan elimination*. For a simplex tableau the process involves a **pivoting operation**.

The *selection* of the pivot elements in the simplex method is not the same as in Gauss–Jordan elimination, as we see in the next section. Example 6, however, reviews the pivoting process. (You might also wish to review the process on page 57.)

COMPUTER APPLICATION You can practice this pivoting process without getting tied up in the arithmetic calculations by using Program 6 on the computer disk accompanying this book.

EXAMPLE 6 Pivot about the circled numbers.

a.
$$\begin{array}{ccccc} x_1 & x_2 & y_1 & y_2 & z \end{array}$$
$$\left[\begin{array}{ccccc|c} ② & 0 & 1 & 0 & 0 & 800 \\ 2 & 3 & 0 & 1 & 0 & 1500 \\ \hline -45 & -30 & 0 & 0 & 1 & 0 \end{array}\right]$$

Step 1
$$\left[\begin{array}{ccccc|c} ① & 0 & \frac{1}{2} & 0 & 0 & 400 \\ 2 & 3 & 0 & 1 & 0 & 1500 \\ \hline -45 & -30 & 0 & 0 & 1 & 0 \end{array}\right]$$ Divide row 1 by 2: $R_1 \div 2$

Step 2
$$\left[\begin{array}{ccccc|c} ① & 0 & \frac{1}{2} & 0 & 0 & 400 \\ 0 & 3 & -1 & 1 & 0 & 700 \\ \hline 0 & -30 & \frac{45}{2} & 0 & 1 & 18{,}000 \end{array}\right]$$

← Pivot row
Add $(-2) \cdot$ row 1 to row 2: $-2R_1 + R_2$
Add $45 \cdot$ row 1 to row 3: $45R_1 + R_3$

b.
$$\begin{array}{cccccc} x_1 & x_2 & x_3 & y_1 & y_2 & z \end{array}$$
$$\left[\begin{array}{cccccc|c} 1 & ② & 3 & 1 & 0 & 0 & 100 \\ 2 & 4 & -1 & 0 & 1 & 0 & 200 \\ \hline -14 & -25 & -10 & 0 & 0 & 1 & 0 \end{array}\right]$$

Step 1
$$\left[\begin{array}{cccccc|c} \frac{1}{2} & ① & \frac{3}{2} & \frac{1}{2} & 0 & 0 & 50 \\ 2 & 4 & -1 & 0 & 1 & 0 & 200 \\ \hline -14 & -25 & -10 & 0 & 0 & 1 & 0 \end{array}\right]$$ Divide row 1 by 2: $\frac{1}{2}R_1$

Step 2
$$\left[\begin{array}{cccccc|c} \frac{1}{2} & ① & \frac{3}{2} & \frac{1}{2} & 0 & 0 & 50 \\ 0 & 0 & -7 & -2 & 1 & 0 & 0 \\ \hline -\frac{3}{2} & 0 & \frac{55}{2} & \frac{25}{2} & 0 & 1 & 1250 \end{array}\right]$$

← Pivot row
Add $(-4) \cdot$ row 1 to row 2: $-4R_1 + R_2$
Add $25 \cdot$ row 1 to row 2: $25R_1 + R_2$ ∎

Problem Set 4.1

Rewrite the linear programming problems (taken from Problem Set 3.3) using the notation introduced in this section. If one of the three conditions for a standard linear programming problem is violated, tell which one.

1. Maximize: $W = 30x + 20y$

Subject to: $\begin{cases} 2x + y \le 12 \\ 5x + 8y \le 40 \\ x \ge 0, \quad y \ge 0 \end{cases}$

2. Maximize: $T = 100x + 10y$

Subject to: $\begin{cases} 2x + 5y \le 20 \\ 2x + y \le 12 \\ x \ge 0, \quad y \ge 0 \end{cases}$

3. Maximize: $P = 100x + 100y$

Subject to: $\begin{cases} 3x + 2y \le 12 \\ x + 2y \le 8 \\ x \ge 0, \quad y \ge 0 \end{cases}$

4. Minimize: $I = 90x + 20y$

Subject to: $\begin{cases} x + y \ge 6 \\ -2x + y \ge -16 \\ y \le 9 \\ x \ge 0, \quad y \ge 0 \end{cases}$

5. Minimize: $X = 5x + 3y$

Subject to: $\begin{cases} 2x + y \ge 8 \\ y \le 5 \\ x - y \le 2 \\ 3x - 2y \ge 5 \\ x \ge 0, \quad y \ge 0 \end{cases}$

6. Maximize: $P = 23x + 46y$

Subject to: $\begin{cases} x + y \le 6 \\ 2x + y \le -16 \\ y \le 9 \\ x \ge 0, \quad y \ge 0 \end{cases}$

7. Maximize: $I = 90x + 20y$

Subject to: $\begin{cases} x + y \le 6 \\ 2x + y \le -16 \\ y \le 9 \end{cases}$

8. Maximize: $P = 6x + 3y$

Subject to: $\begin{cases} x + y \le 9 \\ 2x - 3y \ge -6 \\ x - y \le 3 \\ x \ge 0, \quad y \ge 0 \end{cases}$

9. Maximize: $T = x + 6y$

Subject to: $\begin{cases} 2x + y \ge 8 \\ y \le 5 \\ x - y \le 2 \\ 3x - 2y \ge 5 \\ x \ge 0, \quad y \ge 0 \end{cases}$

10. Maximize: $P = 160x + 130y$

Subject to: $\begin{cases} 2x + 3y \le 8 \\ 4x + 5y \le 8 \end{cases}$

11. Maximize: $K = 140x + 250y$

Subject to: $\begin{cases} 2x + 5y \le 100 \\ 3x + 4y \le 120 \end{cases}$

12. Minimize: $A = 2x - 3y$

Subject to: $\begin{cases} 3x + 2y \le 12 \\ x + 2y \le 8 \\ x \ge 0, \quad y \ge 0 \end{cases}$

Write the initial simplex tableau for the standard programming problems in Problems 13–22.

13. Maximize: $z = 2x_1 + 3x_2$

Subject to: $\begin{cases} 3x_1 + x_2 \le 300 \\ 2x_1 + 2x_2 \le 400 \\ x_1 \ge 0, \quad x_2 \ge 0 \end{cases}$

14. Maximize: $z = 8x_1 + 16x_2$

Subject to: $\begin{cases} 5x_1 + 3x_2 \le 165 \\ 900x_1 + 1200x_2 \le 36{,}000 \\ x_1 \ge 0, \quad x_2 \ge 0 \end{cases}$

15. Maximize: $z = 45x_1 + 35x_2$

Subject to: $\begin{cases} x_1 + x_2 \le 200 \\ 4x_1 + 2x_2 \le 500 \\ x_1 \ge 0, \quad x_2 \ge 0 \end{cases}$

16. Maximize: $z = 5x_1 + 8x_2$

Subject to: $\begin{cases} x_1 + 3x_2 \le 1200 \\ x_1 + 2x_2 \le 1000 \\ x_1 \le 700 \\ x_1 \ge 0, \quad x_2 \ge 0 \end{cases}$

17. Maximize: $z = x_1 + x_2$

Subject to: $\begin{cases} 12x_1 + 150x_2 \le 1200 \\ 6x_1 + 200x_2 \le 1200 \\ 16x_1 + 50x_2 \le 800 \\ x_1 \ge 0, \quad x_2 \ge 0 \end{cases}$

18. Maximize: $z = 4x_1 + 2x_2$

Subject to: $\begin{cases} 2x_1 + x_2 \le 200 \\ 2x_1 + 2x_2 \le 240 \\ 2x_1 + 3x_2 \le 300 \\ x_1 \ge 0, \quad x_2 \ge 0 \end{cases}$

19. Maximize: $z = 12x_1 + 7x_2 + 5x_3$

Subject to: $\begin{cases} 8x_1 + 5x_2 + 3x_3 \le 1000 \\ 5x_1 + x_2 + 3x_3 \le 800 \\ x_1 \ge 0, \quad x_2 \ge 0, \quad x_3 \ge 0 \end{cases}$

20. Maximize: $z = x_1 + x_2 + x_3$

Subject to: $\begin{cases} 60x_1 \le 1800 \\ 60x_2 \le 1500 \\ 60x_3 \le 3000 \\ x_2 + x_3 \le 50 \\ x_1 \ge 0, \quad x_2 \ge 0, \quad x_3 \ge 0 \end{cases}$

21. Maximize: $z = 9x_1 + 7x_2 + 7x_3 + 8x_4$

Subject to: $\begin{cases} x_1 + x_2 \le 90 \\ x_3 + x_4 \le 130 \\ x_1 + x_3 \le 80 \\ x_2 + x_4 \le 110 \\ x_1 \ge 0, \quad x_2 \ge 0, \quad x_3 \ge 0, \quad x_4 \ge 0 \end{cases}$

22. Maximize: $z = 48x_1 + 61x_2 + 39x_3 + 45x_4$

Subject to: $\begin{cases} x_1 + x_2 \le 5 \\ x_2 + x_3 \le 4 \\ x_3 + x_4 \le 8 \\ x_2 + x_4 \le 7 \\ x_1 \ge 0, \quad x_2 \ge 0, \quad x_3 \ge 0, \quad x_4 \ge 0 \end{cases}$

Perform the pivoting process for the circled pivot in Problems 23–34.

23.
$$\begin{array}{ccccc|c} x_1 & x_2 & y_1 & y_2 & z & \\ ③ & 0 & 1 & 0 & 0 & 90 \\ 6 & 2 & 0 & 1 & 0 & 18 \\ \hline -12 & -6 & 0 & 0 & 1 & 0 \end{array}$$

24.
$$\begin{array}{ccccc|c} x_1 & x_2 & y_1 & y_2 & z & \\ ⑤ & 0 & 1 & 0 & 0 & 20 \\ 2 & 3 & 0 & 1 & 0 & 30 \\ \hline -10 & -3 & 0 & 0 & 1 & 0 \end{array}$$

25.
$$\begin{array}{ccccc|c} x_1 & x_2 & y_1 & y_2 & z & \\ 1 & 0 & \frac{1}{3} & 0 & 0 & 30 \\ 0 & ② & -2 & 1 & 0 & -162 \\ \hline 0 & -6 & 4 & 0 & 1 & 360 \end{array}$$

26.
$$\begin{array}{ccccc|c} x_1 & x_2 & y_1 & y_2 & z & \\ 1 & 0 & \frac{1}{5} & 0 & 0 & 4 \\ 0 & ③ & -\frac{2}{5} & 1 & 0 & 22 \\ \hline 0 & -3 & 2 & 0 & 1 & 40 \end{array}$$

27.
$$\begin{array}{ccccc|c} x_1 & x_2 & y_1 & y_2 & z & \\ 8 & ④ & 1 & 0 & 0 & 40 \\ 12 & 6 & 0 & 1 & 0 & 600 \\ \hline -8 & -10 & 0 & 0 & 1 & 0 \end{array}$$

28.
$$\begin{array}{ccccc|c} x_1 & x_2 & y_1 & y_2 & z & \\ ② & 1 & \frac{1}{4} & 0 & 0 & 10 \\ 0 & 0 & -6 & 1 & 0 & 360 \\ \hline -12 & 0 & \frac{5}{2} & 0 & 1 & 100 \end{array}$$

29.
$$\begin{array}{cccccc|c} x_1 & x_2 & y_1 & y_2 & y_3 & z & \\ 2 & 3 & 1 & 0 & 0 & 0 & 200 \\ ② & 6 & 0 & 1 & 0 & 0 & 100 \\ 3 & 2 & 0 & 0 & 1 & 0 & 300 \\ \hline -10 & -5 & 0 & 0 & 0 & 1 & 0 \end{array}$$

30.
$$\begin{array}{cccccc|c} x_1 & x_2 & y_1 & y_2 & y_3 & z & \\ 20 & ⑩ & 1 & 0 & 0 & 0 & 200 \\ 2 & 4 & 0 & 1 & 0 & 0 & 100 \\ 3 & 2 & 0 & 0 & 1 & 0 & 300 \\ \hline -10 & -30 & 0 & 0 & 0 & 1 & 0 \end{array}$$

31.
$$\begin{array}{cccccc|c} x_1 & x_2 & y_1 & y_2 & y_3 & z & \\ 0 & -5 & 1 & 0 & 0 & 0 & 300 \\ 1 & ② & 0 & 1 & 0 & 0 & 20 \\ 0 & 8 & 0 & -2 & 1 & 0 & 180 \\ \hline 0 & -4 & 0 & 6 & 0 & 1 & 300 \end{array}$$

32.
$$\begin{array}{cccccc|c} x_1 & x_2 & y_1 & y_2 & y_3 & z & \\ 2 & 1 & \frac{1}{2} & 0 & 0 & 0 & 30 \\ -3 & 0 & 0 & 1 & 0 & 0 & 50 \\ ⊖2 & 0 & 3 & 0 & 1 & 0 & 20 \\ \hline -4 & 0 & 2 & 0 & 0 & 1 & 140 \end{array}$$

33.
$$\begin{array}{ccccccc|c} x_1 & x_2 & x_3 & y_1 & y_2 & y_3 & z & \\ 1 & 2 & 4 & 1 & 0 & 0 & 0 & 80 \\ 1 & 4 & 3 & 0 & 1 & 0 & 0 & 60 \\ 8 & 4 & ② & 0 & 0 & 1 & 0 & 10 \\ \hline -2 & -3 & -4 & 0 & 0 & 0 & 1 & 0 \end{array}$$

34.
$$\begin{array}{ccccccc|c} x_1 & x_2 & x_3 & y_1 & y_2 & y_3 & z & \\ 1 & 2 & 4 & 1 & 0 & 0 & 0 & 80 \\ ① & 4 & 3 & 0 & 1 & 0 & 0 & 20 \\ 8 & 4 & 2 & 0 & 0 & 1 & 0 & 200 \\ \hline -5 & -3 & -4 & 0 & 0 & 0 & 1 & 0 \end{array}$$

4.2 Maximization by the Simplex Method

In Section 4.1 we set the stage for the **simplex method**. This method requires a standard form linear programming problem written in an initial simplex tableau with slack variables. In this section we discuss three ideas that complete the simplex method:

1. How to read a solution from the simplex tableau
2. How to select a pivot
3. How to recognize when a maximum value is found (that is, how to know when you are finished with the simplex method)

Begin by writing Example 5 of Section 3.3 in standard form:

Maximize: $z = 143x_1 + 60x_2$

Subject to:
$$\begin{cases} x_1 + x_2 \le 100 \\ 120x_1 + 210x_2 \le 15{,}000 \\ 110x_1 + 30x_2 \le 4000 \\ x_1 \ge 0, \quad x_2 \ge 0 \end{cases}$$

Write this using slack variables:

$$\begin{cases} x_1 + x_2 + y_1 = 100 \\ 120x_1 + 210x_2 + y_2 = 15{,}000 \\ 110x_1 + 30x_2 + y_3 = 4000 \\ -143x_1 - 60x_2 + z = 0 \end{cases}$$

Now write the initial simplex tableau:

This vertical line separates the left and right sides of the equations

This dashed line separates the objective function from the constraints ⟶

x_1	x_2	y_1	y_2	y_3	z	
1	1	1	0	0	0	100
120	210	0	1	0	0	15,000
110	30	0	0	1	0	4000
−143	−60	0	0	0	1	0

Solutions can be read from this matrix tableau. It represents the set of feasible solutions. The problem asks us to maximize an objective function subject to two variables and three constraints (not counting those that restrict the variables to nonnegative values). This means that we have a system of four equations and six variables (three given variables, x_1, x_2, and z, and three slack variables, y_1, y_2, and y_3). From our work with systems of equations we know that any two of the six variables can be chosen arbitrarily and the other four can then be found by solving the remaining system. Also, from the matrix solution of systems of equations, the matrix

y_1	y_2	y_3	z	
1	0	0	0	100
0	1	0	0	15,000
0	0	1	0	4000
0	0	0	1	0

gives the solution

$$\begin{cases} y_1 = 100 \\ y_2 = 15{,}000 \\ y_3 = 4000 \\ z = 0 \end{cases}$$

Thus, if we choose to assign values to x_1 and x_2, we can solve the remaining system. A **basic solution** of a system such as this is one in which we arbitrarily assign values to two variables. A **basic feasible solution** is one in which all the variables are nonnegative. For example, if we let $x_1 = 0$ and $x_2 = 0$, then a basic feasible solution is the one specified above. Since x_1 and x_2 are both zero, this is sometimes called the **initial basic solution**. Notice that $z = 0$ is hardly a maximum value for z, but nevertheless it is a solution to the system. We sometimes write y_1, y_2, y_3, and z to the right of the matrix to help remind us which value in the last column corresponds to which variable. Thus, for this example, we would write

x_1	x_2	y_1	y_2	y_3	z		
1	1	1	0	0	0	100	y_1
120	210	0	1	0	0	15,000	y_2
110	30	0	0	1	0	4000	y_3
-143	-60	0	0	0	1	0	z

EXAMPLE 1   Consider the following matrix tableau. Find the initial basic solution.

x_1	x_2	x_3	y_1	y_2	y_3	z	
2	3	4	1	0	0	0	40
4	5	3	0	1	0	0	30
6	9	9	0	0	1	0	15
-3	-10	-5	0	0	0	1	0

Solution   If $x_1 = 0, x_2 = 0$, and $x_3 = 0$, then $y_1 = 40, y_2 = 30, y_3 = 15$, and $z = 0$ (you can see this by inspection). From this information *you begin* the simplex method by labeling the right-hand column as shown:

x_1	x_2	x_3	y_1	y_2	y_3	z		
2	3	4	1	0	0	0	40	y_1
4	5	3	0	1	0	0	30	y_2
6	9	9	0	0	1	0	15	y_3
-3	-10	-5	0	0	0	1	0	z

∎

You do not need to choose values for x_1, x_2, and x_3 as in Example 1, but you can arbitrarily choose values for any convenient variables. For example, to find a basic solution for

x_1	x_2	x_3	y_1	y_2	y_3	z	
0	8	0	1	4	6	0	30
1	4	0	0	3	9	0	30
0	6	1	0	2	3	0	90
0	9	0	0	4	12	1	4000

you can arbitrarily choose some values to be zero. But which ones and how many? Find the columns with a 1 and with all other entries 0, as shown by the shading below:

$$
\begin{array}{ccccccc}
x_1 & x_2 & x_3 & y_1 & y_2 & y_3 & z \\
\end{array}
$$

$$
\left[\begin{array}{ccccccc|c}
0 & 8 & 0 & 1 & 4 & 6 & 0 & 30 \\
1 & 4 & 0 & 0 & 3 & 9 & 0 & 30 \\
0 & 6 & 1 & 0 & 2 & 3 & 0 & 90 \\
\hline
0 & 9 & 0 & 0 & 4 & 12 & 1 & 4000
\end{array}\right]
$$

If you now look at the remaining columns and choose $x_2 = 0$, $y_2 = 0$, and $y_3 = 0$, then the resulting matrix is

$$
\begin{array}{cccc}
x_1 & x_3 & y_1 & z \\
\end{array}
$$

$$
\left[\begin{array}{cccc|c}
0 & 0 & 1 & 0 & 30 \\
1 & 0 & 0 & 0 & 30 \\
0 & 1 & 0 & 0 & 90 \\
\hline
0 & 0 & 0 & 1 & 4000
\end{array}\right]
$$

From this matrix you can see that $y_1 = 30$, $x_1 = 30$, $x_3 = 90$, and $z = 4000$. Now you can label the column of the original matrix at the right to reflect this information:

$$
\begin{array}{ccccccc}
x_1 & x_2 & x_3 & y_1 & y_2 & y_3 & z \\
\end{array}
$$

$$
\left[\begin{array}{ccccccc|c}
0 & 8 & 0 & 1 & 4 & 6 & 0 & 30 \\
1 & 4 & 0 & 0 & 3 & 9 & 0 & 30 \\
0 & 6 & 1 & 0 & 2 & 3 & 0 & 90 \\
\hline
0 & 9 & 0 & 0 & 4 & 12 & 1 & 4000
\end{array}\right]
\begin{array}{c}
y_1 \\
x_1 \\
x_3 \\
z
\end{array}
$$

EXAMPLE 2   Find a basic feasible solution for the given simplex tableau by labeling the column at the right.

$$
\begin{array}{ccccccc}
x_1 & x_2 & x_3 & y_1 & y_2 & y_3 & z \\
\end{array}
$$

$$
\left[\begin{array}{ccccccc|c}
8 & 0 & 0 & 9 & 1 & 3 & 0 & 90 \\
5 & 0 & 1 & 19 & 0 & 6 & 0 & 80 \\
4 & 1 & 0 & 12 & 0 & 9 & 0 & 110 \\
\hline
-12 & 0 & 0 & 5 & 0 & 6 & 1 & 850
\end{array}\right]
$$

Solution   Look for the columns with a 1 and 0s:

$$
\begin{array}{ccccccc}
x_1 & x_2 & x_3 & y_1 & y_2 & y_3 & z \\
\end{array}
$$

$$
\left[\begin{array}{ccccccc|c}
8 & 0 & 0 & 9 & 1 & 3 & 0 & 90 \\
5 & 0 & 1 & 19 & 0 & 6 & 0 & 80 \\
4 & 1 & 0 & 12 & 0 & 9 & 0 & 110 \\
\hline
-12 & 0 & 0 & 5 & 0 & 6 & 1 & 850
\end{array}\right]
\begin{array}{c}
y_2 \\
x_3 \\
x_2 \\
z
\end{array}
$$

A basic feasible solution is $y_2 = 90$, $x_3 = 80$, $x_2 = 110$, and $z = 850$. [This assumes that you let $x_1 = 0$, $y_1 = 0$, and $y_3 = 0$ (the nonshaded columns).]   ∎

Is the basic feasible solution in Example 2 an **optimal solution**? Not as long as the last row has any negative values. This can be seen by writing the last row of the tableau in equation form. If the z value has any negative coefficients, it is not yet maximized.

Optimal Solution

> The z value in a simplex tableau is maximized when all the entries in the bottom row are nonnegative.

The *process* by which we obtain an optimal solution is called the **pivoting process** and is summarized below.

Pivoting Process

> **Rule 1.** The **pivot column** is the column that has the most negative number as the bottom entry.
>
> **Rule 2.** The **pivot row** is the row above the line that has the smallest positive ratio when the entry in the last column is divided by the corresponding positive entry from the same row of the pivot column.
>
> The **pivot element** is the entry in the intersection of the pivot row and pivot column. Using this pivot element, pivoting is completed in two steps:
>
> **1.** Divide the pivot row by the pivot element so that the pivot element becomes a 1.
>
> **2.** Obtain 0s above and below the pivot element.

EXAMPLE 3 Carry out the pivoting process on the given initial simplex tableau.

$$
\begin{array}{ccccc|c}
x_1 & x_2 & y_1 & y_2 & z & \\
2 & 4 & 1 & 0 & 0 & 20 \\
3 & 2 & 0 & 1 & 0 & 18 \\
\hline
-6 & -5 & 0 & 0 & 1 & 0
\end{array}
\quad
\begin{array}{l}
y_1 \\
y_2 \\
z
\end{array}
$$

$\uparrow$
Pivot column; this is the most
negative entry

Solution The variable at the top of the pivot column (x_1 in this example) is sometimes called the **entering variable** because when the pivoting process is finished, the x_1 will appear at the right and will no longer be a zero variable.

We now find the pivot row:

$$
\begin{array}{ccccc|c}
x_1 & x_2 & y_1 & y_2 & z & \\
2 & 4 & 1 & 0 & 0 & 20 \\
3 & 2 & 0 & 1 & 0 & 18 \\
\hline
-6 & -5 & 0 & 0 & 1 & 0
\end{array}
\quad
\begin{array}{l}
y_1 \\
y_2 \\
z
\end{array}
\quad
\begin{array}{l}
20 \div 2 = 10 \\
18 \div 3 = 6 \leftarrow
\end{array}
\left\{
\begin{array}{l}
\text{Pivot row; this} \\
\text{is the smallest} \\
\text{positive entry}
\end{array}
\right.
$$

$\uparrow$

The variable labeling the pivot row (y_2 in this example) is sometimes called the **departing variable** because when the pivoting process is finished, the y_2 will no longer appear at the right.

Circle the pivot element and carry out the pivot process:

$$\begin{array}{ccccc} x_1 & x_2 & y_1 & y_2 & z \end{array}$$

$$\left[\begin{array}{ccccc|c} 2 & 4 & 1 & 0 & 0 & 20 \\ ③ & 2 & 0 & 1 & 0 & 18 \\ \hline -6 & -5 & 0 & 0 & 1 & 0 \end{array}\right] \begin{array}{l} y_1 \\ y_2 \leftarrow \\ z \end{array}$$

$$\uparrow$$

Step 1   Divide to make the pivoting element 1:

$$\begin{array}{ccccc} x_1 & x_2 & y_1 & y_2 & z \end{array}$$

$$\left[\begin{array}{ccccc|c} 2 & 4 & 1 & 0 & 0 & 20 \\ ① & \frac{2}{3} & 0 & \frac{1}{3} & 0 & 6 \\ \hline -6 & -5 & 0 & 0 & 1 & 0 \end{array}\right] \quad \leftarrow \text{Each entry is divided by 3}$$

$$\uparrow$$

Step 2   Perform elementary row operations to obtain 0s:

$$\begin{array}{ccccc} x_1 & x_2 & y_1 & y_2 & z \end{array}$$

$$\left[\begin{array}{ccccc|c} 0 & \frac{8}{3} & 1 & -\frac{2}{3} & 0 & 8 \\ ① & \frac{2}{3} & 0 & \frac{1}{3} & 0 & 6 \\ \hline 0 & -1 & 0 & 2 & 1 & 36 \end{array}\right] \begin{array}{l} y_1 \\ x_1 \\ z \end{array}$$

⌐— Relabel these according to
the 1s and 0s in the columns
after the pivoting process

Note: x_1 entered and y_2
departed

Repeat the pivoting process until there are no negative values in the last row.

$$\begin{array}{ccccc} x_1 & x_2 & y_1 & y_2 & z \end{array}$$

$$\left[\begin{array}{ccccc|c} 0 & \frac{8}{3} & 1 & -\frac{2}{3} & 0 & 8 \\ 1 & \frac{2}{3} & 0 & \frac{1}{3} & 0 & 6 \\ \hline 0 & -1 & 0 & 2 & 1 & 36 \end{array}\right] \begin{array}{l} y_1 \\ x_1 \\ z \end{array} \begin{array}{l} 8 \div \frac{8}{3} = 3 \leftarrow \\ 6 \div \frac{2}{3} = 9 \end{array}$$

$$\uparrow$$

Step 1

$$\begin{array}{ccccc} x_1 & x_2 & y_1 & y_2 & z \end{array}$$

$$\left[\begin{array}{ccccc|c} 0 & ① & \frac{3}{8} & -\frac{1}{4} & 0 & 3 \\ 1 & \frac{2}{3} & 0 & \frac{1}{3} & 0 & 6 \\ \hline 0 & -1 & 0 & 2 & 1 & 36 \end{array}\right] \quad \begin{array}{l} \leftarrow \text{Divide the entries of this} \\ \text{row by } \frac{8}{3} \end{array}$$

$$\uparrow$$

Step 2

$$\begin{array}{ccccc} x_1 & x_2 & y_1 & y_2 & z \end{array}$$

$$\left[\begin{array}{ccccc|c} 0 & 1 & \frac{3}{8} & -\frac{1}{4} & 0 & 3 \\ 1 & 0 & -\frac{1}{4} & \frac{1}{2} & 0 & 4 \\ \hline 0 & 0 & \frac{3}{8} & \frac{7}{4} & 1 & 39 \end{array}\right] \begin{array}{l} x_2 \\ x_1 \\ z \end{array}$$

All values in the last
row are nonnegative, so the
← process is complete

Thus the maximum value of z is 39 and it occurs when

$$x_2 = 3$$
$$x_1 = 4$$
$$z = 39$$

There are three equations with five unknowns, so we can arbitrarily choose two variables. We choose $y_1 = 0$ and $y_2 = 0$; the remaining variables are then found by inspection of the last matrix by looking at the columns where a single 1 and 0s elsewhere are found

■

We can now summarize the simplex method.

Simplex Method

In a standard linear programming problem:

Step 1. Write the initial simplex tableau using slack variables.

Step 2. **Test for maximality:** If all entries in the last row are nonnegative, then the tableau is the final tableau; interpret the solution.

Step 3. Select the pivot element.
  a. The **pivot column** is the column that has the most negative entry at the bottom.*
  b. The **pivot row** is the row that has the smallest positive ratio. If there are no positive ratios, then there is no solution.

Step 4. Carry out the pivoting process and return to step 2.

These steps are written in flowchart form in Figure 4.1.

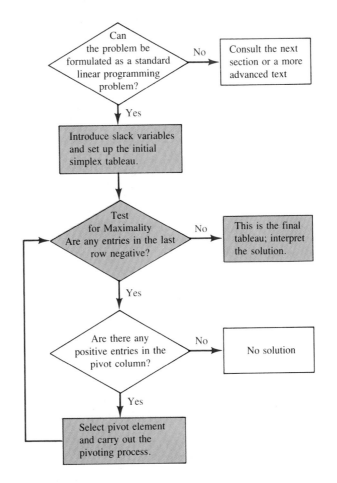

Figure 4.1 The simplex method can easily be written as a computer program. See Program 6 in the computer supplement accompanying this text for an example. (Appendix C provides a complete listing of the programs available in the computer supplement.)

---

* If two columns have the same entry and this is the most negative entry, then either can be chosen.

EXAMPLE 4   Maximize: $z = 4x_1 + 5x_2$

Subject to: $\begin{cases} 2x_1 + 5x_2 \leq 25 \\ 6x_1 + 5x_2 \leq 45 \\ x_1 \geq 0, \quad x_2 \geq 0 \end{cases}$

Solution   We solved this problem in Examples 1–3 of Section 3.3. The maximum value from the graphical method was 35 at (5, 3). We now solve this problem using the complex method.

$$
\begin{array}{ccccc|c}
x_1 & x_2 & y_1 & y_2 & z & \\
\end{array}
$$

$$
\left[
\begin{array}{ccccc|c}
2 & ⑤ & 1 & 0 & 0 & 25 \\
6 & 5 & 0 & 1 & 0 & 45 \\
\hline
-4 & -5 & 0 & 0 & 1 & 0
\end{array}
\right]
\quad
\begin{array}{l}
y_1 \\
y_2 \\
z
\end{array}
\quad
\begin{array}{l}
25 \div 5 = 5 \leftarrow \text{Departing variable} \\
45 \div 5 = 9
\end{array}
$$

↑
Entering variable

$$
\left[
\begin{array}{ccccc|c}
\frac{2}{5} & ① & \frac{1}{5} & 0 & 0 & 5 \\
6 & 5 & 0 & 1 & 0 & 45 \\
\hline
-4 & -5 & 0 & 0 & 1 & 0
\end{array}
\right]
\quad
\begin{array}{l}
\frac{1}{5}R1
\end{array}
$$

After the pivoting process you obtain:

$$
\left[
\begin{array}{ccccc|c}
\frac{2}{5} & 1 & \frac{1}{5} & 0 & 0 & 5 \\
4 & 0 & -1 & 1 & 0 & 20 \\
\hline
-2 & 0 & 1 & 0 & 1 & 25
\end{array}
\right]
\quad
\begin{array}{l}
\\
-5R1 + R2 \\
5R1 + R3
\end{array}
$$

Select a new pivot:

$$
\begin{array}{ccccc|c}
x_1 & x_2 & y_1 & y_2 & z & \\
\end{array}
$$

$$
\left[
\begin{array}{ccccc|c}
\frac{2}{5} & 1 & \frac{1}{5} & 0 & 0 & 5 \\
④ & 0 & -1 & 1 & 0 & 20 \\
\hline
-2 & 0 & 1 & 0 & 1 & 25
\end{array}
\right]
\quad
\begin{array}{l}
x_2 \\
y_2 \\
z
\end{array}
\quad
\begin{array}{l}
5 \div \frac{2}{5} = 12.5 \\
20 \div 4 = 5 \leftarrow \text{Departing variable}
\end{array}
$$

↑
Entering variable

Pivot again:

$$
\left[
\begin{array}{ccccc|c}
\frac{2}{5} & 1 & \frac{1}{5} & 0 & 0 & 5 \\
① & 0 & -\frac{1}{4} & \frac{1}{4} & 0 & 5 \\
\hline
-2 & 0 & 1 & 0 & 1 & 25
\end{array}
\right]
\quad
\begin{array}{l}
\\
\frac{1}{4}R2
\end{array}
$$

$$
\begin{array}{ccccc|c}
x_1 & x_2 & y_1 & y_2 & z & \\
\end{array}
$$

$$
\left[
\begin{array}{ccccc|c}
0 & 1 & \frac{3}{10} & -\frac{1}{10} & 0 & 3 \\
1 & 0 & -\frac{1}{4} & \frac{1}{4} & 0 & 5 \\
\hline
0 & 0 & \frac{1}{2} & \frac{1}{2} & 1 & 35
\end{array}
\right]
\quad
\begin{array}{l}
x_2 \\
x_1 \\
z
\end{array}
\quad
\begin{array}{l}
-\frac{2}{5}R2 + R1 \\
\\
2R2 + R3
\end{array}
$$

This gives the maximum value of 35 at (5, 3). (We have chosen $y_1 = 0$ and $y_2 = 0$.)   ■

EXAMPLE 5   Maximize: $z = 3x_1 + 9x_2 + 12x_3 - 5x_4$

Subject to: $\begin{cases} x_1 + x_2 \le 40 \\ x_3 + x_4 \le 45 \\ x_1 + x_3 \le 30 \\ x_2 + x_4 \le 35 \\ x_1 \ge 0, \quad x_2 \ge 0, \quad x_3 \ge 0, \quad x_4 \ge 0 \end{cases}$

Solution   Write the initial simplex tableau (this was done as Example 5 of the previous section) and determine the entering and departing variables:

$$
\begin{array}{ccccccccc|c}
x_1 & x_2 & x_3 & x_4 & y_1 & y_2 & y_3 & y_4 & z & \\
1 & 1 & 0 & 0 & 1 & 0 & 0 & 0 & 0 & 40 \\
0 & 0 & 1 & 1 & 0 & 1 & 0 & 0 & 0 & 45 \\
1 & 0 & ① & 0 & 0 & 0 & 1 & 0 & 0 & 30 \\
0 & 1 & 0 & 1 & 0 & 0 & 0 & 1 & 0 & 35 \\
\hline
-3 & -9 & -12 & 5 & 0 & 0 & 0 & 0 & 1 & 0
\end{array}
$$

y_1	$40 \div 0$ Not defined
y_2	$45 \div 1 = 45$
y_3	$30 \div 1 = 30 \leftarrow$ [Departing variable
y_4	$35 \div 0$ Not defined
z	

↑
Entering variable

Next, circle the pivot element and pivot as shown below:

$$
\begin{array}{ccccccccc|c}
x_1 & x_2 & x_3 & x_4 & y_1 & y_2 & y_3 & y_4 & z & \\
1 & 1 & 0 & 0 & 1 & 0 & 0 & 0 & 0 & 40 \\
-1 & 0 & 0 & 1 & 0 & 1 & -1 & 0 & 0 & 15 \\
1 & 0 & 1 & 0 & 0 & 0 & 1 & 0 & 0 & 30 \\
0 & ① & 0 & 1 & 0 & 0 & 0 & 1 & 0 & 35 \\
\hline
9 & -9 & 0 & 5 & 0 & 0 & 12 & 0 & 1 & 360
\end{array}
$$

y_1	$40 \div 1 = 40$
y_2	$15 \div 0$ Not defined
x_3	$30 \div 0$ Not defined
y_4	$35 \div 1 = 35 \leftarrow$ [Departing variable
z	

↑
Entering variable

The new entering and departing variables and pivot element are found in the above tableau. Pivot once again:

$$
\begin{array}{ccccccccc|c}
x_1 & x_2 & x_3 & x_4 & y_1 & y_2 & y_3 & y_4 & z & \\
1 & 0 & 0 & -1 & 1 & 0 & 0 & -1 & 0 & 5 \\
-1 & 0 & 0 & 1 & 0 & 1 & -1 & 0 & 0 & 15 \\
1 & 0 & 1 & 0 & 0 & 0 & 1 & 0 & 0 & 30 \\
0 & 1 & 0 & 1 & 0 & 0 & 0 & 1 & 0 & 35 \\
\hline
9 & 0 & 0 & 14 & 0 & 0 & 12 & 9 & 1 & 675
\end{array}
$$

This is the final tableau. The solution (by inspection) is

$y_1 = 5$

$y_2 = 15$

$x_3 = 30$

$x_2 = 35$

$z = 675$

This solution assumes that we let $x_1 = 0$, $x_4 = 0$, $y_3 = 0$, and $y_4 = 0$; there are five equations with nine unknowns, so four variables are arbitrarily chosen

The answer, however, is given in terms of the original variables in the problem, so we say that the objective function has a maximum value of 675 when $x_1 = 0$, $x_2 = 35$, $x_3 = 30$, and $x_4 = 0$. ∎

Problem Set 4.2

Find the initial basic solution for each matrix tableau in Problems 1–3.

1. a.

x_1	x_2	y_1	y_2	z	
4	8	1	0	0	30
6	5	0	1	0	50
−10	−20	0	0	1	0

b.

x_1	x_2	y_1	y_2	z	
0	1	6	2	0	12
1	0	1	4	0	80
0	0	−3	9	1	120

b.

x_1	x_2	y_1	y_2	z	
9	12	1	0	0	120
5	18	0	1	0	180
−40	−30	0	0	1	0

5. a.

x_1	x_2	y_1	y_2	y_3	z	
1	8	0	3	0	0	10
0	4	1	1	0	0	12
0	1	0	0	1	0	20
0	3	0	2	0	1	32

2. a.

x_1	x_2	y_1	y_2	y_3	z	
5	9	1	0	0	0	18
3	12	0	1	0	0	35
7	21	0	0	1	0	49
−5	−9	0	0	0	1	0

b.

x_1	x_2	x_3	y_1	y_2	y_3	z	
0	2	0	1	2	4	0	20
0	4	1	0	8	8	0	80
1	3	0	0	9	3	0	120
0	−5	0	0	4	−2	1	360

b.

x_1	x_2	y_1	y_2	y_3	z	
9	5	1	0	0	0	19
6	12	0	1	0	0	35
12	1	0	0	1	0	48
−10	−25	0	0	0	1	0

6. a.

x_1	x_2	x_3	y_1	y_2	y_3	z	
0	1	0	4	0	0	0	90
1	0	0	6	1	0	0	70
0	0	1	3	3	0	0	65
0	0	0	5	2	1	0	20
0	0	0	19	10	0	1	250

3. a.

x_1	x_2	x_3	y_1	y_2	y_3	z	
8	12	40	1	0	0	0	60
5	9	12	0	1	0	0	30
6	15	8	0	0	1	0	40
−5	−12	−3	0	0	0	1	0

b.

x_1	x_2	x_3	y_1	y_2	y_3	z	
1	2	1	0	0	1	0	5
1	−1	3	0	1	0	0	10
6	5	−2	1	0	0	0	8
−3	−5	−6	0	0	0	1	0

b.

x_1	x_2	x_3	y_1	y_2	y_3	y_4	z	
5	9	11	1	0	0	0	0	80
6	8	4	0	1	0	0	0	50
9	18	1	0	0	1	0	0	60
1	8	1	0	0	0	1	0	90
−8	−12	−5	0	0	0	0	1	0

Carry out the pivoting process on each initial simplex tableau in Problems 7–12.

7.

x_1	x_2	y_1	y_2	z	
6	3	1	0	0	20
2	4	0	1	0	4
−4	−20	0	0	1	0

Find a basic feasible solution for each simplex tableau in Problems 4–6 and determine whether it is the final tableau.

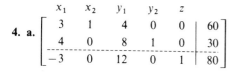

4. a.

x_1	x_2	y_1	y_2	z	
3	1	4	0	0	60
4	0	8	1	0	30
−3	0	12	0	1	80

8.

x_1	x_2	y_1	y_2	z	
3	6	1	0	0	60
5	6	0	1	0	10
−1	−30	0	0	1	0

9.

x_1	x_2	y_1	y_2	y_3	z	
2	3	1	0	0	0	50
5	15	0	1	0	0	10
4	3	0	0	1	0	70
-5	-20	0	0	0	1	0

10.

x_1	x_2	y_1	y_2	y_3	z	
3	15	1	0	0	0	12
5	3	0	1	0	0	50
2	5	0	0	1	0	100
-8	-20	0	0	0	1	0

11.

x_1	x_2	x_3	y_1	y_2	y_3	z	
5	-1	-3	1	0	0	0	20
2	1	9	0	1	0	0	10
1	3	9	0	0	1	0	2
-4	-5	-9	0	0	0	1	0

12.

x_1	x_2	x_3	y_1	y_2	y_3	z	
3	9	6	1	0	0	0	6
5	9	8	0	1	0	0	15
3	6	5	0	0	1	0	30
-5	-30	-25	0	0	0	1	0

In Problems 13–28 solve the linear programming problems, if possible.

13. Maximize: $z = 2x_1 + 3x_2$

Subject to: $\begin{cases} 3x_1 + x_2 \le 300 \\ 2x_1 + 2x_2 \le 400 \\ x_1 \ge 0, \quad x_2 \ge 0 \end{cases}$

14. Maximize: $z = 5x_1 - 3x_2$

Subject to: $\begin{cases} 2x_1 + x_2 \le 200 \\ 5x_1 + 2x_2 \le 100 \\ x_1 \ge 0, \quad x_2 \ge 0 \end{cases}$

15. Maximize: $z = 8x_1 + 16x_2$

Subject to: $\begin{cases} 5x_1 + 3x_2 \le 165 \\ 900x_1 + 1200x_2 \le 36{,}000 \\ x_1 \ge 0, \quad x_2 \ge 2 \end{cases}$

16. Maximize: $z = 240x_1 + 100x_2$

Subject to: $\begin{cases} 300x_1 + 900x_2 \le 36{,}000 \\ 3x_1 + x_2 \le 150 \\ x_1 \ge 0, \quad x_2 \ge 0 \end{cases}$

17. Maximize: $z = 45x_1 + 35x_2$

Subject to: $\begin{cases} x_1 + x_2 \le 200 \\ 4x_1 + 2x_2 \le 500 \\ x_1 \ge 0, \quad x_2 \ge 0 \end{cases}$

18. Maximize: $z = 90x_1 + 55x_2$

Subject to: $\begin{cases} x_1 + x_2 \le 100 \\ 3x_1 + 4x_2 \le 400 \\ x_1 \ge 0, \quad x_2 \ge 0 \end{cases}$

19. Maximize: $z = 5x_1 + 8x_2$

Subject to: $\begin{cases} x_1 + 3x_2 \le 1200 \\ x_1 + 2x_2 \le 1000 \\ x_1 \le 700 \\ x_1 \ge 0, \quad x_2 \ge 0 \end{cases}$

20. Maximize: $z = 8x_1 + 10x_2$

Subject to: $\begin{cases} 2x_1 + x_2 \le 1000 \\ x_1 + 3x_2 \le 1500 \\ x_2 \le 500 \\ x_1 \ge 0, \quad x_2 \ge 0 \end{cases}$

21. Maximize: $z = x_1 + x_2$

Subject to: $\begin{cases} 12x_1 + 150x_2 \le 1200 \\ 6x_1 + 200x_2 \le 1200 \\ 16x_1 + 50x_2 \le 800 \\ x_1 \ge 0, \quad x_2 \ge 0 \end{cases}$

[*Hint:* The first pivot can be selected in the first or second column. Selecting either will give the same answer, but the arithmetic is significantly easier if you select the pivot from the first column.]

22. Maximize: $z = 140x_1 + 80x_2$

Subject to: $\begin{cases} 100x_1 + 5x_2 \le 1500 \\ 200x_1 + 8x_2 \le 1200 \\ 150x_1 + 3x_2 \le 900 \\ x_1 \ge 0, \quad x_2 \ge 0 \end{cases}$

23. Maximize: $z = 4x_1 + 2x_2$

Subject to: $\begin{cases} 2x_1 + x_2 \le 200 \\ 2x_1 + 2x_2 \le 240 \\ 2x_1 + 3x_2 \le 300 \\ x_1 \ge 0, \quad x_2 \ge 0 \end{cases}$

24. Maximize: $z = 3x_1 + 2x_2$

Subject to: $\begin{cases} 2x_1 - 3x_2 \le 80 \\ 2x_1 + x_2 \le 100 \\ 2x_1 + 2x_2 \le 120 \\ x_1 \ge 0, \quad x_2 \ge 0 \end{cases}$

25. Maximize: $z = 12x_1 + 7x_2 + 5x_3$

Subject to: $\begin{cases} 8x_1 + 5x_2 + 3x_3 \le 1000 \\ 5x_1 + x_2 + 3x_3 \le 800 \\ x_1 \ge 0, \quad x_2 \ge 0, \quad x_3 \ge 0 \end{cases}$

26. Maximize: $z = x_1 + x_2 + x_3$

Subject to:
$$\begin{cases} 60x_1 \le 1800 \\ 60x_2 \le 1500 \\ 60x_3 \le 3000 \\ x_2 + x_3 \le 50 \\ x_1 \ge 0, \quad x_2 \ge 0, \quad x_3 \ge 0 \end{cases}$$

27. Maximize: $z = 9x_1 + 7x_2 + 7x_3 + 8x_4$

Subject to:
$$\begin{cases} x_1 + x_2 \le 90 \\ x_3 + x_4 \le 130 \\ x_1 + x_3 \le 80 \\ x_2 + x_4 \le 110 \\ x_1 \ge 0, \quad x_2 \ge 0, \quad x_3 \ge 0, \quad x_4 \ge 0 \end{cases}$$

28. Maximize: $z = 48x_1 + 61x_2 + 39x_3 + 45x_4$

Subject to:
$$\begin{cases} x_1 + x_2 \le 5 \\ x_2 + x_3 \le 4 \\ x_3 + x_4 \le 8 \\ x_2 + x_4 \le 7 \\ x_1 \ge 0, \quad x_2 \ge 0, \quad x_3 \ge 0, \quad x_4 \ge 0 \end{cases}$$

Answer Problems 29–31 by using the simplex method.

APPLICATIONS

29. Alco Company manufactures two products: Alpha and Beta. Each product must pass through two processing operations, and all materials are introduced at the first operation. Alco may produce either one product exclusively or various combinations of both products subject to the constraints given in the table. A shortage of technical labor has limited Alpha production to no more than 700 units per day. There are no constraints on the production of Beta other than the hour constraints in the table. How many of each product should be manufactured in order to maximize the profit?

30. If each Alpha costs $2 to manufacture and each Beta costs $3, rework Problem 29 with the additional restriction that

Product	Hours required to produce 1 unit		Profit per unit
	First process	Second process	
Alpha	1 hr	1 hr	$5
Beta	3 hr	2 hr	$8
Total capacity per day	1200 hr	1000 hr	

only $2,000 per day is available to pay for these manufacturing costs.

31. A shoe company produces three models of athletic shoes. Shoe parts are first produced in the manufacturing shop and then put together in the assembly shop. The number of hours of labor required per pair in each shop is given in the table. The company can sell as many pairs of shoes as it can produce. However, during the next month, no more than 1000 hours can be expended in the manufacturing shop and no more than 800 hours can be expended in the assembly shop. The expected profit from the sale of each pair of shoes is $12 for model 1, $7 for model 2, and $5 for model 3. The company wants to determine how many pairs of each shoe model to produce to maximize total profits over the next month.*

Shop	Model 1	Model 2	Model 3
Manufacturing	8 hr	5 hr	3 hr
Assembly	5 hr	1 hr	3 hr

* This problem is taken from a recent examination administered by the Society of Actuaries. Reprinted by permission of the Society of Actuaries.

4.3 Nonstandard Linear Programming Problems*

The standard linear programming problem is subject to three conditions:

Condition 1.   The objective function is to be maximized.
Condition 2.   All variables are nonnegative.
Condition 3.   All constraints (except those in condition 2) are less than or equal to a nonnegative constant.

_____
* This section is not required for subsequent textual development.

In this section we learn how to handle certain types of nonstandard problems.

Sometimes a linear programming problem is not standard, but can be transformed into standard form by an algebraic process or by substitution, as shown in Examples 1–4.

EXAMPLE 1

Given: Maximize: $z = 14x_1 + 20x_2$

Subject to: $\begin{cases} 3x_1 + 4x_2 \le 120 \\ 5x_1 + 6x_2 \ge -14 \\ x_1 \ge 0, \quad x_2 \ge 0 \end{cases}$

Not standard form

Altered: Maximize: $z = 14x_1 + 20x_2$

Subject to: $\begin{cases} 3x_1 + 4x_2 \le 120 \\ -5x_1 - 6x_2 \le 14 \\ x_1 \ge 0, \quad x_2 \ge 0 \end{cases}$

Standard form

Multiply both sides by -1 (remember to reverse the inequality)

This is now a standard linear programming problem and you can carry out the simplex method:

$$\begin{array}{ccccc|c} x_1 & x_2 & y_1 & y_2 & z & \\ 3 & \boxed{4} & 1 & 0 & 0 & 120 \\ -5 & -6 & 0 & 1 & 0 & 14 \\ \hline -14 & -20 & 0 & 0 & 1 & 0 \end{array}$$

$\uparrow$
Pivot column

$120 \div 4 = 30$

$14 \div (-6)$ is not positive

$$\begin{array}{ccccc|c} x_1 & x_2 & y_1 & y_2 & z & \\ \frac{3}{4} & \boxed{1} & \frac{1}{4} & 0 & 0 & 30 \\ -5 & -6 & 0 & 1 & 0 & 14 \\ \hline -14 & -20 & 0 & 0 & 1 & 0 \end{array}$$

$\longrightarrow$

$$\begin{array}{ccccc|cc} x_1 & x_2 & y_1 & y_2 & z & & \\ \frac{3}{4} & 1 & \frac{1}{4} & 0 & 0 & 30 & x_2 \\ -\frac{1}{2} & 0 & \frac{3}{2} & 1 & 0 & 194 & y_1 \\ \hline 1 & 0 & 5 & 0 & 1 & 600 & z \end{array}$$

The maximum value is 600 when $(x_1, x_2) = (0, 30)$. ∎

EXAMPLE 2

Given: Maximize: $z = 3x_1 + 2x_2$

Altered: Maximize: $z = 3x_1 - 2x_3$

Let $x_3 = -x_2$; then $x_2 = -x_3$ and thus $-x_3 \le 0$ is the same as $x_3 \ge 0$

Subject to: $\begin{cases} x_1 \ge 0 \\ x_2 \le 0 \\ 5x_1 + x_2 \le 100 \\ 3x_1 - 4x_2 \le 50 \end{cases}$

Not standard form

Subject to: $\begin{cases} x_1 \ge 0 \\ x_3 \ge 0 \\ 5x_1 - x_3 \le 100 \\ 3x_1 + 4x_3 \le 50 \end{cases}$

Standard form

Now translate this standard form into tableau form and complete the simplex process:

$$
\begin{array}{ccccc}
x_1 & x_3 & y_1 & y_2 & z \\
\end{array}
$$

$$
\left[
\begin{array}{ccccc|c}
5 & -1 & 1 & 0 & 0 & 100 \\
③ & 4 & 0 & 1 & 0 & 50 \\
\hdashline
-3 & 2 & 0 & 0 & 1 & 0
\end{array}
\right]
\qquad
\begin{array}{l}
100 \div 5 = 20 \\
50 \div 3 = 16.6666\ldots
\end{array}
$$

$$
\begin{array}{c}
\uparrow \\
\text{Pivot column}
\end{array}
$$

$$
\begin{array}{ccccc}
x_1 & x_3 & y_1 & y_2 & z \\
\end{array}
$$

$$
\left[
\begin{array}{ccccc|c}
5 & -1 & 1 & 0 & 0 & 100 \\
① & \frac{4}{3} & 0 & \frac{1}{3} & 0 & \frac{50}{3} \\
\hdashline
-3 & 2 & 0 & 0 & 1 & 0
\end{array}
\right]
\longrightarrow
\left[
\begin{array}{ccccc|c}
0 & -\frac{23}{3} & 1 & -\frac{5}{3} & 0 & \frac{50}{3} \\
1 & \frac{4}{3} & 0 & \frac{1}{3} & 0 & \frac{50}{3} \\
\hdashline
0 & 6 & 0 & 1 & 1 & 50
\end{array}
\right]
\begin{array}{l}
y_1 \\
x_1 \\
z
\end{array}
$$

Now $x_2 = -x_3$, so $x_2 = -0 = 0$ and the solution is $(\frac{50}{3}, 0)$, giving the maximum value of z as 50. ∎

If there are mixed constraints of the type $\leq$ and $\geq$, then you can reverse the inequality by multiplying both sides by -1. This forces all of the constraints to be less than or equal to the constraints, but then the requirement that all variables be nonnegative may be violated. For example, if

$$2x_1 + 3x_2 \geq 4$$

then multiplying both sides by -1 yields

$$-2x_1 - 3x_2 \leq -4$$

This type of constraint violates condition 3:

Linear polynomial $\leq$ nonnegative constant

We will now relax this requirement by allowing

Linear polynomial $\leq$ constant

but in doing this we need to modify the simplex procedure, as illustrated in Example 3.

EXAMPLE 3 Maximize: $z = 8x_1 + 12x_2$

Subject to: $\begin{cases} x_1 + x_2 \leq 10 \\ x_1 - x_2 \geq 5 \\ x_1 \geq 0, \quad x_2 \geq 0 \end{cases}$

Solution Multiply the second constraint by -1:

$$\begin{cases} x_1 + x_2 \leq 10 \\ -x_1 + x_2 \leq -5 \\ x_1 \geq 0, \quad x_2 \geq 0 \end{cases}$$

$$
\begin{array}{ccccc}
x_1 & x_2 & y_1 & y_2 & z \\
\end{array}
$$

$$
\left[
\begin{array}{ccccc|c}
1 & 1 & 1 & 0 & 0 & 10 \\
-1 & 1 & 0 & 1 & 0 & -5 \\
\hdashline
-8 & -12 & 0 & 0 & 1 & 0
\end{array}
\right]
$$

Everything is the same except that there is a negative number in the rightmost column, which violates the simplex method. If you were to try to carry out the simplex method on this matrix you would not arrive at the solution. Why? (For a hint, try using a computer.) All is not lost because we can put the tableau into standard form by pivoting to remove the negative entry in the rightmost column. Use the following procedure:

To Remove a Negative Element in the Rightmost Column

1. Select any negative entry in its row; this is the pivot column.
2. Find the ratios as before, using the pivot column and the entries in the rightmost column (this is the same as before; choose the smallest non-negative ratio).

Pivot:

$$
\begin{array}{ccccc}
x_1 & x_2 & y_1 & y_2 & z \\
\end{array}
$$

$$
\left[\begin{array}{ccccc|c}
1 & 1 & 1 & 0 & 0 & 10 \\
\boxed{-1} & 1 & 0 & 1 & 0 & -5 \\
\hline
-8 & -12 & 0 & 0 & 1 & 0
\end{array}\right]
$$

← Locate any negative elements in the rightmost column (-5 in this example). Next, select *any* negative element in the same row to find the pivot column (the first column in this example shown by box). Finally, consider the ratios: first row: $10 \div 1 = 10$ and second row: $-5 \div (-1) = 5$. The smallest positive ratio gives the pivot element (-1 in this example but the pivot element can be different from the boxed element)

Pivot:

$$
\begin{array}{ccccc}
x_1 & x_2 & y_1 & y_2 & z \\
\end{array}
$$

$$
\left[\begin{array}{ccccc|c}
1 & 1 & 1 & 0 & 0 & 10 \\
① & -1 & 0 & -1 & 0 & 5 \\
\hline
-8 & -12 & 0 & 0 & 1 & 0
\end{array}\right]
$$

You can now begin the simplex process

Pivot:

$$
\begin{array}{ccccc}
x_1 & x_2 & y_1 & y_2 & z \\
\end{array}
$$

$$
\left[\begin{array}{ccccc|c}
0 & 2 & 1 & 1 & 0 & 5 \\
1 & -1 & 0 & -1 & 0 & 5 \\
\hline
0 & -20 & 0 & -8 & 1 & 40
\end{array}\right]
$$

$$
\begin{array}{l}
5 \div 2 = \frac{5}{2} \\
5 \div (-1) \quad \text{not positive}
\end{array}
$$

↑
Pivot column

$$
\left[\begin{array}{ccccc|c}
0 & ① & \frac{1}{2} & \frac{1}{2} & 0 & \frac{5}{2} \\
1 & -1 & 0 & -1 & 0 & 5 \\
\hline
0 & -20 & 0 & -8 & 1 & 40
\end{array}\right]
$$

$$
\begin{array}{l}
\frac{5}{2} \div 1 = \frac{5}{2} \\
5 \div (-1) \quad \text{not positive}
\end{array}
$$

$$
\begin{array}{ccccc}
x_1 & x_2 & y_1 & y_2 & z \\
\end{array}
$$

$$
\left[\begin{array}{ccccc|c}
0 & 1 & \frac{1}{2} & \frac{1}{2} & 0 & \frac{5}{2} \\
1 & 0 & \frac{1}{2} & -\frac{1}{2} & 0 & \frac{15}{2} \\
\hline
0 & 0 & 10 & 2 & 1 & 90
\end{array}\right]
\begin{array}{l}
x_2 \\
x_1 \\
z
\end{array}
$$

The maximum value is $z = 90$, and the optimal solution is $(x_1, x_2) = (\frac{15}{2}, \frac{5}{2})$. ∎

For the rest of this section (as well as the next section), we turn our attention to *minimization* problems. The first, and easiest, type is one in which all of the constraints are of the $\leq$ type. Then we simply need to treat z as we treated x_2 in Example 2. This process is illustrated in Example 4.

EXAMPLE 4 Minimize: $z = -4x_1 - 5x_2$

Solution Subject to: $\begin{cases} 2x_1 + 5x_2 \leq 25 \\ 6x_1 + 5x_2 \leq 45 \\ x_1 \geq 0, \quad x_2 \geq 0 \end{cases}$

This is a minimization problem with $\leq$ constraints. It is *not* a standard linear programming problem. Treat z as we treated x_2 in Example 2. Let $z' = -z$. This means that the smallest value for z is the largest value for z'. To see this more clearly, write down any set of positive numbers, say,

$$S = \{3, 9, 18, 20\}$$

Opposites: $\bar{S} = \{-3, -9, -18, -20\}$

Minimize S: 3 is the smallest element in S

Maximize $\bar{S}$: -3 is the largest element in $\bar{S}$

This means that the objective function for this example can be rewritten as:

Maximize: $z' = -z = 4x_1 + 5x_2$

Subject to: $\begin{cases} 2x_1 + 5x_2 \leq 25 \\ 6x_1 + 5x_2 \leq 45 \\ x_1 \geq 0, \quad x_2 \geq 0 \end{cases}$

Solving the problem using the simplex method is the same as with the standard linear programming problem except that now the coefficient of the z is negative. For this example, the simplex tableau is

$$\begin{bmatrix} x_1 & x_2 & y_1 & y_2 & z & \\ 2 & 5 & 1 & 0 & 0 & 25 \\ 6 & 5 & 0 & 1 & 0 & 45 \\ \hline -4 & -5 & 0 & 0 & -1 & 0 \end{bmatrix} \begin{matrix} \\ y_1 \\ y_2 \\ z \end{matrix}$$

However, since this example has only two variables the most expedient method of solution is by graphing. From Example 1 in Section 3.3, the solution is $z' = 35$ when $x_1 = 5$ and $x_2 = 3$. This means that the minimum value of z is -35 when $x_1 = 5$ and $x_2 = 3$. ∎

The second type of minimization problem in this section is one that has *both* $\leq$ and $\geq$ type constraints. For $\geq$ constraints, we multiply both sides by -1 and treat the tableau like the one in Example 3. We illustrate this procedure with an applied problem from Section 3.2.

EXAMPLE 5 **Transportation Problem**   Sears ships a certain air-conditioning unit from factories in Portland, Oregon, and Flint, Michigan, to distribution centers in Los Angeles,

California, and Atlanta, Georgia. Shipping costs are summarized in the table:

Source	Destination	Shipping cost
Portland	Los Angeles Atlanta	$30 $40
Flint	Los Angeles Atlanta	$60 $50

Supply and demand, in number of units, are:

Supply	Demand
Portland, 200 Flint, 600	Los Angeles, 300 Atlanta, 400

How should shipments be made from Portland and Flint to minimize the shipping cost?

Solution This model was built on page 98 and leads to the following linear programming problem. Let

x_1 = Number shipped from Portland to Los Angeles

x_2 = Number shipped from Portland to Atlanta

x_3 = Number shipped from Flint to Los Angeles

x_4 = Number shipped from Flint to Atlanta

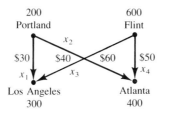

Thus the problem is:

Minimize: $z = 30x_1 + 40x_2 + 60x_3 + 50x_4$

Subject to:
$$\begin{cases} x_1 + x_2 \le 200 \\ x_3 + x_4 \le 600 \\ x_1 + x_3 \ge 300 \\ x_2 + x_4 \ge 400 \end{cases}$$

These last two constraints should be rewritten as

$$-x_1 - x_3 \le -300$$
$$-x_2 - x_4 \le -400$$

Begin with the following initial tableau:

$$
\begin{array}{ccccccccc|c}
x_1 & x_2 & x_3 & x_4 & y_1 & y_2 & y_3 & y_4 & z & \\
\boxed{1} & 1 & 0 & 0 & 1 & 0 & 0 & 0 & 0 & 200 \\
0 & 0 & 1 & 1 & 0 & 1 & 0 & 0 & 0 & 600 \\
-1 & 0 & -1 & 0 & 0 & 0 & 1 & 0 & 0 & -300 \\
0 & -1 & 0 & -1 & 0 & 0 & 0 & 1 & 0 & -400 \\
\hline
30 & 40 & 60 & 50 & 0 & 0 & 0 & 0 & -1 & 0
\end{array}
$$

(the circled 1 is the top-left entry, the boxed -1 is the first entry of the third row)

First we need to work to eliminate the negative elements in the right-hand column. Select the row with right-hand entry -300 and then select one of the negative elements (see the boxed entry). Next we (after checking the ratios) select as a pivot the number circled above and carry out the pivoting process (the details are left for you).

$$
\begin{array}{ccccccccc|c}
x_1 & x_2 & x_3 & x_4 & y_1 & y_2 & y_3 & y_4 & z & \\
1 & 1 & 0 & 0 & 1 & 0 & 0 & 0 & 0 & 200 \\
0 & 0 & 1 & 1 & 0 & 1 & 0 & 0 & 0 & 600 \\
0 & 1 & \boxed{-1} & 0 & 1 & 0 & 1 & 0 & 0 & -100 \\
0 & -1 & 0 & -1 & 0 & 0 & 0 & 1 & 0 & -400 \\
\hline
0 & 10 & 60 & 50 & -30 & 0 & 0 & 0 & -1 & -6000
\end{array}
$$

(the -1 in the third row, x_3 column is circled)

We still have some negative entries, so we select the row with the entry of -100 and use the only negative element in it as a pivot. After pivoting, we have:

$$
\begin{array}{ccccccccc|c}
x_1 & x_2 & x_3 & x_4 & y_1 & y_2 & y_3 & y_4 & z & \\
1 & \boxed{1} & 0 & 0 & 1 & 0 & 0 & 0 & 0 & 200 \\
0 & 1 & 0 & 1 & 1 & 1 & 1 & 0 & 0 & 500 \\
0 & -1 & 1 & 0 & -1 & 0 & -1 & 0 & 0 & 100 \\
0 & \boxed{-1} & 0 & -1 & 0 & 0 & 0 & 1 & 0 & -400 \\
\hline
0 & 70 & 0 & 50 & 30 & 0 & 60 & 0 & -1 & -12{,}000
\end{array}
$$

(the 1 in the first row, x_2 column is circled; the -1 in the fourth row, x_2 column is boxed)

There is one negative entry remaining. Select the negative element in the row shown by the box and use the circled element as the pivot. After pivoting:

$$
\begin{array}{ccccccccc|c}
x_1 & x_2 & x_3 & x_4 & y_1 & y_2 & y_3 & y_4 & z & \\
1 & 1 & 0 & 0 & 1 & 0 & 0 & 0 & 0 & 200 \\
-1 & 0 & 0 & 1 & 0 & 1 & 1 & 0 & 0 & 300 \\
1 & 0 & 1 & 0 & 0 & 0 & -1 & 0 & 0 & 300 \\
1 & 0 & 0 & \boxed{-1} & 1 & 0 & 0 & 1 & 0 & -200 \\
\hline
-70 & 0 & 0 & 50 & -40 & 0 & 60 & 0 & -1 & -26{,}000
\end{array}
$$

Pivot again, using the boxed element:

$$
\begin{array}{ccccccccc|c}
x_1 & x_2 & x_3 & x_4 & y_1 & y_2 & y_3 & y_4 & z & \\
\boxed{1} & 1 & 0 & 0 & 1 & 0 & 0 & 0 & 0 & 200 \\
0 & 0 & 0 & 0 & 1 & 1 & 1 & 1 & 0 & 100 \\
1 & 0 & 1 & 0 & 0 & 0 & -1 & 0 & 0 & 300 \\
-1 & 0 & 0 & 1 & -1 & 0 & 0 & -1 & 0 & 200 \\
\hline
-20 & 0 & 0 & 0 & 10 & 0 & 60 & 50 & -1 & -36{,}000
\end{array}
$$

(the 1 in the first row, x_1 column is circled)

Now we can carry out the simplex method using the circled element above as the pivot:

x_1	x_2	x_3	x_4	y_1	y_2	y_3	y_4	z	
1	1	0	0	1	0	0	0	0	200
0	0	0	0	1	1	1	1	0	100
0	-1	1	0	-1	0	-1	0	0	100
0	1	0	1	0	0	0	-1	0	400
0	20	0	0	30	0	60	50	-1	-32,000

The solution is $x_1 = 200$, $x_2 = 0$, $x_3 = 100$, and $x_4 = 400$. This means that the maximum value is $-32,000$, so the minimum value of the original problem is $32,000 and is achieved by shipping 200 air-conditioning units from Portland to Los Angeles and none from Portland to Atlanta. Flint ships 100 units to Los Angeles and 400 to Atlanta. ∎

A final note on the simplex method. There are some technical complications that were not discussed because they are beyond the scope of this course. For example, when trying certain problems (especially if you randomly choose entries on a computer) the simplex method will break down, and you may at some stage find no nonnegative ratios to consider—which means that you cannot choose a pivot element. But, when the simplex method "breaks down," it simply means that the associated linear programming problem does not have a solution.

Problem Set 4.3

Use the simplex method to solve Problems 1–16.

1. Maximize: $z = 10x_1 + 40x_2$

Subject to: $\begin{cases} 3x_1 + 5x_2 \le 50 \\ 2x_1 + 3x_2 \ge -10 \\ x_1 \ge 0, \quad x_2 \ge 0 \end{cases}$

2. Maximize: $z = 500x_1 + 300x_2$

Subject to: $\begin{cases} 9x_1 + 5x_2 \ge -5 \\ 15x_1 + 3x_2 \le 75 \\ x_1 \ge 0, \quad x_2 \ge 0 \end{cases}$

3. Maximize: $z = 30x_1 + 40x_2$

Subject to: $\begin{cases} 5x_1 + 3x_2 \ge -5 \\ 2x_1 + 3x_2 \le 40 \\ x_1 \ge 0, \quad x_2 \ge 0 \end{cases}$

4. Maximize: $z = 5x_1 + 4x_2$

Subject to: $\begin{cases} 3x_1 + x_2 \le 80 \\ 2x_1 + 5x_2 \le 100 \\ x_1 \ge 0, \quad x_2 \le 0 \end{cases}$

5. Maximize: $z = 30x_1 - 20x_2$

Subject to: $\begin{cases} 2x_1 - x_2 \le 12 \\ -2x_2 \le 9 \\ x_1 \ge 0, \quad x_2 \le 0 \end{cases}$

6. Maximize: $z = 100x_1 - 10x_2$

Subject to: $\begin{cases} 2x_1 - 5x_2 \le 20 \\ 2x_1 - x_2 \le 12 \\ x_1 \ge 0, \quad x_2 \le 0 \end{cases}$

7. Maximize: $z = 50x_1 + 40x_2$

Subject to: $\begin{cases} 3x_1 + 2x_2 \le 8 \\ x_1 + 5x_2 \le 8 \\ x_1 \ge 0, \quad x_2 \ge 0 \end{cases}$

8. Maximize: $z = 3x_1 + 2x_2$

Subject to: $\begin{cases} 2x_1 + 3x_2 \le 105 \\ 5x_1 + 10x_2 \ge 15 \\ x_1 \ge 0, \quad x_2 \ge 0 \end{cases}$

9. Maximize: $z = 5x_1 + 3x_2$

Subject to:
$$\begin{cases} 12x_1 + 4x_2 \leq 16 \\ 2x_1 + 5x_2 \geq 10 \\ x_1 \geq 0, \quad x_2 \geq 0 \end{cases}$$

10. Minimize: $z = -x_1 - 3x_2$

Subject to:
$$\begin{cases} 6x_1 + 2x_2 \leq 5 \\ 2x_1 + 3x_2 \leq 6 \\ x_1 \geq 0, \quad x_2 \geq 0 \end{cases}$$

11. Minimize: $z = -20x_1 - 5x_2$

Subject to:
$$\begin{cases} 3x_1 + 5x_2 \leq 10 \\ 3x_1 + 2x_2 \leq 6 \\ x_1 \geq 0, \quad x_2 \geq 0 \end{cases}$$

12. Minimize: $z = x_1 + 3x_2$

Subject to:
$$\begin{cases} x_1 + x_2 \leq 10 \\ 5x_1 + 2x_2 \geq 20 \\ -x_1 + 2x_2 \geq 0 \\ x_1 \geq 0, \quad x_2 \geq 0 \end{cases}$$

13. Minimize: $z = 35x_1 + 10x_2$

Subject to:
$$\begin{cases} -2x_1 + 3x_2 \geq 0 \\ 8x_1 + x_2 \leq 52 \\ -2x_1 + x_2 \leq 2 \\ x_1 \geq 3 \\ x_1 \geq 0, \quad x_2 \geq 0 \end{cases}$$

14. Minimize: $z = 140x_1 - 60x_2$

Subject to:
$$\begin{cases} -2x_1 + 3x_2 \geq 0 \\ 8x_1 + x_2 \leq 52 \\ -2x_1 + x_2 \leq 2 \\ x_1 \geq 3 \\ x_1 \geq 0, \quad x_2 \geq 0 \end{cases}$$

15. Maximize: $z = 2x_1 + 3x_2 + 5x_3$

Subject to:
$$\begin{cases} x_1 + 3x_2 + x_3 \leq 46 \\ 2x_1 + x_2 + x_3 \leq 40 \\ x_2 + x_3 = 10 \\ x_1 \geq 0, \quad x_2 \geq 0, \quad x_3 \geq 0 \end{cases}$$

[*Hint*: Since $x_2 + x_3 = 10$, write $x_3 = 10 - x_2$ and eliminate the variable x_3 from the problem.]

16. Maximize: $z = x_1 + x_2 + 5x_3$

Subject to:
$$\begin{cases} x_1 - 3x_3 \leq 30 \\ x_2 - 2x_3 \leq 20 \\ x_1 + 5x_2 \leq 40 \\ x_1 + x_2 - x_3 = 8 \\ x_1 \geq 0, x_2 \geq 0, x_3 \geq 0 \end{cases}$$

[*Hint*: Since $x_1 + x_2 - x_3 = 8$, write $x_3 = x_1 + x_2 - 8$ and eliminate x_3 from the problem.]

APPLICATIONS

17. Tony's veterinarian perscribes three food supplements for his horses. Tony must feed them at least 115 kilograms of oats and 50 kilograms of alfalfa, but no more than 200 kilograms of grain. The amounts of each of these in the supplements are summarized in the table.

| | Supplement | | |
	A	B	C
Oats	1 kg	2 kg	3 kg
Alfalfa	2 kg	1 kg	1 kg
Grain	1 kg	0 kg	1 kg
Cost per kg:	$1	$1	$4

How many units of A, B, and C should be mixed to minimize the cost?

18. Helmer's apple farm grows apples. Helmer's cannot grow more than 3000 bins (but can grow less). The crop is divided into two types of apples: eating and applesauce. Helmer's must supply at least 80 bins of eating apples and 800 bins of apples for sauce. The cost of producing these apples is $4 per bin of eating apples and $3.25 per bin of applesauce. How many bins of each should be produced to minimize the cost?

19. Pell, Inc., manufactures two products: tables and shelves. Each product must be processed in each of three departments: machining, assembling, and finishing. The hours needed to produce one unit of product per department and the maximum possible hours per department are shown in the table. Standing orders require that Pell manufacture at least 30 tables and 25 shelves. Pell's net profit is $4 per table and $2 per shelf. How many tables should be manufactured to maximize the profit? Solve this problem graphically.

| Department | Production time per unit (hr) | | Maximum capacity (hr) |
	Tables	Shelves	
Machining	2	1	200
Assembling	2	2	240
Finishing	2	3	300

20. Solve Problem 19 by using the simplex method.

21. The answer to Problem 19 using the graphical method gives some integral coordinates, but the answer to Problem 20 using the simplex method does not. Look at the graphical solution and reconcile the answers to Problems 19 and 20.

4.4 Duality*

The most common type of minimization problem is one in which all of the constraints are of the $\geq$ type. In this section we use a procedure for solving these problems. The process was first introduced by John von Neumann (1903–1957), who has been described as one of the greatest geniuses of this century. It involves solving a related maximization problem called the **dual** of the minimization problem. We begin with a two-dimensional example so that we can work both the original problem and the dual problem graphically to illustrate the plausibility of the method. The dual can be used with any number of variables.

EXAMPLE 1 Minimize: $z = 50x_1 + 60x_2$

Subject to: $\begin{cases} 7x_1 + 2x_2 \geq 14 \\ 3x_1 + 5x_2 \geq 20 \\ x_1 \geq 0, \quad x_2 \geq 0 \end{cases}$

Solution We begin by using the graphical method:

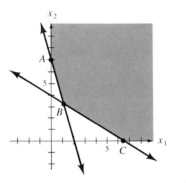

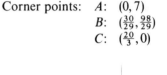

Corner points: A: $(0, 7)$
B: $\left(\frac{30}{29}, \frac{98}{29}\right)$
C: $\left(\frac{20}{3}, 0\right)$

check the corner points and find:

$(0, 7)$: $z = 50(0) + 60(7) = 420$

$\left(\frac{30}{29}, \frac{98}{29}\right)$: $z = 50\left(\frac{30}{29}\right) + 60\left(\frac{98}{29}\right) \approx 254.48$

$\left(\frac{20}{3}, 0\right)$: $z = 50\left(\frac{20}{3}\right) + 60(0) \approx 333.33$

The minimum value is 254.48, which occurs at $\left(\frac{30}{29}, \frac{98}{29}\right)$.

Next, write the problem using an augmented matrix (the objective function is written in the last row):

$$\left[\begin{array}{cc:c} 7 & 2 & 14 \\ 3 & 5 & 20 \\ \hdashline 50 & 60 & 0 \end{array}\right]$$

If we interchange the rows and columns of this matrix (it is called the **transpose** and we will discuss it below), we obtain

$$\left[\begin{array}{cc:c} 7 & 3 & 50 \\ 2 & 5 & 60 \\ \hdashline 14 & 20 & 0 \end{array}\right]$$

---

* This section is not required for subsequent textual development.

John von Neumann discovered that if the problem specified by this matrix is maximized using $\leq$ constraints, the same answer as the corresponding minimum problem will always be obtained. Check it out; use y's instead of x's so we can keep the two problems apart.

Maximize: $z = 14y_1 + 20y_2$

Subject to: $\begin{cases} 7y_1 + 3y_2 \leq 50 \\ 2y_1 + 5y_2 \leq 60 \\ y_1 \geq 0, \quad y_2 \geq 0 \end{cases}$

Here again, solve the linear programming problem by graphing.

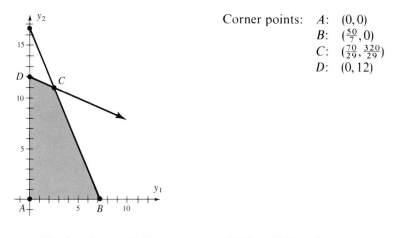

Corner points: A: $(0,0)$
B: $\left(\frac{50}{7},0\right)$
C: $\left(\frac{70}{29},\frac{320}{29}\right)$
D: $(0,12)$

Check values: $(0,0)$: $z' = 14(0) + 20(0) = 0$
$\left(\frac{50}{7},0\right)$: $z' = 14\left(\frac{50}{7}\right) + 20(0) = 100$
$\left(\frac{70}{29},\frac{320}{29}\right)$: $z' = 14\left(\frac{70}{29}\right) + 20\left(\frac{320}{29}\right) \approx 254.48$
$(0,12)$: $z' = 14(0) + 20(12) = 240$

The maximum value is 254.48 at the point $\left(\frac{70}{29},\frac{320}{29}\right)$.

Note that the minimum value of the original problem is the same as the maximum value of the dual: *This is always true.* ■

The feasible regions of the two problems are different, and the corner points are different, but the *extreme values* of the objective functions are the same. You should note an even closer connection between a problem and its dual as we use the simplex method to solve the dual (the maximum problem).

For now, let us use slack variables u and v (you will soon see these turn out to be x_1 and x_2)

y_1	y_2	u	v	z'	
7	3	1	0	0	50
2	⑤	0	1	0	60
-14	-20	0	0	1	0

$50 \div 3 = 16.666\ldots$
$60 \div 5 = 12$

↑
Pivot column

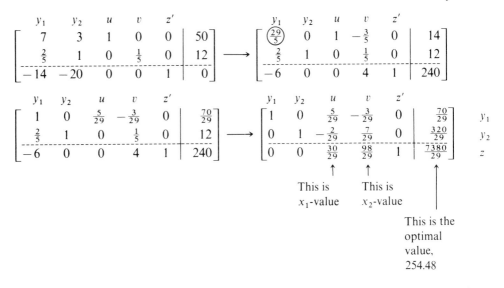

Note that the solution to the *minimum problem* is found in the bottom row of the tableau under the columns u and v (the slack variables); that is, $x_1 = \frac{30}{29}$ and $x_2 = \frac{98}{29}$. This suggests that when solving the dual, you can use the original x-values as slack variables.

The key to changing a minimization problem to its dual maximization problem is the interchanging of the rows and columns in the augmented matrix of the minimization problem. We called this the *transpose*:

Transpose of a Matrix

The *transpose*, M^{T}, of a matrix M is the matrix formed by interchanging the rows and columns of M.

EXAMPLE 2 Find the transpose of the given matrices.

$$A = \begin{bmatrix} 1 & 2 \\ 3 & 4 \end{bmatrix} \qquad B = \begin{bmatrix} 6 & 8 & 9 \\ 4 & 1 & 7 \\ 3 & 2 & 1 \end{bmatrix} \qquad C = \begin{bmatrix} 4 & 8 \end{bmatrix}$$

$$D = \begin{bmatrix} 6 \\ 1 \\ 2 \end{bmatrix} \qquad E = \begin{bmatrix} 1 & 3 & 6 \\ 4 & 9 & 2 \end{bmatrix}$$

Solution

$$A^{\mathsf{T}} = \begin{bmatrix} 1 & 3 \\ 2 & 4 \end{bmatrix} \qquad B^{\mathsf{T}} = \begin{bmatrix} 6 & 4 & 3 \\ 8 & 1 & 2 \\ 9 & 7 & 1 \end{bmatrix} \qquad C^{\mathsf{T}} = \begin{bmatrix} 4 \\ 8 \end{bmatrix}$$

$$D^{\mathsf{T}} = \begin{bmatrix} 6 & 1 & 2 \end{bmatrix} \qquad E^{\mathsf{T}} = \begin{bmatrix} 1 & 4 \\ 3 & 9 \\ 6 & 2 \end{bmatrix}$$

∎

We can now summarize this process.

Von Neumann's Duality Principle

The optimum value of a minimum linear programming problem, if a solution exists, is the same as the optimum value of its dual. That is, the maximum value of z' is the same as the minimum value of z.

To find the dual:

1. The given problem should be a minimization problem consisting only of $\geq$ constraints.* (You may have to use some of the techniques of the last section if the order of some of the constraints is not correct.)
2. Write the simplex problem in matrix form, with each constraint on a different row and with the objective function as the bottom row.
3. Find the transpose of the problem; this gives the dual problem (the dual of a minimum problem is a maximum problem). The constraints of the dual are formed from the rows of the transpose; the objective function is at the bottom; and the inequality signs are the reverse of those in the original problem.
4. The optimal solution is given by the entries in the bottom row of the columns corresponding to the slack variables, and the minimum value of the objective function of the minimization problem is the same as the maximum value of the objective function of the dual.

EXAMPLE 3   Minimize: $z = 50x_1 + 60x_2$

Subject to: $\begin{cases} 7x_1 + 2x_2 \geq 14 \\ 3x_1 + 5x_2 \geq 20 \\ 6x_1 + 10x_2 \geq 30 \\ x_1 \geq 0, \quad x_2 \geq 0 \end{cases}$

Solution   Solve this by solving the dual problem. First, write in augmented matrix form:

$$\begin{bmatrix} 7 & 2 & | & 14 \\ 3 & 5 & | & 20 \\ 6 & 10 & | & 30 \\ \hline 50 & 60 & | & 0 \end{bmatrix}$$

Next, find the transpose:

$$\begin{bmatrix} 7 & 3 & 6 & | & 50 \\ 2 & 5 & 10 & | & 60 \\ \hline 14 & 20 & 30 & | & 0 \end{bmatrix}$$

The transpose leads to the dual problem:

Maximize: $z' = 14y_1 + 20y_2 + 30y_3$

Subject to: $\begin{cases} 7y_1 + 3y_2 + 6y_3 \leq 50 \\ 2y_1 + 5y_2 + 10y_3 \leq 60 \end{cases}$

---

* The dual can also be used to transform a maximization problem into a minimization problem. In this book, however, we will use the dual only to transform minimization problems into maximization ones.

Solve using the simplex method:

$$
\begin{array}{cccccc}
y_1 & y_2 & y_3 & x_1 & x_2 & z'
\end{array}
$$

$$
\left[\begin{array}{cccccc|c}
7 & 3 & 6 & 1 & 0 & 0 & 50 \\
2 & 5 & \boxed{10} & 0 & 1 & 0 & 60 \\
\hdashline
-14 & -20 & -30 & 0 & 0 & 1 & 0
\end{array}\right]
\qquad
\begin{array}{l}
50 \div 6 = 8.333 \\
60 \div 10 = 6
\end{array}
$$

$$
\begin{array}{c}
\uparrow \\
\text{Pivot column}
\end{array}
$$

$$
\begin{array}{cccccc}
y_1 & y_2 & y_3 & x_1 & x_2 & z'
\end{array}
$$

$$
\left[\begin{array}{cccccc|c}
7 & 3 & 6 & 1 & 0 & 0 & 50 \\
\frac{1}{5} & \frac{1}{2} & 1 & 0 & \frac{1}{10} & 0 & 6 \\
\hdashline
-14 & -20 & -30 & 0 & 0 & 1 & 0
\end{array}\right]
$$

$$
\begin{array}{cccccc}
y_1 & y_2 & y_3 & x_1 & x_2 & z'
\end{array}
$$

$$
\left[\begin{array}{cccccc|c}
\boxed{\frac{29}{5}} & 0 & 0 & 1 & -\frac{3}{5} & 0 & 14 \\
\frac{1}{5} & \frac{1}{2} & 1 & 0 & \frac{1}{10} & 0 & 6 \\
\hdashline
-8 & -5 & 0 & 0 & 3 & 1 & 180
\end{array}\right]
\qquad
\begin{array}{l}
14 \div \frac{29}{5} = 2.413\ldots \\
6 \div \frac{1}{5} = 30
\end{array}
$$

$$
\begin{array}{c}
\uparrow \\
\text{Pivot column}
\end{array}
$$

$$
\begin{array}{cccccc}
y_1 & y_2 & y_3 & x_1 & x_2 & z'
\end{array}
$$

$$
\left[\begin{array}{cccccc|c}
1 & 0 & 0 & \frac{5}{29} & -\frac{3}{29} & 0 & \frac{70}{29} \\
\frac{1}{5} & \frac{1}{2} & 1 & 0 & \frac{1}{10} & 0 & 6 \\
\hdashline
-8 & -5 & 0 & 0 & 3 & 1 & 180
\end{array}\right]
$$

$$
\begin{array}{cccccc}
y_1 & y_2 & y_3 & x_1 & x_2 & z'
\end{array}
$$

$$
\left[\begin{array}{cccccc|c}
1 & 0 & 0 & \frac{5}{29} & -\frac{3}{29} & 0 & \frac{70}{29} \\
0 & \boxed{\frac{1}{2}} & 1 & -\frac{1}{29} & \frac{35}{290} & 0 & \frac{160}{29} \\
\hdashline
0 & -5 & 0 & \frac{40}{29} & \frac{63}{29} & 1 & \frac{5780}{29}
\end{array}\right]
$$

$$
\begin{array}{cccccc}
y_1 & y_2 & y_3 & x_1 & x_2 & z'
\end{array}
$$

$$
\left[\begin{array}{cccccc|c}
1 & 0 & 0 & \frac{5}{29} & -\frac{3}{29} & 0 & \frac{70}{29} \\
0 & 1 & 2 & -\frac{2}{29} & \frac{70}{290} & 0 & \frac{320}{29} \\
\hdashline
0 & -5 & 0 & \frac{40}{29} & \frac{63}{29} & 1 & \frac{5780}{29}
\end{array}\right]
$$

$$
\begin{array}{cccccc}
y_1 & y_2 & y_3 & x_1 & x_2 & z'
\end{array}
$$

$$
\left[\begin{array}{cccccc|c}
1 & 0 & 0 & \frac{5}{29} & -\frac{3}{29} & 0 & \frac{70}{29} \\
0 & 1 & 2 & -\frac{2}{29} & \frac{70}{290} & 0 & \frac{320}{29} \\
\hdashline
0 & 0 & 10 & \frac{30}{29} & \frac{98}{29} & 1 & \frac{7380}{29}
\end{array}\right]
$$

The maximum is $z' = \frac{7380}{29} \approx 254.5$, so von Neumann's duality principle tells us that the minimum value of z is also 254.5. In addition to giving the minimum value of z, the duality process also gives the values of x_1 and x_2. These values appear at the bottom of the columns labeled x_1 and x_2, respectively. Thus the objective function is minimized when $x_1 = \frac{30}{29}$ and $x_2 = \frac{98}{29}$. ∎

COMPUTER COMMENT   Most linear programming problems require an extensive amount of tedious arithmetic calculations and for that reason most mathematicians and financial consultants use computer programs of the simplex method to carry out the actual "work" of this method. You might wish to use Program 6 of the computer disk that accompanies this text. One of the things you will notice when using a computer program is that instead of exact results, the work is rounded to a specified number of decimal places. Example 3, for example, looks like this when Program 6 is used:

$$
\begin{array}{ccccccc}
y_1 & y_2 & y_3 & x_1 & x_2 & z' & \\
\begin{bmatrix}
(5.80) & 0 & 0 & 1 & -0.60 & 0 & | & 14 \\
0.20 & 0.50 & 1 & 0 & 0.10 & 0 & | & 6 \\
\hdashline
-8 & -5 & 0 & 0 & 3 & 1 & | & 180
\end{bmatrix}
\end{array}
$$

This is the step in Example 3 where $\frac{29}{5}$ is circled.

$$
\begin{array}{ccccccc}
y_1 & y_2 & y_3 & x_1 & x_2 & z' & \\
\begin{bmatrix}
1 & 0 & 0 & 0.17 & -0.10 & 0 & | & 2.41 \\
0 & (0.50) & 1 & -0.03 & 0.12 & 0 & | & 5.52 \\
\hdashline
0 & -5 & 0 & 1.38 & 2.17 & 1 & | & 199.31
\end{bmatrix}
\end{array}
$$

This is the step in Example 3 where $\frac{1}{2}$ is circled.

$$
\begin{array}{ccccccc}
y_1 & y_2 & y_3 & x_1 & x_2 & z' & \\
\begin{bmatrix}
1 & 0 & 0 & 0.17 & -0.10 & 0 & | & 2.41 \\
0 & 1 & 2 & -0.07 & 0.24 & 0 & | & 11.03 \\
\hdashline
0 & 0 & 10 & 1.03 & 3.38 & 1 & | & 254.48
\end{bmatrix}
\end{array}
$$

This is the final step.

A final note on integer solutions. In this chapter we have been requiring nonnegative variables and have not worked problems requiring integer solutions because linear programming problems that impose integer constraints involve techniques that are beyond the scope of this text. Unfortunately, you cannot simply solve such a problem by the techniques discussed here and then round your results to the nearest integer. The rounded solution *may not* be the best integer solution. However, you can use the following result: Suppose you are maximizing z and

z_0 = Maximum value for the continuous problem (the answer found using the simplex method in this chapter)

z_1 = Maximum found by rounding z_0 to the nearest integer

z_2 = Maximum integer solution (found by techniques not discussed in this text)

The best we can say is that

$$z_1 \leq z_2 \leq z_0$$

Linear programming provides extremely useful mathematical models, but there are many applied applications for which the model developed here is not sufficient. For more advanced study of linear programming, you will need to consult a linear algebra or a linear programming textbook.

Problem Set 4.4

Find the transpose of the matrices in Problems 1–4.

1. a. $A = \begin{bmatrix} 6 & 9 \\ 4 & 8 \end{bmatrix}$

   b. $B = \begin{bmatrix} 5 & 6 \\ 3 & 8 \end{bmatrix}$

   c. $C = \begin{bmatrix} 4 & 9 & 1 \\ 6 & 1 & 4 \end{bmatrix}$

2. a. $D = \begin{bmatrix} 8 & 1 \\ 6 & 2 \\ 4 & 5 \end{bmatrix}$ **b.** $E = \begin{bmatrix} 6 & 8 & 1 \\ 2 & 3 & 4 \\ 5 & 9 & 7 \end{bmatrix}$

   c. $F = \begin{bmatrix} 1 & 0 & 3 \\ 4 & 9 & 7 \\ 8 & 6 & 5 \end{bmatrix}$

3. a. $G = \begin{bmatrix} 1 & 3 & 5 \end{bmatrix}$ **b.** $H = \begin{bmatrix} 4 \\ 9 \\ 6 \end{bmatrix}$

   c. $J = \begin{bmatrix} 1 \\ 0 \\ 3 \\ 2 \end{bmatrix}$

4. a. $K = \begin{bmatrix} 1 & 8 & 7 & 4 \end{bmatrix}$

   b. $L = \begin{bmatrix} 4 & 8 & 0 & 3 & 2 \\ 6 & 1 & 4 & 7 & 9 \end{bmatrix}$

   c. $M = \begin{bmatrix} 1 & 4 & 3 \\ 6 & 9 & 2 \\ 4 & 7 & 1 \\ 5 & 7 & 11 \end{bmatrix}$

Write the dual of Problems 5–8.

5. Minimize: $z = 3x_1 + 4x_2$

  Subject to: $\begin{cases} 2x_1 + 8x_2 \geq 10 \\ 3x_1 + 5x_2 \geq 30 \\ x_1 \geq 0, \quad x_2 \geq 0 \end{cases}$

6. Minimize: $z = 50x_1 + 30x_2$

  Subject to: $\begin{cases} 5x_1 + 5x_2 \geq 25 \\ 6x_1 - 2x_2 \geq 10 \\ x_1 \geq 0, \quad x_2 \geq 0 \end{cases}$

7. Minimize: $z = 3x_1 + 2x_2 + 5x_3$

  Subject to: $\begin{cases} x_1 + x_2 + x_3 \geq 10 \\ 2x_1 + 3x_3 \geq 2 \\ x_2 + 2x_3 \geq 1 \\ 3x_1 + 5x_3 \geq 15 \\ x_1 \geq 0, \quad x_2 \geq 0, \quad x_3 \geq 0 \end{cases}$

8. Minimize: $z = x_1 + 2x_2 + 3x_3$

  Subject to: $\begin{cases} x_1 + x_2 + x_3 \geq 50 \\ 6x_1 + 5x_3 \geq 10 \\ 7x_2 + 5x_3 \geq 35 \\ 9x_1 + 4x_2 \geq 18 \\ x_1 \geq 0, \quad x_2 \geq 0, \quad x_3 \geq 0 \end{cases}$

Solve Problems 9–22 by using the dual.

9. Minimize: $z = 3x_1 + 4x_2$

  Subject to: $\begin{cases} 2x_1 + 8x_2 \geq 10 \\ 3x_1 + 5x_2 \geq 30 \\ x_1 \geq 0, \quad x_2 \geq 0 \end{cases}$

10. Minimize: $z = 50x_1 + 30x_2$

  Subject to: $\begin{cases} 5x_1 + 5x_2 \geq 25 \\ 6x_1 - 2x_2 \geq 10 \\ x_1 \geq 0, \quad x_2 \geq 0 \end{cases}$

11. Minimize: $z = 3x_1 + 2x_2 + 5x_3$

  Subject to: $\begin{cases} x_1 + x_2 + x_3 \geq 10 \\ 2x_1 + 3x_3 \geq 2 \\ x_2 + 2x_3 \geq 1 \\ 3x_1 + 5x_3 \geq 15 \\ x_1 \geq 0, \quad x_2 \geq 0, \quad x_3 \geq 0 \end{cases}$

12. Minimize: $z = x_1 + 2x_2 + 3x_3$

  Subject to: $\begin{cases} x_1 + x_2 + x_3 \geq 50 \\ 6x_1 + 5x_3 \geq 10 \\ 7x_2 + 5x_3 \geq 35 \\ 9x_1 + 4x_2 \geq 18 \\ x_1 \geq 0, \quad x_2 \geq 0, \quad x_3 \geq 0 \end{cases}$

13. Minimize: $z = 50x_1 + 10x_2 + 20x_3$

  Subject to: $\begin{cases} x_1 + x_2 + x_3 \geq 25 \\ 2x_1 + x_2 + 3x_3 \geq 100 \\ x_1 \geq 0, \quad x_2 \geq 0, \quad x_3 \geq 0 \end{cases}$

14. Minimize: $z = 10x_1 + 20x_2 + 30x_3$

  Subject to: $\begin{cases} x_1 + 3x_2 + x_3 \geq 75 \\ 2x_1 + x_2 + 5x_3 \geq 80 \\ x_1 \geq 0, \quad x_2 \geq 0, \quad x_3 \geq 0 \end{cases}$

15. Minimize: $z = 24x_1 + 12x_2$

Subject to:
$$\begin{cases} x_1 \leq 10 \\ x_2 \leq 8 \\ 3x_1 + 2x_2 \geq 12 \\ x_1 \geq 0, \quad x_2 \geq 0 \end{cases}$$

16. Minimize $z = 6x_1 + 18x_2$

Subject to:
$$\begin{cases} x_1 \leq 10 \\ x_2 \leq 8 \\ 3x_1 + 2x_2 \geq 12 \\ x_1 \geq 0, \quad x_2 \geq 0 \end{cases}$$

17. Minimize: $z = 90x_1 + 20x_2$

Subject to:
$$\begin{cases} x_1 + x_2 \geq 6 \\ -2x_1 + x_2 \geq -16 \\ x_2 \leq 9 \\ x_1 \geq 0, \quad x_2 \geq 0 \end{cases}$$

18. Minimize: $z = 400x_1 + 100x_2$

Subject to:
$$\begin{cases} x_1 + x_2 \geq 6 \\ -2x_1 + x_2 \geq -16 \\ x_2 \leq 9 \\ x_1 \geq 0, \quad x_2 \geq 0 \end{cases}$$

19. Minimize: $z = 5x_1 + 3x_2$

Subject to:
$$\begin{cases} 2x_1 + x_2 \geq 8 \\ x_2 \leq 5 \\ x_1 - x_2 \leq 2 \\ 3x_1 - 2x_2 \geq 5 \\ x_1 \geq 0, \quad x_2 \geq 0 \end{cases}$$

20. Minimize: $z = 2x_1 - 3x_2$

Subject to:
$$\begin{cases} 2x_1 + x_2 \geq 8 \\ x_2 \leq 5 \\ x_1 - x_2 \leq 2 \\ 3x_1 - 2x_2 \geq 5 \\ x_1 \geq 0, \quad x_2 \geq 0 \end{cases}$$

21. Minimize: $z = 140x_1 + 250x_2$

Subject to:
$$\begin{cases} 2x_1 + x_2 \geq 8 \\ x_1 - x_2 \leq 7 \\ x_1 - x_2 \geq -3 \\ x_1 \leq 9 \\ x_1 \geq 0, \quad x_2 \geq 0 \end{cases}$$

22. Minimize: $z = 640x_1 - 130x_2$

Subject to:
$$\begin{cases} 2x_1 + x_2 \geq 8 \\ x_1 - x_2 \leq 7 \\ x_1 - x_2 \geq -3 \\ x_1 \leq 9 \\ x_1 \geq 0, \quad x_2 \geq 0 \end{cases}$$

APPLICATIONS

23. Karlin Enterprises manufactures two electronic games. Standing orders require that at least 24,000 space-battle games and 5000 football games be produced. The company has two factories: the Gainesville plant can produce 600 space-battle games and 100 football games per day; the Sacramento plant can produce 300 space-battle games and 100 football games per day. If the Gainesville plant costs $20,000 per day to operate and the Sacramento factory costs $15,000 per day, find the number of days per month each factory should operate to minimize the cost. (Assume that each month has 30 days.)

24. The Cosmopolitan Oil Company requires at least 8000 barrels of low-grade oil per month; it also needs at least 12,000 barrels of medium-grade oil per month and at least 2000 barrels of high-grade oil per month. Oil is produced at one of two refineries. The daily production of these refineries is summarized in the table. If it costs $17,000 per day to operate refinery I and $15,000 per day to operate refinery II, how many days per month should each refinery be operated to satisfy the requirements and at the same time minimize the costs? (Assume that each month has 30 days.)

Refinery	Oil production (barrels)		
	Low grade	Medium grade	High grade
I	200	400	100
II	200	300	100

25. A woman wants to design a weekly exercise schedule that involves jogging, handball, and aerobic dance. She decides to jog at least 3 hours per week, play handball at least 2 hours per week, and dance at least 5 hours per week. She also wants to devote at least as much time to jogging as to handball because she does not like handball as much as she likes the other exercises. She also knows that jogging consumes 900 calories per hour, handball 600 calories per hour, and dance 800 calories per hour. Her physician told her that she must burn up a total of at least 9000 calories per week in this exercise program. How many hours should she devote to each exercise if she wishes to minimize her exercise time?

26. Brown Brothers is an investment company analyzing the pension fund of a certain company. A maximum of $10 million is available to invest in two places. No more than $8 million can be invested in stocks yielding 12%, and at least $2 million can be invested in long-term bonds yielding 8%. The stock-to-bond investment ratio cannot be more than 1 to 3. How should Brown Brothers advise their client so that the pension fund will receive the maximum yearly return on investment?

27. To utilize costly equipment efficiently, an assembly line with a capacity of 50 units per shift must operate 24 hours per day. This requires scheduling three shifts with varying labor costs as follows:

Day shift:	Labor costs are $100 per unit
Swing shift:	Labor costs are $150 per unit
Graveyard shift:	Labor costs are $180 per unit

Each unit produced uses 60 linear feet of sheeting material, and there is a total of 1800 feet available for the day shift, 1500 feet for the swing shift, and 3000 feet for the graveyard shift. The day shift must produce at least 20 units and the other shifts must produce at least 50 units together. How many units should be produced on each shift to maximize the revenue if the units are sold for $850 each?

28. Repeat Problem 27 except minimize labor costs rather than maximize revenue.

29. Suppose a computer dealer has stores in Hillsborough and Palo Alto and warehouses in San Jose and Burlingame. The cost of shipping a computer from one location to another as well as the supply and demand at each location are shown on the following "map":

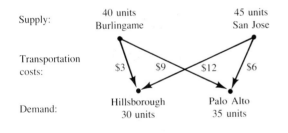

How should the dealer ship the computers to minimize the shipping costs?

30. Camstop, an importer of computer chips, can ship case lots of components to stores in Dallas and Chicago from ports in either Los Angeles or Seattle. The costs and demand are given in the table. How many cases should be shipped from each port to minimize shipping costs?

Store	Cost to ship (each case)		
	Los Angeles port	Seattle port	Store demand
Chicago	$9	$7	80 cases
Dallas	$7	$8	110 cases
Available inventory at each port	90 cases	130 cases	

31. A Texas appliance dealer has stores in Fort Worth and in Houston and warehouses in Dallas and San Antonio. The cost of shipping a refrigerator from Dallas to Fort Worth is $14, and from Dallas to Houston shipping is $15; from San Antonio to Fort Worth the cost is $10, and from San Antonio to Houston it is $18. Suppose that the Fort Worth store has orders for 15 refrigerators and the Houston store has orders for 35. Also suppose that there are 25 refrigerators in stock at Dallas and 45 at San Antonio. What is the most economical way to supply the requested refrigerators to the two stores?

4.5 Summary and Review

IMPORTANT TERMS

Basic feasible solution [4.2]
Basic solution [4.2]
Departing variable [4.2]
Dual [4.4]
Duality principle [4.4]
Entering variable [4.2]
Initial basic solution [4.2]
Initial simplex tableau [4.1]
Maximization [4.2]
Minimization [4.3]
Nonstandard linear programming problem [4.3]
Optimal solution [4.2]

Pivot [4.2]
Pivoting process [4.2]
Pivot row [4.2]
Simplex method [4.1, 4.2]
Simplex tableau [4.1]
Slack variable [4.1]
Standard linear programming problem [4.1]
Test for maximality [4.2]
Transpose of a matrix [4.4]
Von Neumann's duality principle [4.4]

For additional practice there are a large number of review problems categorized by objective in the Student Solutions Manual. The following sample test (40 minutes) is intended to review the main ideas of the chapter.

In Problems 1–4 set up the initial tableau for the linear programming problems and indicate the location of the first pivot. You do not need to solve.*

1. Maximize: $z = 6x_1 + 25x_2 + 3x_3$

Subject to: $\begin{cases} 2x_1 + x_3 \le 50 \\ 4x_2 + x_3 \le 90 \\ 3x_1 + 4x_2 \le 100 \\ x_1 \ge 0, \quad x_2 \ge 0, \quad x_3 \ge 0 \end{cases}$

2. Maximize: $z = 5x_1 + 3x_2$

Subject to: $\begin{cases} 7x_1 - 3x_2 \ge 3 \\ 2x_1 + x_2 \le 12 \\ x_1 \ge 0, \quad x_2 \ge 0 \end{cases}$

3. Minimize: $z = 50x_1 + 80x_2$

Subject to: $\begin{cases} 9x_1 + x_2 \ge 18 \\ 3x_1 + 12x_2 \ge 36 \\ 2x_1 + 3x_2 \ge 30 \\ x_1 \ge 0, \quad x_2 \ge 0 \end{cases}$

4. Maximize: $z = x_1 + 3x_2 + 2x_3$

Subject to: $\begin{cases} x_1 + x_2 + x_3 \le 10 \\ 4x_1 + 2x_2 + 3x_3 \le 32 \\ x_1 + 2x_2 + x_3 \le 16 \\ x_1 \ge 0, \quad x_2 \ge 0, \quad x_3 \ge 0 \end{cases}$

Problems 5–10 are reprinted with the permission of the American Institute of Certified Public Accountants, Inc. The questions are multiple-choice questions taken from recent CPA examinations. Choose the best, or most, appropriate response for each question.

The Ball Company manufactures three types of lamps: A, B, and C. Each lamp is processed in two departments—I and II. Total available hours of labor per day for departments I and II are 400 and 600, respectively. No additional labor is available. Time requirements and profit per unit for each lamp type are:

	A	B	C
Hours required in department I	2	3	1
Hours required in department II	4	2	3
Profit per unit (sales price less all variable costs)	$5	$4	$3

The company has assigned you, as the accounting member of its profit planning committee, to determine the number of types of A, B, and C lamps that it should produce in order to maximize its total profit from the sale of lamps. The following questions concern the linear programming model your group has developed.

5. The coefficients of the objective function would be

   A. 4, 2, 3   **B.** 2, 3, 1   **C.** 5, 4, 3   **D.** 400, 600

6. The constraints in the model would be

   A. 2, 3, 1   **B.** 5, 4, 3   **C.** 4, 2, 3   **D.** 400, 600

7. The constraint imposed by the available hours in department I could be expressed as

   A. $4x_1 + 2x_2 + 3x_3 \le 400$ **B.** $4x_1 + 2x_2 + 3x_3 \ge 400$

   C. $2x_1 + 3x_2 + 1x_3 \le 400$ **D.** $2x_1 + 3x_2 + 1x_3 \ge 400$

8. The most types of lamps that would be included in the optimal solution would be

   A. 2   **B.** 1   **C.** 3   **D.** 0

9. The Pauley Company plans to expand its sales force by opening several new branch offices. Pauley has $10,400,000 in capital available for new branch offices. Pauley will

---

* For extra practice you can solve Problems 1–4, but to do so will require more time.

consider opening only two types of branches: 20-person branches (type A) and 10-person branches (type B). Expected initial cash outlays are $1,300,000 for a type A branch and $670,000 for a type B branch. The expected annual profit is $92,000 for a type A branch and $36,000 for a type B branch. Pauley will hire no more than 200 employees for the new branch offices and will not open more than 20 branch offices. Use linear programming to help decide how many branch offices should be opened.

In a system of equations for a linear programming model, which of the following equations would **not** represent a constraint (restriction)?

A. $A + B \leq 20$
B. $20A + 10B \leq 200$
C. $\$92,000A + \$36,000B \leq \$128,000$
D. $\$1,300,000A + \$670,000B \leq \$10,400,000$

CPA Exam
November 1975

10. Patsy, Inc., manufactures two products, X and Y. Each product must be processed in each of three departments: machining, assembling, and finishing. The hours needed to produce one unit of product per department and the maximum possible hours per department are:

Department	Production hours per unit X	Production hours per unit Y	Maximum capacity in hours
Machining	2	1	420
Assembling	2	2	500
Finishing	2	3	600

Other restrictions are:

$$X \geq 50$$
$$Y \geq 50$$

The objective function is to maximize profits where profit $= \$4X + \$2Y$. Given the objective and constraints, what is the most profitable number of units of X and Y, respectively, to manufacture?

4.6 Cumulative Review for Chapters 2–4

In Problems 1–4, let

$$A = \begin{bmatrix} 4 & 0 & -1 \\ 3 & -2 & 1 \\ 0 & 3 & -2 \end{bmatrix} \quad B = \begin{bmatrix} 4 & -1 & 3 \\ -3 & 1 & 2 \\ 2 & -3 & 0 \end{bmatrix} \quad I = \begin{bmatrix} 1 & 0 & 0 \\ 0 & 1 & 0 \\ 0 & 0 & 1 \end{bmatrix}$$

Perform the indicated operations.

1. AB **2.** $A(B + I)$ **3.** $I(2A + B)$ **4.** A^{-1}

Solve the systems of equations or inequalities in Problems 5–10.

5. $\begin{cases} x + y = 3 \\ 2x - y = 9 \end{cases}$ **6.** $\begin{cases} y = 2x + 1 \\ y = 3x - 5 \\ y = x + 7 \end{cases}$ **7.** $\begin{cases} x + 2y + 3z = 2 \\ 2x - y + z = -1 \end{cases}$

8. $\begin{cases} 3x + y + z = -6 \\ x + 2y - z = 9 \\ 5x + y - 3z = -4 \end{cases}$ **9.** $\begin{cases} 3x + 2y < -3 \\ x - y > 0 \\ x \qquad \le 0 \end{cases}$ **10.** $\begin{cases} 2x + 3y \le 60 \\ x + 2y \le 36 \\ 3x - 2y \le 24 \\ x \ge 0, \quad y \ge 0 \end{cases}$

11. Maximize: $23x_1 + 24x_2$

Subject to: $\begin{cases} 4x_1 + 2x_2 \le 1800 \\ 2x_1 + 3x_2 \le 1200 \\ 5x_1 + 4x_2 \le 2400 \\ x_1 \ge 0, \quad x_2 \ge 0 \end{cases}$

Solve the linear programming problems in Problems 12–15.

12. A hospital wishes to provide a diet for its patients that has a minimum of 50 grams of carbohydrates, 30 grams of proteins, and 40 grams of fats per day. These requirements can be met with two foods, A and B. Food A costs 14¢ per ounce and supplies 6 grams of carbohydrates, 3 grams of proteins, and 1 gram of fats per ounce, and food B costs 6¢ per ounce and supplies 2 grams of carbohydrates, 2 grams of proteins, and 2 grams of fats per ounce. How many ounces of each food should be bought for each patient per day in order to meet the minimum requirements at the lowest cost?

13. An island is inhabited by three species of animals, A, B, and C, which feed on three types of food, F1, F2, and F3. The amount of each type of food (in ounces) to sustain one animal of each type is summarized in the table:

Animal		Food	
	F1	F2	F3
A	12	14	16
B	15	20	5
C	5	8	6

If the island contains 2000 ounces of food F1, 12,000 ounces of food F2, and 8000 ounces of food F3, what is the maximum number of animals that can be supported?

14. Karlin Manufacturing produces two types of children's toys, wagons and carts. Each product must be processed in each of three departments: machining, assembling, and finishing. The hours needed to produce one unit of each product per department and the maximum possible hours per department are shown in the table:

Department	Production Time Per Unit (hr)		Maximum Capacity (hr)
	Wagons	Carts	
Machining	2	1	2000
Assembling	2	2	2400
Finishing	1	2	3000

Karlin's net profit is $5 per wagon and $4 per cart. How many wagons and carts should be manufactured to maximize the profit?

15. Of the twenty or so amino acids making up protein, eight cannot be synthesized by humans and are known as the *essential amino acids*. A diet that provides adequate amounts of the three amino acids tryptophan (TRP), lysine (LYS), and methionine (MET) will also generally provide enough of the other essential amino acids. The table shows three sources of these amino acids and the approximate number of grams of each per pound.

| | Amino Acid | | |
Food	TRP	LYS	MET
Beef	1.2	8.0	2.4
Peanuts	1.5	5.0	1.2
Cashews	2.0	3.0	1.5

If peanuts cost $2.50 per pound and cashews cost $5.00 per pound, find the cheapest combination of peanuts and cashews that contains an amount of the three amino acids equal to or greater than the amount in a pound of beef.

The following two questions were taken from recent CPA examinations given by the American Institute of Certified Public Accountants, Inc. The questions are reproduced with permission.

**CPA Exam
November 1974**   The Golden Hawk Manufacturing Company wants to maximize the profits on products A, B, and C. The contribution margin for each product follows:

Product	Contribution margin
A	$2
B	$5
C	$4

The production requirements and departmental capacities, by departments, are as follows:

Department	Production requirements by product (hr)			Departmental capacity (total hr)
	A	B	C	
Assembling	2	3	2	30,000
Painting	1	2	2	38,000
Finishing	2	3	1	28,000

16. What is the profit maximization formula for the Golden Hawk Company?
   A. $2A + $5B + $4C = x$,   where $x =$ profit
   B. $5A + 8B + 5C \leq 96{,}000$
   C. $2A + $5B + $4C \leq x$,   where $x =$ profit
   D. $2A + $5B + $4C = 96{,}000$

17. What is the constraint for the painting department of the Golden Hawk Company?
   A. $1A + 2B + 2C \geq 38{,}000$ **B.** $2A + $5B + $4C \geq 38{,}000$
   C. $1A + 2B + 2C \leq 38{,}000$ **D.** $2A + 3B + 2C \leq 30{,}000$

The last three questions are taken from an exam given by the Institute of Management Accounting of the National Association of Accountants. The questions are reproduced by permission.

**CMA Exam
December 1979**   The Elon Co. manufactures two industrial products: X-10, which sells for $90 a unit, and Y-12, which sells for $85 a unit. Each product is processed through both of the company's manufacturing departments. The limited availability of labor, material, and equipment capacity has restricted the ability of the firm to meet the demand for its products. The production department believes that linear programming can be used to routinize the production schedule for the two products.

The following data are available to the production department:

	Amount required per unit	
	X-10	Y-12
Direct material: Weekly supply is limited to 1800 pounds at $12 per pound	4 lb	2 lb
Direct labor:		
Department 1—weekly supply is limited to 10 people at 40 hours each at an hourly cost of $6	$\frac{2}{3}$ hr	1 hr
Department 2—weekly supply is limited to 15 people at 40 hours each at an hourly rate of $8	$1\frac{1}{4}$ hr	1 hr
Machine time:		
Department 1—weekly capacity is limited to 250 hours	$\frac{1}{2}$ hr	$\frac{1}{2}$ hr
Department 2—weekly capacity is limited to 300 hours	0 hr	1 hr

The overhead costs for Elon are accumulated on a plantwide basis. The overhead is assigned to products on the basis of the number of direct labor hours required to manufacture the product. This base is appropriate for overhead assignment because most of the variable overhead costs vary as a function of labor time. The estimated overhead cost per direct labor hour is:

Variable overhead cost	$ 6
Fixed overhead cost	6
Total overhead cost per direct labor hour	$12

The production department formulated the following equations for the linear programming statement of the problem:

A = number of units of X-10 to be produced

B = number of units of Y-12 to be produced

Objective function to minimize costs:

Minimize: $Z = 85A + 62B$

Constraints:

Material	$4A + 2B \leq 1800$ lb
Department 1 labor	$\frac{2}{3}A + 1B \leq 400$ hr
Department 2 labor	$1\frac{1}{4}A + 1B \leq 600$ hr
Nonnegativity	$A \geq 0, \quad B \geq 0$

18. The formulation of the linear programming equations as prepared by Elon Co.'s production department is incorrect. Explain what errors have been made in the formulation prepared by the production department.

19. Formulate and label the proper equations (i.e., the simplex tableau) for the linear programming statement of Elon Co.'s production problem.

20. Explain how linear programming could help Elon Co. determine how large a change in the price of direct materials would have to be to change the optimum production mix of X-10 and Y-12.

Modeling Application 4

Ecology: Maintaining a Profitable Ecological Balance*

In the Edwards Plateau country of west Texas the vegetation is easily modified by grazing animals from a mixed vegetation to a dominance of grasses, or forbs, or browse, or various combinations of these. It is common to see in the pastures herds of cattle, bands of sheep, and flocks of mohair goats. White-tail deer and wild turkeys are common if there is sufficient browse and mixed vegetation. Catfish thrive in the ponds where there is suitable vegetation cover to prevent siltation. Ranchers in the area sell beef, wool, mutton, and mohair, and lease deer and turkey hunting rights, as well as catfishing rights. The relative monetary income value per animal are:

Cattle	10	Goats	1
Deer	0.5	Turkeys	0.05
Sheep		1 (wool and mutton combined)	
Fish	0.001		

A rancher owns 10 sections of land with a maximum carrying capacity of ten animal units per section per year. Suppose that the animal unit equivalents for the various species per animal are:

Cattle	1	Goats	0.25	Sheep	0.2
Deer	0.3	Turkeys	0	Fish	0

To properly organize the operation for livestock production, the rancher has to have at least 20 cattle and at least 20 goats on the ranch. The rancher also wants some sheep and some deer. The desired vegetation cover for turkey and fish can be maintained if there are: (1) cattle, sheep, goats, and deer; or (2) cattle, sheep, and goats; or (3) cattle, goats, and deer—but no more than 75% of the grazing load (measured in animal units) may be due to cattle and goats combined.

Of course, the rancher wants the total stocking rate of all organisms combined to be equal to or less than the carrying capacity of the range.

Other constraints are that there can be no more than one pond per section, each of which will support no more than 500 fish. Only 25% of the catfish per pond can be harvested each year. There cannot be more than two flocks of 10 wild turkeys per flock per section (with no more than 20% of the turkeys harvested each year).

---

* This modeling application is adapted from G. M. Van Dyne and K. R. Rebman, "Maintaining a Profitable Ecological Balance," in Robert M. Thrall, ed., *Some Mathematical Models in Biology* (Ann Arbor: University of Michigan Press, 1967).

Only castrated male goats are kept for mohair, and they are sheared once a year. The cattle, sheep, and deer harvests, which maintain each population, are about 25%, 35%, and 15% of the population per year, respectively. (The 35% for sheep includes both wool and mutton).

Write a paper discussing how many of each animal should be stocked by the rancher in order to maximize profit. For general guidelines, see the commentary for Modeling Application 1 on page 36.

CHAPTER 5
Combinatorics

APPLICATIONS

Management (*Business, Economics, Finance, and Investments*)

Analysis of sales representatives (5.1, Problem 55)
Comparison of inflation and unemployment rates (5.1, Problem 56)
Different categories of employees within the Ampex Corp. (5.1, Problem 58)
Comparison of number of executives with business degrees and MBA degrees (5.1, Problem 61)
Ways of rating businesses (5.3, Problem 34)
Grouping stocks (5.3, Problem 37)
Likelihood of an IRS investigation (5.3, Problem 39)

Life sciences (*Biology, Ecology, Health, and Medicine*)

Surveys of exercise preferences (5.1, Problems 62–64)
Survey of shampoo preferences (5.1, Problem 63)
Menu selections (5.2, Problem 43)
Human blood types (5.5, Problem 15)

Social sciences (*Demography, Political Science, Population, Psychology, Society, and Sociology*)

Survey of voter preferences (5.1, Problem 65)
Survey of TV show preferences (5.1, Problem 66)
Election problem (5.2, Problems 34–36; 5.5, Problems 10–11)
Social Security numbers (5.2, Problem 44)
License plate problem (5.2, Problems 45–49, 51)
Committee problem (5.3, Problems 13, 32–33, 42–43)
Grouping individuals in a psychology experiment (5.3, Problem 35)
Subset problem (5.4, Problems 34–35)

General interest

Likelihood of purchasing a defective tire (5.1, Problem 59)
Historical questions dealing with John Venn (5.1, Problem 71)
Selecting outfits from a wardrobe (5.2, Problems 37–38)
Transfinite numbers (5.2, Problem 53)
Ways of obtaining different poker hands (5.3, Problems 14–17, 22)
Pascal's triangle (5.4, Problems 39–42)
Number of ways of answering a true/false and a multiple-choice test (5.5, Problems 13–14)

Modeling application—Earthquake Epicenter

CHAPTER OVERVIEW

This chapter focuses on sets and counting techniques. In addition, you will learn about the fundamental counting principle and two very useful counting models, permutations and combinations. This chapter concludes with the binomial theorem.

PREVIEW

The idea of a set is one of those very simple, and profound, concepts that is used to simplify and unify mathematics. After looking at sets, set operations, and set relationships in the first section, we move to the notion of counting and develop the ideas of permutations, combinations, and the fundamental counting principle. Make a special note of the box on page 184 because being able to tell *how* to proceed is just as important as carrying out the actual process.

PERSPECTIVE

We all learned to count in a straightforward "one, two, three, . . ." procedure. But in mathematics, especially when working with probabilities, it is necessary to count the number of elements in involved sets, and straightforward counting is often difficult or even impossible. This chapter involves preliminary, but very necessary, work for the next chapter, on probability.

5.1 Sets and Set Operations

The concept of sets and set theory is attributed to Georg Cantor (1845–1918), a great German mathematician. Cantor's ideas were not adopted as fundamental underlying concepts in mathematics until the twentieth century, but today the idea of set is used to unify and explain nearly every other concept in mathematics. Recall that a counting number is an element of the set $\{1, 2, 3, \ldots\}$.

Sets are specified by roster or by description. In the **roster method**, sets are specified by listing the elements and enclosing those elements between braces: { }. In the **description method**, sets are described in words, such as

> The set of subscribers to the *Wall Street Journal*

or by using **set-builder notation**, such as

> $\{s \mid 0 < s < 100, s \text{ a counting number}\}$

The proper way to read the set-builder notation here is: "the set of all counting numbers, s, such that s is between 0 and 100."

If S is a set, we write $a \in S$ if a is a member of the set S and $b \notin S$ if b is not a member of S. For example, let C be the set of cities in California. If a represents the city of Anaheim and b stands for the city of Berlin, Germany, we would write

> $a \in C$ and $b \notin C$

Two special sets require attention. The first is the set that contains no elements and the second is a set that contains every element under consideration:

Empty and Universal Sets

> The **empty set** contains no elements and is denoted by { } or $\varnothing$.
>
> The **universal set** contains all the elements under consideration in a given discussion and is denoted by U.

For example, if we agree that $U = \{1, 2, 3, 4, 5, 6, 7, 8, 9\}$, then all sets that are considered would have elements only among the elements of U. No set could contain the number 10 since 10 is not included in what we have decided is the universe. For every problem, a universal set must be specified or implied, and it must remain fixed for that problem. However, when a new problem is begun, a new universal set can be specified.

EXAMPLE 1 Let U be the set of counting numbers. Write the given sets by roster.

$$A = \{x \,|\, (x - 2)(x - 3) = 0\}$$
$$B = \{y \,|\, 3 \le y < 7\}$$
$$C = \{y \,|\, y^2 = 4\}$$

Solution By roster,

$$A = \{2, 3\}$$
$$B = \{3, 4, 5, 6\}$$
$$C = \{2\} \qquad \textit{Note:} \quad \text{The number } -2 \text{ also makes the statement } y^2 = 4 \text{ true, but } -2$$
$$\text{is not in the universe for this problem} \qquad \blacksquare$$

Let us now take a look at certain relationships among sets.

Subset

> A set A is a **subset** of a set B, denoted by $A \subseteq B$, if every element of A is an element of B.

Consider the following sets:

$$U = \{1, 2, 3, 4, 5, 6, 7, 8, 9\}$$
$$A = \{2, 4, 6, 8\}$$
$$B = \{1, 3, 5, 7\}$$
$$C = \{5, 7\}$$

A, B, and C are subsets of the universal set U (all sets considered are subsets of the universe, by definition of the universal set). Also, $C \subseteq B$ since every element in C is also an element of B.

However, C is not a subset of A. To substantiate this claim, we must show that there is some member of C that is not a member of A. In this case we merely note that $5 \in C$ and $5 \notin A$.

Do not confuse the notions of "element" and "subset". That is, 5 is an *element* of C since it is listed in C; $\{5\}$ is a *subset* of C but is not an element since we do not find $\{5\}$ contained in C—if we did, C might look like $\{5, \{5\}, 7\}$.

To find all the possible subsets of the set $C = \{5, 7\}$ we use what is called a **tree diagram**, which is a device that helps us enumerate a list of

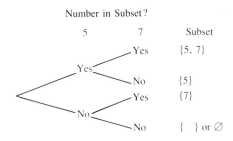

possibilities: The subsets of C are: $\{5, 7\}, \{5\}, \{7\}, \varnothing$. Note that the set C has only two elements, but it has four subsets. Certainly, then, you must be careful to distinguish between a subset and an element.

The subsets of C can be classified into two categories: **proper** and **improper**.

Proper subsets of C	Improper subset of C
$\{5\}$ $\{7\}$ $\{\ \}$	$\{5, 7\}$

Every set has just one improper subset, and that is the set itself. All other subsets are proper subsets. The notation for a proper subset (as distinguished from an improper subset) does not include the small equality symbol; that is,

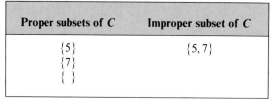

$A \subset B$ Used to denote the idea that A is a *proper subset* of B

The notion of subset can be used to define the equality of two sets. Two sets are equal if they have exactly the same elements:

Equality of Sets   Sets A and B are *equal*, denoted by $A = B$, if $A \subseteq B$ and $B \subseteq A$.

A useful way to depict relationships among sets is to represent the universal set by a rectangle, with the proper subsets in the universe represented by circular or oval-shaped regions, as in Figure 5.1.

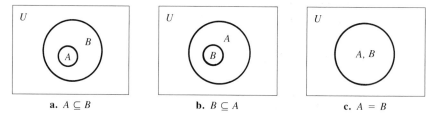

Figure 5.1 Venn diagrams showing subset and equal relationships

a. $A \subseteq B$ **b.** $B \subseteq A$ **c.** $A = B$

These figures are called **Venn diagrams**, after John Venn (1834–1923). The Swiss mathematician Leonhard Euler (1707–1783) also used circles to illustrate principles of logic, so sometimes these diagrams are called *Euler circles*. However, Venn was the first person to use them in a general way.

We can also illustrate other relationships between two sets: A and B may have no elements in common, in which case they are **disjoint**, or they may overlap and have some elements in common (Figure 5.2).

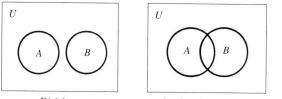

Figure 5.2 Venn diagrams for disjoint and overlapping sets

a. Disjoint sets   **b.** Overlapping sets

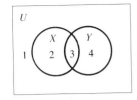

Figure 5.3 Venn diagram showing two general sets

Sometimes we are given two sets X and Y and we know nothing about how they are related. In this situation we draw a general figure, such as the one shown in Figure 5.3.

Note that these circles divide the universe into four disjoint regions. When sets are drawn in this manner it does not mean that the only possibility is overlapping sets. For example:

If $X \subseteq Y$, then region 2 is empty.
If $Y \subseteq X$, then region 4 is empty.
If $X = Y$, then regions 2 and 4 are both empty.
If X and Y are disjoint, then region 3 is empty.

Three common operations performed on sets are union, intersection, and complementation.

Union is an operation for sets A and B in which a set is formed that consists of all the elements in A or B or both. The symbol for the operation of union is $\cup$, and we write $A \cup B$.

Union of Sets

The *union* of sets A and B, denoted by $A \cup B$, is the set consisting of all elements of A or B or both.
$$A \cup B = \{x \mid x \in A \text{ or } x \in B \text{ or } x \in (\text{both } A \text{ and } B)\}$$

EXAMPLE 2   Let $U = \{1, 2, 3, 4, 5, 6, 7, 8, 9\}$, $A = \{2, 4, 6, 8\}$, $B = \{1, 3, 5, 7\}$, $C = \{5, 7\}$. Then:

a. $A \cup C = \{2, 4, 5, 6, 7, 8\}$; that is, the union of A and C is the set consisting of all elements in A or in C or in both.
b. $B \cup C = \{1, 3, 5, 7\}$; note that, even though the elements 5 and 7 appear in both sets, they are listed only once. That is, the sets $\{1, 3, 5, 7\}$ and $\{1, 3, 5, 5, 7, 7\}$ are equal (exactly the same).
c. $A \cup B = (1, 2, 3, 4, 5, 6, 7, 8)$.
d. $(A \cup B) \cup \{9\} = \{1, 2, 3, 4, 5, 6, 7, 8, 9\} = U$; here we are considering the union of three sets; however, the parentheses indicate the operation that should be performed first. Note also that, because the solution $\{1, 2, 3, 4, 5, 6, 7, 8, 9\}$ has a name, we write down the name rather than the set. That is, $(A \cup B) \cup \{9\} = U$ is the simpler representation.   ∎

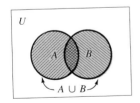

Figure 5.4 Venn diagram showing the union of two sets (color shading)

We can use Venn diagrams to illustrate union. In Figure 5.4 we first shade A and then shade B. *The union is all parts that have been shaded at least once.*

A second operation for sets is called **intersection**.

Intersection of Sets

The *intersection* of sets A and B, denoted by $A \cap B$, is the set consisting of all elements common to A and B.
$$A \cap B = \{x \mid x \in A \quad and \quad x \in B\}$$

EXAMPLE 3   Let $U = \{a, b, c, d, e\}$, $A = \{a, c, e\}$, $B = \{c, d, e\}$, $C = \{a\}$, and $D = \{e\}$. Then:

a. $A \cap B = \{c, e\}$; that is, the intersection of A and B is the set consisting of elements in both A and B.

b. $A \cap C = C$ since $A \cap C = \{a\}$ and $\{a\} = C$; we write down the name for the set that is the intersection.

c. $B \cap C = \varnothing$ since B and C have no elements in common (they are disjoint).

d. Parentheses tell us which operation to do first:

$$(A \cap B) \cap D = \{c, e\} \cap \{e\}$$
$$= \{e\} \qquad \text{\{e\} is the set of elements common to the sets \{c, e\}}$$
$$= D \qquad \text{and \{e\}}$$

∎

Intersection of sets can also be easily shown in a Venn diagram. To find the intersection of two sets A and B, first shade A and then shade B; *the intersection is all parts shaded twice* (Figure 5.5).

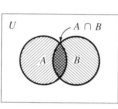

Figure 5.5 Venn diagram showing the intersection of two sets (color shading)

Complementation is an operation on a set that must be performed in relation to the universal set.

Complementation

> The *complement* of a set A, denoted by $\bar{A}$, is the set of all elements of U that are not in the set A.
>
> $$\bar{A} = \{x \mid x \notin A\}$$

EXAMPLE 4   Let $U = \{\text{People in California}\}$, $A = \{\text{People who are over 30}\}$, $B = \{\text{People who are 30 or under}\}$, and $C = \{\text{People who own a car}\}$. Then:

a. $\bar{C} = \{\text{Californians who do not own a car}\}$

b. $\bar{A} = B$

c. $\bar{B} = A$

d. $\bar{U} = \varnothing$

e. $\bar{\varnothing} = U$

∎

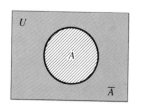

Figure 5.6 Venn diagram showing the complement of a set A (color shading)

Complementation can be shown in a Venn diagram. The color shading in Figure 5.6 shows the complement of A. In a Venn diagram, the complement is everything in U that is not in the given set (in this case, everything not in A).

The real payoff for studying these relationships comes when we have combined operations or several sets at the same time.

EXAMPLE 5   Illustrate $\overline{A \cup B}$ in a Venn diagram.

Solution   This is a combined operation. *First,* find $A \cup B$; *then* find the complement.

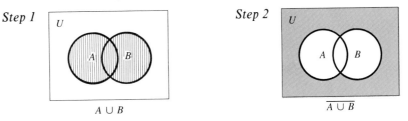

Step 1 $A \cup B$ *Step 2* $\overline{A \cup B}$

These steps are generally combined into one diagram:

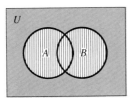

The answer is shown by color shading. The vertical lines show the preliminary step. It is generally a good idea to show the final answer (the part in color) by using a highlighter pen. ■

EXAMPLE 6   Illustrate $\overline{A} \cup \overline{B}$ in a Venn diagram.

Solution   Compare this statement with Example 5. Here, we *first* find $\overline{A}$ and $\overline{B}$; *then* find the union.

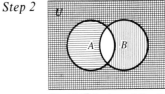

Step 1 $\overline{A}$ (vertical lines)   *Step 2* $\overline{A}$ with $\overline{B}$ (horizontal lines)

Your work will normally show the above steps with one diagram.

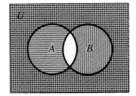

$\overline{A} \cup \overline{B}$ (color shading; the union is all parts that have horizontal or vertical lines or both) ■

Notice that $\overline{A \cup B} \neq \overline{A} \cup \overline{B}$. If they were equal, the final color portions of the Venn diagrams from Examples 5 and 6 would have been the same.

The *order* of operations for sets is left to right, unless parentheses indicate other-wise. Operations within parentheses are performed first, as shown by Example 7.

EXAMPLE 7   Illustrate $(A \cup C) \cap \bar{C}$ in a Venn diagram.

Solution

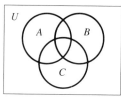

$(A \cup C) \cap \bar{C}$

Detail:

$A \cup C$:   Vertical lines
$\bar{C}$:   Horizontal lines
$(A \cup C) \cap \bar{C}$:   The intersection of vertical and horizontal lines is the part shaded in color. ■

Sometimes you will be asked to consider relationships among three sets. The general Venn diagram for this is shown in Figure 5.7. Note that three sets divide the universe into eight regions. Can you number each?

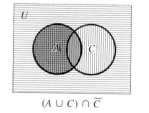

Figure 5.7 Three general sets

EXAMPLE 8   Using Figure 5.7 as a guide, shade in each of the following sets:
a. $A \cup B$ **b.** $\overline{A \cap C}$ **c.** $B \cap C$
d. $\bar{A}$ **e.** $\overline{A \cup B}$ **f.** $A \cap B \cap C$ **g.** $A \cup B \cup C$

Solution

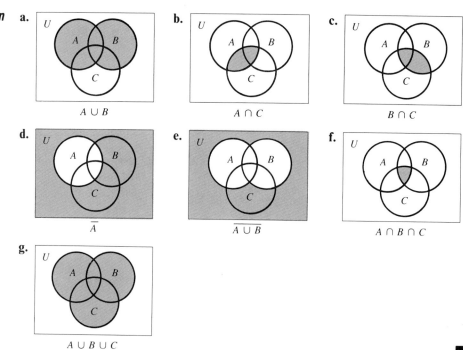

■

Venn diagrams are also used for survey problems. Suppose W. B. Blaire, a marketing research firm, conducts a survey of people buying electronic calculators during a week. It is found that 48 people bought machines with paper tape printout and 71 bought machines without tape printout. How many people purchased machines?

The "obvious" answer of $48 + 71 = 119$ is not so obvious if you consider that some people who were surveyed may have purchased both types of machines. So, let

$A = \{x \mid x$ purchased an electronic calculator with paper tape$\}$

$B = \{x \mid x$ purchased an electronic calculator without paper tape$\}$

Also, **let $n(A)$ denote the number of elements in set A.** Thus $n(A) = 48$. Also, $n(B) = 71$. Now, $n(A \cup B) = 48 + 71 = 119$ if $A \cap B = \varnothing$. But suppose it is found that 19 persons purchased both types of machines. That is, $n(A \cap B) = 19$. Consider the Venn diagram shown in Figure 5.8.

Figure 5.8 Venn diagram for sets A and B. Note that since 19 is in the intersection, then $48 - 19 = 29$ is in the remaining part of A; similarly, there is $71 - 19 = 52$ in the remaining part of B

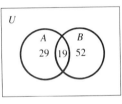

We see that if we seek $n(A \cup B)$, we need to count the number of elements in A plus the number of elements in B. But when we do this, we have counted the number of elements in the intersection twice. Therefore

$$n(A \cup B) = n(A) + n(B) - n(A \cap B)$$
$$= 48 + 71 - 19$$
$$= 100$$

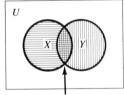

This part has been counted twice

Figure 5.9

In general, for *any* sets X and Y, the number of elements in the union is the sum of the number of elements in each of the individual sets, *minus* the number of elements in the intersection, as shown in Figure 5.9.

Number of Elements in the Union of Sets

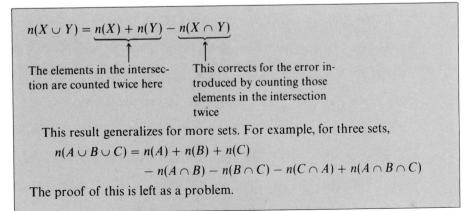

$$n(X \cup Y) = \underline{n(X) + n(Y)} - \underline{n(X \cap Y)}$$

The elements in the intersection are counted twice here

This corrects for the error introduced by counting those elements in the intersection twice

This result generalizes for more sets. For example, for three sets,

$$n(A \cup B \cup C) = n(A) + n(B) + n(C)$$
$$- n(A \cap B) - n(B \cap C) - n(C \cap A) + n(A \cap B \cap C)$$

The proof of this is left as a problem.

EXAMPLE 9   **Survey**   A survey of 470 students gives the following information:

45 students are taking finite math
41 students are taking statistics
40 students are taking computer programming
15 students are taking finite math and statistics
18 students are taking finite math and computer programming
17 students are taking statistics and computer programming
7 students are taking all three

a. How many students are taking only finite math?
b. How many students are taking only statistics?
c. How many students are taking only computer programming?
d. How many students are not taking any of these courses?

Solution   The method of solution is to draw a Venn diagram and fill in the various regions. Let

$F = \{x \mid x \text{ is taking finite math}\}$
$S = \{x \mid x \text{ is taking statistics}\}$
$C = \{x \mid x \text{ is taking computer programming}\}$

Step 1   Fill in $F \cap S \cap C$, which is given: $n(F \cap S \cap C) = 7$.

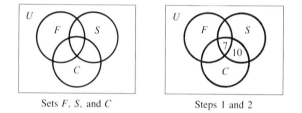

Sets F, S, and C Steps 1 and 2

Step 2   Fill in $S \cap C$, which is given: $n(S \cap C) = 17$. But 7 has previously been accounted for in this region, so we need only account for 10 additional members.

Step 3   Fill in $n(F \cap C) = 18$. Need to fill in only 11 additional members.

Step 4   Fill in $n(F \cap S) = 15$. Need to fill in only 8 additional members.

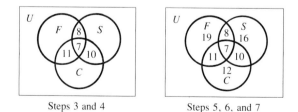

Steps 3 and 4   Steps 5, 6, and 7

Step 5   Fill in $n(C) = 40$. Note that 28 has previously been filled in, so we need an additional 12 members.

Step 6   Fill in $n(S) = 41$. Need to fill in 16 members.

Step 7   Fill in $n(F) = 45$. Need to fill in 19 members.

Step 8   Adding all the numbers in the diagram, we have accounted for 83 members. Since 470 students were surveyed, we see that 387 are not taking any of the three courses.

We now have all the answers to the questions directly from the Venn diagram:

a. 19   **b.** 16   **c.** 12   **d.** 387 ■

Problem Set 5.1

1. Explain the difference between *element of a set* and *subset of a set*. Give examples.
2. Explain the difference between *subset* and *proper subset*.
3. Explain the difference between *universal set* and *empty set*.
4. In your own words, define *union, intersection,* and *complement*. Illustrate with examples.

List all subsets of each set given in Problems 5–12.

5. $\{m, y\}$
6. $\{4, 5\}$
7. $\{y, o, u\}$
8. $\{6, 9, 0\}$
9. $\{3, 6, 9\}$
10. $\{m, a, t, h\}$
11. $\{1, 2, 3, 4\}$
12. $\{a, b, c, d\}$

Perform the set operations in Problems 13–22. Let $U = \{1, 2, 3, 4, 5, 6, 7, 8, 9, 10\}$.

13. $\{2, 6, 8\} \cup \{6, 8, 10\}$
14. $\{2, 6, 8\} \cap \{6, 8, 10\}$
15. $\{1, 2, 3, 4, 5\} \cap \{3, 4, 5, 6, 7\}$
16. $\{1, 2, 3, 4, 5\} \cup \{3, 4, 5, 6, 7\}$
17. $\{2, 5, 8\} \cup \{3, 6, 9\}$
18. $\{2, 5, 8\} \cap \{3, 6, 9\}$
19. $\overline{\{2, 8, 9\}}$
20. $\overline{\{1, 2, 5, 7, 9\}}$
21. $\overline{\{8\}}$
22. $\overline{\{6, 7, 8, 9, 10\}}$

Let $U = \{1, 2, 3, 4, 5, 6, 7\}$, $A = \{1, 2, 3, 4\}$, $B = \{1, 2, 5, 6\}$, and $C = \{3, 5, 7\}$. List all the members of each of the sets in Problems 23–38.

23. $A \cup B$
24. $A \cup C$
25. $B \cup C$
26. $A \cap B$
27. $A \cap C$
28. $B \cap C$
29. $\bar{A}$
30. $\bar{B}$
31. $\bar{C}$
32. $\bar{U}$
33. $(A \cup B) \cup C$
34. $(A \cap B) \cap C$
35. $A \cup (B \cap C)$
36. $\overline{A \cap (B \cup C)}$
37. $\bar{A} \cup (B \cap C)$
38. $\overline{A \cup (B \cap C)}$

39. Determine whether each of the following is true or false:
   a. $\varnothing \in \varnothing$ **b.** $\varnothing \subseteq \varnothing$ **c.** $\varnothing \in \{\varnothing\}$
   d. $\varnothing = \varnothing$ **e.** $\varnothing = \{\varnothing\}$
40. Determine whether each of the following is true or false:
   a. $\varnothing \subseteq \{\varnothing\}$ **b.** $\varnothing \in A$ for all sets A
   c. $\varnothing = 0$ **d.** $\varnothing = \{0\}$ **e.** $\varnothing \subseteq \{0\}$

Draw a Venn diagram for each relationship in Problems 41–54.

41. $X \cup Y$
42. $Y \cup Z$
43. $X \cap Z$
44. $X \cap Y$
45. $\bar{X}$
46. $\bar{Z}$
47. $X \cup \bar{Y}$
48. $\bar{X} \cap Z$
49. $A \cap (B \cup C)$
50. $A \cup (B \cup C)$
51. $\overline{(A \cup B) \cup C}$
52. $\overline{(A \cap B) \cup C}$
53. $A \cap \overline{B \cup C}$
54. $\overline{A \cup B \cup C}$

APPLICATIONS

55. Let A represent the top sales representatives of the year 1987:

   {Bob Wisner, Joan Marsh, Craig Barth, Phyllis Niklas}

   and let B represent the top representatives of the year 1988:

   {Phyllis Niklas, Craig Barth, Shannon Smith, Christy Anton}

   a. What is the set of top sales representatives for 1987 or 1988? What set operation should be used to find this answer?
   b. What is the set of top sales representatives for the 2 years in a row? What set operation should be used to find this answer?

56. The following table gives inflation and unemployment rates for the years 1978–1985:

Year	Inflation rate	Unemployment rate
1978	7.6	6.0
1979	11.3	5.8
1980	13.6	7.1
1981	10.4	7.6
1982	6.1	9.7
1983	3.2	9.6
1984	4.2	7.5
1985	2.8	7.3

   Let $A = \{$years from 1978–1985 in which unemployment was at least 7%$\}$
   $B = \{$years from 1978–1985 in which inflation was at least 6%$\}$
   a. Find: $A, B, A \cup B, A \cap B$.
   b. Describe, in words, the sets $A \cup B$ and $A \cap B$.

57. Let $U = \{$All people in California$\}$
$A = \{$All people over 30$\}$
$B = \{$All people 30 or less$\}$
$C = \{$All people who own a car$\}$

a. Draw a Venn diagram showing how these sets are related.

Determine whether each of the following is true or false:

b. $B \subseteq A$ **c.** $C \subseteq U$ **d.** $B \subseteq U$
e. $C \in U$ **f.** $A \cap B = \emptyset$ **g.** $\bar{A} = B$
h. $A \cap C = \emptyset$ **i.** $\emptyset = U$

58. Let

$U = \{$All employees of Ampex Corporation$\}$
$A =$ Set of executives $= \{$Employees with salaries of $60,000 or more$\}$
$B =$ Set of junior executives $= \{$Employees with salaries between $40,000 and $60,000$\}$
$C =$ Set of nonexecutives $= \{$Employees with salaries of $20,000 and less$\}$
$D =$ Set of white-collar workers $= \{$Employees with salaries of $12,000 or more$\}$

a. Draw a Venn diagram showing how these sets are related.

Determine whether each of the following is true or false:

b. $A \subseteq U$ **c.** $A \cup B = U$ **d.** $D \subseteq A$
e. $\emptyset \subseteq A$ **f.** $C \cap D = \emptyset$ **g.** $\emptyset \in A$
h. $B \cap C = \emptyset$ **i.** $A \cup B \cup C = D$

59. In a sample of defective tires, 72 have defects in materials, 89 have defects in workmanship, and 17 have defects of both types. How many tires are in the sample of defective tires (the sample consists of only tires with defects in materials or workmanship)?

60. In a survey of the executive board of a national charity, it is found that 15 members earn more than $100,000 per year and 9 own more than $1,000,000 in negotiable securities. If 7 earn more than $100,000 and own more than $1,000,000 in negotiable securities, how many members are on the executive board, if all the members fall into one of the two categories?

61. A survey of executives of Fortune 500 companies finds that 520 have MBA degrees, 650 have business degrees, and 450 have both degrees. How many executives with MBAs have nonbusiness undergraduate degrees? (*Note:* The MBA is a graduate degree and requires an undergraduate degree.)

62. A survey of 100 persons at Plimbo Corporation shows that 40 jog, 25 swim, and 15 both swim and jog. How many do neither?

63. A survey of 100 women finds that 59 use shampoo A, 41 use shampoo B, 35 use shampoo C, 24 use shampoos A and B, 19 use shampoos A and C, 13 use shampoos B and C,

and 11 use all three. Let

$A = \{$Women who use shampoo A$\}$
$B = \{$Women who use shampoo B$\}$
$C = \{$Women who use shampoo C$\}$

Use a Venn diagram to show how many are in each of the eight possible categories.

64. A survey of 100 persons at Better Widgets finds that 40 jog, 25 swim, 16 cycle, 15 swim and jog, 10 swim and cycle, 8 jog and cycle, and 3 jog, swim, and cycle. Let

$J = \{$People who jog$\}$
$S = \{$People who swim$\}$
$C = \{$People who cycle$\}$

Use a Venn diagram to show how many are in each of the eight possible categories.

65. In an interview of 50 students: 12 liked Proposition 8 and Proposition 13, 18 liked Proposition 8 but not Proposition 2, 4 liked Proposition 8, Proposition 13, and Proposition 2, 25 liked Proposition 8, 15 liked Proposition 13, 10 liked Proposition 2 but not Proposition 8 or Proposition 13, and 1 liked Proposition 13 and Proposition 2 but not Proposition 8.

a. Of those surveyed, how many did not like any of the three propositions?
b. How many liked Proposition 8 and Proposition 2?
c. Show the completed Venn diagram.

66. To see if Shannon Smith can handle the job he has applied for, the personnel manager sends him out to poll 100 people about their favorite types of TV shows. Shannon obtains the following data: 59 prefer comedies, 38 prefer variety shows, 42 prefer serious drama, 18 prefer comedies or variety programs, 12 prefer variety or serious drama, 16 prefer comedies or serious drama, 7 prefer all types, and 2 do not like any TV shows. If you were the personnel manager, would you hire Shannon on the basis of this survey?

67. If $A \subseteq B$, describe:
a. $A \cup B$ **b.** $A \cap B$ **c.** $\bar{A} \cup B$ **d.** $A \cap \bar{B}$

68. What is the millionth number that is not a perfect square or a perfect cube?

69. What is the millionth number that is not a perfect square, a perfect cube, or a perfect fifth power?

70. Show that

$$n(A \cup B \cup C) = n(A) + n(B) + n(C) - n(A \cap B)$$
$$- n(A \cap C) - n(B \cap C) + n(A \cap B \cap C)$$

71. Historical Question John Venn (1834–1923) used diagrams to illustrate logic in his *Symbolic Logic*, published in 1881. Venn was an ordained priest, but he left the clergy in 1883 to spend all his time teaching and studying logic. In the text we noted that three intersecting circles divide the universe into eight regions. Venn also considered the case of four intersecting circles. Show what this case might look like, and tell how many disjoint regions are produced by four general intersecting circles.

5.2 Permutations

Consider the set

$$A = \{a, b, c, d, e\}$$

Remember, when set symbols are used, the order in which the elements are listed is not important. Suppose now that we wish to select elements from A by taking them *in a certain order*. For example, suppose set A is a club and the members are holding an election for president and secretary. If the first person selected is the president, and the second person selected is the secretary, then selecting members a and b is not the same as selecting members b and a for these positions. Since set notation signifies that the order of the elements is not important, another notation must be used when the order is important. This notation uses parentheses rather than braces. For this example, (a, b) or (b, a) represents the offices of (president, secretary) and $(a, b) \neq (b, a)$. If two elements are selected, then the notation (a, b) is called an **ordered pair**; if three are selected, it is called an **ordered triplet**; if four are selected, it is called an **ordered four-tuple**; and so on. These selections are called **arrangements** of the given set, A in this example. Arrangements are said to be selections from the given set *without repetitions*, since a symbol cannot be selected from the set twice.

EXAMPLE 1 List all possible arrangements of the elements a, c, and d selected from A.

Solution (a, c, d), (a, d, c), (c, a, d), (c, d, a), (d, a, c), (d, c, a) ∎

Permutation

> A *permutation* of r elements of a set S with n elements is an ordered arrangement of those r elements selected without repetitions.

EXAMPLE 2 How many permutations of two elements can be selected from a set of six elements?

Solution Let $B = \{a, b, c, d, e, f\}$ and select two elements:

$$
\left.
\begin{array}{ccccc}
(a, b) & (a, c) & (a, d) & (a, e) & (a, f) \\
(b, a) & (b, c) & (b, d) & (b, e) & (b, f) \\
(c, a) & (c, b) & (c, d) & (c, e) & (c, f) \\
(d, a) & (d, b) & (d, c) & (d, e) & (d, f) \\
(e, a) & (e, b) & (e, c) & (e, d) & (e, f) \\
(f, a) & (f, b) & (f, c) & (f, d) & (f, e)
\end{array}
\right\}
$$

There are 30 permutations of two elements selected from a set of six elements ∎

Example 2 brings up two difficulties. The first is the lack of notation for the phrase

*the number of permutations of two elements
selected from a set of six elements*

and the second is the inadequacy of relying on direct counting, especially if the sets are very large.

Notation for Permutations

$_nP_r$ is a symbol used to denote the *number of permutations* of r elements selected from a set of n elements.

Example 2 can now be shortened by writing

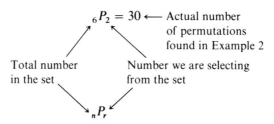

$_6P_2 = 30$ ⟵ Actual number of permutations found in Example 2

Total number in the set

Number we are selecting from the set

$_nP_r$

Next, to find a formula for $_nP_r$, we turn to a general result called the *fundamental counting principle*. We begin by considering the methods of solution illustrated in the following example.

EXAMPLE 3   Consider a club with five members:

{Alfie, Bogie, Calvin, Doug, Ernie}

In how many ways can they elect a president and secretary?

Solution   There are at least two ways to solve this problem. The first, and perhaps the easiest, is by drawing a tree diagram:

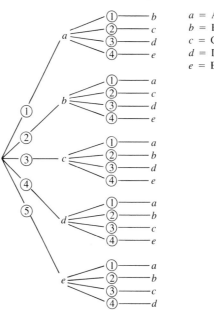

President   Secretary

a = Alfie
b = Bogie
c = Calvin
d = Doug
e = Ernie

By counting the number of branches in the tree, we see that there are 20 possibilities. This method is effective for small numbers, but the technique quickly gets out of hand. For example, if we wished to see how many ways the club could elect a president, secretary, and treasurer, this technique would be very lengthy.

A second method of solution is by using boxes or "pigeonholes" to represent each choice separately. For example:

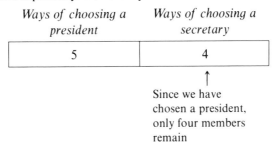

If we multiply the numbers in the pigeonholes, we get

$$5 \cdot 4 = 20$$

and we see that the result is the same as that from the tree diagram. This pigeonhole procedure is called the **fundamental counting principle**. ∎

Fundamental Counting Principle	If task *A* can be performed in *m* ways, and, after task *A* is performed, a second task *B* can be performed in *n* ways, then task *A* followed by task *B* can be performed in *m* · *n* ways.

The fundamental counting principle can be used repeatedly in a particular problem so that it can be applied not only to two tasks, but also to three tasks, four tasks, or any number of tasks.

EXAMPLE 4 How many ways can the club in Example 3 select a president, secretary, and treasurer?

Solution Continue from where we left off in Example 3 and use the fundamental counting principle a second time:

$$\underset{\text{President}}{5} \cdot \underset{\text{Secretary}}{4} \cdot \underset{\text{Number of ways of selecting a treasurer}}{3} = 60$$

From Example 3

Using permutation notation, we write

$$_5P_3 = 5 \cdot 4 \cdot 3 = 60$$
∎

Can you draw a tree diagram for this example?

EXAMPLE 5 $_7P_3 = \underset{\text{Three factors}}{7 \cdot 6 \cdot 5} = 210 \qquad _{10}P_4 = \underset{\text{Four factors}}{10 \cdot 9 \cdot 8 \cdot 7} = 5040$ ∎

In general, the number of permutations of n objects taken r at a time is given by the following formula:

$$_nP_r = \underbrace{n \cdot (n-1) \cdot (n-2) \cdot \cdots \cdot (n-r+1)}_{r \text{ factors}}$$

EXAMPLE 6 $_4P_4 = 4 \cdot 3 \cdot 2 \cdot 1 = 24$ $_4P_4$ is a permutation of 4 objects

$_5P_5 = 5 \cdot 4 \cdot 3 \cdot 2 \cdot 1 = 120$ $_5P_5$ is a permutation of 5 objects

$_6P_6 = 6 \cdot 5 \cdot 4 \cdot 3 \cdot 2 \cdot 1 = 720$ $_6P_6$ is a permutation of 6 objects ∎

In our work with permutations, we will frequently encounter products such as

$6 \cdot 5 \cdot 4 \cdot 3 \cdot 2 \cdot 1$

$10 \cdot 9 \cdot 8 \cdot 7 \cdot 6 \cdot 5 \cdot 4 \cdot 3 \cdot 2 \cdot 1$

$52 \cdot 51 \cdot 50 \cdot 49 \cdot \cdots \cdot 4 \cdot 3 \cdot 2 \cdot 1$

Since these are tedious to write out, **factorial notation** is used:

Definition of Factorial

$n!$ is called n *factorial* and is defined by

$$n! = n(n-1)(n-2) \cdot \cdots \cdot 3 \cdot 2 \cdot 1$$

for n a natural number. Also, $0! = 1$.

Thus, we see from Example 6 that there are $n!$ permutations of n objects.

EXAMPLE 7 Find the factorial of each whole number up to 10.

$0! = \mathbf{1}$

$1! = \mathbf{1}$

$2! = 1 \cdot 2 = \mathbf{2}$

$3! = 1 \cdot 2 \cdot 3 = \mathbf{6}$

$4! = 1 \cdot 2 \cdot 3 \cdot 4 = \mathbf{24}$

$5! = 1 \cdot 2 \cdot 3 \cdot 4 \cdot 5 = \mathbf{120}$

$6! = 1 \cdot 2 \cdot 3 \cdot 4 \cdot 5 \cdot 6 = \mathbf{720}$

$7! = 1 \cdot 2 \cdot 3 \cdot 4 \cdot 5 \cdot 6 \cdot 7 = \mathbf{5040}$

$8! = 1 \cdot 2 \cdot 3 \cdot 4 \cdot 5 \cdot 6 \cdot 7 \cdot 8 = \mathbf{40{,}320}$

$9! = 1 \cdot 2 \cdot 3 \cdot \cdots \cdot 8 \cdot 9 = \mathbf{362{,}880}$

$10! = 1 \cdot 2 \cdot 3 \cdot \cdots \cdot 9 \cdot 10 = \mathbf{3{,}628{,}800}$ ∎

Calculator note: Look for a $\boxed{!}$ or $\boxed{x!}$ key on your calculator. This is a factorial key. Verify one of the entries above by using your calculator.

EXAMPLE 8 $5! - 4! = 120 - 24$ PRESS: $\boxed{5}\ \boxed{x!}\ \boxed{-}\ \boxed{4}\ \boxed{x!}\ \boxed{=}$

$= 96$ ∎

EXAMPLE 9 $(5 - 4)! = 1!$ PRESS: $\boxed{5}\ \boxed{-}\ \boxed{4}\ \boxed{=}\ \boxed{x!}$

$= 1$ ∎

EXAMPLE 10 $(2 \cdot 3)! = 6!$

$= 720$ ∎

EXAMPLE 11 $2!3! = 2 \cdot 1 \cdot 3 \cdot 2 \cdot 1$

$= 12$ ∎

EXAMPLE 12 $\dfrac{8!}{4!} = \dfrac{8 \cdot 7 \cdot 6 \cdot 5 \cdot \not{4} \cdot \not{3} \cdot \not{2} \cdot \not{1}}{\not{4} \cdot \not{3} \cdot \not{2} \cdot \not{1}}$

$= 8 \cdot 7 \cdot 6 \cdot 5$

$= 1680$ ∎

EXAMPLE 13 $\dfrac{100!}{98!} = \dfrac{100 \cdot 99 \cdot \not{98!}}{\not{98!}}$ Note that $100! = 100 \cdot 99!$

$= 100 \cdot 99 \cdot 98!$

$= 9900$ ∎

As the problems use larger numbers, you will find it necessary to simplify first and then use your calculator, if necessary. Example 13, for example, cannot be worked on a calculator without first simplifying.

EXAMPLE 14 $\dfrac{8!}{3!(8 - 3)!} = \dfrac{8!}{3!5!}$

$= \dfrac{8 \cdot 7 \cdot \not{6} \cdot \not{5}!}{\not{3} \cdot \not{2} \cdot 1 \cdot \not{5}!}$

$= 56$ ∎

Examples 13 and 14 illustrate a useful property of factorial.

Multiplication Property of Factorial

$$n! = n(n - 1)!$$

This property was used in Example 13:

$$100! = 100 \cdot 99! \quad \text{and} \quad 99! = 99 \cdot 98!$$

Thus $100! = 100 \cdot 99 \cdot 98!$. This means that when using factorial notation, we can "count down" to any convenient number and then affix a factorial symbol. See Example 14, where we used $8! = 8 \cdot 7 \cdot 6 \cdot 5!$.

Using factorials, the formula for $_nP_r$ can be written more simply as follows:

$$_nP_r = n(n-1)(n-2)\cdots(n-r+1)$$

$$= n(n-1)(n-2)\cdots(n-r+1)\cdot\frac{(n-r)!}{(n-r)!}$$

$$= \frac{n(n-1)(n-2)\cdots(n-r+1)(n-r)(n-r-1)\cdots\cdot 3\cdot 2\cdot 1}{(n-r)!}$$

$$= \frac{n!}{(n-r)!}$$

This is the general formula for $_nP_r$.

Permutation Formula

$$_nP_r = \frac{n!}{(n-r)!}$$

EXAMPLE 15
$$_{10}P_2 = \frac{10!}{(10-2)!}$$
$$= \frac{10\cdot 9\cdot 8!}{8!}$$
$$= 90 \qquad \blacksquare$$

EXAMPLE 16
$$_nP_0 = \frac{n!}{(n-0)!}$$
$$= 1 \qquad \blacksquare$$

EXAMPLE 17 Find the number of license plates possible in a state using only three letters, if none of the letters can be repeated.

Solution This is a permutation of 26 objects taken 3 at a time. Thus the solution is given by

$$_{26}P_3 = \underbrace{26\cdot 25\cdot 24}_{\text{Three factors}} = 15{,}600 \qquad \blacksquare$$

EXAMPLE 18 Repeat Example 17 if repetitions are allowed.

Solution Remember, permutations do not allow repetitions, so this is *not* a permutation. However, the fundamental counting principle still applies:

$$26\cdot 26\cdot 26 = 17{,}576 \qquad \blacksquare$$

EXAMPLE 19 Find the number of arrangements of letters in the word MATH.

Solution This is a permutation of four objects taken 4 at a time:

$$_4P_4 = 4\cdot 3\cdot 2\cdot 1 = 24 \qquad \blacksquare$$

EXAMPLE 20 How many permutations are there of the letters in the word HATH?

Solution If you try to solve this problem as you did Example 19, you would have

$$_4P_4 = 4! = 24$$

However, if you list the possibilities, you would find only *12 different permutations.* The difficulty here is that two of the letters in the word HATH are indistinguishable. If you label them as H_1ATH_2, you would find such possibilities as

$$H_1ATH_2 \qquad H_2ATH_1 \qquad H_1AH_2T$$

If you complete this list, you would indeed find $4! = 24$ possibilities. But since there are *two indistinguishable* letters, you divide the total, 4!, by 2 to find the result:

$$\frac{4!}{2} = \frac{24}{2} = 12 \qquad\qquad \blacksquare$$

EXAMPLE 21 How many permutations are there of the letters in the word ASSIST?

Solution There are six letters, and if you consider the letters as distinguishable, as in

$$AS_1S_2IS_3T$$

there are $_6P_6 = 6! = 720$ possibilities. However,

$$AS_1S_2IS_3T \qquad AS_1S_3IS_2T \qquad AS_2S_1IS_3T$$
$$AS_2S_3IS_1T \qquad AS_3S_1IS_2T \qquad AS_3S_2IS_1T$$

are all indistinguishable, so you must divide the total by $3! = 6$.

$$\frac{6!}{3!} = \frac{6 \cdot 5 \cdot 4 \cdot 3!}{3!} = 120$$

There are 120 permutations of the letters in the word ASSIST. $\qquad \blacksquare$

Examples 20 and 21 suggest a general result:

Number of Distinguishable Permutations

> The number of **distinguishable permutations** of n objects, of which r are alike and the remaining objects are different from each other, is
>
> $$\frac{n!}{r!}$$

This result generalizes to include several subcategories.

EXAMPLE 22 The number of permutations of the letters in the word ATTRACT is 7! (since there are seven letters) divided by factorials of the number of subcategories of repeated letters.

$$\frac{7!}{3!\,2!\,1!\,1!} \quad \leftarrow \text{Total number of objects}$$

Letters R and C occur once

Letter T occurs three times

Letter A occurs twice

This number can now be simplified:

$$\frac{7 \cdot 6 \cdot 5 \cdot \overset{2}{\cancel{4}} \cdot \cancel{3!}}{\cancel{3!} \cdot \cancel{2}} = 420 \qquad\qquad \blacksquare$$

General Formula for the Number of Distinguishable Permutations

> The number of *distinguishable permutations* of n objects in which n_1 are of one kind, n_2 are of another kind,..., and n_k are of a further kind so that
>
> $$n = n_1 + n_2 + \cdots + n_k$$
>
> is given by the formula
>
> $$\frac{n!}{n_1! n_2! \cdots n_k!}$$

Problem Set 5.2

Evaluate each expression in Problems 1–18.

1. $6! - 4!$ **2.** $7! - 3!$ **3.** $8! - 5!$

4. $(6 - 4)!$ **5.** $(7 - 3)!$ **6.** $(8 - 5)!$

7. $4! \, 3!$ **8.** $(4 \cdot 3)!$ **9.** $(5 \cdot 2)!$

10. $5! \, 2!$ **11.** $\dfrac{9!}{7!}$ **12.** $\dfrac{10!}{6!}$

13. $\dfrac{12!}{8!}$ **14.** $\dfrac{10!}{4! \, 6!}$ **15.** $\dfrac{9!}{5! \, 4!}$

16. $\dfrac{12!}{3!(12-3)!}$ **17.** $\dfrac{11!}{4!(11-4)!}$ **18.** $\dfrac{52!}{3!(52-3)!}$

Evaluate each of the numbers in Problems 19–24.

19. a. $_9P_1$ **b.** $_9P_2$ **c.** $_9P_3$ **d.** $_9P_4$ **e.** $_9P_0$

20. a. $_5P_4$ **b.** $_{52}P_3$ **c.** $_7P_2$ **d.** $_4P_4$ **e.** $_{100}P_1$

21. a. $_{12}P_5$ **b.** $_5P_3$ **c.** $_8P_4$ **d.** $_8P_0$ **e.** $_gP_h$

22. a. $_{92}P_0$ **b.** $_{52}P_1$ **c.** $_7P_5$ **d.** $_{16}P_3$ **e.** $_nP_4$

23. a. $_7P_3$ **b.** $_5P_5$ **c.** $_{50}P_{48}$ **d.** $_{25}P_1$ **e.** $_mP_3$

24. a. $_8P_3$ **b.** $_{12}P_0$ **c.** $_{10}P_2$ **d.** $_{11}P_4$ **e.** $_nP_5$

How many permutations are there in the words in Problems 25–33?

25. HOLIDAY   **26.** ANNEX

27. ESCHEW   **28.** OBFUSCATION

29. MISSISSIPPI   **30.** CONCENTRATION

31. BOOKKEEPING   **32.** GRAMMATICAL

33. APOSIOPESIS

Answer the questions in Problems 34–52 by using the fundamental counting principle, the permutation formula, or both.

APPLICATIONS

34. In how many ways can a group of five people elect a president, vice president, secretary, and treasurer?

35. In how many ways can a group of 15 people elect a president and a vice president?

36. In how many ways can a group of 10 people elect a president, a vice president, and a secretary?

37. How many outfits consisting of a skirt and a blouse can a woman select if she has three skirts and five blouses?

38. How many outfits consisting of a suit and a tie can a man select if he has two suits and eight ties?

39. In how many different ways can eight books be arranged on a shelf?

40. In how many ways can you select and read three books from a shelf of eight books?

41. In how many ways can a row of 3 contestants for a TV game show be selected from an audience of 362 people?

42. How many seven-digit telephone numbers are possible if the first two digits cannot be ones or zeros?

43. Foley's Village Inn offers the following menu in its restaurant:

Main course	Dessert	Beverage
Prime rib	Ice cream	Coffee
Steak	Sherbet	Tea
Chicken	Cheesecake	Milk
Ham		Sanka
Shrimp		

In how many different ways can someone order a meal consisting of one choice from each category?

44. A typical Social Security number is 576-38-4459; the first digit cannot be zero. How many Social Security numbers are possible?

45. California license plates consist of one digit, followed by three letters, followed by three digits. How many such license plates are possible?

46. If a state issues license plates that consist of one letter followed by five digits, how many different plates are possible?

47. Repeat Problem 46 if the first letter cannot be *O*, *Q*, or *I*.

48. Repeat Problem 46 without repetition of digits.

49. New York license plates consist of three letters followed by three digits. It is also known that 245 specific arrangements of three letters are not allowed because they are considered obscene. How many license plates are possible?

50. A certain lock has five tumblers, and each tumbler can assume six positions. How many different positions are possible?

51. Texas license plates have three letters followed by three digits. If we assume that all arrangements are allowed, how many Texas license plates have a repeating letter?

52. Suppose you flip a coin and keep a record of the results. In how many ways could you obtain at least one head if you flip the coin six times?

53. **Historical Question** Georg Cantor (1845–1918) was the originator of set theory and the study of transfinite numbers. When he first published his paper on the theory of sets in 1874 it was very controversial because it differed from current mathematical thinking. One of Cantor's former teachers, Leopold Kronecker (1823–1891), was particularly strong in his criticism not only of Cantor's work, but also of Cantor himself. Although Cantor believed that "the essence of mathematics lies in its freedom," he tragically had a series of mental breakdowns and died in a mental hospital in 1918. Cantor proved that the set of counting numbers contains precisely as many members as its proper subset $\{2, 4, 6, \ldots\}$. This turned out to be the defining property for an infinite set. Prove that the set $\{1, 2, 3, 4, \ldots\}$ contains precisely the same number of elements as the proper subset $\{2, 4, 6, \ldots\}$.

5.3 Combinations

Consider the set

$$A = \{a, b, c, d, e\}$$

If two elements are selected from *A* in a certain order, they are represented by an *ordered pair* and the ordered pair is called a *permutation*. On the other hand, if two elements are selected from *A without regard to the order in which they are selected*, they are represented as a subset of *A*.

EXAMPLE 1 Select two elements from *A*:

Permutations—Order important

(a, b)	(a, c)	(a, d)	(a, e)
(b, a)	(b, c)	(b, d)	(b, e)
(c, a)	(c, b)	(c, d)	(c, e)
(d, a)	(d, b)	(d, c)	(d, e)
(e, a)	(e, b)	(e, c)	(e, d)

Notation: $_5P_2 = 20$

There are 20 permutations— order is important.

Subsets—Order not important

$\{a, b\}$	$\{a, c\}$	$\{a, d\}$	$\{a, e\}$
	$\{b, c\}$	$\{b, d\}$	$\{b, e\}$
		$\{c, d\}$	$\{c, e\}$
			$\{d, e\}$

Do not list $\{b, a\}$ since $\{b, a\} = \{a, b\}$

There are 10 subsets; note that set notation is used. ∎

We use the following name and notation when listing different subsets of a given set.

Definition of Combination

A **combination** of *r* elements of a finite set *S* is a subset of *S* that contains *r* distinct elements.

Remember when listing the elements of a subset that the order in which those elements are listed is not important. A notation similar to that used for permutations is used to denote the number of combinations.

Notation for Combinations

$\dbinom{n}{r}$ and $_nC_r$ are symbols used to denote the number of combinations of r elements selected from a set of n elements $(r \le n)$.

The notation $_nC_r$ is similar to the notation used for permutations, but since $\dbinom{n}{r}$ is used in later work, we will use that notation for combinations.

The formula for the number of permutations leads directly to a formula for the number of combinations since each subset of r elements has $r!$ permutations of its members. Thus

$$_nP_r = \binom{n}{r} \cdot r!$$

$$\binom{n}{r} = \frac{_nP_r}{r!}$$

$$= \frac{n!}{r!(n-r)!}$$

Combination Formula

$$\binom{n}{r} = \frac{n!}{r!(n-r)!}$$

EXAMPLE 2
$$\binom{10}{3} = \frac{10!}{3!\,7!}$$
$$= \frac{\overset{5}{\cancel{10}} \cdot \overset{3}{\cancel{9}} \cdot 8 \cdot \cancel{7!}}{\cancel{3} \cdot \cancel{2} \cdot \cancel{7!}}$$
$$= 120$$ ∎

EXAMPLE 3
$$\binom{n}{0} = \frac{n!}{0!(n-0)!}$$
$$= 1$$ ∎

EXAMPLE 4
$$\binom{m-1}{2} = \frac{(m-1)!}{2!(m-1-2)!}$$
$$= \frac{(m-1)(m-2)(m-3)!}{2 \cdot 1 \cdot (m-3)!}$$
$$= \frac{(m-1)(m-2)}{2}$$ ∎

EXAMPLE 5 In how many ways can a club of five members select a three-person committee?

Solution $\binom{5}{3} = \frac{5!}{3!2!} = \frac{5 \cdot 4 \cdot 3}{3!}$

$= 10$ ∎

EXAMPLE 6 Find the number of five-card hands that can be drawn from an ordinary deck of cards.

Solution $\binom{52}{5} = \frac{52!}{5!47!}$

$= \frac{52 \cdot 51 \cdot 50 \cdot 49 \cdot 48}{5 \cdot 4 \cdot 3 \cdot 2 \cdot 1} = 2,598,960$ ∎

EXAMPLE 7 In how many ways can a diamond flush be drawn in poker? (A diamond flush is a hand of five diamonds.)

Solution This is a combination of 13 objects (diamonds) taken 5 at a time. Thus, the solution is given by

$\binom{13}{5} = \frac{13!}{5!8!}$

$= \frac{13 \cdot 12 \cdot 11 \cdot 10 \cdot 9}{5!} = 1287$ ∎

EXAMPLE 8 In how many ways can a flush be drawn in poker?*

Solution Begin with the fundamental counting principle:

$$\underbrace{4}_{\substack{\text{Number} \\ \text{of suits}}} \cdot \underbrace{1287}_{\substack{\text{Number of ways of} \\ \text{drawing a flush in} \\ \text{a particular suit} \\ \text{(from Example 7)}}} = 5148$$ ∎

Ordered Partitions

As we have seen, not all counting problems can be solved as permutation or combination problems. But suppose we generalize the idea of filling pigeonholes. We wish to divide a set *S* into *r* subsets. There are several ways that we might do this. For example, suppose

$S = \{x \mid x \text{ is a sales representative of the Sharp Investment Company}\}$

By roster,

$S = \{\text{Ralph, Ron, Lim, Shannon, Mark, Todd, Tim, Greg}\}$

---

* A flush in poker is five cards from one suit (see Figure 6.1 on page 203.)

We divide the set S in three different ways:

1. By geographical areas covered:

North	South	East	West
Ron	Ralph	Lim	Mark
Todd		Shannon	Greg
		Tim	

2. By amount of sales:

Over $100,000	$50,000–$100,000	Under $50,000
Ron	Todd	Tim
Lim	Shannon	
Mark	Greg	

3. By musical instruments each plays:

Piano	Guitar	Clarinet	Accordion
Ralph	Ralph	Ron	Ralph
Todd	Mark		Todd
Greg			

If no two of the subsets have any elements in common, and if all the elements of the original set are in some subset, we call such a set of subsets a **partition** of the original set. This corresponds to our everyday usage of the word when we say a house is partitioned into several rooms. In the division by geographical area, the four sets {Ron, Todd}, {Ralph}, {Lim, Shannon, Tim}, and {Mark, Greg} form a partition of S. On the other hand, the divisions by sales and instrument are not partitions since Ralph does not appear in any of the subsets of the former, and the subsets in the latter are not disjoint.

Partition of a Set

A set S is said to be *partitioned* if S is divided into r subsets satisfying the following two conditions:

1. If A and B are any two subsets, then $A \cap B = \emptyset$; that is, the members of the subsets are pairwise disjoint.
2. The union of all the subsets is S; that is, there are no elements of S that are not included in one of the subsets.

EXAMPLE 9 Form a partition of the set $\{1, 2, 3, 4\}$.

Solution Answers vary; some possible ones are:

$$\{1, 2\}, \{3, 4\}$$

or

$\{1\}, \{2\}, \{3\}, \{4\}$

or

$\{1, 2, 3\}, \{4\}$

However, $\{1, 2, 3\}, \{3, 4\}$ is *not* a partition.   ∎

EXAMPLE 10   Any set and its complement form a partition of a universal set.   ∎

Suppose the eight sales representatives of Sharp Investment Company are to be reassigned into three geographical areas:

Area #1	*Area #2*	*Area #3*
North	East	West

We are interested in knowing how many ways we can assign two sales representatives to area #1, three sales representatives to area #2, and three sales representatives to area #3. Use the fundamental counting principle to fill the pigeonholes below:

Area #1	*Area #2*	*Area #3*
$\binom{8}{2}$	$\binom{6}{3}$	$\binom{3}{3}$

There are $\binom{8}{2}$ ways of filling area #1. This leaves six sales representatives to distribute among areas #2 and #3. For area #2, we see there are $\binom{6}{3}$ ways, and this leaves us with three sales representatives to be distributed to area #3: $\binom{3}{3}$.

Instead of calculating these numbers separately and then taking the product, let us do all the calculations at once:

$$\binom{8}{2}\binom{6}{3}\binom{3}{3} = \frac{8!}{2!\,6!} \cdot \frac{6!}{3!\,3!} \cdot \frac{3!}{3!\,0!}$$

$$= \frac{8!}{2!\,3!\,3!}$$

$$= \frac{8 \cdot 7 \cdot 6 \cdot 5 \cdot 4 \cdot 3!}{2 \cdot 1 \cdot 3 \cdot 2 \cdot 1 \cdot 3!}$$

$$= 560 \text{ ways}$$

This is an example of an **ordered**, or **distinguishable**, **partition** (two sales representatives for area #1 and three for area #2 is different from three sales representatives for area #1 and two for area #2—order is important).

The above example can be written in condensed form as

$$\binom{8}{2,3,3} = \frac{8!}{2!\,3!\,3!}$$

You can see that what we have done here is add a new notation to the formula for the number of distinguishable permutations that was stated in Section 5.2.

Number of Ordered Partitions

For any set of positive integers $\{n_1, n_2, \ldots, n_k\}$, where $n_1 + n_2 + \cdots + n_k = n$, we have

$$\underbrace{\binom{n}{n_1, n_2, \ldots, n_k}}_{\substack{\text{Partition} \\ \text{notation}}} \text{ for } \underbrace{\frac{n!}{n_1! n_2! \cdots \cdots n_k!}}_{\substack{\text{Number of} \\ \text{distinguishable} \\ \text{permutations}}}$$

EXAMPLE 11   Suppose there are 15 sales representatives and they are to be divided among the three areas as follows: 4 in area #1, 5 in area #2, and 6 in area #3.

Solution   Use pigeonholes to find

$$\binom{15}{4}\binom{11}{5}\binom{6}{6} = \frac{15!}{4!\,\cancel{11!}} \cdot \frac{\cancel{11!}}{5!\,\cancel{6!}} \cdot \frac{\cancel{6!}}{6!\,0!}$$

$$= \frac{15!}{4!\,5!\,6!}$$

This can be rewritten as

$$\binom{15}{4,5,6} = \frac{15!}{4!\,5!\,6!}$$ ∎

Which Method?

We have now looked at several counting schemes: tree diagrams, the fundamental counting principle, permutations, combinations, and ordered partitions. In practice, you will generally not be told what type of counting problem you are dealing with — you will need to decide. Table 5.1 should help with that decision.

Table 5.1

Fundamental counting principle	Permutations		Combinations	Ordered partitions
Counting total of separate tasks	Number of ways of selecting r items out of n items			Partition n elements into k categories
Repetitions allowed	Repetitions are not allowed			Repetitions are not allowed
If tasks $1, 2, 3, \ldots, k$ can be performed in $n_1, n_2, n_3, \ldots, n_k$ ways, respectively, then the total number of ways the k tasks can be performed is $n_1 \cdot n_2 \cdot n_3 \cdot \cdots \cdot n_k$ ways	Order is important (select and arrange)		Order is not important (just select)	Order is not important
	Arrangements of r items from a set of n items		Subsets of r items from a set of n items	Place n items into k categories where $n = n_1 + n_2 + n_3 + \cdots + n_k$
	$_nP_r = \dfrac{n!}{(n-r)!}$		$\dbinom{n}{r} = \dfrac{n!}{r!(n-r)!}$	$\dbinom{n}{n_1, n_2, \ldots, n_k} = \dfrac{n!}{n_1! n_2! \cdots n_k!}$

COMPUTER APPLICATION   If you have access to a computer, Program 7 on the computer disk accompanying this book involves practice with counting problems.

EXAMPLE 12   What is the number of license plates possible in Florida if each license plate consists of three letters followed by three digits, and we add the condition that repetition of letters or digits is not permitted?

Solution   This is a permutation problem since the *order* in which the elements are arranged is important. That is, CWB072 and BCW072 are different plates. The number is found by using permutations along with the fundamental counting principle:

$$_{26}P_3 \cdot {}_{10}P_3 = 26 \cdot 25 \cdot 24 \cdot 10 \cdot 9 \cdot 8 \quad \leftarrow \text{This answer is acceptable}$$
$$= 11{,}232{,}000 \quad \blacksquare$$

EXAMPLE 13   Find the number of three-letter "words" that can be formed using three letters from $\{m, a, t, h\}$.

Solution   Finding these words is also a permutation problem since *mat* is different from *tam*.

$$_4P_3 = 4 \cdot 3 \cdot 2 = 24 \quad \blacksquare$$

EXAMPLE 14   The football team at UCLA plays 11 games each season. In how many ways can the team complete the season with 6 wins, 4 losses, and 1 tie?

Solution   This is an ordered partition $(6 + 4 + 1 = 11)$, so

$$\binom{11}{6,4,1} = \frac{11!}{6!4!1!}$$
$$= \frac{11 \cdot 10 \cdot 9 \cdot 8 \cdot 7 \cdot \cancel{6!}}{\cancel{6!}4 \cdot 3 \cdot 2 \cdot 1 \cdot 1} = 2310 \quad \blacksquare$$

EXAMPLE 15   Find the number of bridge hands (13 cards) consisting of six hearts, four spades, and three diamonds.

Solution   The order in which the cards are received is unimportant, so finding the number of bridge hands is a combination problem. Also use the fundamental counting principle:

$$\underbrace{\binom{13}{6}}_{\substack{\text{Number of ways of} \\ \text{obtaining hearts}}} \cdot \underbrace{\binom{13}{4}}_{\substack{\text{Number of ways of} \\ \text{obtaining spades}}} \cdot \underbrace{\binom{13}{3}}_{\substack{\text{Number of ways of} \\ \text{obtaining diamonds}}}$$

$$= \frac{13!}{6!(13-6)!} \cdot \frac{13!}{4!(13-4)!} \cdot \frac{13!}{3!(13-3)!}$$
$$= \frac{13!13!13!}{6!7!4!9!3!10!} \quad \leftarrow \text{This answer is acceptable}$$
$$= 350{,}904{,}840 \quad \blacksquare$$

EXAMPLE 16 A club with 42 members wants to elect a president, a vice president, and a treasurer. From the other members, an advisory committee of 5 people is to be selected. In how many ways can this be done?

Solution This is both a permutation and a combination problem, with the final result calculated by using the fundamental counting principle:

$$\underbrace{\begin{array}{c} \text{Number of ways} \\ \text{of selecting} \\ \text{officers} \end{array}}_{{}_{42}P_3} \cdot \underbrace{\begin{array}{c} \text{Number of ways} \\ \text{of selecting} \\ \text{committee} \end{array}}_{\dbinom{39}{5}}$$

$$= 42 \cdot 41 \cdot 40 \quad \cdot \quad \frac{39!}{5!(39 - 5)!}$$

$$= 39{,}658{,}142{,}160 \qquad \leftarrow \text{Found with a calculator} \qquad \blacksquare$$

Problem Set 5.3

Evaluate each of the numbers in Problems 1–11.

1. a. $\dbinom{9}{1}$ **b.** $\dbinom{9}{2}$ **c.** $\dbinom{9}{3}$ **d.** $\dbinom{9}{4}$ **e.** $\dbinom{9}{0}$

2. a. $\dbinom{5}{4}$ **b.** $\dbinom{52}{3}$ **c.** $\dbinom{7}{2}$ **d.** $\dbinom{4}{4}$ **e.** $\dbinom{100}{1}$

3. a. $\dbinom{7}{3}$ **b.** $\dbinom{5}{5}$ **c.** $\dbinom{50}{48}$ **d.** $\dbinom{25}{1}$ **e.** $\dbinom{g}{h}$

4. a. ${}_7P_3$ **b.** ${}_5P_5$ **c.** ${}_{52}P_2$ **d.** ${}_{25}P_1$ **e.** ${}_gP_h$

5. a. ${}_7P_5$ **b.** $\dbinom{8}{0}$ **c.** $\dbinom{10}{2}$ **d.** ${}_nP_4$ **e.** $\dbinom{n}{4}$

6. a. $\dbinom{92}{0}$ **b.** ${}_{52}P_3$ **c.** ${}_mP_n$ **d.** $\dbinom{16}{3}$ **e.** $\dbinom{m}{n}$

7. a. ${}_{92}P_0$ **b.** $\dbinom{53}{3}$ **c.** $\dbinom{7}{7}$ **d.** ${}_7P_7$ **e.** ${}_nP_5$

8. a. $\dbinom{4}{1,1,2}$ **b.** $\dbinom{6}{3,2,1}$

9. a. $\dbinom{5}{1,3,1}$ **b.** $\dbinom{8}{2,4,2}$

10. a. $\dbinom{9}{1,4,4}$ **b.** $\dbinom{7}{1,4,2}$

11. a. $\dbinom{10}{6,2,1,1}$ **b.** $\dbinom{8}{2,3,1,2}$

12. A bag contains 12 pieces of candy. In how many ways can 5 pieces be selected?

13. If the Senate is to form a new committee of 5 members, in how many different ways can the committee be chosen if all 100 senators are available to serve on this committee?

14. In how many ways can three aces be drawn from a deck of cards?

15. In how many ways can two kings be drawn from a deck of cards?

16. In how many ways can a heart flush be obtained? (A heart flush is a hand of five hearts.)

17. In how many ways can a spade flush be obtained? (A spade flush is a hand of five spades.)

In Problems 18–31 decide whether you would use the fundamental counting principle, a permutation, a combination, an ordered partition, or none of these. Next, write the solution using permutation, combination or ordered partition notation, if possible, and finally, answer the question asked.

18. How many different arrangements are there of letters in the word CORRECT?

19. At Artist's Dance Studio, every man must dance the last dance. If there are five men and eight women, in how many ways can dance couples be formed for the last dance?

20. Martin's Ice Cream Store sells sundaes with chocolate, strawberry, butterscotch, or marshmallow toppings, nuts, and whipped cream. If you can choose exactly three of these extras, how many possible sundaes are there?

21. Five people are to dine together at a rectangular table, but the host cannot decide on a seating arrangement. In how many ways can the guests be seated, assuming fixed chair positions?

22. In how many ways can three hearts be drawn from a deck of cards?

23. A shipment of a hundred TV sets is received. Six sets are to be chosen at random and tested for defects. In how many ways can the six sets be chosen?

24. A night watchman visits 15 offices every night. To prevent others from knowing when he will be at a particular office, he varies the order of his visits. In how many ways can this be done?

25. A certain manufacturing process calls for six chemicals to be mixed. One liquid is to be poured into the vat, and then the others are to be added in turn. All possible combinations must be tested to see which gives the best results. How many tests must be performed?

26. There are three boys and three girls at a party. In how many ways can they be seated in a row if they can sit down four at a time?

27. There are three boys and three girls at a party. In how many ways can they be seated in a row if they want to sit alternating boy, girl, boy, girl?

28. If there are ten people in a club, in how many ways can they choose a dishwasher and a bouncer?

29. How many subsets of size three can be formed from a set of five elements?

30. In how many ways can you be dealt two cards from an ordinary deck of cards?

31. In how many ways can five taxi drivers be assigned to six cars?

APPLICATIONS

32. A fraternity contains 8 sophomores, 15 juniors, and 12 seniors. In how many ways can a committee consisting of 1 sophomore, 3 juniors, and 3 seniors be chosen?

33. A sorority has 2 freshmen, 5 sophomores, 25 juniors, and 18 seniors as members. In how many ways can a committee consisting of 2 sophomores, 5 juniors, and 5 seniors be chosen?

34. A rating service rates 25 businesses as A, AA, or AAA. Suppose that 5 companies will be rated AAA, 4 companies will be rated A, and the rest rated AA. In how many ways can this be done?

35. A psychology experiment will group 50 persons into one of three categories: dominant, passive, neutral. In how many ways can 15 be dominant, 10 passive, with the rest neutral? (You can leave your answer in factorial notation.)

36. In a state lottery, 20 people will be chosen as winners. Of these 20, 5 will be awarded $3 million, 5 will win $100,000, and the rest will win $10,000. In how many ways can this be done?

37. Suppose that in a study of 50 stocks, 7 are expected to go up, 5 to go down, and the rest will stay about the same. In how many different ways (to the nearest billion) can the 50 stocks be divided up so that the conditions of this problem are satisfied?

38. A group of students found that on a certain multiple-choice test of 15 questions, the answer sheet shows response "a" occurs 3 times, "b" occurs 4 times, "c" occurs 3 times, "d" occurs 2 times, and "e" occurs 3 times. In how many ways can the questions be answered? Leave your answer in factorial notation.

39. The investigative division of the Internal Revenue Service has a list of 10 companies cited for violations of proper tax procedures. A spy for one of the companies found out that the IRS intends to bring 4 of these companies to trial, will refer 5 cases for further investigation, and the other case will be dropped. In how many ways can the disposition of the 10 cases be made according to the known information?

40. A sample of 5 watches is selected at random from a lot of 20 watches. The entire lot will be rejected if 2 or more watches are found to be defective.
   a. How many different samples can be selected?
   b. If 2 of the watches are defective, how many of the samples will lead to rejection?

41. A group of 10 Americans, 6 Russians, and 5 Finns must choose a committee of 3 persons. In how many ways can a committee be chosen that consists of:
   a. All Americans?
   b. 2 Americans?
   c. 1 American?
   d. 1 of each nationality?
   e. 2 Russians and 1 Finn?

42. How many committees of 2 men and 3 women can be chosen from 10 men and 8 women?

43. How many committees of 3 Americans and 3 Russians can be chosen from 10 Americans and 12 Russians?

44. Show that $\dbinom{n}{n-r} = \dbinom{n}{r}$.

45. Show that $\dbinom{n}{k-1} + \dbinom{n}{k} = \dbinom{n+1}{k}$.

46. Prove the formula for the number of ordered partitions for the case where $k = 3$.

5.4 Binomial Theorem

Combinations can be used to help us form a pattern for the expansion of a binomial $(a + b)^n$. This problem occurs frequently in mathematics, and direct calculation is too tedious for a very large exponent n. Also, we sometimes need to find only one term in the expansion, and a pattern will help us find only that term without having to find all the others.

Consider the powers of $(a + b)$ listed below, which are found by direct multiplication:

$$(a + b)^0 = \qquad\qquad\qquad 1$$
$$(a + b)^1 = \qquad\qquad\qquad 1 \cdot a + 1 \cdot b$$
$$(a + b)^2 = \qquad\qquad 1 \cdot a^2 + 2 \cdot ab + 1 \cdot b^2$$
$$(a + b)^3 = \qquad\qquad 1 \cdot a^3 + 3 \cdot a^2 b + 3 \cdot ab^2 + 1 \cdot b^3$$
$$(a + b)^4 = \qquad 1 \cdot a^4 + 4 \cdot a^3 b + 6 \cdot a^2 b^2 + 4 \cdot ab^3 + 1 \cdot b^4$$
$$(a + b)^5 = 1 \cdot a^5 + 5 \cdot a^4 b + 10 \cdot a^3 b^2 + 10 \cdot a^2 b^3 + 5 \cdot ab^4 + 1 \cdot b^5$$
$$\vdots \qquad\qquad\qquad \vdots$$

First, ignore the coefficients (shown in color) and focus on the variables:

$$(a + b)^1: \quad a \quad b$$
$$(a + b)^2: \quad a^2 \quad ab \quad b^2$$
$$(a + b)^3: \quad a^3 \quad a^2 b \quad ab^2 \quad b^3$$
$$(a + b)^4: \quad a^4 \quad a^3 b \quad a^2 b^2 \quad ab^3 \quad b^4$$
$$(a + b)^5: \quad a^5 \quad a^4 b \quad a^3 b^2 \quad a^2 b^3 \quad ab^4 \quad b^5$$
$$\vdots$$

Do you see a pattern? As you read from left to right, the powers of a decrease and the powers of b increase. Note that the sum of the exponents for each term is the same as the original exponent:

$$(a + b)^n: \quad a^n b^0 \quad a^{n-1} b^1 \quad a^{n-2} b^2 \quad \cdots \quad a^{n-r} b^r \quad \cdots \quad a^2 b^{n-2} \quad a^1 b^{n-1} \quad a^0 b^n$$

Next, consider the numerical coefficients:

$$(a + b)^0: \qquad\qquad\qquad\qquad 1$$
$$(a + b)^1: \qquad\qquad\qquad 1 \qquad 1$$
$$(a + b)^2: \qquad\qquad 1 \qquad 2 \qquad 1$$
$$(a + b)^3: \qquad 1 \qquad 3 \qquad 3 \qquad 1$$
$$(a + b)^4: \quad 1 \qquad 4 \qquad 6 \qquad 4 \qquad 1$$
$$(a + b)^5: 1 \qquad 5 \qquad 10 \qquad 10 \qquad 5 \qquad 1$$
$$\vdots$$

Do you see the pattern? This arrangement of numbers is called **Pascal's triangle**. The rows and columns of Pascal's triangle are usually numbered, as shown in Figure 5.10.

Note that the rows of Pascal's triangle are numbered to correspond to the exponent n on $(a + b)^n$. There are many relationships associated with this pattern, but, for now, we are concerned with an expression representing the entries in the pattern. Do you see how to generate additional rows of the triangle? Look at Figure 5.10 and note the following:

1. Each row begins and ends with a 1.
2. We began counting the rows with row 0. This is because after row 0, the second entry in the row is the same as the row number. Thus, row 7 begins 1   7....

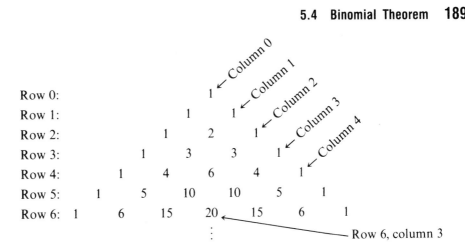

Figure 5.10 Pascal's triangle

3. The triangle is symmetric about the middle. This means that the entries of each row are the same at the beginning and the end. Thus row 7 ends with...7 1. (This property is proved in Problem 41.)
4. To find new entries, we simply add the two entries just above in the preceding row. Thus, row 7 is found by looking at row 6:

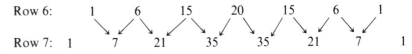

(This property is proved in Problem 40.)

Also note that the entries in Pascal's triangle are the combinations we computed in Section 5.3. That is, the entry in the third column of the fourth row is $\binom{4}{3}$, as shown below:

$$\binom{0}{0} = 1$$

$$\binom{1}{0} = 1 \qquad \binom{1}{1} = 1$$

$$\binom{2}{0} = 1 \qquad \binom{2}{1} = 2 \qquad \binom{2}{2} = 1$$

$$\binom{3}{0} = 1 \qquad \binom{3}{1} = 3 \qquad \binom{3}{2} = 3 \qquad \binom{3}{3} = 1$$

$$\binom{4}{0} = 1 \qquad \binom{4}{1} = 4 \qquad \binom{4}{2} = 6 \qquad \binom{4}{3} = 4 \qquad \binom{4}{4} = 1$$

EXAMPLE 1 Find $\binom{3}{2}$ both by Pascal's triangle and by formula.

Solution Third row, second column; entry in Pascal's triangle is 3. By formula,

$$\binom{3}{2} = \frac{3!}{2!(3-2)!} = \frac{3!}{2!1!} = 3$$

■

EXAMPLE 2 Find $\binom{6}{4}$ both by Pascal's triangle and by formula.

Solution Row 6, column 4; entry is 15. By formula,

$$\binom{6}{4} = \frac{6!}{4!(6-4)!}$$

$$= \frac{6!}{4!2!}$$

$$= \frac{6 \cdot 5 \cdot 4!}{2 \cdot 1 \cdot 4!}$$

$$= 15 \qquad \blacksquare$$

Using this notation we can now state a very important theorem in mathematics—the **binomial theorem**.

Binomial Theorem

For any positive integer n,

$$(a + b)^n = \binom{n}{0}a^n + \binom{n}{1}a^{n-1}b + \binom{n}{2}a^{n-2}b^2 + \cdots$$

$$+ \binom{n}{r}a^{n-r}b^r + \cdots + \binom{n}{n-2}a^2b^{n-2} + \binom{n}{n-1}ab^{n-1} + \binom{n}{n}b^n$$

Pascal's triangle is efficient for finding the numerical coefficients for exponents that are relatively small, as shown in Figure 5.11. However, for larger exponents we need to use the binomial theorem to find the coefficients.

EXAMPLE 3 Find $(x + y)^8$.

Solution Use Pascal's triangle (Figure 5.11) to obtain the coefficients in the expansion. Thus

$$(x + y)^8 = x^8 + 8x^7y + 28x^6y^2 + 56x^5y^3 + 70x^4y^4$$

$$+ 56x^3y^5 + 28x^2y^6 + 8xy^7 + y^8 \qquad \blacksquare$$

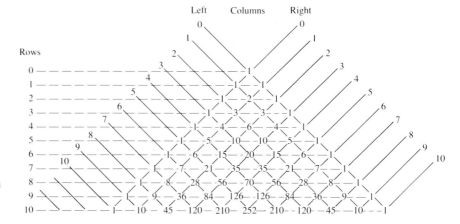

Figure 5.11 Pascal's triangle—an extended version can be found in Table 1, Appendix E

EXAMPLE 4   Find $(x - 2y)^4$.

Solution   In this example, $a = x$ and $b = -2y$ and the coefficients are shown in Figure 5.11.

$$(x - 2y)^4 = [x + (-2y)]^4$$
$$= 1 \cdot x^4 + 4 \cdot x^3(-2y) + 6 \cdot x^2(-2y)^2 + 4 \cdot x(-2y)^3 + 1 \cdot (-2y)^4$$
$$= x^4 - 8x^3y + 24x^2y^2 - 32xy^3 + 16y^4$$ ∎

EXAMPLE 5   Find $(a + b)^{15}$.

Solution   The power is rather large, so use the binomial theorem or Table 1, Appendix E, to find the coefficients:

$$(a + b)^{15} = \binom{15}{0}a^{15} + \binom{15}{1}a^{14}b + \binom{15}{2}a^{13}b^2 + \cdots + \binom{15}{14}ab^{14} + \binom{15}{15}b^{15}$$
$$= \frac{15!}{0!15!}a^{15} + \frac{15!}{1!14!}a^{14}b + \frac{15!}{2!13!}a^{13}b^2 + \cdots + \frac{15!}{14!1!}ab^{14} + \frac{15!}{15!0!}b^{15}$$
$$= a^{15} + 15a^{14}b + 105a^{13}b^2 + \cdots + 15ab^{14} + b^{15}$$ ∎

EXAMPLE 6   Find the coefficient of the term x^2y^{10} in the expansion of $(x + 2y)^{12}$.

Solution   Here, $n = 12, r = 10, a = x$, and $b = 2y$; thus

$$\binom{12}{10}x^2(2y)^{10} = \frac{12!}{10!2!}(2)^{10}x^2y^{10}$$

The coefficient is $66(1024) = 67,584$. ∎

COMPUTER APPLICATION   If you have access to a computer, Options 1 and 2 of Program 11 on the computer disk accompanying this text will help with problems like Examples 5 and 6.

EXAMPLE 7   How many subsets of $\{1, 2, 3, 4, 5\}$ are there?

Solution   In Section 5.1 we found that a set containing two elements has four subsets, and we did this by direct enumeration with a tree diagram. To follow the same procedure for this example would be too tedious. Instead, consider the following argument: There are $\binom{5}{5}$ subsets of five elements; $\binom{5}{4}$ subsets of four elements; $\binom{5}{3}$ subsets of three elements; $\binom{5}{2}$ subsets of two elements; $\binom{5}{1}$ subsets of a single element; and $\binom{5}{0}$ subsets of zero elements (the empty set). Thus the *total* number of subsets is

$$\binom{5}{0} + \binom{5}{1} + \binom{5}{2} + \binom{5}{3} + \binom{5}{4} + \binom{5}{5}$$

Now note that

$$(a + b)^5 = \binom{5}{0}a^5 + \binom{5}{1}a^4b + \binom{5}{2}a^3b^2 + \binom{5}{3}a^2b^3 + \binom{5}{4}ab^4 + \binom{5}{5}b^5$$

so that if we let $a = 1$ and $b = 1$,

$$\underbrace{(1 + 1)^5}_{2^5} = \underbrace{\binom{5}{0} + \binom{5}{1} + \binom{5}{2} + \binom{5}{3} + \binom{5}{4} + \binom{5}{5}}_{\text{Total number of subsets}}$$

We see that there are 2^5 subsets of a set containing five elements. ∎

EXAMPLE 8 How many subsets are there of a set containing 10 elements?

Solution Using the ideas of Example 7, there are 2^{10} subsets. ∎

Examples 7 and 8 suggest a general result:

Number of Subsets of a Finite Set | A set of n distinct elements has 2^n subsets.

Problem Set 5.4

Evaluate the expressions in Problems 1–12 using both Pascal's triangle and the binomial theorem.

1. $\binom{8}{1}$ **2.** $\binom{5}{4}$ **3.** $\binom{8}{2}$ **4.** $\binom{7}{5}$

5. $\binom{8}{3}$ **6.** $\binom{9}{5}$ **7.** $\binom{12}{1}$ **8.** $\binom{15}{0}$

9. $\binom{20}{20}$ **10.** $\binom{32}{31}$ **11.** $\binom{18}{2}$ **12.** $\binom{46}{2}$

Expand the binomial expressions in Problems 13–27 using the binomial theorem or Pascal's triangle.

13. $(x + y)^5$ **14.** $(x + y)^6$ **15.** $(x + y)^4$

16. $(x - y)^4$ **17.** $(x - y)^5$ **18.** $(x - y)^6$

19. $(x + 2)^5$ **20.** $(x - 3)^5$ **21.** $(2x + 3y)^4$

22. $(1 - x)^8$ **23.** $(1 - x)^{10}$ **24.** $(x - 2y)^8$

25. $(x - y)^{15}$ **26.** $(x + 1)^{18}$ **27.** $(x + y)^{20}$

Find the requested coefficient in Problems 28–33.

28. a^5b^6 term in $(a + b)^{11}$ **29.** a^4b^7 term in $(a - b)^{11}$

30. $a^{14}b$ term in $(a - 2b)^{15}$ **31.** $a^{10}b^4$ term in $(a + 3b)^{14}$

32. $x^{14}y^2$ term in $(2x^2 + y)^9$ **33.** $x^{14}y$ term in $(x^2 - 2y)^8$

APPLICATIONS

34. How many different subsets can be chosen from a set of seven elements?

35. How many different subsets can be chosen from the U.S. Senate? (There are 100 U.S. senators.)

36. Suppose a coin is tossed eight times. How many possible outcomes will there be of five heads and three tails? [*Hint:* Denote the possibilities of a single toss by H + T. Since there are eight tosses, we can denote the set of all possibilities by $(H + T)^8$. In this problem, you are looking for the coefficients of H^5T^3; that is, heads five times (H^5) and tails three times (T^3).]

37. Suppose a coin is tossed ten times. How many possible outcomes will there be of four heads and six tails? [See the hint in Problem 36.]

38. Suppose a coin is tossed nine times. How many possible outcomes will there be of two heads and seven tails? [See the hint in Problem 36.]

39. Show that $\binom{n}{0} + \binom{n}{1} + \binom{n}{2} + \cdots + \binom{n}{n-1} + \binom{n}{n} = 2^n$.

(This says that the sum of the entries of the nth row of Pascal's triangle is 2^n.)

40. Show that $\binom{n-1}{r-1} + \binom{n-1}{r} = \binom{n}{r}$.

[This says that to find any entry in Pascal's triangle (except the first and last), simply add the two entries directly above.]

41. Show that Pascal's triangle is symmetric.

42. Historical Question Blaise Pascal (1623–1662) has been described as "the greatest 'might-have-been' in the history of mathematics." Pascal was frail, and, because he needed to conserve his energy, he was forbidden to study mathematics. This aroused his curiosity and forced him to acquire most of his knowledge of the subject by himself. At 16, he wrote an essay on conic sections that astounded the mathematicians of his time. At 18, he had already invented one of the first calculating machines. However, at 27, because of his health, he promised God that he would abandon mathematics and spend his time in religious study. Three years later he broke this promise and wrote *Traité du Triangle Arithmétique*, in which he investigated what we today call Pascal's triangle. The very next year he was almost killed when his runaway horse jumped an embankment. He took this to be a sign of God's displeasure with him and again gave up mathematics—this time permanently. Answer the following questions about Pascal's triangle:

a. What is the sum of the entries in the nth row?

b. How are the powers of 11 related to Pascal's triangle?

c. Find a formula for the largest number in the nth row.

5.5 Summary and Review

IMPORTANT TERMS

Arrangement [5.2]
Binomial theorem [5.4]
Combination [5.3]
Complementation [5.1]
Description method [5.1]
Disjoint sets [5.1]
Distinguishable permutations [5.2]
Empty set [5.1]
Equality of sets [5.1]
Factorial [5.2]
Fundamental counting principle [5.2]
Improper subset [5.1]
Intersection [5.1]

Ordered pair [5.2]
Ordered partition [5.3]
Partition [5.3]
Pascal's triangle [5.4]
Permutation [5.2]
Proper subset [5.1]
Roster method [5.1]
Set [5.1]
Set-builder notation [5.1]
Subset [5.1]
Tree diagram [5.1]
Union [5.1]
Universal set [5.1]
Venn diagram [5.1]

SAMPLE TEST

For additional practice there are a large number of review problems categorized by objective in the Student Solutions Manual. The following sample test (40 minutes) is intended to review the main ideas of this chapter.

Let $U = \{2, 3, 4, 5, 6, 7, 8, 9, 10, 11, 12\}$, $A = \{5, 6, 7\}$, $B = \{7, 11\}$, $C = \{2, 7, 12\}$. *Find the requested sets in Problems 1–4.*

1. $\overline{A \cup B}$ **2.** $\overline{A} \cup \overline{B}$ **3.** $A \cap \overline{(B \cup C)}$ **4.** $\overline{A \cap (B \cup C)}$

Evaluate the expressions in Problems 5–7.

5. a. $7!$ **b.** $7! - 4!$ **c.** $(7 - 4)!$ **d.** $\dfrac{100!}{98!}$

6. a. $_6P_3$ **b.** $_6C_3$ **7. a.** $\dbinom{8}{2}$ **b.** $\dbinom{8}{5,2,1}$

8. In how many ways can a deck of 52 cards be dealt to four bridge players? Leave your answer in factorial notation.

9. In how many ways can you choose two books to read from a bookshelf containing seven books?

10. A club consists of 17 men and 19 women. In how many ways can a president, vice president, treasurer, and secretary be chosen, along with an advisory committee of 6 people, if no 2 persons can serve in more than one capacity?

11. Repeat Problem 10, but now add that exactly two of the officers must be women.

12. In how many ways can the letters in the word WORRY be arranged?

13. A mathematics test contains ten questions. In how many ways can the test be answered if the possible answers are true and false?

14. Answer Problem 13 if the possible answers are a, b, c, d, and e; that is, if the test is a multiple-choice test.

15. Human blood can be classified into types A, B, AB, or O. Furthermore, each of these types is classified as positive or negative depending on whether the antigen called Rh is present. These possibilities are summarized in the Venn diagram. Note that O blood type refers to the absence of either the A or B blood types.

Region	Blood type
1	A^-
2	B^-
3	O^+
4	A^+
5	AB^-
6	B^+
7	AB^+
8	O^-

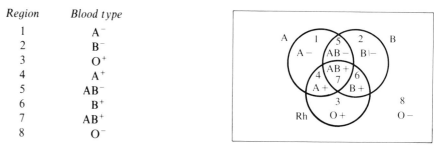

All people with the A antigen are put in circle A, those with the B antigen are put in circle B, and those with the Rh antigen are put in circle Rh. Now suppose that a hospital reports the following data:

    60 patients have the A antigen
    50 patients have the B antigen
    65 patients have the Rh antigen
    15 patients have both the A and B antigens
    20 patients have both the B and Rh antigens
    30 patients have both the A and Rh antigens
     5 patients have all three antigens
    10 patients have none of the antigens

How many patients
a. Were surveyed?
b. Have AB^+ blood?
c. Have A^+ blood?
d. Have O^- blood?
e. Have exactly one antigen?

Modeling Application 5

Earthquake Epicenter

During an earthquake, vibrations emanate in all directions from an origin called the *epicenter*. Two kinds of vibrations of importance in the study of earthquakes are called *primary* (P waves) and *secondary* (S waves). These names are due to the waves' order of arrival at a seismographic laboratory. The P wave is the fastest type of wave generated by an earthquake, with an average speed of about 8 kilometers per second. Concurrently, an S wave is produced that travels at approximately two-thirds the velocity of a P wave, so that it arrives at the recording station after the initial P wave. The greater the distance between the epicenter of the earthquake and the seismograph, the greater the time interval between the arrival of the two waves. Suppose an earthquake is recorded at three recording stations, one in Tokyo, another in Melbourne, and a third in San Francisco. The arrival times of the P and S waves are given in the following table:

| Recording station | Arrival times | |
	P wave	S wave
I. Tokyo	6:25:15 P.M.	6:32:19 P.M.
II. Melbourne	7:25:15 P.M.	7:34:07 P.M.
III. San Francisco	7:25:15 A.M.	7:29:15 A.M.

Write a paper that develops a mathematical model that determines the probable location of the epicenter by using Venn diagrams.* For general guidelines about writing this essay, see the commentary for Modeling Application 1 on page 36.

---

* The idea and material for this extended application are derived from Joseph Di Carlucci, "Earthquakes and Venn Diagrams," *The Mathematics Teacher*, September 1979, pp. 428–433.

CHAPTER 6
Probability

CHAPTER CONTENTS

APPLICATIONS

Management (*Business, Economics, Finance, and Investments*)

Probability of a defective light bulb (6.4, Problem 35; 6.7, Problems 8–9)
Analysis of accident risk (6.6, Problem 28)
Probability that a shipment has a defect (6.6, Problem 29)
Manufacturing analysis (6.7, Problems 8–9)

Life sciences (*Biology, Ecology, Health, and Medicine*)

Fruit fly experiment (6.4, Problems 19–28)
Medical diagnosis (6.6, Problem 30)
Test for glaucoma (6.7, Problem 10)

Social science (*Demography, Political Science, Population, Psychology, Society, and Sociology*)

Sample space for a study of sibling relationships (6.1, Problem 9)
Public opinion polling (6.1, Problems 8, 17–19)
Odds of a person lying (6.3, Problem 16)
Survey of persons dropping a college course (6.4, Problem 36)
Qualifying for graduate school (6.6, Problem 24)
Probability that a criminal is lying (6.6, Problem 25)
Demographics of college students (6.6, Problems 26–27)

General interest

Sample space for the results of the World Series (6.1, Problem 5)
Probability of an A on an examination (6.2, Problem 17)
Dice probabilities (6.2, Problems 23–37; 6.3, Problem 33; 6.4, Problems 15–18)
Dungeons and Dragons (6.2, Problem 41)
Historical dice games (6.2, Problems 42–43)
Probability of winning a raffle (6.2, Problem 19)
Racetrack odds (6.3, Problem 32)
Historical probability problems (6.3, Problem 62)
System players for roulette (6.4, Problem 38)
Birthday problem (6.4, Problems 55–56)
Roulette probabilities (6.7, Problem 5)

Modeling application—Medical Diagnosis

CHAPTER OVERVIEW We begin by discussing the notion of a probability model; then consider the probability of simple, equally likely events; and finally construct more complicated models by relating them back to the simpler cases.

PREVIEW This chapter introduces probabilistic models. Real world events can either be predicted with certainty or not. Of course, very little in the real world is *certain*, but events can be predicted with more or less certainty. Mathematical models are used to measure the amount of certainty, or probability. As you can imagine, it is an extremely important idea, not only in mathematics, but in almost all fields of study.

PERSPECTIVE The first major topic of finite mathematics was solving systems (both equations and inequalities) and the mathematics of matrices was used for such solutions. The second major topic of finite mathematics is probability, and is introduced in this chapter.

6.1 Probability Models

The models we have considered so far have been deterministic models. We now turn to a different type of model, called a **probabilistic model**. This model is used with situations that are random in character and attempts to predict the outcomes of events with a certain stated or known degree of accuracy. For example, if we toss a coin, it is impossible to predict in advance whether the outcome will be a head or a tail. Our intuition tells us that it is equally likely to be a head or a tail, and somehow we sense that if we repeat the experiment of tossing a coin a large number of times, heads will occur "about half the time." To check this out, I recently flipped a coin 1000 times and obtained 460 heads and 540 tails. The percentage of heads is $\frac{460}{1000} = .46 = 46\%$, which is called the **relative frequency**:

Relative Frequency of a Repeated Experiment

> If an experiment is repeated n times and an event occurs m times, then
>
> $$\frac{m}{n}$$
>
> is called the *relative frequency* of the event.

Our task in this chapter is to create a model that will assign a number p, called the *probability of an event*, which will predict the relative frequency. This means that for a *sufficiently large number of repetitions* of an experiment

$$p \approx \frac{m}{n}$$

Probabilities can be obtained in one of three ways:

1. **Theoretical probabilities** (also called *a priori models*) are obtained by logical reasoning. For example, the probability of rolling a die and obtaining a 3 is $\frac{1}{6}$ because there are six possible outcomes, each with an equal chance of occurring, so a 3 should appear $\frac{1}{6}$ of the time.

2. **Empirical probabilities** (also called *a posteriori models*) are obtained from experimental data. For example, an assembly line producing brake assemblies for General Motors produces 1500 items per day. The probability of a defective brake can be obtained by experimentation. Suppose the 1500 brakes are tested and 3 are found to be defective. Then the relative frequency, or probability, is

$$\frac{3}{1500} = .002 \qquad \text{or} \qquad .2\%$$

3. **Subjective probabilities** are obtained by experience and indicate a measure of "certainty." For example, a TV reporter studies the satellite maps and issues a prediction about tomorrow's weather based on past experience under similar circumstances: 80% chance of rain tomorrow.

We must define a probability measure so that it conforms to these different ways of using the word *probability*.

We begin by formalizing our terminology. We have been speaking about experiments and events. An **experiment** is the observation of any physical occurrence. A **sample space** of an experiment is the set of all possible outcomes. An **event** is a subset of the sample space. If an event is the empty set, it is called the **impossible event**; and if it has only one element, it is called a **simple event**.

EXAMPLE 1   The sample space S for the experiment of simultaneously tossing a coin and rolling a die is

$$S = \{1H, 1T, 2H, 2T, 3H, 3T, 4H, 4T, 5H, 5T, 6H, 6T\}$$

An example of an event E is "obtaining an even number or a head." That is, $E = \{1H, 2H, 2T, 3H, 4H, 4T, 5H, 6H, 6T\}$.   ∎

EXAMPLE 2   A computer chip is operated until it fails, and its lifetime, t (in hours), is recorded. The sample space S is

$$S = \{t \mid t \geq 0\}$$

An example of an event E is that the chip lasts more than 1000 hours; that is, $E = \{t \mid t > 1000\}$.   ∎

EXAMPLE 3   A UCLA alumnus records the results of a UCLA football game. The sample space S is

$$S = \{w, l, t\}$$

where w, l, and t denote win, lose, and tie, respectively. An example of an event E is that UCLA wins; that is, $E = \{w\}$. This is also an example of a simple event, whereas the events in Examples 1 and 2 are not simple events.   ∎

Two events E and F are said to be **mutually exclusive** if $E \cap F = \varnothing$.

EXAMPLE 4   Suppose you perform an experiment of rolling a die. Then the sample space S is

$$S = \{1, 2, 3, 4, 5, 6\}$$

Let $E = \{1, 3, 5\}$, $F = \{2, 4, 6\}$, $G = \{1, 3, 6\}$, and $H = \{2, 4\}$. Then:

E and F are mutually exclusive
G and H are mutually exclusive

But

F and H are *not* mutually exclusive
E and G are *not* mutually exclusive ■

We can now define a probability measure:

Probability Measure

> Let S be a sample space associated with some experiment. With each event E we associate a real number called the *probability of E*, denoted by P(E), satisfying the following properties:
> 1. $0 \leq P(E) \leq 1$
> 2. $P(S) = 1$
> 3. If E and F are mutually exclusive events, then
> $$P(E \cup F) = P(E) + P(F)$$

Note that this definition does not tell us how to compute P(E); it simply gives some general properties of P(E) from which we can build a variety of probability models. Also note that set notation is used in property 3. There is a very direct relationship between the language used to describe probabilistic situations and set notation. This relationship is shown in Table 6.1.

TABLE 6.1

$\varnothing$	Empty set	$A \cup B$	Union of sets A and B
U	Universal set	$A \cap B$	Intersection of sets A and B
S	Sample space	$\bar{A}$	Complement of a set A

Probabilistic statement	Set notation
Events A and B	A, B
A and B are mutually exclusive	$A \cap B = \varnothing$
A and B occur	$A \cap B$
A or B occurs	$A \cup B$
A does not occur	$\bar{A}$
Neither A nor B occurs	$\overline{A \cup B}$, or (equivalently) $\bar{A} \cap \bar{B}$
A and B are equally likely	$P(A) = P(B)$
A is more likely than B	$P(A) > P(B)$
A is less likely than B	$P(A) < P(B)$

In this book our discussion will be limited to experiments with a finite number of outcomes. Example 2 illustrates an infinite sample space. When the sample space is finite, it can be written as

$$S = \{s_1, s_2, s_3, \ldots, s_k\}$$

for some counting number k. Here $s_1, s_2, \ldots, s_k$ are the simple events or outcomes of our experiment. It follows from property 2 (above) that

$$P(s_1) + P(s_2) + P(s_3) + \cdots + P(s_k) = 1$$

which means that all the probabilities of the simple events in a probabilistic model must add up to 1.

EXAMPLE 5 Suppose a die is rolled and the number of the top face is recorded. Then $S = \{1, 2, 3, 4, 5, 6\}$. There are many possible models:

Model 1 $P(1) = \frac{1}{6}$, $P(2) = \frac{1}{6}$, $P(3) = \frac{1}{6}$, $P(4) = \frac{1}{6}$, $P(5) = \frac{1}{6}$, and $P(6) = \frac{1}{6}$. This is a model consisting of equally likely outcomes.

Model 2 $P(1) = \frac{1}{12}$, $P(2) = \frac{2}{12}$, $P(3) = \frac{3}{12}$, $P(4) = \frac{3}{12}$, $P(5) = \frac{2}{12}$, and $P(6) = \frac{1}{12}$. This is a model for a die loaded to favor the outcomes of 3 and 4.

Model 3 $P(1) = \frac{1}{10}$, $P(2) = \frac{2}{10}$, $P(3) = \frac{3}{10}$, $P(4) = \frac{3}{10}$, $P(5) = \frac{2}{10}$, and $P(6) = \frac{1}{10}$. This is *not* a probability model because the sum of the probabilities of the simple events is not 1.

Other models for this experiment are also possible. ∎

If an event is not a simple event, but has finitely many elements, then its probability can be found using the **addition principle**:

Addition Principle

If $E = \{s_1, s_2, \ldots, s_n\}$, then

$$P(E) = P(s_1) + P(s_2) + \cdots + P(s_n)$$

where $P(s_1), P(s_2), \ldots, P(s_n)$ are the probabilities of the simple events.

EXAMPLE 6 Suppose a die is loaded so that the probabilities of the simple events are $P(1) = \frac{1}{12}$, $P(2) = \frac{2}{12}$, $P(3) = \frac{3}{12}$, $P(4) = \frac{3}{12}$, $P(5) = \frac{2}{12}$, and $P(6) = \frac{1}{12}$. Find the probabilities for rolling the die once and obtaining:

a. An even number   **b.** A prime number   **c.** A 2 or a 6
d. A number greater than 3   **e.** A 2 and a 6

Solution **a.** $E = \{2, 4, 6\}$ so $P(E) = P(2) + P(4) + P(6)$
$$= \frac{2}{12} + \frac{3}{12} + \frac{1}{12}$$
$$= \frac{6}{12} = \frac{1}{2}$$

b. $F = \{2, 3, 5\}$ so $P(F) = P(2) + P(3) + P(5)$
$$= \frac{2}{12} + \frac{3}{12} + \frac{2}{12}$$
$$= \frac{7}{12}$$

c. $G = \{2, 6\}$ so $P(G) = P(2) + P(6)$
$$= \frac{2}{12} + \frac{1}{12}$$
$$= \frac{3}{12} = \frac{1}{4}$$

d. $H = \{4, 5, 6\}$ so $P(H) = P(4) + P(5) + P(6)$
$$= \tfrac{3}{12} + \tfrac{2}{12} + \tfrac{1}{12}$$
$$= \tfrac{6}{12} = \tfrac{1}{2}$$

e. $I = \varnothing$ so $P(I) = 0$ ■

In Example 6e, our intuition tells us that the probability of the empty event should be 0, but the following argument proves this fact by using only the definition of a probability measure.

$P(E) = P(E \cup \varnothing)$	Since $E = E \cup \varnothing$ for any set E
$P(E) = P(E) + P(\varnothing)$	Property 3 of a probability measure since E and $\varnothing$ are mutually exclusive
$0 = P(\varnothing)$	Subtract $P(E)$ from both sides

Probability of the Empty Set

$P(\varnothing) = 0$

In the next section we discuss finding a model for simple events, and in Section 6.4 we find a model for the probabilities of events that are not simple.

Problem Set 6.1

Describe the sample space for the experiments in Problems 1–10.

1. A pair of dice is rolled and the sum of the top faces is recorded.

2. A coin is tossed three times and the sequence of heads and tails is recorded.

3. Ten people are asked if they graduated from college and the number of people responding "yes" is recorded.

4. A jar contains five red balls, four white balls, and three green balls. One ball is drawn and the result is recorded.

APPLICATIONS

5. The number of games necessary to complete the World Series for the years 1960–1980 is recorded. (The World Series consists of playing until one team wins four games.)

6. One hundred Eveready Long-Life® batteries are tested and the life of each battery is recorded.

7. The number of words in which an error is made on a typing test consisting of 500 words is recorded.

8. A person is randomly selected for a public opinion poll. The person is asked two questions:

   Sex: male (m), female (f)
   Political party: Democrat (d), Republican (r), Independent (i), not registered (n)

The responses are recorded.

9. A psychologist is studying sibling relationships in families with three children. A family is asked about the sex of their children and the result is recorded. If a family has a girl, then a boy, and finally another boy, this would be recorded as {gbb}.

10. A sample of five radios is selected from an assembly line and tested. The number of defective radios is recorded.

In Problems 11–19 an experiment is described and some events are listed. Write each event using set notation and determine whether the given events are mutually exclusive.

11. A pair of dice is rolled and the sum of the numbers on the top faces is recorded.
   a. Event E is rolling an even number.
   b. Event F is rolling a prime number.

12. A pair of dice is rolled and the sum of the numbers on the top faces is recorded.
   a. Event M is rolling a 7 or an 11.
   b. Event N is rolling a 2, 3, or 12.

13. A pair of dice is rolled and the sum of the numbers on the top faces is recorded.
   a. Event G is rolling a 2 (snake eyes).
   b. Event H is rolling a sum greater than 2.

14. A coin is tossed three times and the sequence of heads and tails is recorded.
   a. Event E is tossing one head.
   b. Event F is tossing two heads.

15. A coin is tossed three times and the sequence of heads and tails is recorded.
    a. Event G is tossing a head on the first roll.
    b. Event H is tossing a head on the second roll.

16. A coin is tossed three times and the sequence of heads and tails is recorded.
    a. Event I is tossing at least one head.
    b. Event J is tossing no heads.

APPLICATIONS

17. Consider the experiment described in Problem 8.
    a. Event F is the selected person is female.
    b. Event D is the selected person is a Democrat.

18. Consider the experiment described in Problem 8.
    a. Event I is the selected person is a registered Independent.
    b. Event N is the selected person is not registered.

19. Consider the experiment described in Problem 8.
    a. Event E is the selected person is either female or Republican.
    b. Event R is the selected person is a registered voter.

In Problems 20–28 consider the experiment of rolling a single die and recording the number on the top face. Let N be the event of rolling an odd number; E be the event of rolling an even number; and L be the event of rolling a number less than 3. Describe each event in words.

20. $N \cup E$ 21. $L \cup E$ 22. $N \cap E$

23. $E \cap L$ 24. $\bar{N}$ 25. $\bar{L}$

26. $\overline{N \cup L}$ 27. $\bar{E} \cap \bar{L}$ 28. $\bar{N} \cap \bar{E}$

In Problems 29–38 suppose a pair of dice are loaded so that the probabilities of the simple events are

$$P(2) = \tfrac{1}{11} \quad P(3) = \tfrac{1}{11} \quad P(4) = \tfrac{1}{11} \quad P(5) = \tfrac{1}{11}$$
$$P(6) = \tfrac{1}{11} \quad P(7) = \tfrac{1}{11} \quad P(8) = \tfrac{1}{11} \quad P(9) = \tfrac{1}{11}$$
$$P(10) = \tfrac{1}{11} \quad P(11) = \tfrac{1}{11} \quad P(12) = \tfrac{1}{11}$$

Let $C = \{2, 3, 12\}$, $E = \{7, 11\}$, and $F = \{8, 9, 10\}$.

29. Is this a probability model? If it is, explain why.

30. Find P(E).   31. Find P(C).

32. Find P(F).   33. Find P($E \cup F$).

34. Find P($E \cup C$).   35. Find P($E \cap C$).

36. Find P($\bar{F}$).   37. Find P($\bar{C}$).

38. Find P($\overline{E \cup F}$).

In Problems 39–48 suppose a pair of dice are loaded so that the probabilities of the simple events are

$$P(2) = \tfrac{1}{36} \quad P(3) = \tfrac{2}{36} \quad P(4) = \tfrac{3}{36} \quad P(5) = \tfrac{4}{36}$$
$$P(6) = \tfrac{5}{36} \quad P(7) = \tfrac{6}{36} \quad P(8) = \tfrac{5}{36} \quad P(9) = \tfrac{4}{36}$$
$$P(10) = \tfrac{3}{36} \quad P(11) = \tfrac{2}{36} \quad P(12) = \tfrac{1}{36}$$

Let $C = \{2, 3, 12\}$, $E = \{7, 11\}$, and $F = \{8, 9, 10\}$.

39. Is this a probability model? If it is, explain why.

40. Find P(E).   41. Find P(C).

42. Find P(F).   43. Find P($E \cup F$).

44. Find P($E \cup C$).   45. Find P($E \cap C$).

46. Find P($\bar{F}$).   47. Find P($\bar{C}$).

48. Find P($\overline{E \cup F}$).

6.2 Probability of Equally Likely Events

In this section we begin to calculate theoretical probabilities. Suppose a sample space is divided into simple events that are **equally likely**. For example, the experiment of flipping a coin has a sample space

$$S = \{H, T\}$$

and the simple events H = {heads}, T = {tails} are equally likely. In this text, coins are considered *fair* (equally likely heads and tails) unless otherwise noted. This means that the coin is perfectly balanced and symmetrical and the events H and T are equally likely to occur.

EXAMPLE 1   Consider two dice, each with faces labeled 1, 2, 3, 4, 5, 6.
For die A, $P(1) = \tfrac{1}{6}$, $P(2) = \tfrac{1}{6}$, $P(3) = \tfrac{1}{6}$, $P(4) = \tfrac{1}{6}$, $P(5) = \tfrac{1}{6}$, and $P(6) = \tfrac{1}{6}$.
For die B, $P(1) = \tfrac{1}{12}$, $P(2) = \tfrac{2}{12}$, $P(3) = \tfrac{3}{12}$, $P(4) = \tfrac{3}{12}$, $P(5) = \tfrac{2}{12}$, and $P(6) = \tfrac{1}{12}$.
For die A, each of the six possible outcomes are equally likely. Die A is called a *fair die*. The outcomes for die B are not equally likely, so die B is called a *loaded die*. All dice used in this book are considered to be fair dice unless otherwise noted. ■

Die A in Example 1 has a sample space consisting of six equally likely simple events, so we will assign each simple event probability $\frac{1}{6}$. In so doing we are creating a probability model for rolling a single die called a **uniform probability model**.

Uniform Probability Model

> If an experiment has a sample space consisting of n mutually exclusive and equally likely simple events, then a *uniform probability model* assigns the probability of $1/n$ to each simple event.

Spades (black cards)

Hearts (red cards)

Clubs (black cards)

Diamonds (red cards)

Figure 6.1 A deck of 52 cards

EXAMPLE 2   Suppose a single card is chosen from a deck of cards. What is the probability it is an ace of spades?

Solution   In this book, when we refer to a deck of cards we are assuming the standard bridge deck shown in Figure 6.1. Since there are 52 equally likely outcomes for this experiment, we assign a probability of $\frac{1}{52}$ to the event of drawing the ace of spades. We write this as

$$P(\text{ace of spades}) = \tfrac{1}{52}$$ ∎

EXAMPLE 3   Find P(ace) for the experiment in Example 2.

Solution   Let $A = \{$ace of spades, ace of clubs, ace of hearts, ace of diamonds$\}$. Then

$$P(A) = P(\text{ace of spades}) + P(\text{ace of clubs}) + P(\text{ace of hearts})$$
$$\quad + P(\text{ace of diamonds})$$
$$= \tfrac{1}{52} + \tfrac{1}{52} + \tfrac{1}{52} + \tfrac{1}{52}$$
$$= \tfrac{4}{52}$$
$$= \tfrac{1}{13} \quad \text{or} \quad .08 \qquad$$ State probability as a reduced fraction or as a decimal rounded to the nearest hundredth   ∎

The procedure illustrated by Example 3 is unnecessarily lengthy. Let us consider a general result. Suppose

$$E = \{k_1, k_2, k_3, \ldots, k_s\}$$

Then

$$P(E) = P(k_1) + P(k_2) + P(k_3) + \cdots + P(k_s)$$

$$= \underbrace{\frac{1}{n} + \frac{1}{n} + \frac{1}{n} + \cdots + \frac{1}{n}}_{s \text{ simple events}} \qquad \textit{Each} \text{ simple event has probability of } 1/n$$

$$= \frac{s}{n}$$

This leads us to the following important probability model:

Probability of an Event that Can Occur in Any One of n Mutually Exclusive and Equally Likely Ways

If an experiment can occur in any of n ($n \geq 1$) mutually exclusive and equally likely ways and if s of these ways are considered favorable, then the probability of the event E, denoted by $P(E)$, is

$$P(E) = \frac{s}{n} = \frac{\text{Number of outcomes favorable to } E}{\text{Number of all possible outcomes}}$$

For Examples 4–8, suppose a single card is selected from a deck of cards (see Figure 6.1).

EXAMPLE 4   Find P(heart).

Solution   There are 52 elements in the sample space and 13 of these are hearts, or successes. Therefore

$$P(\text{heart}) = \tfrac{13}{52}$$

$$= \tfrac{1}{4}$$ ∎

EXAMPLE 5 $P(\text{heart or an ace}) = \tfrac{16}{52}$ 13 hearts + 3 *additional* aces (be careful not to count the ace of hearts twice)

$$= \tfrac{4}{13}$$ ∎

EXAMPLE 6 $P(\text{heart and an ace}) = \tfrac{1}{52}$ The ace of hearts is the only such card   ∎

EXAMPLE 7 $P(\text{ace or a 2}) = \tfrac{8}{52}$

$$= \tfrac{2}{13}$$ ∎

EXAMPLE 8 $P(\text{ace and a 2}) = \tfrac{0}{52}$ There is no way of drawing a single card and obtaining an ace *and* a 2

$$= 0$$ ∎

Finding the Probability
of an Event

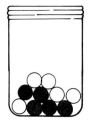

In summary, to find the probability of some event:

1. Describe and identify the sample space, and then count the number of elements (these should be equally likely). Call this number n.
2. Count the number of occurrences that are favorable to the event; call this the *number of successes* and denote it by s.
3. Compute the probability of the event: $P(E) = \dfrac{s}{n}$.

Do not forget that the simple events must be equally likely. Consider the following situation for Examples 9–11: A jar contains three red marbles, two black marbles, and five white marbles. Conduct an experiment of drawing out a single marble from the jar.

EXAMPLE 9   Find P(red).

Solution   A possible sample space is {red, black, white}, but the simple events {red}, {black}, {white} are *not* equally likely. The problem, then, is to create a sample space that *has* equally likely simple events. For this example it seems that the individual marbles each have an equal chance of being chosen. Consider the sample space consisting of the individual marbles labeled as follows:

$$\{R_1, R_2, R_3, B_1, B_2, W_1, W_2, W_3, W_4, W_5\}$$

where R_1, R_2, R_3 represent the red marbles; B_1, B_2 represent the black marbles; and W_1, W_2, W_3, W_4, W_5 represent the white marbles. Then the sample space, as now described, consists of equally likely outcomes; so we use this model. Thus

$$P(\text{red}) = \frac{\text{Number of favorable outcomes}}{\text{Number of all possible outcomes}}$$

$$= \frac{3}{10}$$

EXAMPLE 10 $P(\text{black or white}) = \frac{7}{10}$

EXAMPLE 11 $P(\text{black and white}) = \frac{0}{10} = 0$ The number of favorable outcomes is 0. You cannot draw a black *and* a white marble with only one draw

Sometimes probabilities are found empirically by using the model developed in this section, as shown in Example 12.

EXAMPLE 12   Suppose that in a certain study, 46 out of 155 people showed a certain kind of behavior. Assign a probability to this behavior.

Solution $p = \frac{46}{155}$

$\approx .30$ By calculator

EXAMPLE 13   Suppose we conduct an experiment by rolling a pair of dice 50 times and recording the sum. This experiment is repeated 3 times for a total of 150 rolls of the dice. The results of these experiments are recorded below:

Outcome	First trial	Second trial	Third trial	Total
2	|		|	2
3	|||	||	||	7
4	卌	||||	卌	14
5	||	卌 ||	卌 |	15
6	卌 ||||	卌 |||	卌 |	23
7	卌 |||	卌 卌 |	卌 ||	26
8	卌 |	卌 |||	卌 ||||	23
9	卌 |	|||	卌 ||	16
10	||||	||||	||	10
11	卌	||	卌	12
12	|	|		2
Total	50	50	50	150

The empirical probabilities can now be calculated:

$$P(2) = \tfrac{2}{150} \approx .01 \qquad P(3) = \tfrac{7}{150} \approx .05 \qquad P(4) = \tfrac{14}{150} \approx .09$$
$$P(5) = \tfrac{15}{150} = .10 \qquad P(6) = \tfrac{23}{150} \approx .15 \qquad P(7) = \tfrac{26}{150} \approx .17$$
$$P(8) = \tfrac{23}{150} \approx .15 \qquad P(9) = \tfrac{16}{150} \approx .11 \qquad P(10) = \tfrac{10}{150} \approx .07$$
$$P(11) = \tfrac{12}{150} = .08 \qquad P(12) = \tfrac{2}{150} \approx .01$$

■

EXAMPLE 14   Find the theoretical probabilities for rolling a pair of dice.

Solution   The sample space is $\{2, 3, 4, 5, 6, 7, 8, 9, 10, 11, 12\}$. If we assume that each simple event is equally likely, then

$$P(2) = \tfrac{1}{11} \approx .09$$

This same result should then hold for any of the numbers 2–12. But these theoretical probabilities do not correspond to the probabilities found in Example 13 by actually conducting an experiment of rolling a pair of dice. We know these two numbers should be approximately the same. The difficulty here is our assumption that the outcomes in this sample space are equally likely. You *must* find a sample space of *equally likely possibilities* to use the model given in this section. To find the correct sample space, let us use the fundamental counting principle. Since each of the two dice can be arranged in six ways, the total number of arrangements is $6 \cdot 6 = 36$. If you picture the dice as two different colors, you can see these arrangements in Figure 6.2.

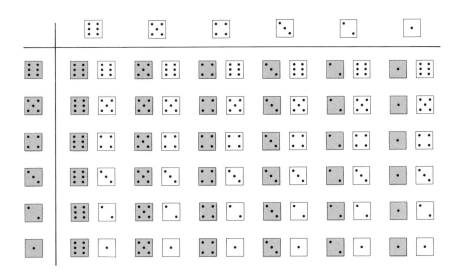

Figure 6.2 Sample space
for tossing a pair of dice

Using this model, we find

$$P(2) = \frac{1}{36}$$ ← Number of successful possibilities in sample space
← Number of equally likely possibilities
$$P(3) = \frac{2}{36}$$ ← Look at Figure 6.2

$$P(4) = \frac{3}{36}$$ ∎

After calculating the other probabilities in Example 14 (you will be asked to do
some of these in Problem Set 6.2), we can compare *these* theoretical probabilities
with the empirical probabilities and find consistent results, as shown in Table 6.2.

TABLE 6.2
Comparison of the empirical
and theoretical probabilities for
rolling a pair of dice

Outcome	Theoretical probability	Empirical probability
2	.0278	.0133
3	.0556	.0467
4	.0833	.0933
5	.1111	.1000
6	.1389	.1533
7	.1667	.1733
8	.1389	.1533
9	.1111	.1067
10	.0833	.0667
11	.0556	.0800
12	.0278	.0133

Problem Set 6.2

Use the spinner shown here for Problems 1–3 and find the requested probabilities. Assume that the pointer can never lie on a border line.

1. P(white)   2. P(red)   3. P(black)

Consider the jar containing marbles shown here. Suppose each marble has an equal chance of being picked from the jar. Find the probabilities in Problems 4–9.

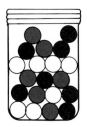

4. P(white)   5. P(red)
6. P(black)   7. P(red or black)
8. P(white or black)   9. P(white and black)

A single card is selected from a deck of 52 cards. Find the probabilities in Problems 10–15.

10. P(5 of clubs)   11. P(5)
12. P(club)   13. P(jack and spade)
14. P(jack)   15. P(jack or spade)

APPLICATIONS

Give the probabilities in Problems 16–19 in decimal form (correct to two decimal places).

16. Last year, 1485 calculators were returned to the manufacturer. If 85,000 were produced, assign a number to specify the probability that a particular calculator would be returned.

17. Last semester, Professor Math gave 13 A's out of 285 grades. If the grades were assigned randomly, what is the probability of an A?

18. Last year, it rained on 85 days in a certain city. What is the probability of rain on a day selected at random?

19. A campus club is having a raffle and needs to sell 1500 tickets. If the people on your floor of the dorm buy 285 of the tickets, what is the probability that someone on your floor will hold the winning ticket?

Perform the experiments described in Problems 20–22, tally your results, and calculate the empirical probabilities for each outcome (to the nearest hundredth).

20. Flip three coins simultaneously 50 times and note the results. The possible outcomes are (1) three heads, (2) two heads and one tail, (3) two tails and one head, and (4) three tails. Do these appear to be equally likely outcomes?

21. Simultaneously toss a coin and roll a die 50 times, and note the results. The possible outcomes are 1H, 1T, 2H, 2T, 3H, 3T, 4H, 4T, 5H, 5T, 6H, and 6T. Do these appear to be equally likely outcomes?

22. Prepare three cards that are identical except for the color. One card is black on both sides, one is white on both sides, and one is black on one side and white on the other. One card is selected at random and placed flat on the table. You will see either a black or a white card; record the color of the face. This is not the event with which we are concerned; rather, we are interested in finding the probability of the *other* side being black or white. Record the color of the underside, as shown in the table:

Color of face	Frequency	Outcome (color of the underside)	Frequency	Probability
White		White		
		Black		
Black		White		
		Black		

Repeat the experiment 50 times and find the probability of occurrence with respect to the known color. Do these appear to be equally likely outcomes?

Use the sample space shown in Figure 6.2 to find the probabilities that the sums of the top faces on the dice are the numbers requested in Problems 23–34.

23. P(5)   24. P(6)   25. P(7)   26. P(8)
27. P(9)   28. P(10)   29. P(11)   30. P(12)
31. P(4 or 5)   32. P(4 and 5)
33. P(even number)   34. P(odd number)

35. Dice is a popular game in gambling casinos. Two dice are tossed, and various amounts are paid according to the outcome. If a 7 or 11 occurs on the first roll, the player wins. What is the probability of winning on the first roll?

36. In the game of dice, the player loses if the outcome of the first roll is a 2, 3, or 12. What is the probability of losing on the first roll?

37. In the game of dice, a pair of 1s is called *snake eyes*. What is the probability of losing a dice game by rolling snake eyes?

38. Consider a die with only four sides, marked 1, 2, 3, and 4. Write out a sample space similar to the one shown in Figure 7.2 for rolling a pair of these dice.

39. Using the sample space you found in Problem 38, find the probability that the sum of the dice is the given number. Assume equally likely outcomes.
    a. P(2)   **b.** P(3)   **c.** P(4)

40. Using the sample space you found in Problem 38, find the probability that the sum of the dice is the given number. Assume equally likely outcomes.
    a. P(5)   **b.** P(6)   **c.** P(7)

41. The game of Dungeons and Dragons uses nonstandard dice. Consider a die with eight sides marked 1, 2, 3, 4, 5, 6, 7, and 8. Write out a sample space similar to the one shown in Figure 7.2 for rolling a pair of these dice.

42. **Historical Question** The Romans played many dice games using a stone with 14 faces marked with the roman numerals I to XIV. Assuming that each face of this die has an equally likely chance of occurring, find the requested probabilities when one such stone is tossed.
    a. P(V)   **b.** P(VII)   **c.** P(X)   **d.** P(XV)

43. **Historical Question** The Romans played a game of chance that required the participants to roll a pair of dice like the one described in Problem 42. Find the probability that the sum of the faces is the given number.
    a. P(20)   **b.** P(2)   **c.** P(15)   **d.** P(25)

6.3 Calculated Probabilities

In Section 6.2 our focus was on deciding whether the events were mutually exclusive and equally likely. Then we applied the formula

$$P(E) = \frac{\text{Number of favorable outcomes}}{\text{Number of all possible outcomes}}$$

In this section we focus on different ways of counting the number of all possible outcomes and the number of favorable outcomes. You will need to use the fundamental counting principle, permutations, and combinations. It will also be helpful to have a calculator to put your answers in decimal form.

In Examples 1–3 assume that in an assortment of 20 electronic calculators there are 5 with defective switches.

EXAMPLE 1 If one machine is selected at random, what is the probability of picking one with a defective switch?

Solution The solution is given by

$$\frac{\text{Number of ways of selecting 1 defective from the 5}}{\text{Number of ways of selecting 1 machine from the 20}} = \frac{5}{20} = \frac{1}{4}$$

Now, to use the terminology of combinations in anticipation of more complicated problems, we can rework this problem:

$$\frac{\binom{5}{1}}{\binom{20}{1}} = \frac{5}{20} = .25$$

EXAMPLE 2   If two machines are selected at random, what is the probability that they both have defective switches?

Solution   The solution is given by

$$\frac{\left(\begin{array}{c}\text{Number of ways of selecting}\\\text{2 defectives from the 5}\end{array}\right)}{\left(\begin{array}{c}\text{Number of ways of selecting}\\\text{2 machines from the 20}\end{array}\right)} = \frac{\binom{5}{2}}{\binom{20}{2}} = \frac{\dfrac{5\cdot 4}{2}}{\dfrac{20\cdot 19}{2}}$$

$$= \frac{1}{19} \approx .05$$ ∎

EXAMPLE 3   If three machines are selected at random, what is the probability that exactly one has a defective switch?

Solution   In this problem we use the fundamental counting principle to determine the number of successes. Picking one machine with a defective switch:

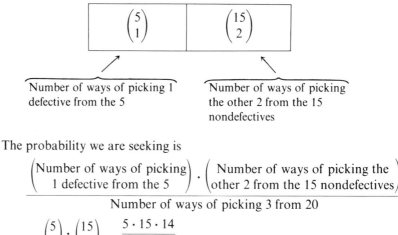

The probability we are seeking is

$$\frac{\left(\begin{array}{c}\text{Number of ways of picking}\\\text{1 defective from the 5}\end{array}\right) \cdot \left(\begin{array}{c}\text{Number of ways of picking the}\\\text{other 2 from the 15 nondefectives}\end{array}\right)}{\text{Number of ways of picking 3 from 20}}$$

$$= \frac{\binom{5}{1}\cdot\binom{15}{2}}{\binom{20}{3}} = \frac{\dfrac{5\cdot 15\cdot 14}{2}}{\dfrac{20\cdot 19\cdot 18}{3\cdot 2\cdot 1}} = \frac{5\cdot 15\cdot 7}{20\cdot 19\cdot 3} = \frac{35}{76} \approx .46$$ ∎

EXAMPLE 4   Extrasensory perception (ESP) can be tested by using five colored cards. The subject is asked to arrange the red, blue, green, black, and white cards in the same order in which a person in another room has arranged them. What is the probability that the subject would arrange the cards in the same order by chance?

Solution   Let $E = \{$proper arrangement$\}$.

$$P(E) = \frac{\text{Number of proper arrangements}}{\text{Number of possible arrangements}} = \frac{1}{{}_5P_5}$$

$$= \frac{1}{5!} = \frac{1}{120} \approx .008$$ ∎

EXAMPLE 5 What is the probability of being dealt a flush in poker?

Solution A flush is 5 cards (from a deck of 52 cards) in the same suit (hearts, diamonds, clubs, or spades).

$$P(\text{flush}) = \frac{\text{Number of ways of obtaining a flush}}{\text{Number of possible poker hands}}$$

$$= \frac{(\text{Number of suits}) \cdot (\text{Number of flushes of a particular suit})}{\text{Number of ways of drawing 5 cards from 52}}$$

$$= \frac{\binom{4}{1} \cdot \binom{13}{5}}{\binom{52}{5}} \qquad \text{Fundamental counting principle}$$

$$= \frac{4 \cdot 1287}{2{,}598{,}960} \qquad \text{Detail of work:} \quad \binom{4}{1} = 4$$

$$\approx .00198 \qquad\qquad \binom{13}{5} = \frac{13!}{5!(13-5)!} = 1287 \qquad \blacksquare$$

$$\binom{52}{5} = \frac{52!}{5!(52-5)!} = 2{,}598{,}960$$

EXAMPLE 6 What is the probability that a family that has five children has exactly two girls?

Solution Assume that the probability of a boy or a girl is the same.

$$P(\text{exactly 2 girls}) = \frac{\text{Number of families of 5 having exactly 2 girls}}{\text{Number of all possible 5-children families}}$$

$$= \frac{\binom{5}{2}}{2^5} \qquad \begin{array}{l} \leftarrow \text{Ways of selecting 2 out of 5} \\[6pt] \leftarrow \text{Fundamental counting principle;} \\ \quad \text{2 possibilities for each of the} \\ \quad \text{5 children} \end{array}$$

$$= \frac{10}{32}$$

$$\approx .3125 \qquad\qquad\qquad\qquad\qquad\qquad\qquad \blacksquare$$

An easily calculated probability often involves **complementary events**. Events are complementary if they are mutually exclusive and together make up the entire sample space. The complement of any event E is denoted by $\bar{E}$. Given a sample space S and any event E,

$$E \cup \bar{E} = S \qquad \text{and} \qquad E \cap \bar{E} = \varnothing$$

so that

$$1 = P(S) = P(E \cup \bar{E}) = P(E) + P(\bar{E})$$

Therefore

$$P(E) = 1 - P(\bar{E})$$

This formula is used when it is easier to find $P(\bar{E})$ than it is to find $P(E)$.

EXAMPLE 7 Find the probability (to the nearest hundredth) that a poker hand (5 cards) has at least one ace.

Solution Direct calculation involves finding the probabilities of having one ace, two aces, three aces, or four aces in a hand. But if $E = \{$at least one ace$\}$, then $\bar{E} = \{$not at least one ace$\} = \{$no aces$\}$. It is easier to find $P(\bar{E})$ than it is to find $P(E)$:

$$P(\bar{E}) = P(\text{no aces}) = \frac{\dbinom{48}{5}}{\dbinom{52}{5}} \quad \begin{array}{l} \leftarrow \text{5 out of 48 cards that are not aces} \\[2em] \leftarrow \text{5 out of 52 cards in deck} \end{array}$$

$$\approx .6588 \qquad \text{Using a calculator}$$

Thus

$$P(E) = 1 - P(\bar{E}) \approx 1 - .6588 \approx .34 \qquad \blacksquare$$

Probabilities are often stated in terms of **odds**. There are two ways of stating odds: *odds in favor* and *odds against*. If the odds against your winning are 2 to 5, then the odds in favor of your winning are 5 to 2. In general:

Odds in Favor

The *odds in favor of an event E*, where $P(E)$ is the probability that E will occur, are

$$P(E) \quad \text{to} \quad 1 - P(E) \qquad \text{or} \qquad \frac{P(E)}{1 - P(E)} = \frac{P(E)}{P(\bar{E})}$$

Odds Against

The *odds against an event E* are

$$1 - P(E) \quad \text{to} \quad P(E) \qquad \text{or} \qquad \frac{1 - P(E)}{P(E)} = \frac{P(\bar{E})}{P(E)}$$

EXAMPLE 8 If a contest has 1000 entries and you purchase 10 tickets, what are the odds against your winning?

Solution $$P(\text{winning}) = \frac{10}{1000} = \frac{1}{100} \qquad 1 - P(\text{winning}) = \frac{99}{100}$$

Odds against:

$$\frac{\frac{99}{100}}{\frac{1}{100}} = \frac{99}{100} \times \frac{100}{1} = \frac{99}{1}$$

The odds against winning are 99 to 1. $\qquad \blacksquare$

In Example 8 we found the odds by first calculating the probability. Suppose, instead, we are given the odds and wish to calculate the probability.

Procedure for Finding Probability Given the Odds

If the odds in favor of an event E are s to b, then the probability of E is given by

$$P(E) = \frac{s}{b + s}$$

EXAMPLE 9 If the odds in favor of some event are 2 to 5, what is the probability?

Solution In this example, $s = 2$ and $b = 5$, so $P(E) = \frac{2}{7}$. ∎

EXAMPLE 10 If the odds against you are 100 to 1, what is the probability?

Solution First change (mentally) to odds in favor: 1 to 100. Then, $P(E) = \frac{1}{101}$. ∎

Problem Set 6.3

APPLICATIONS

It is known that a company has 10 pieces of machinery with no defects, 4 with minor defects, and 2 with major defects. Thus each of the 16 machines falls into one of these three categories. The inspector is about to come, and it is her policy to choose one machine and check it. Find the requested probabilities in Problems 1–6.

1. P(no defects)
2. P(major defect)
3. P(minor defect)
4. P(no major defects)
5. P(no minor defects)
6. P(defect)

Suppose a jar contains six red balls, four white balls, and three black balls. One ball is drawn at random. Find the requested probabilities in Problems 7–12.

7. P(red)
8. P(white)
9. P(black)
10. P(black or white)
11. P(red or white)
12. P(black and white)

13. What are the odds in favor of drawing a heart from an ordinary deck of 52 cards?
14. What are the odds against a family with four children having four boys?
15. Suppose the odds that a man will be bald by the time he is 60 are 9 to 1. State this as a probability.
16. Suppose the odds are 33 to 1 that someone will lie to you at least once in the next 7 days. State this as a probability.
17. What are the odds in favor of drawing an ace from an ordinary deck of 52 cards?

A certain magic trick with cards requires that five cards be randomly placed side by side in a line. Find the requested probabilities in Problems 18–23.

18. P(first card at the left is the ace of spades)
19. P(middle card is a heart)
20. P(first and last cards are diamonds)
21. P(left 3 cards are clubs and the next two are not)
22. P(second and fourth cards are kings)
23. P(middle 3 cards are red)

Suppose that a family has four children. Find the requested probabilities in Problems 24–28.

24. P(exactly 2 boys)
25. P(exactly 1 girl)
26. P(all boys)
27. P(2 girls and 2 boys)
28. P(3 girls and 1 boy)
29. What is the probability of flipping a coin five times and obtaining exactly two heads?
30. What is the probability of flipping a coin five times and obtaining three heads and two tails?
31. What is the probability of flipping a coin six times and obtaining exactly three heads?
32. Racetracks quote the approximate odds for a particular race on a large display board called a tote board. Some examples from a particular race are:

Horse number	Odds
1	2 to 1
2	15 to 1
3	3 to 2
4	7 to 5
5	1 to 1

The odds stated are for the horse losing. Thus

$$P(\text{horse 1 losing}) = \frac{2}{2+1} = \frac{2}{3}$$

$$P(\text{horse 1 winning}) = 1 - \frac{2}{3} = \frac{1}{3}$$

What would be the probability of winning for each of these horses?

33. What are the odds in favor of rolling a 7 or 11 on a single roll of a pair of dice?

For Problems 34–39 assume that the inspector described in Problems 1–6 selects 2 machines at random. Find the requested probabilities as decimals rounded to the nearest hundredth.

34. P(both defective)
35. P(both nondefective)
36. P(both have major defects)
37. P(both have minor defects)
38. P(one with no defects and one with major defects)
39. P(one with a major defect and one with a minor defect)

For Problems 40–45 suppose two balls are drawn at random from the jar described in Problems 7–12. Find the requested probabilities as decimals rounded to the nearest hundredth.

40. P(both red)

41. P(both white)

42. P(both black)

43. P(1 red and 1 white)

44. P(1 black and 1 white)

45. P(1 red and 1 black)

For Problems 46–51 assume that the inspector described in Problems 1–6 selects 3 machines at random. Find the requested probabilities as decimals rounded to the nearest hundredth.

46. P(all have major defects)   **47.** P(all have minor defects)

48. P(all are nondefective)

49. P(exactly one major defect)

50. P(exactly one minor defect)

51. P(one of each type)

For Problems 52–58 suppose three balls are drawn at random from the jar described in Problems 7–12. Find the requested probabilities as decimals rounded to the nearest hundredth.

52. P(3 red)   **53.** P(3 white)   **54.** P(3 black)

55. P(2 red and 1 white)   **56.** P(2 white and 1 black)

57. P(2 black and 1 red)   **58.** P(one of each color)

Find the requested probabilities in Problems 59–61 as decimals correct to eight decimal places.

59. P(royal flush). (*Note:* A royal flush is an ace, king, queen, jack, and 10 of one suit.)

60. P(full house of three aces and a pair of 2s)

61. P(pair). [*Note:* A hand better than a pair (such as three of a kind) is not called a pair.]

62. **Historical Question**   The mathematical theory of probability arose in France in the seventeenth century when a gambler, the Chevalier de Méré, was interested in adjusting the stakes so that he could be certain of winning if he played long enough. He was betting that he could get at least one 6 in four rolls of a die. He also bet that in 24 tosses of a pair of dice, he would get at least one 12. He found that he won more often than he lost with the first bet, but not with the second. He did not know why, so he wrote to the mathematician Blaise Pascal (1623–1662) to find out why. Pascal sent these questions to another mathematician, Pierre de Fermat (1601–1665), and together they developed the first theory of probability. What are the probabilities for winning in the two games described by de Méré?

6.4 Conditional Probability

The probability of an event depends on what information is known about that event. For example, suppose you know that a family has two children and you are interested in the probability that both the children are boys. This probability depends on additional information, as illustrated by Example 1.

EXAMPLE 1   What is the probability that a family with two children has two boys if:

a. You have no additional information.

b. You know that there is at least one boy.

c. You know that the youngest child is a boy.

Solution   Let

$$D = \{2 \text{ boys}\}$$

$$E = \{\text{at least 1 boy}\}$$

$$F = \{\text{youngest is a boy}\}$$

a. Consider the sample space:

Sample space

BB   ← Success ⎫
BG   ⎬ Four possibilities
GB   ⎪
GG   ⎭

Thus $P(D) = \frac{1}{4}$.

b. The sample space from part a is *altered* because of the additional information:

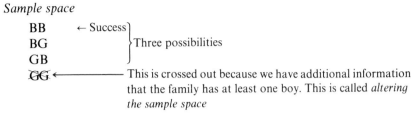

Sample space

BB   ← Success ⎫
BG   ⎬ Three possibilities
GB   ⎭
~~GG~~ ←———————— This is crossed out because we have additional information that the family has at least one boy. This is called *altering the sample space*

We write $P(D|E)$ to mean the probability of D *given the additional information* that E has occurred. Thus $P(D|E) = \frac{1}{3}$.

c. Again consider the altered sample space:

Sample space

BB
BG
~~GB~~ ⎫ These are crossed out because you know that the youngest child
~~GG~~ ⎭ is a boy

Thus $P(D|F) = \frac{1}{2}$. ∎

In Example 1 we needed a notation to represent the probability of an event E *given that an event F has occurred*. This is the idea of **conditional probability**. We write $P(E|F)$ to denote the probability of an event E *given* that an event F has occurred. One way of finding a conditional probability is to consider an altered sample space.

EXAMPLE 2   Consider rolling a pair of dice.

a. What is the probability that the sum is 6?
b. What is the probability that the sum is 6 if you know that one of the dice shows a 5?
c. What is the probability that at least one of the numbers showing is a 6?
d. What is the probability of rolling a 2, 3, or 12 (this is known as *craps*) if you know that at least one of the dice shows a 6?

Solution   The sample space is shown in Figure 6.2 (page 207).

a. Out of a sample space consisting of 36 equally likely pairs (x, y), there are five ways of obtaining a 6: $(5, 1), (4, 2), (3, 3), (2, 4),$ and $(1, 5)$. Thus $P(6) = \frac{5}{36}$.
b. The sample space is now reduced to 11 possibilities: $(6, 5), (5, 5), (4, 5), (3, 5), (2, 5), (1, 5), (5, 6), (5, 4), (5, 3), (5, 2),$ and $(5, 1)$. Of these, two are considered success. Thus $P(6|\text{one is a 5}) = \frac{2}{11}$.
c. There are 11 ways to have at least one 6: $(6, 6), (5, 6), (4, 6), (3, 6), (2, 6), (1, 6), (6, 5), (6, 4), (6, 3), (6, 2),$ and $(6, 1)$. There are 36 equally likely pairs (x, y), so $P(\text{at least one number shown is a 6}) = \frac{11}{36}$.
d. The sample space is reduced to those pairs shown in part c, and a 2, 3, or 12 can be obtained as follows: $(1, 1), (1, 2), (2, 1),$ and $(6, 6)$. There is one possibility out of the sample space, so $P(2, 3, \text{or } 12 | \text{at least one is a 6}) = \frac{1}{11}$. ∎

EXAMPLE 3 Before an advertising campaign for a new product is launched, a survey is conducted to determine the present usage for a certain brand of hair dryer. The results are shown in the table. Assume that one respondent is chosen at random from this sample of 1400. Let

$$M = \{\text{male}\} \qquad D = \{\text{daily use}\}$$
$$F = \{\text{female}\} \qquad C = \{\text{occasional use}\}$$

	Daily use	Occasional use	Total
Male	142	258	400
Female	619	381	1000
Total	761	639	1400

Find each of the following (to the nearest hundredth) and interpret:

a. $P(M)$ **b.** $P(D)$ **c.** $P(M|D)$ **d.** $P(D|M)$

Solution **a.** $P(M)$ is the probability that the respondent is male:

$$P(M) = \frac{400}{1400} \qquad \begin{array}{l} \leftarrow \text{Number of males} \\ \leftarrow \text{Total number of respondents} \end{array}$$
$$\approx .29$$

b. $P(D)$ is the probability that the respondent uses a hair dryer daily:

$$P(D) = \frac{761}{1400} \qquad \begin{array}{l} \leftarrow \text{Number of daily users} \\ \leftarrow \text{Total number of respondents} \end{array}$$
$$\approx .54$$

c. $P(M|D)$ is the probability that the respondent is a male if it is known that the respondent uses a hair dryer daily:

$$P(M|D) = \frac{142}{761} \qquad \left. \begin{array}{l} \leftarrow \text{Number of males} \\ \leftarrow \text{Total number} \end{array} \right\} \text{Reduced sample space; daily users}$$
$$\approx .19$$

d. $P(D|M)$ is the probability that the respondent uses a hair dryer daily if it is known that the respondent is a male:

$$P(D|M) = \frac{142}{400} \qquad \left. \begin{array}{l} \leftarrow \text{Number of daily users} \\ \leftarrow \text{Total number} \end{array} \right\} \text{Reduced sample space; males}$$
$$\approx .36 \qquad\qquad\qquad\qquad\qquad\qquad\qquad\qquad \blacksquare$$

Examples 2 and 3 were designed to help you understand the idea of conditional probability, but these are simple examples with simple reduced sample spaces. It is not always practical to list the sample space and then the reduced sample space, so we seek to define $P(E|F)$ in such a way as to tell us how to calculate $P(E|F)$ without listing the sample space.

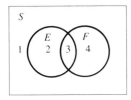

Figure 6.3

Consider the Venn diagram for two events E and F in Figure 6.3. We seek $P(E\,|\,F)$. Let $n(E)$, $n(F)$, and $n(E \cap F)$ be the number of times that events E, F, and $E \cap F$ have occurred among the n repetitions of the experiment. Now $P(E\,|\,F)$ can be found by focusing our attention on region 3. We calculate the number

$$\frac{n(E \cap F)}{n(F)}$$

which represents the number of times E has occurred among those outcomes in which F has occurred. Now

$$\frac{n(E \cap F)}{n(F)} = \frac{\dfrac{n(E \cap F)}{n}}{\dfrac{n(F)}{n}} = \frac{P(E \cap F)}{P(F)}$$

This leads to the following definition:

Conditional Probability

> The *conditional probability* that an event E has occurred given that event F has occurred is denoted by $P(E\,|\,F)$ and defined by
>
> $$P(E\,|\,F) = \frac{P(E \cap F)}{P(F)} \qquad \text{provided} \quad P(F) \neq 0$$

EXAMPLE 4 A tire manufacturer has found that 10% of the tires produced have cosmetic defects and 2% have both cosmetic and structural defects. What is the probability that one tire selected at random is structurally defective if it is known that it has a cosmetic defect?

Solution Let $C = \{\text{cosmetic defect}\}$ and $S = \{\text{structural defect}\}$; find $P(S\,|\,C)$:

$$P(S\,|\,C) = \frac{P(S \cap C)}{P(C)}$$

$$= \frac{\frac{2}{100}}{\frac{10}{100}}$$

$$= \tfrac{1}{5} \qquad \blacksquare$$

The conditional probability $P(E\,|\,F)$ is the probability that E occurs when it is known that F occurs. Sometimes, though, the occurrence of F does not affect $P(E)$. In such circumstances we say that E and F are **independent**.

Independent Events

> The events E and F are said to be *independent* if
>
> $$P(E\,|\,F) = P(E)$$
>
> That is, the occurrence of E is not affected by the occurrence or nonoccurrence of F.

EXAMPLE 5 Suppose a coin is tossed twice. Let

$$H_1 = \{\text{head on first toss}\}$$
$$H_2 = \{\text{head on second toss}\}$$

These events seem to be independent since the occurrence of one event does not affect the occurrence of the other. We can verify this by using the definition.

Sample space

HH	$P(H_1) = \frac{2}{4} = \frac{1}{2}$
HT	$P(H_2) = \frac{2}{4} = \frac{1}{2}$
TH	$P(H_1 \cap H_2) = \frac{1}{4}$
TT	

These probabilities are found by looking at the sample space

Thus

$$P(H_1 | H_2) = \frac{P(H_1 \cap H_2)}{P(H_2)} = \frac{\frac{1}{4}}{\frac{1}{2}} = \frac{1}{2} = P(H_1)$$

Since $P(H_1 | H_2) = P(H_1)$, we know that H_1 and H_2 are independent. ∎

EXAMPLE 6 An experiment consists of drawing two consecutive cards from an ordinary deck of cards. Let $E = \{\text{first card is red}\}$ and $F = \{\text{second card is black}\}$. Are these events independent if:

a. The cards are drawn with replacement?
b. The cards are drawn without replacement?

Solution To verify independence we need to calculate $P(F | E)$ and compare this with $P(F)$; the events are independent if

$$P(F | E) = P(F)$$

a. $P(F | E) = \dfrac{26}{52}$ ← Number of black cards in deck
   ← Number of cards in deck

$P(F) = \dfrac{1}{2}$ There are the same number of outcomes with the second card black as there are with the second card red

Thus $P(F | E) = P(F)$, so these are independent events. This should be consistent with your intuition about these events.

b. $P(F | E) = \dfrac{26}{51}$ ← Number of black cards in deck
   ← Number of cards in deck (remember, this time the card is not replaced)

$P(F) = \dfrac{1}{2}$ This event, by itself, has the same probability whether or not the experiment is done with replacement

Thus $P(F | E) \neq P(F)$, so these are not independent events. ∎

There is an alternate way to check for independence. Suppose that E and F are independent. Then

$$P(E | F) = P(E)$$

Suppose we rewrite this expression:

$$P(E) = P(E \mid F) = \frac{P(E \cap F)}{P(F)}$$

Therefore, by multiplication of both sides of the equation,

$$P(E) \cdot P(F) = P(E \cap F)$$

provided E and F are independent. We can also carry out the same calculation in reverse to arrive at the following test for independence:

Test for Independent Events

Events E and F are independent if and only if
$$P(E \cap F) = P(E) \cdot P(F)$$

EXAMPLE 7 Toss a single die twice. Let

$E = \{$first toss is a prime$\}$
$F = \{$first toss is a 3$\}$
$G = \{$second toss is a 2$\}$
$H = \{$second toss is a 3$\}$

Decide which of these events are independent.

Solution The sample space for both tosses has 36 possibilities and is shown in Figure 6.2 (page 207). Using this sample space, we find:

$P(E) = \frac{18}{36} = \frac{1}{2}$

$P(F) = \frac{6}{36} = \frac{1}{6}$ $P(E \cap F) = \frac{6}{36} = \frac{1}{6}$

$P(G) = \frac{6}{36} = \frac{1}{6}$ $P(E \cap G) = \frac{3}{36} = \frac{1}{12}$ $P(F \cap G) = \frac{1}{36}$

$P(H) = \frac{6}{36} = \frac{1}{6}$ $P(E \cap H) = \frac{3}{36} = \frac{1}{12}$ $P(F \cap H) = \frac{1}{36}$ $P(G \cap H) = 0$

Therefore:

E and F are not independent since $P(E \cap F) \neq P(E) \cdot P(F)$.
E and G are independent since $P(E \cap G) = P(E) \cdot P(G)$.
F and G are independent since $P(F \cap G) = P(F) \cdot P(G)$.
E and H are independent since $P(E \cap H) = P(E) \cdot P(H)$.
F and H are independent since $P(F \cap H) = P(F) \cdot P(H)$.
G and H are not independent since $P(G \cap H) \neq P(G) \cdot P(H)$. ∎

Problem Set 6.4

In Problems 1–8 consider a family with three children and find the requested probabilities.

1. P(exactly 2 boys)

2. P(exactly 1 girl)

3. P(exactly 3 girls)

4. P(exactly 2 boys given that at least 1 child is a boy)

5. P(exactly 1 girl given that at least 1 child is a girl)

6. P(exactly 3 girls given that at least 1 child is a girl)

7. P(at least 1 girl)

8. P(all girls given that there is at least 1 girl)

Suppose a coin is flipped four times. Find the requested probabilities in Problems 9–14.

9. P(exactly 3H)   10. P(exactly 2H)

11. P(exactly 3T)

12. P(exactly 3H given that there are at least 2H)

13. P(exactly 2H given that there is at least 1H)

14. P(exactly 3T given that there is at least 1T)

Consider rolling a pair of dice. Find the probabilities requested in Problems 15–18.

15. P(sum is 7)

16. P(at least one 5 is rolled)

17. P(sum is 7 | 2 is on one of the dice)

18. P(sum is 2 or 3 | at least one 5 is rolled)

APPLICATIONS

In an experiment it is necessary to examine fruit flies and determine their sex and whether they have mutated after exposure to radiation. For 1000 fruit flies examined, there are 643 females and 357 males. Also, 403 of the females are normal and 240 are mutated, while 190 of the males are normal and 167 are mutated. In Problems 19–28 assume a single fruit fly is chosen at random from the sample of 1000 and calculate the requested probabilities to the nearest hundredth.

19. P(male)   20. P(female)

21. P(normal male)   22. P(mutated male)

23. P(normal | male)   24. P(mutated | male)

25. P(male | normal)   26. P(male | mutated)

27. P(female | mutated)   28. P(mutated | female)

In Problems 29–34 suppose that E and F are events with the given probabilities. Use the definition of conditional probability to calculate both P(E | F) and P(F | E).

29. $P(E) = .5$, $P(F) = .2$, $P(E \cap F) = .1$

30. $P(E) = .85$, $P(F) = .45$, $P(E \cap F) = .3$

31. $P(E) = \frac{1}{3}$, $P(F) = \frac{1}{2}$, $P(E \cap F) = \frac{1}{6}$

32. $P(E) = \frac{3}{5}$, $P(F) = \frac{7}{10}$, $P(E \cap F) = \frac{1}{2}$

33. $P(E) = .5$, $P(F) = .8$, $P(E \cap F) = .4$

34. $P(E) = \frac{6}{7}$, $P(F) = \frac{2}{3}$, $P(E \cap F) = \frac{4}{7}$

35. The Ross Light Bulb Company has found that 5% of the bulbs it manufactures have defective filaments and 3% have both defective filaments and defective workmanship. What is the probability of defective workmanship in a particular bulb if you know it has a defective filament?

36. At Southeastern University a survey of students taking both college algebra and statistics found that 25% dropped college algebra, 30% dropped statistics, and 10% dropped both courses. If a person dropped college algebra,

what is the probability that the person also dropped statistics?

37. The probability of tossing a coin four times and obtaining four heads in a row is $P(4H) = \frac{1}{2} \cdot \frac{1}{2} \cdot \frac{1}{2} \cdot \frac{1}{2} = \frac{1}{16}$. What is the probability of tossing a coin and obtaining a head if we know that heads have occurred on the previous four flips of the coin?

38. One "system" used by roulette players is to watch a game of roulette until a large number of reds occur in a row (say, 10). After 10 successive reds, they reason that black is "due" to occur and begin to bet large sums on black. Where is the fallacy in the reasoning of this "system" betting?

Use the definition in Problems 39–42 to determine whether events E and F are independent.

39. $P(E) = .5$, $P(F) = .2$, $P(E \cap F) = .1$

40. $P(E) = \frac{3}{5}$, $P(F) = \frac{7}{10}$, $P(E \cap F) = \frac{1}{2}$

41. $P(E) = \frac{6}{7}$, $P(F) = \frac{2}{3}$, $P(E \cap F) = \frac{4}{7}$

42. $P(E) = .5$, $P(F) = .8$, $P(E \cap F) = .4$

Suppose a die is rolled twice. Let

$$A = \{ \text{first toss is an even} \}$$
$$B = \{ \text{first toss is a } 6 \}$$
$$C = \{ \text{second toss is a } 2 \}$$
$$D = \{ \text{second toss is a } 3 \}$$

Determine whether the events in Problems 43–48 are independent.

43. A and B 44. A and C 45. A and D

46. B and C 47. B and D 48. C and D

In Problems 49–54, consider a family with four children. Let

$$E = \{ 2 \text{ boys and 2 girls} \}$$
$$F = \{ \text{exactly 1 boy} \}$$
$$G = \{ \text{at most 1 boy} \}$$
$$H = \{ \text{at least 1 child of each sex} \}$$

Determine whether the events are independent.

49. E and F 50. E and G 51. E and H

52. F and G 53. F and H 54. G and H

55. **Birthday Problem** Consider the set of people in your classroom or the set of U.S. presidents. What would you guess is the probability that two persons of the group are born on the same day of the year? As it turns out, Polk and Harding were both born on November 2. If any 24 or more people are selected at random, the probability that two or more of them will have the same birthday is greater than 50%! This is a seemingly paradoxical situation that will

fool most people. The actual probabilities for the birthday problem are shown in the figure. Verify a value in this figure by experimentation. Use your class, a biographical dictionary, or other list of birth dates to see if 2 persons in a group of 24 persons have the same birthday.

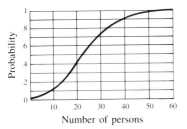

56. Birthday Problem (continued) To calculate the probabilities graphed in the figure, suppose you have a group of

n people. Let E be the event that at least two of the people have the same birthday. It is very difficult to find $P(E)$, but not too difficult to find $P(\bar{E})$. If we assume that each year has 365 days (ignore leap years), then

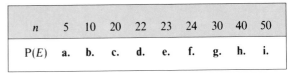

$$P(\bar{E}) = \frac{{}_{365}P_n}{365^n} = \frac{365 \cdot 364 \cdot 363 \cdots (365 - n + 1)}{365^n}$$

Complete the following table of probabilities using the above formula:

n	5	10	20	22	23	24	30	40	50
$P(E)$	a.	b.	c.	d.	e.	f.	g.	h.	i.

6.5 Probability of Intersections and Unions

Some probability models can be built by breaking down the events being considered into simpler ones by using the words *and* (intersection) or *or* (union).

For a formula for the **probability of the intersection of two events**, recall that events E and F are independent if the occurrence of one of these events in no way affects the occurrence of the other; that is, $P(E|F) = P(E)$. Therefore, from the conditional probability formula,

$$P(E|F) = \frac{P(E \cap F)}{P(F)}$$

$$P(E|F) \cdot P(F) = P(E \cap F)$$

If E and F are independent, $P(E \cap F) = P(E)P(F)$. This formula is now used to derive a formula for the **probability of a union of two events**. From the definition of a probability measure (Section 6.1), we obtain a formula for

$$P(E \cup F) = P(E) + P(F) \qquad \text{if } E \text{ and } F \text{ are mutually exclusive}$$

If E and F are not mutually exclusive, then Figure 6.3 (page 217) leads us to a generalization of this formula:

$$P(E \cup F) = P(E) + P(F) - P(E \cap F)$$

We can now summarize these formulas for the probability of the intersection or union of two events.

Probability of Intersections and Unions

INTERSECTION:	$P(E \cap F) = P(E) \cdot P(F)$ Provided E and F are independent
UNION:	$P(E \cup F) = P(E) + P(F) - P(E \cap F)$

Suppose a coin is tossed and a die is simultaneously rolled. Let $T = \{\text{tail is tossed}\}$, $F = \{4 \text{ is rolled}\}$, and $N = \{\text{odd number is rolled}\}$. Find the probabilities in Examples 1 and 2.

EXAMPLE 1 $P(T \cap F) = P(T) \cdot P(F)$ This is the probability of a tail *and* a 4. Note that T and F

$\qquad\qquad\quad = \frac{1}{2} \cdot \frac{1}{6}$ are independent

$\qquad\qquad\quad = \frac{1}{12}$ ∎

EXAMPLE 2 $P(T \cup F) = P(T) + P(F) - P(T \cap F)$ This is the probability of a tail *or* a 4

$\qquad\qquad\quad = \frac{1}{2} + \frac{1}{6} - \frac{1}{12}$

$\qquad\qquad\quad = \frac{7}{12}$ ∎

In Examples 3–6 suppose a die is rolled twice and let

$A = \{\text{first toss is a prime}\}$ $B = \{\text{first toss is a 3}\}$

$C = \{\text{second toss is a 2}\}$ $D = \{\text{second toss is a 3}\}$

The sample space for this experiment has 36 possible outcomes (see Figure 6.2, page 207). By considering this sample space, we see that

$\qquad P(A) = \frac{18}{36} = \frac{1}{2}$

$\qquad P(B) = P(C) = P(D) = \frac{6}{36} = \frac{1}{6}$

$\qquad P(A \cap B) = \frac{6}{36} = \frac{1}{6}$ Events A and B are not independent because $P(A \cap B) \neq P(A) \cdot P(B)$

$\qquad P(A \cap C) = P(A \cap D) = \frac{3}{36} = \frac{1}{12}$ Events A and C *are* independent because $P(A \cap C) = P(A) \cdot P(C)$

$\qquad P(B \cap C) = P(B \cap D) = \frac{1}{36}$

$\qquad P(C \cap D) = 0$ This means that C and D are mutually exclusive

Can you name other pairs that are or are not independent?

EXAMPLE 3 $P(A \cup B) = P(A) + P(B) - P(A \cap B)$ This is the probability that a prime *or*

$\qquad\qquad\quad = \frac{1}{2} + \frac{1}{6} - \frac{1}{6}$ a 3 is obtained on the first toss

$\qquad\qquad\quad = \frac{1}{2}$ ∎

EXAMPLE 4 $P(A \cup C) = P(A) + P(C) - P(A \cap C)$ This is the probability that the first toss is a

$\qquad\qquad\quad = \frac{1}{2} + \frac{1}{6} - \frac{1}{12}$ prime *or* the second toss is a 2

$\qquad\qquad\quad = \frac{7}{12}$ ∎

EXAMPLE 5 $P(B \cup D) = P(B) + P(D) - P(B \cap D)$ This is the probability that the first toss

$\qquad\qquad\quad = \frac{1}{6} + \frac{1}{6} - \frac{1}{36}$ is a 3 *or* the second toss is a 3

$\qquad\qquad\quad = \frac{11}{36}$ ∎

EXAMPLE 6 $P(C \cup D) = P(C) + P(D) - P(C \cap D)$ This is the probability that the second toss

$\qquad\qquad\quad = \frac{1}{6} + \frac{1}{6} - 0$ is a 2 *or* a 3. Note that C and D are

$\qquad\qquad\quad = \frac{1}{3}$ mutually exclusive   ∎

Some experiments are said to be performed *with replacement* and others *without replacement*. Consider the following examples:

Suppose cards are drawn from a deck of 52 cards. Let

$S_1 = \{$draw a spade on the first draw$\}$

$H_1 = \{$draw a heart on the first draw$\}$

$H_2 = \{$draw a heart on the second draw$\}$

To draw **with replacement** means that the first card is drawn, the result is noted, and then the card is replaced in the deck before the second card is drawn. To draw **without replacement** means that one card is drawn, the result is noted, and then a second card is drawn without replacing the first card. Find the probabilities in Examples 7–10.

EXAMPLE 7 Find $P(S_1 \cap H_2)$, with replacement.

Solution S_1 and H_2 are independent since the cards are drawn with replacement.

$$P(S_1 \cap H_2) = P(S_1) \cdot P(H_2) = \tfrac{1}{4} \cdot \tfrac{1}{4} = \tfrac{1}{16}$$ ∎

EXAMPLE 8 Find $P(S_1 \cap H_2)$, without replacement.

Solution In this experiment the events are not independent since the probability of drawing the second card depends on what was drawn on the first card. This problem is equivalent to drawing two cards from a deck of cards. The order in which the cards are drawn is important since the question specifies that the *first* card is a spade and the *second* card is a heart; thus this is a permutation problem.

$$P(S_1 \cap H_2) = \frac{_{13}P_1 \cdot {}_{13}P_1}{_{52}P_2}$$

$$= \frac{13 \cdot 13}{52 \cdot 51} = \frac{13}{204}$$ ∎

EXAMPLE 9 What is the probability of drawing two hearts with replacement?

Solution H_1 and H_2 are independent.

$$P(H_1 \cap H_2) = P(H_1) \cdot P(H_2) = \tfrac{1}{4} \cdot \tfrac{1}{4} = \tfrac{1}{16}$$ ∎

EXAMPLE 10 What is the probability of drawing two hearts without replacement?

Solution H_1 and H_2 are not independent, but the order in which the cards are drawn is not important since the question simply asks for two hearts; thus this is a combination problem.

$$P(\text{two hearts}) = \frac{_{13}C_2}{_{52}C_2}$$

$$= \frac{\dfrac{13 \cdot 12}{2}}{\dfrac{52 \cdot 51}{2}} = \tfrac{1}{17}$$ ∎

We conclude this section by summarizing the probability formulas:

Summary of Probability Formulas

1. $P(E) = \dfrac{s}{n}$ where event E can occur in n mutually exclusive and equally likely ways, and where s of them are considered favorable
2. $P(S) = 1$ where S is the sample space
3. $P(\varnothing) = 0$
4. $P(E) = 1 - P(\bar{E})$
5. $P(E \mid F) = \dfrac{P(E \cap F)}{P(F)}$ provided $P(F) \neq 0$
6. $P(E \text{ and } F) = P(E \cap F) = P(E) \cdot P(F)$ provided E and F are independent
7. $P(E \text{ or } F) = P(E \cup F) = P(E) + P(F) - P(E \cap F)$
8. $P(E \text{ or } F) = P(E \cup F) = P(E) + P(F)$ provided E and F are mutually exclusive

Note: When doing probability problems, assume that experiments are performed *without replacement* unless it is otherwise stated.

Problem Set 6.5

Suppose that events A, B, and C are all independent so that

$$P(A) = \tfrac{1}{2} \qquad P(B) = \tfrac{1}{3} \qquad P(C) = \tfrac{1}{6}$$

Find the probabilities in Problems 1–16.

1. $P(\bar{A})$
2. $P(\bar{B})$
3. $P(\bar{C})$
4. $P(A \cap B)$
5. $P(A \cap C)$
6. $P(B \cap C)$
7. $P(A \cup B)$
8. $P(A \cup C)$
9. $P(B \cup C)$
10. $P(\overline{A \cap B})$
11. $P(\overline{A \cap C})$
12. $P(\overline{B \cap C})$
13. $P(\overline{A \cup B})$
14. $P(\overline{A \cup C})$
15. $P(\overline{B \cup C})$
16. $P(A \cap B \cap C)$

In Problems 17–21 suppose a coin is tossed twice. Find the requested probabilities.

17. $P(2H)$
18. $P(2T)$
19. $P(3T)$
20. $P(1H \text{ and } 1T)$
21. $P(\text{match})*$

Suppose A, B, and C are independent events so that

$$P(A) = \tfrac{1}{2} \qquad P(B) = \tfrac{2}{3} \qquad P(C) = \tfrac{5}{6}$$

Find the probabilities in Problems 22–25.

22. a. $P(\bar{A})$ b. $P(\bar{B})$ c. $P(\bar{C})$
23. a. $P(A \cap B)$ b. $P(A \cap C)$ c. $P(B \cap C)$
24. a. $P(A \cup B)$ b. $P(A \cup C)$ c. $P(B \cup C)$
25. a. $P(\overline{A \cap C})$ b. $P(\overline{A \cup B})$ c. $P(\overline{B \cup C})$

---

* This means P (both the same).

In Problems 26–38 suppose a coin is tossed and simultaneously a die is rolled. Let $H = \{head \text{ is } tossed\}$, $S = \{6 \text{ is } rolled\}$, and $E = \{even \text{ number is } rolled\}$. Find the requested probabilities.

26. a. $P(H)$ b. $P(S)$ c. $P(E)$
27. $P(H \cap S)$ 28. $P(H \cap E)$ 29. $P(S \cap E)$
30. $P(S \cup E)$ 31. $P(H \cup E)$ 32. $P(H \cup S)$
33. $P(H \mid S)$ 34. $P(S \mid H)$ 35. $P(S \mid E)$
36. $P(E \mid S)$ 37. $P(H \mid E)$ 38. $P(E \mid H)$

In Problems 39–43 assume a box has five red cards and three black cards, and two cards are drawn with replacement. Find the requested probabilities.

39. P(2 red cards)   40. P(2 black cards)
41. P(1 red and 1 black card)
42. P(red on first draw and black on second draw)
43. P(red on first draw or black on second draw)

Problems 44–48 repeat the experiment of Problems 39–43 except that the cards are drawn without replacement.

44. P(2 red cards)   45. P(2 black cards)
46. P(1 red and 1 black card)
47. P(red on first draw and black on second draw)
48. P(red on first draw or black on second draw)

6.6 Stochastic Processes and Bayes' Theorem

In Section 6.4 we developed the following formula for conditional probability:

$$P(E\,|\,F) = \frac{P(E \cap F)}{P(F)}$$

This formula can be rewritten as

$$P(E \cap F) = P(E\,|\,F)P(F)$$
$$= P(F\,|\,E)P(E)$$

We now use this form to develop a formula for the probability of an event in which the possible outcomes depend on the outcomes of a preceding experiment. For example, in Section 6.3 we discussed an experiment of selecting 2 calculators from a sample space consisting of 20 calculators of which 5 had defective switches. Instead of selecting 2 at the same time, we now select them one at a time and let

$A = \{$first calculator picked has a defective switch$\}$

$B = \{$second calculator picked has a defective switch$\}$

Now, $P(A) = \frac{5}{20} = \frac{1}{4}$, but what is $P(B)$?

This sequence of experiments is called a **stochastic process**, which means that the outcome of one event depends on the outcome of the previous experiment.

EXAMPLE 1 Find $P(B)$ for the experiment just described.

Solution To find $P(B)$ consider the following tree diagram:

Step 1 Label the paths with appropriate probabilities. The sum of these should be 1.

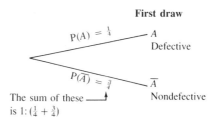

Step 2

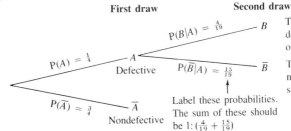

Step 3

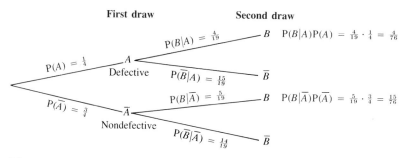

First draw   Second draw

$P(A) = \frac{1}{4}$ A
Defective

$P(B|A) = \frac{4}{19}$ B

$P(\overline{B}|A) = \frac{15}{19}$ $\overline{B}$

$P(\overline{A}) = \frac{3}{4}$ $\overline{A}$
Nondefective

$P(B|\overline{A}) = \frac{5}{19}$ B

$P(\overline{B}|\overline{A}) = \frac{14}{19}$ $\overline{B}$

This is $P(B|\overline{A})$; there are 5 defectives left in a sample of 19.

This is $P(\overline{B}|\overline{A})$; there are 14 nondefectives left in a sample of 19.

The sum of these should be 1: $(\frac{5}{19} + \frac{14}{19})$

Now, to find the probability of B you must take all the possibilities into account. Use the product rule for probability along a single path and the addition principle for different paths.

First draw   Second draw

$P(A) = \frac{1}{4}$ A
Defective

$P(B|A) = \frac{4}{19}$ B $P(B|A)P(A) = \frac{4}{19} \cdot \frac{1}{4} = \frac{4}{76}$

$P(\overline{B}|A) = \frac{15}{19}$ $\overline{B}$

$P(\overline{A}) = \frac{3}{4}$ $\overline{A}$
Nondefective

$P(B|\overline{A}) = \frac{5}{19}$ B $P(B|\overline{A})P(\overline{A}) = \frac{5}{19} \cdot \frac{3}{4} = \frac{15}{76}$

$P(\overline{B}|\overline{A}) = \frac{14}{19}$ $\overline{B}$

Thus

$$P(B) = P(B \mid A)P(A) + P(B \mid \overline{A})P(\overline{A}) = \frac{4}{76} + \frac{15}{76}$$
$$= \frac{19}{76}$$
$$= \frac{1}{4}$$ ∎

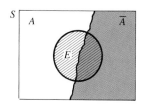

Figure 6.4

To formalize the results discussed in Example 1, note that A and $\overline{A}$ form a **partition of the sample space**, as illustrated in Figure 6.4.

It can be shown that for any event E, $E = (E \cap A) \cup (E \cap \overline{A})$. Thus

$$P(E) = P[(E \cap A) \cup (E \cap \overline{A})]$$
$$= P(E \cap A) + P(E \cap \overline{A})$$ Since $E \cap A$ and $E \cap \overline{A}$ are mutually exclusive

$$P(E) = P(E \mid A)P(A) + P(E \mid \overline{A})P(\overline{A})$$ Formula for conditional probability

This result (shown in color in Figure 6.4) is the same result we obtained from the tree diagram in Example 1. It generalizes to any partition of the sample space.

Probability of a Partitioned Event

If $A_1, A_2, \ldots, A_n$ form a partition of a sample space and E is any event, then

$$P(E) = P(E \mid A_1)P(A_1) + P(E \mid A_2)P(A_2) + \cdots + P(E \mid A_n)P(A_n)$$

EXAMPLE 2 Three cards are drawn (without replacement) from a deck of cards. What is the probability that the third one drawn is a heart?

Solution Construct the following tree diagram. Be sure you understand how the probabilities at each step are found.

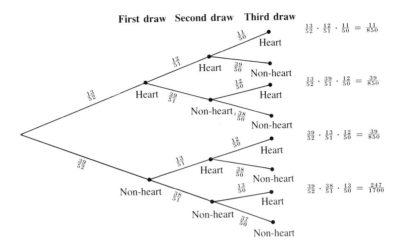

Thus

$$P(\text{heart on the third draw}) = \tfrac{11}{850} + \tfrac{39}{850} + \tfrac{39}{850} + \tfrac{247}{1700}$$

$$= \tfrac{425}{1700}$$

$$= \tfrac{1}{4}$$

∎

EXAMPLE 3 Three cards are drawn from a deck of cards. What is the probability that exactly two cards are hearts?

Solution Use the tree diagram from Example 2, and trace out different paths to indicate success.

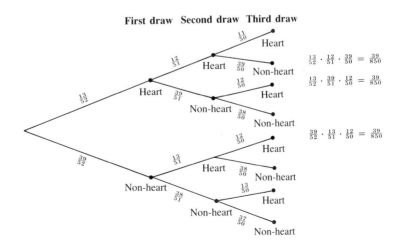

Thus

$$P(\text{exactly two hearts}) = \tfrac{39}{850} + \tfrac{39}{850} + \tfrac{39}{850}$$
$$= \tfrac{117}{850}$$
$$\approx .138 \qquad\blacksquare$$

There is an important result relating $P(E|F)$ and $P(F|E)$. This formula is important because we are frequently concerned with finding $P(F|E)$ when we know (or it is easy to find) $P(E|F)$.

$$P(F|E) = \frac{P(F \cap E)}{P(E)} \qquad\qquad \text{Conditional probability formula}$$

$$= \frac{P(E \cap F)}{P(E)} \qquad\qquad P(F \cap E) = P(E \cap F)$$

$$= \frac{P(E|F)P(F)}{P(E)} \qquad\qquad \begin{array}{l}\text{Probability of intersection:}\\ P(E \cap F) = P(E|F)P(F)\end{array}$$

$$= \frac{P(E|F)P(F)}{P(E|F)P(F) + P(E|\bar{F})P(\bar{F})} \qquad \begin{array}{l}\text{Events } F \text{ and } \bar{F} \text{ form a partition}\\ \text{of the sample space}\end{array}$$

Note that $P(F|E)$ is now written in terms of $P(E|F)$. This formula was first published in 1763 by Thomas Bayes (1702–1763).

Bayes' Theorem

If $A_1, A_2, \ldots, A_n$ form a partition of the sample space and E is an event associated with S, then

$$P(A_i|E) = \frac{P(E|A_i)P(A_i)}{P(E|A_1)P(A_1) + P(E|A_2)P(A_2) + \cdots + P(E|A_n)P(A_n)}$$

$$i = 1, 2, \ldots, n$$

EXAMPLE 4 Thompson Associates, a management consulting firm, has agreed to run a quality control check on three assembly lines at Karlin Manufacturing. In checking the inventory, Thompson found a defective item. What is the probability that it came from a particular assembly line?

Analysis In checking the records, Thompson Associates gathers the following information:

Number of items in inventory	*Previous testing results*
50% from assembly line A	2% of items from assembly line A are defective
30% from assembly line B	3% of items from assembly line B are defective
20% from assembly line C	4% of items from assembly line C are defective

Let

$A = \{$item tested is from assembly line A$\}$

$B = \{$item tested is from assembly line B$\}$

$C = \{$item tested is from assembly line C$\}$

$D = \{$item tested is defective$\}$

Thompson is now able to calculate the following probabilities from the given information:

$$P(A) = .5 \qquad P(B) = .3 \qquad P(C) = .2$$

$$P(D\,|\,A) = .02 \qquad P(D\,|\,B) = .03 \qquad P(D\,|\,C) = .04$$

Find $P(A\,|\,D)$, $P(B\,|\,D)$, and $P(C\,|\,D)$.

Solution   Find $P(A\,|\,D)$ first by formula:

$$P(A\,|\,D) = \frac{P(D\,|\,A)P(A)}{P(D\,|\,A)P(A) + P(D\,|\,B)P(B) + P(D\,|\,C)P(C)}$$

$$= \frac{(.02)(.5)}{(.02)(.5) + (.03)(.3) + (.04)(.2)}$$

$$= \frac{.01}{.027} \approx .37$$

Since it can often be confusing to use the formula, a variation of a tree diagram, called a *Bayes' tree*, may be drawn:

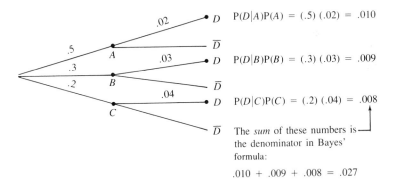

Using Bayes' tree, find the quotient of the desired path (shown in color) divided by the sum of all path possibilities:

$$P(A\,|\,D) = \frac{.010}{.027} \approx .37$$

Use Bayes' tree to find the other probabilities:

$$P(B\,|\,D) = \frac{.009}{.027} \approx .33 \qquad P(C\,|\,D) = \frac{.008}{.027} \approx .30$$

EXAMPLE 5   Suppose a test for AIDS is given to a patient. Let

$$A = \{\text{person has AIDS}\}$$
$$R = \{\text{test shows a positive result}\}$$

If the person has AIDS, the test shows a positive result 95% of the time; if the person does not have AIDS, the test shows a positive result 5% of the time. It is also known that .1% of the population has undetected AIDS. Find the probability that if the test shows a positive result, the person actually has AIDS.

Solution   Given: $P(R\,|\,A) = .95$, $P(R\,|\,\bar{A}) = .05$, and $P(A) = .001$.
Find $P(A\,|\,R)$.

$$P(A\,|\,R) = \frac{P(R\,|\,A)P(A)}{P(R\,|\,A)P(A) + P(R\,|\,\bar{A})P(\bar{A})}$$

$$= \frac{(.95)(.001)}{(.95)(.001) + (.05)(.999)} = \frac{.00095}{.00095 + .04995} \approx .0186640472$$

This means that even though the test is highly reliable and will detect AIDS in 95% of the cases, a positive result indicates that the person actually has AIDS in only 1.8% of the cases.   ∎

Problem Set 6.6

In Problems 1–6 consider the experiment of selecting 2 items (without replacement) from a sample space of 100, of which 5 items are defective. Let A = { first item selected is defective} and B = {second item selected is defective}. Find the probabilities correct to two decimal places.

1. $P(A)$ **2.** $P(\bar{A})$ **3.** $P(B\,|\,A)$
4. $P(B\,|\,\bar{A})$ **5.** $P(B)$ **6.** $P(\bar{B})$

Two cards are drawn in succession from a deck of 52 cards (without replacement). Let D_1 = {diamond is drawn on first draw} and D_2 = {diamond is drawn on second draw}. Find the probabilities in Problems 7–12.

7. $P(D_1)$ **8.** $P(\bar{D_1})$ **9.** $P(D_2\,|\,D_1)$
10. $P(D_2\,|\,\bar{D_1})$ **11.** $P(D_2)$ **12.** $P(\bar{D_2})$

For the experiment in Problems 1–6, draw 3 items and let C = {third item selected is defective}. Find the probabilities in Problems 13–18 correct to the nearest hundredth.

13. $P(C\,|\,B)$ **14.** $P(C\,|\,\bar{B})$ **15.** $P(C\,|\,A)$
16. $P(C\,|\,\bar{A})$ **17.** $P(C)$ **18.** $P(\bar{C})$

For the experiment in Problems 7–12, draw three cards and let D_3 = {diamond is drawn on third draw}. Find the probabilities in Problems 19–23 correct to the nearest hundredth.

19. $P(D_3)$ **20.** $P(D_3\,|\,D_2)$
21. P(exactly 1 diamond is drawn)
22. P(3 diamonds) **23.** P(at least 1 diamond)

Write your answers to Problems 24–30 correct to three decimal places.

APPLICATIONS
24. A test is given to candidates for graduate school. If the person is qualified, the probability of passing the test is 95%; if the person is not qualified, the probability of passing the test is 5%. Assume that the probability that a person taking the test is actually qualified is 70%. What is the probability that a person who passes the test is actually qualified?

25. Two persons are questioned by the police. One man tells the truth half of the time and the other always tells the truth. One man is chosen and asked one test question (a question to which the answer is known). The man answers truthfully. What is the probability that this man is the one who always tells the truth?

26. The student body of Oak Ridge University consists of 55% men and 45% women. A survey finds that 60% of the men and 30% of the women have outside jobs. What is the probability that a student who has a job is:
a. A man? **b.** A woman?

27. At Oak Ridge University 60% of the students come from the west, 15% from the midwest, and 25% from the east. A surveys finds that 80% of the students from the west, 40% of the students from the midwest, and 60% of the students from the east had attended at least one rock concert. What is the probability that a student who has attended a rock concert comes from:

a. The east?   b. The midwest?   c. The west?

28. Evergreen Insurance Company classifies its policyholders according to the following table:

Policyholder	Percent of policyholders	Probability of an accident-free year
AA(good risk)	60%	.999
A(average)	30%	.995
B(poor risk)	10%	.99
X(bad risk)	0%	.75

If Cliff Wilson buys a policy and subsequently has an accident, what is the probability he is a class B policyholder?

29. An electronic calculator has eight essential components that a manufacturer purchases from a supplier. Three components are selected at random and are tested. One is found to be defective. A search of the records gives the following information about the supplier:

Supplier's name	Number of shipments	Number of defects per shipment			
		0	1	2	More than 2
Abdullah	30	21	6	3	0

Given the sample result, what is the probability that the shipment has two defects?

30. Suppose a test for cancer is given. If a person has cancer, the test will detect it in 96% of the cases; and if the person does not have cancer, the test will show a positive result 1% of the time. If we assume that 12% of the population taking the test actually has cancer, what is the probability that a person taking the test and obtaining a positive result actually has cancer?

6.7 Summary and Review

IMPORTANT TERMS

Addition principle [6.1]
Bayes' theorem [6.6]
Complementary events [6.3]
Conditional probability [6.4]
Empirical probability [6.1]
Equally likely events [6.2]
Event [6.1]
Experiment [6.1]
Impossible event [6.1]
Independent events [6.4]
Intersection of events [6.5]
Mutually exclusive [6.1]

Odds [6.3]
Partition of a sample space [6.6]
Probabilistic model [6.1]
Probability measure [6.1]
Relative frequency [6.1]
Sample space [6.1]
Simple event [6.1]
Stochastic process [6.6]
Subjective probability [6.1]
Theoretical probability [6.1]
Uniform probability model [6.2]
Union of events [6.5]

COMPUTER APPLICATION If you have access to a computer, you might practice the concepts of this chapter by using Programs 8 and 9 on the computer disk accompanying this book. These programs review probability problems and probability properties.

SAMPLE TEST *For additional practice there are a large number of review problems. Categorized by objective in the Student Solutions Manual. The following sample test (40 minutes) is intended to review the main ideas of the chapter.*

1. a. Define a probability measure.
   b. What is meant by mutually exclusive events?
   c. What is meant by independent events?

APPLICATIONS

2. Suppose a box contains five flavors of jelly beans: 10 raspberry, 20 coconut, 25 blueberry, 30 watermelon, and 15 mint. The outcomes of selecting one jelly bean at random from the box are: $A = \{coconut\}$; $B = \{raspberry,\ blueberry\}$; and $C = \{watermelon,\ mint\}$. Find:
   a. $P(A)$ b. $P(B)$ c. $P(\bar{C})$ d. $P(A \cup B)$ e. $P(\overline{A \cap B})$

3. Suppose you flip two coins simultaneously 100 times and note the results according to the following list of possibilities: $A = \{2H\}$, $B = \{2T\}$, and $C = \{1H\ and\ 1T\}$. Will the experiment show these to be equally likely events? Why or why not?

4. Suppose two cards are drawn from a deck of cards (without replacement). Find the probabilities of the following events.
   a. P(both spaces)   b. P(spade first, heart second)
   c. P(1 space and 1 heart)   d. P(not both spades)

5. In the American version of roulette, there are 38 individual slots into which a steel ball has an equal chance of falling: 18 red, 18 black, and 2 green. Suppose you decide to bet $100 on red, but instead of just placing your bet you wait until black has occurred five consecutive times and then place your bet on red on the sixth play. Given these conditions, what is the probability of red on the sixth play after black occurs on the first five plays?

6. The following table shows the results of an examination of 1000 fruit flies exposed to massive amounts of radiation.

Sex of fruit fly	Normal, N	Mutated, $\bar{N}$	Total
Male, M	125	475	600
Female, $\bar{M}$	250	150	400
Total	375	625	1000

Find:
   a. $P(M)$ b. $P(\bar{N})$ c. $P(M \cup N)$ d. $P(M \cap \bar{N})$

7. Using the information in Problem 6, find:
   a. $P(M|N)$ b. $(N|M)$ c. $P(\bar{M}|\bar{N})$ d. $P(N|\bar{M})$

8. A shipment of six Broncos and four Thunderbirds arrives at a dealership, and two are selected at random and tested for defects. Let

   $B_1 = \{first\ choice\ is\ a\ Bronco\}$
   $B_2 = \{second\ choice\ is\ a\ Bronco\}$
   $T_1 = \{first\ choice\ is\ a\ Thunderbird\}$
   $T_2 = \{second\ choice\ is\ a\ Thunderbird\}$

   Compute the requested probabilities by drawing with replacement.
   a. P(both Broncos)   b. $P(T_1 \cap B_2)$ c. $P(T_1 \cup B_2)$

9. Repeat Problem 8 by drawing without replacement.

10. A test for glaucoma will detect the disease in 95% of the cases and give a false positive result 2% of the time. If we assume that 10% of the population taking the test actually has glaucoma, what is the probability that a person taking the test and obtaining a positive result actually has glaucoma?

Modeling Application 6

Medical Diagnosis

After a patient with observable symptoms sees a doctor, the doctor must decide which of several possible diseases is the most probable cause of the symptoms. The doctor may order further tests before making a diagnosis. Suppose a new test for detecting renal disease is being developed. The test will be given in a double-blind study to 24 patients with confirmed renal disease and 36 patients who are free of the disease. The results are as follows: of those with renal disease, 19 test positively and 5 test negatively; for those free of the disease, 10 test positively and 26 show negative results.*

Write a paper using Bayes' theorem to develop a model to answer the question: Are the results of this test conclusive enough for a doctor to make a diagnosis? The development of this model should show that two probabilities are small. One of these is a *false positive result*, which is obtaining a positive test result when in fact the patient does not have the disease. The other is a *false negative result*, which is obtaining a negative test result when in fact the patient does have the disease. For general guidelines about writing this essay, see the commentary for Modeling Application 1 on page 36.

* This modeling application is adapted from J. S. Milton and J. J. Corbet, "Conditional Probability and Medical Tests: An Exercise in Conditional Probability," *The UMAP Journal*, Vol. III, No. 2, 1982. ©1982 Education Development Center, Inc.

CHAPTER 7
Statistics and Probability Applications

APPLICATIONS

Management (*Business, Economics, Finance, and Investments*)

Quality control (7.1, Problems 14–15; 7.3, Problems 1–6, 38–39)
Salary comparisons (7.2, Problems 15 and 34)
Analysis of telephone usage (7.2, Problem 29)
Number of customers arriving at a bank (7.2, Problem 30)
Mean life of light bulbs (7.4, Problems 31–36)
Reliability of a computer system (7.5, Problem 50)
Overbooking of airline flights (7.5, Problem 51)
Actuary Exam questions (7.7, Problems 12–15; 7.8, Problems 22–28)

Life sciences (*Biology, Ecology, Health, and Medicine*)

Cure rate of a drug (7.1, Problems 16–17)
Probability of white-faced calves (7.3, Problems 7–12)
Physical characteristics (7.3, Problems 30–31; 7.4, Problems 17–21)
Germination rate of a seed (7.3, Problem 41)
Death rate from cancer (7.5, Problems 48–49)
Probability of a 310-day pregnancy (7.7, Problem 9)
Radioactive contamination in Oregon (7.7, Problems 10–11)
Ecology on an island (7.8, Problem 21)

Social sciences (*Demography, Political Science, Population, Psychology, Society, and Sociology*)

Analysis of test scores (7.2, Problems 17–19, 32–33; 7.3, Problem 40)
Probability of a missile reaching target (7.3, Problem 37)
Testing for ESP (7.3, Problems 42–43)
Grading on a curve (7.4, Problems 22–26)
Parapsychology experiment (7.5, Problem 46)
Correlation between IQ and GPA (7.6, Problems 26–28)

General interest

Probability that a team will win the World Series (7.3, Problems 34–36)
Game of odd person out (7.3, Problems 44–47)
Breaking strength of materials (7.4, Problems 27–28, 38–40)

Modeling application—World Running Records

CHAPTER OVERVIEW   We are now introduced to the concept of a random variable. We use random variables to discuss two important probability distributions: the binomial distribution and the normal distribution. We also discuss some statistical measures needed to analyze data. These are the measures of central tendency (often called *averages*)—the mean, median, and mode. We also discuss the measures of dispersion—the range, variance, and standard deviation. Knowledge of these statistical measures enables us to make general statements and comparisons when dealing with sets of data or large populations.

PREVIEW   When an experiment with two possible outcomes is repeated, the binomial distribution is used to analyze the results. For example, it is used to determine the failure rate on an assembly line, to determine the likelihood of certain occurrences in genetics, and in determining the reliability of predictions based on psychology experiments. The normal distribution is a continuous distribution that is used to measure variability within a population or as an approximation to certain binomial distributions. Together these distributions comprise a powerful tool in making probabilistic statements in a great variety of circumstances.

PERSPECTIVE   The central thread tying the material of this course together is that of a mathematical model. One of the difficult aspects of building a model is being able to handle the large amounts of data available in real world settings. Matrices are one type of model used for analyzing data. In this chapter other types of mathematical models are considered: those used to measure central tendency and dispersion and those used to describe binomial and normal distributions.

7.1   Random Variables

The management at Chrysler wants to introduce a "zero defects" advertising campaign on a new line of automobiles, so they decide to do extensive testing of parts coming off the assembly line. Suppose 150 parts come off the assembly line every day, and inspectors keep track of the number of items with any type of defect. If X is the number of defective items, then X can obviously assume any of the values $0, 1, 2, 3, \ldots, 149, 150$. The variable X is called a **discrete random variable**.

Random Variable

> A **random variable** X associated with a probability space S is a function that assigns a real number to each simple event in S. If S has a finite number of outcomes, then X is called a **discrete random variable**; and if X can assume any real value on an interval, then it is called a **continuous random variable**.

In finite mathematics we focus on discrete random variables, but we can often convert continuous values to discrete values by rounding off to a given number of decimal places. For example, if X represents the heights of individuals, then X is a continuous random variable. However, if we round to the nearest eighth of an inch, then it is a discrete random variable. We will look at relationships such as this and other continuous random variables in Sections 7.4 and 7.5. In the meantime, we limit our discussion to discrete random variables.

In statistics, random variables are represented by using capital letters, such as X or Y, while lowercase letters, such as x or y, are used to denote a particular value of a random variable.

EXAMPLE 1   The waiting time for a marriage license by state is given by the following 1982 list:

Alabama—None	Louisiana—72 hours	Ohio—5 days
Alaska—3 days	Maine—5 days	Oklahoma—None
Arizona—None	Maryland—48 hours	Oregon—3 days
Arkansas—3 days	Massachusetts—3 days	Pennsylvania—3 days
California—None	Michigan—3 days	Rhode Island—None
Colorado—None	Minnesota—5 days	South Carolina—24 hours
Connecticut—4 days	Mississippi—3 days	South Dakota—None
Delaware—24 hours	Missouri—3 days	Tennessee—3 days
Florida—3 days	Montana—8 days	Texas—None
Georgia—3 days	Nebraska—2 days	Utah—None
Hawaii—None	Nevada—None	Vermont—3 days
Idaho—3 days	New Hampshire—5 days	Virginia—None
Illinois—1 day	New Jersey—72 hours	Washington—3 days
Indiana—3 days	New Mexico—None	West Virginia—3 days
Iowa—3 days	New York—None	Wisconsin—5 days
Kansas—3 days	North Carolina—None	Wyoming—None
Kentucky—3 days	North Dakota—None	

These data can be summarized in a table called a **frequency distribution** as follows. Make three columns; put the number of days' wait (notice that some times are given in hours—convert these to days) into the first column, put the tally into the second column, and put the frequency into the third column:

Wait for marriage license (days)	Tally	Frequency (number of states)
0													17
1					3
2				2
3															21
4			1
5					5
6		0
7		0
8			1
Total		50

If X is the number of days necessary to wait for a marriage license, then X is a discrete random variable where $X = 0, 1, 2, \ldots, 8$.   ∎

We are often interested in the probability that a random variable assumes a given value. For Example 1, we would not expect $P(X = 3)$ to be the same as $P(X = 8)$. In order to find these probabilities, we divide the frequency by the total to find the **relative frequency**.

EXAMPLE 2 Find the relative frequencies for Example 1.

Solution

Time (days)	Frequency	Relative frequency
0	17	$\frac{17}{50} = .34$
1	3	$\frac{3}{50} = .06$
2	2	$\frac{2}{50} = .04$
3	21	$\frac{21}{50} = .42$
4	1	$\frac{1}{50} = .02$
5	5	$\frac{5}{50} = .10$
6	0	$\frac{0}{50} = 0$
7	0	$\frac{0}{50} = 0$
8	1	$\frac{1}{50} = .02$
Total	50	1.00

■

Probability Distribution

A **probability distribution** is the collection of all values that a random variable assumes along with the probabilities that correspond to these values. Furthermore,

1. $P(X = 1) + P(X = 2) + \cdots + P(X = n) = 1$

2. $0 \le P(X = x_i) \le 1$ for every $1 \le i \le n$

It is often useful to display the information in a probability distribution graphically in a special kind of bar graph called a **histogram**.

EXAMPLE 3 Represent the information from Example 1 in a histogram.

Solution Use the horizontal axis to delineate the values of the random variable and the vertical axis to represent the relative frequencies.

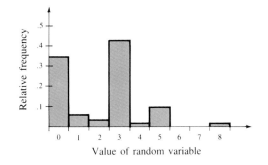

There is a very important relationship between the area of the rectangles in a histogram and the probabilities. Note that since the width of each bar is 1 unit, the **area of each bar is the probability of occurrence for that random variable**. Therefore the **total area of the rectangles is 1**. This property is very important in our study of probability distributions.

EXAMPLE 4   Draw a histogram for the probability distribution of rolling a single die and noting the number on top. Shade the portion that represents $P(X \geq 5)$.

Solution   Let $X = 1, 2, 3, 4, 5, 6$; then $P(X = x_i)$, for $1 \leq i \leq 6$, and

$$P(X \geq 5) = P(X = 5) + P(X = 6)$$

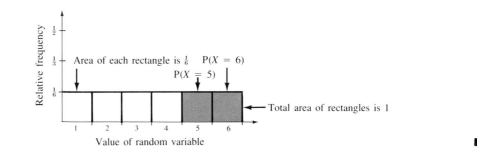

Problem Set 7.1

Give the frequency distribution and define the random variables in Problems 1–3.

1. The heights of 30 students are as follows (figures rounded to the nearest inch):

```
66  68  64  70  67  67  68  64  65  66
64  70  72  71  69  64  63  70  71  63
68  67  67  65  69  65  67  66  69  69
```

2. The number of years several leading batters played in the major leagues are as follows:

Ty Cobb	24	Paul Waner	20
Rogers Hornsby	23	Stan Musial	22
Ed Delehanty	16	Henie Manush	17
Dan Brouthers	19	Honus Wagner	21
Sam Thompson	15	Willie Keeler	19
Al Simmons	20	Ted Williams	19
Joe DiMaggio	13	Tris Speaker	22
Jimmy Foxx	20	Billy Hamilton	14
Lou Gehrig	17	Harry Heilmann	17
George Sisler	16	Babe Ruth	22
Nap Lajoie	21	Jesse Burkett	16
Cap Anson	22	Bill Terry	14
Eddie Collins	25		

3. A pair of dice are rolled and the sum of the spots on the tops of the dice are recorded as follows:

```
3   2   6   5   3   8   8   7  10   9
7   5  12   9   6  11   8  11  11   8
```

```
7   7   7  10  11   6   4   8   8   7
6   4  10   7   9   7   9   6   6   9
4   4   6   3   4  10   6   9   6  11
```

Find the probability distribution (that is, the relative frequencies) and draw a histogram for Problems 4–9.

4. Use the data in Problem 1.

5. Use the data in Problem 2.

6. Use the data in Problem 3.

APPLICATIONS

7. Blane, Inc., a consulting firm, is employed to perform an efficiency study at National City Bank. As part of the study, the number of times per day that people are waiting in line is summarized.

Number waiting	Frequency
2	20
3	15
4	7
5	5
6	2
7	1
8 or more	0

8. Blane, Inc., also notes the transaction times (rounded to the nearest minute) for customers at National City Bank, as listed in the table:

Time	Frequency
1	10
2	12
3	18
4	25
5	16
6	10
7	6
8	1
9	0
10	2

9. Three coins are tossed onto a table and the following frequencies are noted:

Number of heads	Frequency
3	18
2	56
1	59
0	17

Use your knowledge of probability from Chapter 6 to find the probability distributions in Problems 10–19. Let X be the random variable.

10. Three coins are tossed onto a table and the number of tails is noted.

11. Four coins are tossed onto a table and the number of heads is noted.

12. Two dice are rolled and the total number of spots on the top faces is recorded.

13. Two cards are simultaneously drawn from a deck of 52 cards and the number of aces is noted.

14. From a lot containing 25 items, 5 of which are defective, 3 items are chosen at random and the number of defectives is noted.

15. Repeat Problem 14 except choose the 3 items one at a time, note the result, and then return the chosen item before choosing again.

16. A drug is known to effect a cure in 60% of the patients to whom it is administered. This drug is administered to three patients and the number of cures is noted.

17. Repeat Problem 16 for four patients.

18. A bag of Halloween candy contains 25 Hershey bars, 10 Rocky Roads, and 15 Almond Joys. Two pieces of candy are randomly selected and the number of Rocky Roads selected is noted.

19. Repeat Problem 18 except note the number of Almond Joys.

Draw a histogram in Problems 20–29 and shade in the region indicated for each problem.

20. $P(X \geq 2)$ for the experiment in Problem 10.

21. $P(X$ is at least 3) for the experiment in Problem 11.

22. $P(X = 7$ or $X = 11)$ for the experiment in Problem 12.

23. $P(X = 1)$ for the experiment in Problem 13.

24. $P(X \geq 1)$ for the experiment in Problem 14.

25. $P(X = 2)$ for the experiment in Problem 15.

26. $P(X = 3)$ for the experiment in Problem 16.

27. $P(2 \leq X \leq 3)$ for the experiment in Problem 17.

28. $P(X = 2)$ for the experiment in Problem 18.

29. $P(X$ at least 1) for the experiment in Problem 19.

30. Draw a histogram for $P(X = x_i)$, where $P(X = x_i) = \dfrac{x_i}{10}$ for $0 \leq i \leq 4$, i a counting number. For example, if $x_i = 4$, then $P(X = 4) = \frac{4}{10} = \frac{2}{5}$.

31. Draw a histogram for $P(X = x_i)$, where $P(X = x_i) = \dfrac{x_i}{20}$ for $2 \leq i \leq 6$, i a counting number. For example, if $x_i = 2$, then $P(X = 2) = \frac{2}{20} = \frac{1}{10}$.

7.2 Analysis of Data

The word *average* is frequently used in everyday language, but in mathematics it can take on a variety of meanings. For example, suppose we compare the "averages" of two bowlers in a tournament consisting of seven games:

Andrew: 185 average
Bob: 170 average

It would appear that Andrew did better in the tournament, but suppose the winner is found by looking at the individual games, as listed below:

	Andrew	Bob
Game 1	175	180
Game 2	150	130
Game 3	160	161
Game 4	180	185
Game 5	160	163
Game 6	183	185
Game 7	287	186
Total	1295	1190

To find each average, we divide the total by the number of games:

Andrew: $\dfrac{1295}{7} = 185$

Bob: $\dfrac{1190}{7} = 170$

However, if we consider the games separately, Bob beat Andrew in *five out of seven games.* Clearly, the "average" we calculated above does not completely describe the situation.

The first statistical measures we consider are called **measures of central tendency**, or **averages**. In statistics Σx is used to mean *the sum of all values that x can assume.* Similarly, Σx^2 means *square each value that x can assume and then add the results.*

Averages, or Measures of Central Tendency

Given a set of data, three common *measures of central tendency* are:
1. The **mean** (or **arithmetic mean**) is found by adding the data and then dividing by the number of data items, n:

$$\text{Mean} = \frac{\Sigma x}{n}$$

The mean of a set of sample scores is denoted by $\bar{x}$.
2. The **median** is the middle number when the data are arranged in order of size. If there is an even number of data items, then the median is the mean of the two middle numbers.
3. The **mode** is the value that occurs most frequently. If there is no number that occurs more than once, there is no mode. It is possible to have more than one mode.

EXAMPLE 1 Find the mean, median, and mode for Andrew's and Bob's bowling scores.

Solution Mean:

	Andrew	Bob
	$\bar{x} = \dfrac{1295}{7}$	$\bar{x} = \dfrac{1190}{7}$
	$= 185$	$= 170$

Median:

Andrew		Bob
150		130
160		161
160		163
175	← The middle number →	180
180	is the median	185
183		185
287		186

Mode:

Andrew		Bob
150		130
160⎫		161
160⎭	The mode is the	163
175	number that occurs	180
180	most frequently	⎧185
183		⎩185
287		186

Summary:

	Andrew	Bob
Mean	185	170
Median	175	180
Mode	160	185

Find the mean, median, and mode for the sets of data in Examples 2–4.

EXAMPLE 2 3, 5, 5, 8, 9

Solution Mean: $\dfrac{\text{Sum of terms}}{\text{Number of terms}} = \dfrac{3 + 5 + 5 + 8 + 9}{5}$

$$= \dfrac{30}{5}$$

$$= 6$$

Median: Arrange in order: 3, 5, 5, 8, 9
The middle term, 5, is the median.

Mode: The term that occurs most frequently is the mode, which is 5.

EXAMPLE 3 4, 10, 9, 8, 9, 4, 5

Solution Mean: $\dfrac{4 + 10 + 9 + 8 + 9 + 4 + 5}{7} = \dfrac{49}{7} = 7$

Median: 4, 4, 5, 8, 9, 9, 10
         The median is 8.

Mode: This set of data is *bimodal*; that is, it has two modes.
       The modes are 4 and 9. ∎

EXAMPLE 4 6, 5, 4, 7, 1, 9

Solution Mean: $\dfrac{6 + 5 + 4 + 7 + 1 + 9}{6} = \dfrac{32}{6} \approx 5.33$

Median: 1, 4, 5, 6, 7, 9

$$\frac{5 + 6}{2} = \frac{11}{2} = 5.5$$

Mode: There is no mode. ∎

Suppose the data are presented in a frequency distribution and there are more data than can be handled conveniently by the method used in Examples 2–4. Consider Example 5, which shows how you might proceed.

EXAMPLE 5 In Section 7.1 we set up a frequency distribution for the number of days one must wait for a marriage license. This information is repeated below. Find the mean, median, and mode.

Wait for marriage license (days)	Frequency (number of states)
0	17
1	3
2	2
3	2
4	1
5	5
6	0
7	0
8	1
Total	50

Solution Mean: To find the mean, we could, of course, add all 50 individual numbers. But, instead, notice that

0 occurs 17 times, so we write $0 \cdot 17$
1 occurs 3 times, so we write $1 \cdot 3$
2 occurs 2 times, so we write $2 \cdot 2$
3 occurs 21 times, so we write $3 \cdot 21$
⋮

Thus the mean is

$$\bar{x} = \frac{0 \cdot 17 + 1 \cdot 3 + 2 \cdot 2 + 3 \cdot 21 + 4 \cdot 1 + 5 \cdot 5 + 6 \cdot 0 + 7 \cdot 0 + 8 \cdot 1}{50}$$

$$= \frac{0 + 3 + 4 + 63 + 4 + 25 + 0 + 0 + 8}{50}$$

$$= \frac{107}{50}$$

$$= 2.14$$

Median: Since there are 50 values, the mean of the twenty-fifth and twenty-sixth largest values is the median. From the table, we see that the twenty-fifth term is 3 and the twenty-sixth term is 3, so the median is

$$\frac{3 + 3}{2} = \frac{6}{2} = 3$$

Mode: The mode is the value that occurs most frequently, which is 3. ∎

The mean from a frequency distribution is called the **weighted mean**.

Weighted Mean

> If a list of scores $x_1, x_2, x_3, \ldots, x_n$ occurs $w_1, w_2, \ldots, w_n$ times, respectively, then
>
> $$\bar{x} = \frac{\Sigma w \cdot x}{\Sigma w}$$

We have been using $\bar{x}$ to denote the mean of a set of *sample scores*. Now we let μ denote the mean of *all scores* in some population. A **population** is the complete and entire collection of elements to be studied, so a sample must be a subset of a population. In the formula for a weighted mean, if we consider the entire population, then $\Sigma wx / \Sigma w$ can be rewritten as

$$\mu = \Sigma x \cdot P(x)$$

if we consider $P(x)$ to be the relative frequency with which x occurs.

The measures we have been discussing can help us interpret information, but they do not give the whole story. For example, consider the following two sets of data:

Data set 1: 8, 10, 9, 9, 9
Data set 2: 9, 9, 2, 12, 13

For both these data sets the mean, median, and mode are 9. It would seem that some additional analysis is in order. Note that the data in the second set are more spread out than the data in the first set. The amount that the data are spread out is called the **dispersion**. We now consider three measures of dispersion: the **range**, **variance**, and standard deviation. The simplest measure of dispersion is the range:

Range

> The *range* in a set of data is the difference between the largest and the smallest numbers in the set.

The ranges for the above data sets are:

Range of data set 1: $10 - 8 = 2$
Range of data set 2: $13 - 2 = 11$

Note that the range is determined only by the largest and the smallest numbers in the set; it does not give us any information about the other numbers. It thus seems reasonable to invent some other measures of dispersion that take into account all the numbers in the data. The variance and standard deviation are measures that give information about the dispersion. The **variance** uses all the numbers in the data set to measure the dispersion. When finding the variance, we must make a distinction between the variance of the entire population and the variance of a random sample from that population. When the variance is based on a set of sample scores, it is denoted by s^2; and when it is based on all scores in a population, it is denoted by σ^2 (σ is the lowercase Greek letter sigma). The variance is found by

$$s^2 = \frac{\Sigma(x - \bar{x})^2}{n - 1} \qquad \sigma^2 = \frac{\Sigma(x - \mu)^2}{n}$$

If n is large (say, greater than 30), s^2 and σ^2 will be almost the same. The formula for s^2 looks a little intimidating, but it is based on simple arithmetic procedures, which, if taken one at a time, are quite simple. Understanding the variance formula is easy if you systematically deal with the data as indicated by the following procedure:

Procedure for Finding the Variance

> *Step 1*   Determine the mean of the numbers. (Find $\bar{x}$.)
> *Step 2*   Subtract the mean from each number. (Find $x - \bar{x}$.)
> *Step 3*   Square each of these differences. [This is $(x - \bar{x})^2$.]
> *Step 4*   Find the sum of the squares of these differences. [This is $\Sigma(x - \bar{x})^2$.]
> *Step 5*   Divide this sum by 1 less than the number of pieces of data.

EXAMPLE 6   Find the variance for each set of data:

a. Data set 1:   8, 9, 9, 9, 10   **b.** Data set 2:   2, 9, 9, 12, 13

Solution   Remember that the mean, median, and mode for both these examples is the same (9). The range for data set 1 is 2, and for data set 2 it is 11. Now we want to find the second measure of dispersion, the variance. The procedure is lengthy, so make sure to follow each step carefully.

Step 1   Find the mean.
   a. $\bar{x} = 9$ **b.** $\bar{x} = 9$

Step 2 Subtract each number from the mean.

a. Data, x	Difference from the mean, $x - \bar{x}$		b. Data, x	Difference from the mean, $x - \bar{x}$
8	−1		2	−7
9	0		9	0
9	0		9	0
9	0		12	3
10	1		13	4

Step 3 Square each of these differences. Notice that some of the differences in step 2 are positive and others are negative. Remember, we wish to find a measure of total dispersion. But if we add all these differences, we will not obtain the total variability. Indeed, if we simply add the differences for either example, the sum is zero. But we do not wish to say there is no dispersion. To resolve this difficulty with positive and negative differences, we square each difference so the result will always be nonnegative.

a. Data, x	Difference from the mean, $x - \bar{x}$	Square of the difference, $(x - \bar{x})^2$
8	−1	1
9	0	0
9	0	0
9	0	0
10	1	1

b. Data, x	Difference from the mean, $x - \bar{x}$	Square of the difference, $(x - \bar{x})^2$
2	−7	49
9	0	0
9	0	0
12	3	9
13	4	16

Step 4 Find the sum of these squares.
a. $1 + 0 + 0 + 0 + 1 = 2$ **b.** $49 + 0 + 0 + 9 + 16 = 74$

Step 5 Divide this sum by 1 less than the number of terms. In each of these examples there are five pieces of data, so to find the variance, divide by 4:
a. Variance $= \frac{2}{4} = .5$ **b.** Variance $= \frac{74}{4} = 18.5$ ■

The larger the variance, the more dispersion there is in the original data, and a small variance means that the data are more closely clustered around the mean.

The **standard deviation** for a sample is s and for the entire population is σ. That is,

$$\text{Standard deviation} = \sqrt{\text{variance}}$$

EXAMPLE 7 Find the standard deviations for the data of Example 6.

a. Data set 1: 8, 9, 9, 9, 10 **b.** Data set 2: 2, 9, 9, 12, 13

Solution We did most of the work in finding the standard deviation in Example 6. Now, all we need to do is to find the square root of the answers to Example 6.

a. $\sqrt{.5} \approx .71$ **b.** $\sqrt{18.5} \approx 4.30$ ∎

If the standard deviation is based on a random variable of a probability distribution, it can be written as

$$\sigma^2 = \Sigma(X - \mu)^2 \cdot P(X)$$

This formula can be manipulated into an equivalent form to help us carry out the calculations more easily:

$$\sigma^2 = [\Sigma X^2 \cdot P(X)] - \mu^2$$

EXAMPLE 8 Roll a pair of dice and let the random variable represent the total on the tops of the dice. Find the mean, variance, and standard deviation of the population.

Solution Organize the computations in table form, as shown.

X	$P(X)$	$X \cdot P(X)$	X^2	$X^2 \cdot P(X)$
2	$\frac{1}{36}$	$\frac{2}{36}$	4	$\frac{4}{36}$
3	$\frac{2}{36}$	$\frac{6}{36}$	9	$\frac{18}{36}$
4	$\frac{3}{36}$	$\frac{12}{36}$	16	$\frac{48}{36}$
5	$\frac{4}{36}$	$\frac{20}{36}$	25	$\frac{100}{36}$
6	$\frac{5}{36}$	$\frac{30}{36}$	36	$\frac{180}{36}$
7	$\frac{6}{36}$	$\frac{42}{36}$	49	$\frac{294}{36}$
8	$\frac{5}{36}$	$\frac{40}{36}$	64	$\frac{320}{36}$
9	$\frac{4}{36}$	$\frac{36}{36}$	81	$\frac{324}{36}$
10	$\frac{3}{36}$	$\frac{30}{36}$	100	$\frac{300}{36}$
11	$\frac{2}{36}$	$\frac{22}{36}$	121	$\frac{242}{36}$
12	$\frac{1}{36}$	$\frac{12}{36}$	144	$\frac{144}{36}$
Total		$\frac{252}{36} = 7$		$\frac{1974}{36}$

Mean: $\mu = \Sigma X \cdot P(X) = 7$

Variance: $\sigma^2 = [\Sigma X^2 \cdot P(X)] - \mu^2$

$$= \frac{1974}{36} - 7^2 = 5.8$$

Standard deviation: $\sigma = \sqrt{5.8} \approx 2.4$ ∎

Many calculators have keys to calculate the mean, variance, and standard deviation. Check to see if you can carry out these operations on your calculator. If your calculator does not have these keys, you can arrange the calculations as shown in Example 8 and use the square root table given in Appendix E.

COMPUTER APPLICATION   For more real life examples, or for additional practice, use Program 10 on the computer disk accompanying this text. It gives practice with the mean, median, mode, standard deviation, and histograms.

Problem Set 7.2

Find the mean, median, mode, range, variance, and standard deviation for each set of values in Problems 1–12.

1. 1, 2, 3, 4, 5

2. 17, 18, 19, 20, 21

3. 103, 104, 105, 106, 107

4. 765, 766, 767, 768, 769

5. 4, 7, 10, 7, 5, 2, 7

6. 15, 13, 10, 7, 6, 9, 10

7. 3, 5, 8, 13, 21

8. 1, 4, 9, 16, 25

9. 79, 90, 95, 95, 96

10. 70, 81, 95, 79, 85

11. 1, 2, 3, 3, 3, 4, 5

12. 0, 1, 1, 2, 3, 4, 16

13. Compare Problems 1–4. What do you notice about the mean and standard deviation?

14. By looking at Problems 1–4 and discovering a pattern, find the mean and standard deviation of the following set of numbers:

217,850,   217,851,   217,852,   217,853,   217,854

APPLICATIONS

15. Find the mean, median, and mode of the following salaries of the employees of Green Lawn Landscaping Company:

Salary	Frequency
$10,000	4
16,000	3
20,000	2
30,000	1

16. Linda Foley, the leading salesperson for the Green Lawn Landscaping Company, turned in the following summary of sales contacts for the week of October 23–28. Find the mean, median, and mode.

Date	Number of clients contacted
Oct. 23	12
Oct. 24	9
Oct. 25	10
Oct. 26	16
Oct. 27	10
Oct. 28	21

17. Find the mean, median, and mode of the following test scores:

Test score	Frequency
90	1
80	3
70	10
60	5
50	2

18. A class obtained the following scores on a test:

Score	Frequency
90	1
80	6
70	10
60	4
50	3
40	1

Find the mean, median, mode, and range for the class.

19. A class obtained the following scores on a test:

Score	Frequency
90	2
80	4
70	9
60	5
50	3
40	1
30	2
0	4

Find the mean, median, mode, and range for the class.

Find the variance and standard deviation for Problems 20–24.

20. Problem 17   **21.** Problem 18   **22.** Problem 19

23. Problem 15   **24.** Problem 16

Find the mean, variance, and standard deviation of the probability distributions in Problems 25–28.

25.

X	P(X)
0	$\frac{1}{2}$
1	$\frac{1}{4}$
2	$\frac{1}{4}$

26.

X	P(X)
1	.1
2	.8
3	.1

27.

X	P(X)
2	.2
3	.3
4	.4
5	.1

28.

X	P(X)
5	$\frac{1}{6}$
10	$\frac{1}{3}$
20	$\frac{1}{3}$
30	$\frac{1}{6}$

29. A survey shows that the number of times a nonbusiness telephone rings before it is answered is:

X	P(X)
0	.00
1	.20
2	.31
3	.16
4	.18
5	.10
6	.02
7 or more	.03

Find the mean, variance, and standard deviation for the number of rings.

30. The probability for the number of customers arriving at a bank in a given period of time is summarized below:

X	P(X)
0	.25
1	.43
2	.18
3	.09
4	.03
5	.01
6 or more	.01

Find the mean, variance, and standard deviation for the number of arrivals.

31. Suppose a variance is zero. What can you say about the data?

32. A professor gives five exams. The scores for two students have the same mean, although one student seemed to do better on all the tests except one. Give an example of such scores.

33. A professor gives six exams. The scores of two students have the same mean, although one student's scores have a small standard deviation and the other student's scores have a large standard deviation. Give an example of such scores.

34. The salaries for the executives of a small company are shown below. Find the mean, median, and mode. Which measure seems to best describe the average executive salary for the company?

Position	Salary
President	$90,000
1st VP	40,000
2nd VP	40,000
Supervising manager	34,000
Accounting manager	30,000
Personnel manager	30,000

35. Roll a pair of dice 20 times. Find the mean, variance, and standard deviation for your data. Compare your results with Example 8.

36. Repeat Problem 35 for 100 rolls of the dice.

37. Roll a pair of dice until all 11 numbers occur at least once. Repeat the experiment 20 times. Find the mean, variance, and standard deviation for the number of tosses.

38. The *harmonic mean* (or H.M.) is found by dividing the number of scores, n, by the sum of the reciprocals of all scores:

$$\text{H.M.} = \frac{n}{\sum \frac{1}{x}}$$

Find the harmonic mean for the data in Problem 5.

39. Repeat Problem 38 for the data in Problem 6.

40. The *geometric mean* (or G.M.) is found by taking the nth root of the product of the n scores. Find the geometric mean for the data in Problem 5.

41. Repeat Problem 40 for the data in Problem 6.

42. The *quadratic mean*, or *root mean square* (R.M.S.), is found by squaring each score; adding the results; dividing by the number of scores, n; and then taking the square root of the result:

$$\text{R.M.S.} = \sqrt{\frac{\sum x^2}{n}}$$

Find the quadratic mean for the data in Problem 5.

43. Repeat Problem 42 for the data in Problem 6.

7.3 The Binomial Distribution

Consider now a common type of experiment—one with only two outcomes, A and $\bar{A}$. Suppose that $P(A) = p$ and $P(\bar{A}) = q = 1 - p$. We are interested in n repetitions of the experiment. If $P(A)$ remains the same for each repetition and we let X represent the number of times that event A has occurred, then we call X a **binomial random variable**.

EXAMPLE 1 Toss a coin four times. The sample space is shown below:

Number of heads	Outcomes					
4	HHHH					
3	HHHT	HHTH	HTHH	THHH		
2	HHTT	HTHT	HTTH	THTH	THHT	TTHH
1	TTTH	TTHT	THTT	HTTT		
0	TTTT					

Let the random variable X represent the number of heads that have occurred. That is, $X = 0$ if we obtain no heads; $X = 1$ means one head is obtained; $X = 2$ means two heads are obtained; $X = 3$, $X = 4$ mean three and four heads are obtained, respectively. Since there are 16 possibilities,

$$P(X = 4) = \tfrac{1}{16}$$
$$P(X = 3) = \tfrac{4}{16} = \tfrac{1}{4}$$
$$P(X = 2) = \tfrac{6}{16} = \tfrac{3}{8}$$
$$P(X = 1) = \tfrac{4}{16} = \tfrac{1}{4}$$
$$P(X = 0) = \tfrac{1}{16}$$

Note that the sum is one:

$$\tfrac{1}{16} + \tfrac{4}{16} + \tfrac{6}{16} + \tfrac{4}{16} + \tfrac{1}{16} = 1$$

∎

This example illustrates a common discrete probability distribution called the **binomial distribution**, which is a list of outcomes and probabilities for a **binomial experiment**.

Binomial Experiment

A **binomial experiment** is an experiment that meets four conditions:

1. There must be a fixed number of trials. Denote this number by n.
2. There must be two possible mutually exclusive outcomes for each trial. Call them *success* and *failure*.
3. Each trial must be independent. That is, the outcome of a particular trial is not affected by the outcome of any other trial.
4. The probability of success and failure must remain constant for each trial.

Consider a manufacturer of transistor radios. Suppose three items are chosen at random from a day's production and are classified as defective (F) or nondefective (S). We are interested in the number of successes obtained. (A "success" is the occurrence of the event we are considering; in this case, nondefectives.) Suppose that an item has a probability of .1 of being defective and therefore a probability of .9 of being nondefective. We will assume that these probabilities remain the same throughout the experiment and that the classification of any particular item is independent of the classification of any other item. The sample space, along with the probabilities, is listed below:

Sample space	*Associated probabilities*
1. FFF	$(.1)(.1)(.1) = (.1)^3$
2. FFS	$(.1)(.1)(.9) = (.9)(.1)^2$
3. FSF	$(.1)(.9)(.1) = (.9)(.1)^2$
4. SFF	$(.9)(.1)(.1) = (.9)(.1)^2$
5. FSS	$(.1)(.9)(.9) = (.9)^2(.1)$
6. SFS	$(.9)(.1)(.9) = (.9)^2(.1)$
7. SSF	$(.9)(.9)(.1) = (.9)^2(.1)$
8. SSS	$(.9)(.9)(.9) = (.9)^3$

If we let X = the number of successes obtained, then $P(X = 0)$ is found on line 1 above:

$$P(X = 0) = (.1)^3$$

$P(X = 1)$ is found from lines 2, 3, and 4:

2. FFS	$(.9)(.1)^2$
3. FSF	$(.9)(.1)^2$
4. SFF	$(.9)(.1)^2$

Total: $\quad P(X = 1) = 3(.9)(.1)^2$

$P(X = 2)$ is found by

5. FSS	$(.9)^2(.1)$
6. SFS	$(.9)^2(.1)$
7. SSF	$(.9)^2(.1)$

Total: $\quad P(X = 2) = 3(.9)^2(.1)$

$P(X = 3)$ is found in line 8:

$$P(X = 3) = (.9)^3$$

Note that the same results can be achieved by simply considering

$$(.1 + .9)^3 = (.1)^3 + 3(.1)^2(.9) + 3(.1)(.9)^2 + (.9)^3$$

This leads us to the following theorem:

Binomial Distribution Theorem

Let X be a random variable for the number of successes in n independent and identical repetitions of an experiment with two possible outcomes, success and failure. If p is the probability of success, then

$$P(X = k) = \binom{n}{k} p^k (1 - p)^{n-k} \qquad k = 0, 1, \ldots, n$$

A sequence of independent trials for which there are only two possible outcomes is also sometimes called a sequence of **Bernoulli trials**, after Jacob Bernoulli (1654–1705).

EXAMPLE 2   Suppose a sociology teacher always gives true-false tests with 10 questions.

a. What is the probability of getting exactly 70% by guessing?
b. What is the probability of getting 70% or better by guessing?
c. If a student can be sure of getting five questions correct, but must guess at the others, what is the probability of getting 70% or better?

Solution   **a.** $P(X = 7) = \binom{10}{7}\left(\dfrac{1}{2}\right)^7\left(\dfrac{1}{2}\right)^3 = \dfrac{10!}{7!3!} \cdot \dfrac{1}{2^{10}} = \dfrac{120}{1024} = \dfrac{15}{128}$

b. $P(X = 8) = \binom{10}{8}\left(\dfrac{1}{2}\right)^8\left(\dfrac{1}{2}\right)^2 = \dfrac{45}{1024}$

$P(X = 9) = \binom{10}{9}\left(\dfrac{1}{2}\right)^9\left(\dfrac{1}{2}\right)^1 = \dfrac{10}{1024} = \dfrac{5}{512}$

$P(X = 10) = \binom{10}{10}\left(\dfrac{1}{2}\right)^{10} = \dfrac{1}{1024}$

$P(X \geq 7) = \dfrac{120}{1024} + \dfrac{45}{1024} + \dfrac{10}{1024} + \dfrac{1}{1024}$

$= \dfrac{176}{1024} = \dfrac{11}{64}$

c. $P(X \geq 2) = \binom{5}{2}\left(\dfrac{1}{2}\right)^5 + \binom{5}{3}\left(\dfrac{1}{2}\right)^5 + \binom{5}{4}\left(\dfrac{1}{2}\right)^5 + \binom{5}{5}\left(\dfrac{1}{2}\right)^5$

$= 10 \cdot \dfrac{1}{32} + 10 \cdot \dfrac{1}{32} + 5 \cdot \dfrac{1}{32} + 1 \cdot \dfrac{1}{32}$

$= \dfrac{5}{16} + \dfrac{5}{16} + \dfrac{5}{32} + \dfrac{1}{32}$

$= \dfrac{13}{16}$ ∎

EXAMPLE 3   A missile has a probability of $\frac{1}{10}$ of penetrating enemy defenses and reaching its target. If five missiles are aimed at the same target, what is the probability that exactly one will hit its target? What is the probability that at least one will hit its target?

Solution $P(X = 1) = \binom{5}{1}\left(\frac{1}{10}\right)^1\left(\frac{9}{10}\right)^4$

$$= \frac{5 \cdot 9^4}{10^5} = \frac{6561}{20{,}000} = .32805$$

$$P(X \geq 1) = 1 - P(X = 0)$$

$$= 1 - \binom{5}{0}\left(\frac{9}{10}\right)^5$$

$$= 1 - \frac{9^5}{10^5} = 1 - .59049$$

$$= .40951 \qquad\blacksquare$$

As you can see from the above examples, the calculations for binomial probabilities can become rather tedious. For this reason, tables of binomial distributions have been compiled. If you wish to calculate the probability of X successes in n independent Bernoulli trials with the probability of success on a single trial equal to p, you can use Table 3 in Appendix E. For example, the probability

$$P(X = 1) = \binom{5}{1}\left(\frac{1}{10}\right)^1\left(\frac{9}{10}\right)^4$$

can be found in Table 3, where $n = 5$, $k = 1$, and $p = .1$:

$$P(X = 1) = .328$$

Also, we find $P(X = 0) = .590$.

EXAMPLE 4   A typist can type a page accurately (with no errors) with a probability of .25. What is the probability that there will be exactly one error in a report of five pages?

Solution   The probability of one error on one page is .75, so the solution is given by

$$P(X = 1) = \binom{5}{1}(.75)^1(.25)^4$$

But Table 3 in Appendix E only has values for $p \leq .5$. However,

$$\binom{n}{k}p^k(1 - p)^{n-k} = \binom{n}{n-k}(1 - p)^{n-k}p^k$$

This means that we can interchange the second and third terms (one of these must be less than or equal to .5); p now has the value .25 and we can use the table for $X = 1$ replaced by $X = 4$ (this is $n - X$) since the exponent on the new "p term" is now 4. This means that

$$P(X = 1) = P(X = 4) = \binom{5}{4}(.25)^4(.75)^1$$

The entry in the table where $n = 5$, $k = 4$, and $p = .25$ is $P(X = 1) = .015$. $\qquad\blacksquare$

COMPUTER APPLICATION Program 11 (options 3 and 4) on the computer disk accompanying this text will help you find binomial probabilities as well as draw the accompanying histograms.

We can also find the mean, variance, and standard deviation for a binomial distribution.

EXAMPLE 5 Find the mean, variance, and standard deviation for the number of heads obtained from tossing a coin four times.

Solution Use the information in Example 1 along with the formulas for mean and variance from Section 7.2:

$$\mu = \Sigma X \cdot P(X) \qquad \text{and} \qquad \sigma^2 = [\Sigma X^2 \cdot P(X)] - \mu^2$$

X	$P(X)$	$\overset{\mu}{X \cdot P(X)}$	X^2	$X^2 \cdot P(X)$
4	$\frac{1}{16} = .0625$	.25	16	1
3	$\frac{1}{4} = .25$	.75	9	2.25
2	$\frac{3}{8} = .375$	.75	4	1.5
1	$\frac{1}{4} = .25$	.25	1	.25
0	$\frac{1}{16} = .0625$	0	0	0
Total		2		5

From the table, $\mu = 2$, and
$$\sigma^2 = 5 - 4 = 1 \qquad \sigma = \sqrt{1} = 1 \qquad \blacksquare$$

The computations in Example 5 use the formulas for mean, variance, and standard deviation given in Section 7.2. However, if we apply some complicated algebraic manipulations to these formulas (the details are too involved to show here), we arrive at some very simple formulas for the mean, variance, and standard deviation of a binomial distribution.

Mean, Variance, and Standard Deviation of a Binomial Distribution

A binomial distribution of n independent trials, each with a probability of success p and failure $q = 1 - p$, has mean μ, variance σ^2, and standard deviation σ, given by
$$\mu = np \qquad \sigma^2 = npq \qquad \sigma = \sqrt{npq}$$

We can verify these formulas by comparison with the results found in Example 5. We had four repetitions of tossing a coin, so $n = 4$; $p = \frac{1}{2}$ since the probability of success (a head) is $\frac{1}{2}$; and $q = 1 - p = \frac{1}{2}$. Thus

$$
\begin{aligned}
\mu &= np & \sigma^2 &= npq & \sigma &= \sqrt{npq} \\
&= 4 \cdot \tfrac{1}{2} & &= 4 \cdot \tfrac{1}{2} \cdot \tfrac{1}{2} & &= \sqrt{1} \\
&= 2 & &= 1 & &= 1
\end{aligned}
$$

EXAMPLE 6   The probability of a transmission failure in the first year on a new automobile is .005. Find the mean number of failures for 10,000 cars if we assume that the failures are independent. Also find the standard deviation.

Solution $n = 10,000; q = .005; p = 1 - .005 = .995$

Mean number of *failures*

$$\mu = nq \qquad\qquad \sigma = \sqrt{npq}$$
$$= 10,000(.005) \qquad\qquad = \sqrt{(10,000)(.995)(.005)}$$
$$= 50 \qquad\qquad = \sqrt{49.75}$$
$$\approx 7.05 \qquad\qquad\qquad\blacksquare$$

Problem Set 7.3

APPLICATIONS

A manufacturer of disk drives chooses three items from the day's production and classifies these as defective or nondefective. Suppose that an item has a probability of .2 of being defective. In Problems 1–5 find the requested probabilities.

1. No defective disk drives are found.

2. All three are defective.

3. Two are found to be defective.

4. At least two are defective.

5. One or more are defective.

6. Numerically, how are your answers for Problems 1 and 5 related?

All the cows in a certain herd are white-faced. The probability that a white-faced calf will be born from the mating with a certain bull is .9. Suppose four cows are bred to the same bull. Find the probabilities in Problems 7–12.

7. Four white-faced calves

8. Exactly three white-faced calves

9. No white-faced calves

10. One white-faced calf

11. At least three white-faced calves

12. Not more than two white-faced calves

Use Table 3 (Appendix E) to find the binomial probabilities in Problems 13–20.

13. $n = 5, \quad X = 3, \quad p = .30$

14. $n = 4, \quad X = 3, \quad p = .25$

15. $n = 12, \quad X = 6, \quad p = .65$

16. $n = 10, \quad X = 4, \quad p = .80$

17. $n = 6, \quad X = 6, \quad p = \frac{1}{2}$

18. $n = 8, \quad X = 8, \quad p = \frac{3}{4}$

19. $n = 7, \quad X = 5, \quad p = \frac{1}{10}$

20. $n = 15, \quad X = 13, \quad p = \frac{2}{5}$

In Problems 21–29 find the mean, variance, and standard deviation for the given values of n and p. Assume that the binomial conditions are satisfied in each case.

21. $n = 5, \quad p = .3$

22. $n = 4, \quad p = .25$

23. $n = 12, \quad p = .65$

24. $n = 10, \quad p = .8$

25. $n = 6, \quad p = \frac{1}{2}$

26. $n = 8, \quad p = \frac{3}{4}$

27. $n = 7, \quad p = \frac{1}{10}$

28. $n = 15, \quad p = \frac{2}{5}$

29. $n = 10, \quad p = \frac{3}{5}$

30. The probability that children in a specific family will have blond hair is $\frac{3}{4}$. If there are four children in the family, what is the probability that two of them are blond?

31. The probability that the children in a specific family will have brown eyes is $\frac{3}{8}$. If there are four children in the family, what is the probability that at least two of them have brown eyes?

32. It is known that 85% of the graduates of Foley's School of Motel Management are placed in a job within 6 months of graduation. If a class has 20 graduates, what is the probability that 15 will be placed within 6 months?

33. If a graduating class of Foley's School in Problem 32 has 10 graduates, what is the probability that at least 9 will be placed within 6 months?

34. Suppose the National League team has a probability of $\frac{3}{5}$ of winning the World Series and the American League team has a probability of $\frac{2}{5}$. The series is over as soon as one team wins four games. Find the probability that the series is over in four games.

35. In Problem 34 what is the probability that the series will go seven games?

36. The batting average of the star baseball player Brian Thompson is .300 (that is, P(hit) = .3). What is the probability that Thompson will get at least three hits for four times at bat?

37. Suppose that research has shown that the probability that a missile penetrates enemy defenses and reaches its target is $\frac{1}{10}$. Find the smallest number of identical missiles that are necessary in order to be 80% certain of hitting the target at least once.

38. Eighty percent of the widgets of the Ampex Widget Company meet the specifications of its customers. If a sample of six widgets is tested, what is the probability that three or more of them would fail to meet the specifications?

39. Find the mean number of failures for the widgets described in Problem 38. Also find the standard deviation.

40. Ten percent of the students fail the final exam in Math 9. If the class size is 30, find the mean and standard deviation for the number of failures in each class.

41. A certain type of seed has a 60% germination rate. If the seeds are planted in rows of 50 seeds each, find the mean and standard deviation for the number of seeds that germinate in each row.

42. Suppose a person claims to have ESP and says he can read mental images 75% of the time. We set up an experiment where he must determine whether we are looking at a picture of a circle or a square. We agree that if he can correctly identify the mental image at least five out of six times we will grant his claim.
  a. What is the probability of granting his claim if he is in fact only guessing?
  b. What is the probability of denying his claim if he really does have the ability he claims?

43. Repeat Problem 42 where the person claims to read mental images 90% of the time.

44. In a certain office, three men determine who will pay for coffee by each flipping a coin. If one of them has an outcome that is different from the other two, he must pay for the coffee. What is the probability that in any play of the game there will be an odd man out?

45. In the game of Problem 44, what is the probability that there will be an odd man out in a particular play if there are five players?

46. In Problem 44, what is the probability that it will require more than two tosses to produce an odd man out?

47. Generalize Problem 44 for n men playing the game of odd man out.

7.4 The Normal Distribution

Suppose we survey the results of 20 children's scores on an IQ test. The scores (rounded to the nearest 5 points) are: 115, 90, 100, 95, 105, 105, 95, 105, 105, 95, 125, 120, 110, 100, 100, 90, 110, 100, 115, and 80. The mean is 103, the standard deviation is 10.93, and the frequency is shown in Figure 7.1a.

If we consider 100,000 IQ scores instead of only 20, we might obtain the frequency distribution shown in Figure 7.1b. As you can see, the frequency distribution

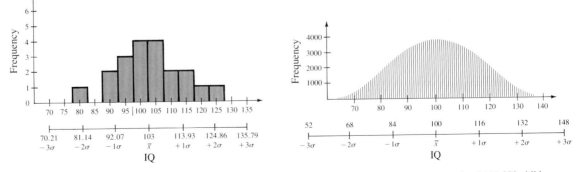

a. Frequencies of IQs for a sample of 20 children   **b.** Frequencies of IQs for a sample of 100,000 children

Figure 7.1 Frequencies of IQ scores

approximates a curve. If we connect the endpoints of the bars in Figure 7.1b, we obtain a curve that is very close to a curve called the **normal distribution curve**, or simply the **normal curve**, as shown in Figure 7.2.

The normal distribution is a **continuous** (rather than finite) **distribution**, and it extends indefinitely in both directions, never touching the x-axis. It is symmetric about a vertical line drawn through the mean, μ. The equation of this curve is

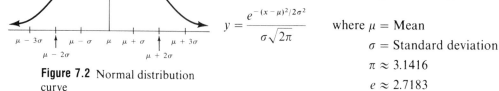

$$y = \frac{e^{-(x-\mu)^2/2\sigma^2}}{\sigma\sqrt{2\pi}}$$

where μ = Mean

σ = Standard deviation

$\pi \approx 3.1416$

$e \approx 2.7183$

Figure 7.2 Normal distribution curve

Graphs of this curve for several choices of σ are shown in Figure 7.3. Using calculus, it can be shown that the "curvature" of the normal curve changes at $\mu + \sigma$ and $\mu - \sigma$. In calculus this point is called a *point of inflection*.

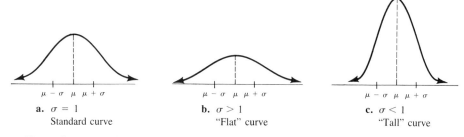

Figure 7.3 Variations of normal curves

a. $\sigma = 1$
Standard curve

b. $\sigma > 1$
"Flat" curve

c. $\sigma < 1$
"Tall" curve

Since the normal distribution is a probability distribution, we know the area under this curve is 1. Therefore we can relate the area to probabilities:

Probabilities of a Normal Probability Distribution

Let X be a random variable with a normal probability distribution. Then:

$P(a \leq X \leq b)$ is the area under the associated normal curve between a and b.

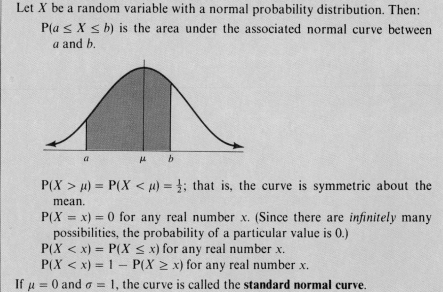

$P(X > \mu) = P(X < \mu) = \frac{1}{2}$; that is, the curve is symmetric about the mean.

$P(X = x) = 0$ for any real number x. (Since there are *infinitely* many possibilities, the probability of a particular value is 0.)

$P(X < x) = P(X \leq x)$ for any real number x.

$P(X < x) = 1 - P(X \geq x)$ for any real number x.

If $\mu = 0$ and $\sigma = 1$, the curve is called the **standard normal curve**.

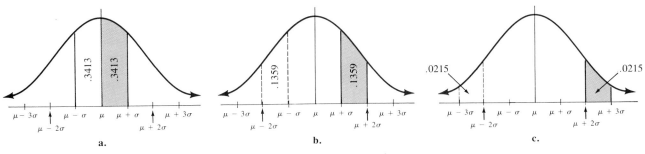

Figure 7.4 Areas under a normal curve

Since it requires calculus to find particular areas under the standard normal curve, extensive tables have been compiled to determine the area under this curve without the necessity of going through actual computations. We know, for example, that approximately 68% (.6826) of the area lies between $\mu + \sigma$ and $\mu - \sigma$. Also, approximately 14% (.1359) is between $\mu + \sigma$ and $\mu + 2\sigma$, and 2% (.0215) between $\mu + 2\sigma$ and $\mu + 3\sigma$.

For additional areas, use Table 4 in Appendix E. The table is arranged to give the area under the standard normal curve to the left of a vertical line through some number z, where $\mu = 0$ and $\sigma = 1$. Thus, to verify the area between $\mu + \sigma$ and $\mu + 2\sigma$ (shaded in color in Figures 7.4b and 7.5), you need to find $z = 2$ and then subtract $z = 1$. From the table, the area to the left of $z = 2$ is .9772 and the area to the left of $z = 1$ is .8413. Therefore the area between $z = 2$ and $z = 1$ is

$$.9772 - .8413 = .1359$$

or about 14% of the area under the curve (as stated earlier).

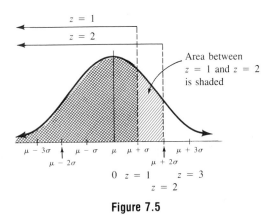

Figure 7.5

EXAMPLE 1 Find the area under the standard normal curve to the left of $z = .57$.

Solution From Table 4 in Appendix E, it is .7157. ∎

EXAMPLE 2 Find the area under the standard normal curve to the right of $z = -.13$.

Solution From Table 4, the area to the *left* of $-.13$ is .4483, so the area to the right is

$$1 - .4483 = .5517$$ ∎

EXAMPLE 3 Find the area under the standard normal curve between $z = -.05$ and $z = .93$.

Solution The area to the left of $-.05$ is .4801 and to the left of .93 it is .8238, so the area between those values is

$$.8238 - .4801 = .3437$$ ∎

What if we are not working with a standard normal curve? That is, suppose the curve is a normal curve, but it does not have a mean of 0. We can use the information in Figure 7.4, which, for convenience, is summarized in Figure 7.6.

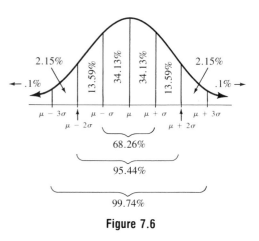

Figure 7.6

EXAMPLE 4   A teacher claims to grade "on a curve." This means the teacher believes that the scores on a given test are normally distributed. If 200 students take the exam, with mean 73 and standard deviation 9, how would the teacher grade the students?

Solution   First, draw a normal curve with a mean 73 and standard deviation 9, as shown below:

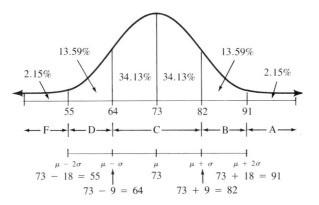

The range of 73 to 82 will contain about 34% of the class, and the range of 82 to 91 will contain about 14% of the class. Finally, about 2% of the class will score higher than 91. The teacher would therefore give grades according to the following table:

Grade on final	Letter grade	Number receiving grade	Percentage of class
Above 91	A	4	2%
83–91	B	28	14%
64–82	C	136	68%
55–63	D	28	14%
54 or below	F	4	2%

EXAMPLE 5   The Ridgemont Light Bulb Company tested a new line of light bulbs and found them to be normally distributed, with a mean life of 98 hours and a standard deviation of 13.

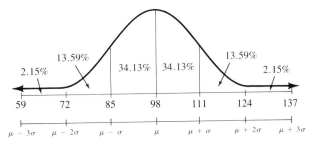

a. What percentage of bulbs will last less than 72 hours?
b. What is the probability that a bulb selected at random will last more than 111 hours?

Solution   Draw a normal curve with mean 98 and standard deviation 13.

a. About 2% will last less than 72 hours.
b. We see that about 16% of the bulbs will last longer than 111 hours, so
$$P(X > 111) \approx .16$$ ■

If you need to use a finer division of standard deviations than 1, 2, or 3, you will have to use calculus, or have a table available for that particular μ and σ, or use a number called a **z-score**. The z-score essentially translates any normal curve into a standard normal curve by use of the simple calculation given below. (This is why Table 4 in Appendix E uses z values.)

z-Score   The area to the left of a value x under a normal curve with mean μ and standard deviation σ is the same as the area under a standard normal curve to the left of the following value of z:

$$z = \frac{x - \mu}{\sigma}$$

EXAMPLE 6   Find the probability that one of the light bulbs described in Example 5 will last between 110 and 120 hours.

Solution   From Example 5, $\mu = 98$ and $\sigma = 13$. Let $x = 110$; then
$$z = \frac{110 - 98}{13} \approx .92$$

Now, look up $z \approx .92$ in Table 4 (Appendix E). The area to the left of $x = 110$ is .8212.
For $x = 120$,
$$z = \frac{120 - 98}{13} \approx 1.69$$

Again, from Table 4, the area to the left of $x = 120$ is .9545 (find $z \approx 1.69$ in the table). The area between $x = 110$ and $x = 120$ is

$$.9545 - .8212 = .1333$$

Thus $P(110 < X < 120) \approx .13$.   ∎

Problem Set 7.4

Find the area under the standard normal curve satisfying the conditions in Problems 1–16.

1. Left of -2

2. Left of 0

3. Left of 1

4. Left of 1.23

5. Left of $-.61$

6. Left of $.81$

7. Right of -2

8. Right of 1

9. Right of -1

10. Right of -1.73

11. Right of 1.69

12. Right of $-.11$

13. Between $.5$ and 1.61

14. Between $-.4$ and $.4$

15. Between -1.03 and 1.59

16. Between -2.8 and $-.46$

APPLICATIONS

In Problems 17–21 suppose that people's heights (in centimeters) are normally distributed, with a mean of 170 and a standard deviation of 5. We take a sample of 50 persons.

17. How many would you expect to be between 165 and 175 centimeters tall?

18. How many would you expect to be taller than 160 centimeters?

19. How many would you expect to be taller than 175 centimeters?

20. If a person is selected at random, what is the probability that he or she is taller than 165 centimeters?

21. What is the variance (s^2) for this sample?

In Problems 22–26 suppose that, for a certain exam, a teacher grades on a curve. It is known that the mean is 50 and the standard deviation is 5. There are 45 students in the class.

22. How many students would receive a C?

23. How many students would receive an A?

24. What score would be necessary to obtain an A?

25. If an exam paper is selected at random, what is the probability that it will be a failing paper?

26. What is the variance for this exam?

27. The breaking strength of a rope (in pounds) is normally distributed, with a mean of 100 pounds and a standard deviation of 16. What is the probability that a certain rope will break with a force of 132 pounds?

28. The diameter of an electric cable is normally distributed, with a mean of .9 inch and a standard deviation of .01. What is the probability that the diameter will exceed .91 inch?

29. The annual rainfall in Ferndale, California, is known to be normally distributed, with a mean of 35.5 inches and a standard deviation of 2.5. About 2.15% of the time, how many inches will the rainfall exceed?

30. In Problem 29, what is the probability that the rainfall will exceed 30.5 inches?

A light bulb is normally distributed with a mean life of 250 hours and a standard deviation of 25 hours. Find the probabilities requested in Problems 31–36.

31. $P(X > 250)$

32. $P(X > 300)$

33. $P(X < 220)$

34. $P(200 < X < 300)$

35. $P(220 \leq X \leq 320)$

36. $P(230 \leq X \leq 240)$

37. The diameter of a pipe is normally distributed, with a mean of .4 inch and a variance of .0004. What is the probability that the diameter will exceed .44 inch?

38. The breaking strength of a certain new synthetic material is normally distributed, with a mean of 165 pounds and a variance of 9. The material is considered defective if the breaking strength is less than 159 pounds. What is the probability that a sample chosen at random will be defective?

39. Repeat Problem 37 to find the probability that the diameter will exceed .43 inch.

40. Repeat Problem 38 to find the probability that the breaking strength of a sample is between 170 and 180 pounds.

7.5 Normal Approximation to the Binomial

The normal distribution in Section 7.4 was introduced by looking at a histogram showing IQ frequencies. In this section we will look at the relationship between the binomial and normal distributions.

EXAMPLE 1 Consider an experiment of tossing a coin 16 times. What is the probability of obtaining exactly x heads for $x = 0, 1, 2, \ldots, 15, 16$?

Solution This is a binomial distribution for which $n = 16$, $p = \frac{1}{2}$, and $q = \frac{1}{2}$. The model is

$$P(X = x) = \binom{n}{x} p^x q^{n-x}$$

$$= \binom{16}{x}\left(\frac{1}{2}\right)^x \left(\frac{1}{2}\right)^{16-x}$$

X	P(X)
0	.00002
1	.0002
2	.0018
3	.0085
4	.0278
5	.0667
6	.1222
7	.1746
8	.1964
9	.1746
10	.1222
11	.0667
12	.0278
13	.0085
14	.0018
15	.0002
16	.00002

The table in the margin shows the results of all the calculations for $x = 0, 1, 2, \ldots, 15$, and 16. Let us consider the procedure for one of these, say, $x = 9$:

$$P(X = 9) = \binom{16}{9}\left(\frac{1}{2}\right)^9 \left(\frac{1}{2}\right)^7$$

$$= \frac{16!}{(16-9)!9!}\left(\frac{1}{2}\right)^{16}$$

$$= \frac{16!}{7!9!2^{16}}$$

$$\approx .1746 \qquad \blacksquare$$

We can construct a histogram showing the results for Example 1. In Figure 7.7, which shows this, we use $2^{16} = 65,536$ for the total number of possibilities.

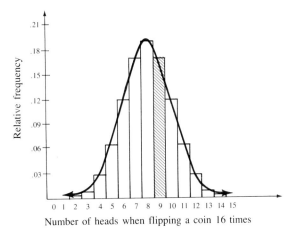

Figure 7.7 Histogram for the binomial distribution in Example 1

Note that we have superimposed a normal curve over the histogram in Figure 7.7. Under certain conditions a normal curve can be used to approximate a binomial distribution. This relationship was first noted by the mathematician Abraham De Moivre in 1718. Suppose we use the normal curve to approximate the binomial distribution in Example 1. We first find the mean and the standard deviation:

$$\mu = np \qquad\qquad \sigma = \sqrt{npq}$$
$$= 16(\tfrac{1}{2}) \qquad\qquad = \sqrt{16(\tfrac{1}{2})(\tfrac{1}{2})}$$
$$= 8 \qquad\qquad\qquad = \sqrt{4}$$
$$\qquad\qquad\qquad\qquad = 2$$

The normal curve shown in Figure 7.7 has mean 8 and standard deviation 2. The important difference between the binomial and normal distributions is that the binomial is *discrete* (or finite) and the normal is *continuous* (or infinite). If we were to find the probability of $x = 9$ for a normal distribution, we would find it to be 0. However, take a closer look at Figure 7.7, as shown in Figure 7.8.

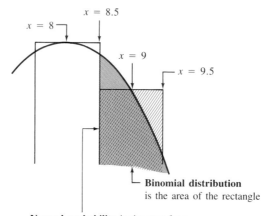

Figure 7.8 Detail of Figure 7.7

$x = 8.5$

$x = 8$

$x = 8.5$

$x = 9$

$x = 9.5$

Binomial distribution is the area of the rectangle

Normal probability is the area from $x = 8.5$ to $x = 9.5$ under the curve. This is used to approximate the area of the rectangle.

The z-score for $x = 9.5$ is

$$z = \frac{x - \mu}{\sigma}$$

$$= \frac{9.5 - 8}{2}$$

$$= .75$$

The Table 4 entry for $z = .75$ is .7734

The z-score for $x = 8.5$ is

$$z = \frac{8.5 - 8}{2}$$

$$= .25$$

The Table 4 entry for $z = .25$ is .5987

The area between $x = 9.5$ and $x = 8.5$ is therefore

$$.7734 - .5987 = .1747$$

Note that this is a *very* good approximation for $P(X = 9)$ in the binomial distribution found in Example 1.

The Normal Distribution as an Approximation to the Binomial Distribution

If X is a binomial random variable of n independent trials, each with probability of success p and failure $q = 1 - p$, and if

$$np \geq 5 \quad \text{and} \quad nq \geq 5$$

then the binomial random variable is approximated by a normal distribution with mean and standard deviation given by

$$\mu = np \qquad \sigma = \sqrt{npq}$$

To show the tremendous value of this theorem, consider another, more realistic example.

EXAMPLE 2   Experimental data indicate that the cure rate of a new drug is 85%. If the drug is given to 100 patients, find the probability that at least 90 of them will be cured.

Solution   This is a binomial distribution with $p = .85$, $q = .15$, and $n = 100$. We want to find $P(X \geq 90)$.

$$P(X \geq 90) = P(X = 90) + P(X = 91) + \cdots + P(X = 99) + P(X = 100)$$

Now,

$$P(X = 90) = \binom{100}{90}(.85)^{90}(.15)^{10}$$

$$P(X = 91) = \binom{100}{91}(.85)^{91}(.15)^{9}$$

$$\vdots$$

$$P(X = 100) = \binom{100}{100}(.85)^{100}(.15)^{0}$$

These calculations are certainly out of hand (they are even out of the range of what most calculators can handle). However,

$$np = 100(.85) = 85 \geq 5 \quad \text{and} \quad nq = 100(.15) = 15 \geq 5$$

so we can use the normal distribution as an approximation.

$$\mu = np \qquad \sigma = \sqrt{npq}$$
$$= 85 \qquad\quad = \sqrt{100(.85)(.15)}$$
$$\approx 3.571$$

Now, $P(X \geq 90) = 1 - P(X < 90)$, so for $x = 90$, we use the z-score and Table 4 (Appendix E):

$$z = \frac{x - \mu}{\sigma}$$

$$= \frac{90 - 85}{3.571}$$

$$\approx 1.40 \qquad \text{From Table 4, } P(X < 90) \approx .9192$$

Then

$$P(X \geq 90) \approx 1 - .9192$$
$$= .0808 \qquad\qquad\qquad\qquad\qquad \blacksquare$$

In Example 2 we calculated np and nq in order to see if we could apply a normal approximation. Table 7.1 provides a quick reference as to when we may use this approximation. For Example 2, we could have used $p = .9$ and the table to find that n must be at least 50 to conclude that the normal is an acceptable approximation.

TABLE 7.1
Minimum sample required to approximate a binomial distribution by a normal distribution

p	n must be at least
.001	5000
.01	500
.1	50
.2	25
.3	17
.4	13
.5	10
.6	13
.7	17
.8	25
.9	50
.99	500
.999	5000

EXAMPLE 3 In certain grades in elementary school, students are given a vision test. Experience has shown that 18% of the students do not pass the test. Find the probability that if the test is given to 100 students, *exactly* 18 will fail the test.

Solution This is a binomial distribution with $n = 100$ and $p = .18$.

$$P(X = 18) = \binom{100}{18}(.18)^{18}(.82)^{82}$$

This is a formidable calculation (even with a calculator). However, Table 7.1 indicates that we can use a normal approximation with

$$\mu = np \qquad\qquad \sigma = \sqrt{npq}$$
$$= (100)(.18) \qquad = \sqrt{100(.18)(.82)}$$
$$= 18 \qquad\qquad \approx 3.84$$

We cannot find $x = 18$ exactly, so we find the difference of $x = 18.5$ and $x = 17.5$:

For x = 18.5: *For x = 17.5:*

$$z = \frac{18.5 - 18}{3.84} \qquad z = \frac{17.5 - 18}{3.84}$$

$$\approx .13 \qquad\qquad \approx -.13$$

From Table 4,

$$P(X \le 18.5) \approx .5517 \qquad P(X \le 17.5) \approx .4483$$

Thus

$$P(17.5 \le X \le 18.5) \approx .5517 - .4483$$
$$= .1034 \qquad\qquad\blacksquare$$

Problem Set 7.5

For the given values associated with a binomial experiment in Problems 1–9, determine whether the normal distribution is a suitable approximation.

1. $n = 1000$, $p = .01$
2. $n = 20$, $p = .5$
3. $n = 10$, $p = .6$
4. $n = 15$, $p = .6$
5. $n = 100$, $p = .4$
6. $n = 50$, $p = .1$
7. $n = 1000$, $p = .04$
8. $n = 12$, $p = .6$
9. $n = 40$, $p = .09$

A coin is tossed 15 times. Let X be a random variable representing the number of tails. Find the probabilities in Problems 10–21.

10. $P(X < 8)$
11. $P(X < 10)$
12. $P(X < 11)$
13. $P(X > 13)$
14. $P(X > 6)$
15. $P(X > 9)$
16. $P(7 < X < 9)$
17. $P(4 \le X \le 8)$
18. $P(5 \le X \le 7)$
19. $P(X = 5)$
20. $P(X = 11)$
21. $P(X = 14)$

A die is tossed 1000 times. Let X be the number of 6s tossed. Find the probabilities in Problems 22–33.

22. $P(X < 200)$
23. $P(X \le 170)$
24. $P(X \le 190)$
25. $P(X \ge 150)$
26. $P(X \ge 140)$
27. $P(X > 160)$
28. $P(140 \le X \le 150)$
29. $P(150 < X < 160)$
30. $P(140 < X < 150)$
31. $P(X = 140)$
32. $P(X = 150)$
33. $P(X = 160)$

APPLICATIONS

A new drug has a cure rate of 92%. The drug is administered to 1000 patients. Let X be the number who are cured by the drug. Find the probabilities in Problems 34–45.

34. $P(X < 900)$
35. $P(X < 950)$
36. $P(X \le 850)$
37. $P(X \ge 920)$
38. $P(X > 900)$
39. $P(X > 910)$
40. $P(900 \le X \le 1000)$
41. $P(850 < X < 950)$
42. $P(800 < X < 950)$
43. $P(X = 900)$
44. $P(X = 920)$
45. $P(X = 930)$

46. An experiment in parapsychology consists of correctly identifying one of five shapes. If the test is repeated 10 times, what is the probability that the subject will correctly identify the object more than 8 times?

47. A baseball player's batting average is .250. What is the probability of more than six hits in the next ten times at bat?

48. In 1980 the yearly death rate from cancer was 186.3 per 100,000. If a person associates with roughly 500 people, estimate the probability that more than 1 out of these 500 will die of cancer in a year.

49. Repeat Problem 48 to find the probability that more than 2 of the 500 will die in a year.

50. A computer system is made up of 100 components, each with a reliability of .95 (that is, the probability that the component operates properly is .95). If these components function independently of one another, and the computer requires at least 80 components to function properly, what is the reliability of the whole system?

51. An airline is penalized for overbooking, but loses money with empty seats. Suppose the records show that for a certain flight, 8% of the advance reservations do not show up. If the plane seats 300 people, find the probability that if the airline takes 315 advance reservations, more than 300 will show up.

7.6 Correlation and Regression*

Mathematical modeling relates numerical data to assumptions about the relationship between two variables. For example:

IQ and salary
Study time and grades
Age and heart disease
Runner's speed and runner's shoes
Teacher's salaries and beer consumption

---

* The material of this section is not required for subsequent material.

All are attempts to relate two variables in some way or another. If a **correlation** is established, then the next step in the modeling process is to identify the nature of the relationship. This is called **regression analysis**. We now turn our attention to finding a *linear relationship* that is called the *best-fitting line* by a technique called the **least squares method**. The derivations of the results and the formulas given in this section are, for the most part, based on calculus and are therefore beyond the scope of this book. We will, however, focus on how to use the formulas as well as discuss what they mean and how they are interpreted.

We begin with correlation. We want to know if two variables are related—that is, depend on one another. Let us call one variable x and the other y. These variables can be represented as ordered pairs (x, y) in a graph called a **scatter diagram**.

EXAMPLE 1 A survey of 20 students compared the grade received on an examination with the length of time the student studied. Draw a scatter diagram to represent the data in the table.

Student Number	1	2	3	4	5	6	7	8	9	10
Length of Study Time (to nearest 5 minutes)	30	40	30	35	45	15	15	50	30	0
Grade (100 possible)	72	85	75	78	89	58	71	94	78	10
Student Number	11	12	13	14	15	16	17	18	19	20
Length of Study Time (to nearest 5 minutes)	20	10	25	25	25	30	40	35	20	15
Grade (100 possible)	75	43	68	60	70	68	82	75	65	62

Solution Let x be the study time (in minutes) and let y be the grade (in points). The graph is shown below:

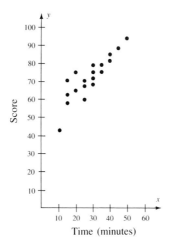

Correlation is a measure to determine whether there is a statistically significant relationship between two variables. Intuitively, it should assign a measure consistent with the scatter diagram shown in Figure 7.9.

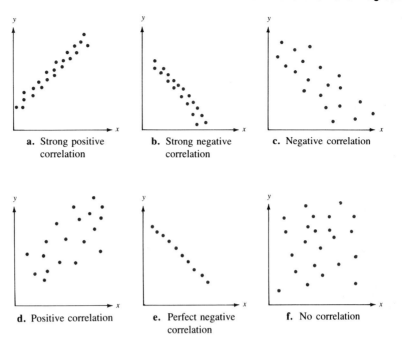

Figure 7.9 Examples of correlation

a. Strong positive correlation

b. Strong negative correlation

c. Negative correlation

d. Positive correlation

e. Perfect negative correlation

f. No correlation

Such a measure, called the **linear correlation coefficient**, **r**, is defined so that it has the following properties:

1. r measures the correlation between x and y.
2. r is between -1 and 1.
3. If r is close to 0, it means there is little correlation.
4. If r is close to 1, it means there is a strong positive correlation.
5. If r is close to -1, it means there is a strong negative correlation.

To write a formula for r, we let n denote the number of pairs of data present; and, as before:

Σx denotes the sum of the x values
Σx^2 means square the x values and then sum
$(\Sigma x)^2$ means sum the x values and then square
Σxy means multiply each x value by the corresponding y value and then sum
$n\Sigma xy$ means multiply n times Σxy
$(\Sigma x)(\Sigma y)$ means multiply Σx and Σy

Correlation Coefficient

The *linear correlation coefficient r* is

$$r = \frac{n\Sigma xy - (\Sigma x)(\Sigma y)}{\sqrt{n(\Sigma x^2) - (\Sigma x)^2}\ \sqrt{n(\Sigma y^2) - (\Sigma y)^2}}$$

EXAMPLE 2 Find r for the data in Example 1.

Solution

Study time, x	Score, y	xy	x^2	y^2
30	72	2160	900	5184
40	85	3400	1600	7225
30	75	2250	900	5625
35	78	2730	1225	6084
45	89	4005	2025	7921
15	58	870	225	3364
15	71	1065	225	5041
50	94	4700	2500	8836
30	78	2340	900	6084
0	10	0	0	100
20	75	1500	400	5625
10	43	430	100	1849
25	68	1700	625	4624
25	60	1500	625	3600
25	70	1750	625	4900
30	68	2040	900	4624
40	82	3280	1600	6724
35	75	2625	1225	5625
20	65	1300	400	4225
15	62	930	225	3844
Total 535	1378	40,575	17,225	101,104
↑ Σx	↑ Σy	↑ Σxy	↑ Σx^2	↑ Σy^2

$$r = \frac{n\Sigma xy - (\Sigma x)(\Sigma y)}{\sqrt{n(\Sigma x^2) - (\Sigma x)^2}\sqrt{n(\Sigma y^2) - (\Sigma y)^2}}$$

$$= \frac{20(40,575) - (535)(1378)}{\sqrt{20(17,225) - (535)^2}\sqrt{20(101,104) - (1378)^2}}$$

$$= \frac{74,270}{\sqrt{58,275}\sqrt{123,196}}$$

$$\approx .8765$$

 ∎

Example 2 shows a very strong positive correlation. But if r for Example 2 had been .46, would we still be able to assume that there is a strong correlation? This question is a topic of major concern in statistics. The term **significance level** is used to denote the cutoff between results attributed to chance and the results attributed to significant differences. Table 7.2 gives **critical values** for determining whether two variables are correlated. If r is greater than the given table value, then we may assume that a correlation exists between the variables. If we use the column labeled

$\alpha = .05$, then the significance level is 5%. This means that the probability is .05 that we will say the variables are correlated when, in fact, the results are attributed to chance. This is also true for a significance level of 1% ($\alpha = .01$). For Example 2, since $n = 20$, we see in Table 7.2 that $r = .46$ would show a linear correlation at a 5% significance level, but not at a 1% level.

TABLE 7.2
Correlation coefficient

n	$\alpha = .05$	$\alpha = .01$	n	$\alpha = .05$	$\alpha = .01$
4	.950	.999	18	.468	.590
5	.878	.959	19	.456	.575
6	.811	.917	20	.444	.561
7	.754	.875	25	.396	.505
8	.707	.834	30	.361	.463
9	.666	.798	35	.335	.430
10	.632	.765	40	.312	.402
11	.602	.735	45	.294	.378
12	.576	.708	50	.279	.361
13	.553	.684	60	.254	.330
14	.532	.661	70	.236	.305
15	.514	.641	80	.220	.286
16	.497	.623	90	.207	.269
17	.482	.606	100	.196	.256

Note: The derivation of this table is beyond the scope of this course. It shows the critical values of the *Pearson correlation coefficient.*

EXAMPLE 3 Find the critical value of the linear correlation coefficient for 10 pairs of data and a significance level of .05.

Solution From Table 7.2, the critical value is $r = .632$. For $n = 10$, any value greater than $r = .632$ is considered linearly correlated. ■

EXAMPLE 4 If $r = -.85$ and $n = 10$, are the variables correlated at a significance level of 1%?

Solution For $n = 10$ and $\alpha = .01$, the Table 7.2 entry is .765. Since r is negative and since $|r| > .765$, we see that there is a negative linear correlation. ■

EXAMPLE 5 The following table is a sample of some past annual mean salaries for teachers in elementary and secondary schools, along with the annual per capita beer consumption (in gallons) for Americans. Find the correlation coefficient.*

---

* Example 5 and the paragraph following are from Mario F. Triola, *Elementary Statistics*, 2nd ed. (Menlo Park, Calif.: Benjamin/Cummings, 1983). © 1983 by The Benjamin/Cummings Publishing Company, Inc. Reprinted by permission.

Year	1960	1965	1970	1972	1973	1983
Mean Teacher Salary	$5,000	$6,200	$8,600	$9,700	$10,200	$16,400
Per Capita Beer Consumption (gal)	24.02	25.46	28.55	29.43	29.68	35.2

Solution

$n = 6$

$\Sigma x = 56,100$

$\Sigma y = 172.34$

$\Sigma x^2 = 604,490,000$

$\Sigma y^2 = 5026.3418$

$(\Sigma x)^2 = 3,147,210,000$

$(\Sigma y)^2 = 29,701.0756$

$\Sigma xy = 1,688,969$

$$r = \frac{6(1,688,969) - (56,100)(172.34)}{\sqrt{6(604,490,000) - (56,100)^2} \ \sqrt{6(5026.3418) - (172.34)^2}}$$

$$\approx .994$$

The number $r \approx .994$ is so close to 1 that we hardly need to consult Table 7.2 to know that there is a strong positive linear correlation between teachers' salaries and per capita beer drinking. ∎

The significance of the correlation implies that teachers use their raises to buy more beer, right? Wrong. Perhaps increases in teachers' salaries precipitate higher taxes, which in turn cause taxpayers to drown their sorrows and forget their financial difficulties by drinking more beer. Or perhaps higher teachers' salaries and greater beer consumption are both manifestations of some other factor, such as a general improvement in the standard of living. In any event, the techniques in this chapter can be used only to establish a *statistical* linear relationship. *We cannot establish the existence or absence of any inherent cause-and-effect relationship.*

The final step in our discussion is to find the best-fitting line. That is, we want to find a line $y' = mx + b$ so that the sum of the distances of the data points from this line will be as small as possible. (We use y' instead of y to distinguish between the actual second component, y, and the predicted y value, y'.) Since some of these distances may be positive and some negative and since we do not want large opposites to "cancel each other out," we minimize the sum of the *squares* of these distances. Therefore, the **regression line** is sometimes called the **least squares line**.

Least Squares, or Regression, Line

The *least squares*, or *regression*, line $y' = mx + b$ is the line of best fit when

$$m = \frac{n(\Sigma xy) - (\Sigma x)(\Sigma y)}{n(\Sigma x^2) - (\Sigma x)^2} \qquad b = \frac{\Sigma y - m(\Sigma x)}{n}$$

EXAMPLE 6 Find the best-fitting line for the data in Example 1.

Solution From Example 2, $n = 20$, and

$$\Sigma xy = 40{,}575$$
$$\Sigma x = 535$$
$$\Sigma y = 1378$$
$$\Sigma x^2 = 17{,}225$$
$$(\Sigma x)^2 = (535)^2$$

Since m is used as part of the formula for b, you must first find m:

$$m = \frac{20(40{,}575) - (535)(1378)}{20(17{,}225) - (535)^2}$$

$$= \frac{74{,}270}{58{,}275} \approx 1.27447$$

Now you can use this m when finding b:

$$b = \frac{1378 - 1.27447(535)}{20} \approx 34.8078$$

Thus we may approximate the least squares line as $y' = 1.3x + 35$. This line is shown in the figure in the margin. ■

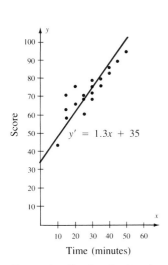

Score — y-axis; Time (minutes) — x-axis

$y' = 1.3x + 35$

Regression line for the data in Example 1

EXAMPLE 7 Use the regression line of Example 6 to predict the score of a person who studied $\frac{1}{2}$ hour.

Solution $x = 30$ minutes, so $y' = 1.3(30) + 35 = 74$ ■

A final word of caution: Use the regression line only if r indicates that there is a significant linear correlation, as given in Table 7.2.

> COMPUTER APPLICATION Program 12 on the computer disk accompanying this book will find the linear relationship between two variables using the least squares method, and it will also find the correlation between those variables.

Problem Set 7.6

In Problems 1–12 a sample of paired data gives a linear coefficient r. In each case use Table 7.2 to determine whether there is a significant linear correlation.

1. $n = 10$, $r = .7$, significance level 5%

2. $n = 10$, $r = .7$, significance level 1%

3. $n = 30$, $r = .4$, significance level 1%

4. $n = 30$, $r = .4$, significance level 5%

5. $n = 15$, $r = -.732$, significance level 5%

6. $n = 35$, $r = -.4127$, significance level 1%

7. $n = 50$, $r = -.3416$, significance level 1%

8. $n = 100$, $r = -.41096$, significance level 5%

9. $n = 23$, $r = .501$, significance level 1%

10. $n = 38$, $r = .416$, significance level 5%

11. $n = 28$, $r = -.214$, significance level 5%

12. $n = 55$, $r = -.14613$, significance level 1%

Draw a scatter diagram and find r for the data in each table in Problems 13–18 and determine whether there is a linear correlation at either the 5% or 1% level.

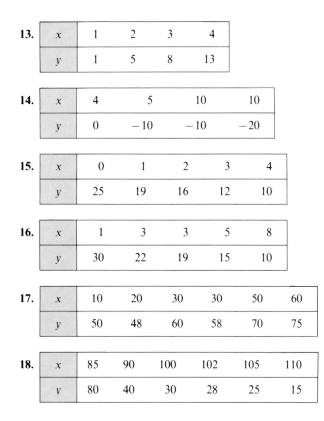

13.

x	1	2	3	4
y	1	5	8	13

14.

x	4	5	10	10
y	0	−10	−10	−20

15.

x	0	1	2	3	4
y	25	19	16	12	10

16.

x	1	3	3	5	8
y	30	22	19	15	10

17.

x	10	20	30	30	50	60
y	50	48	60	58	70	75

18.

x	85	90	100	102	105	110
v	80	40	30	28	25	15

In Problems 19–24 find the regression line for the indicated problem.

19. Problem 13   **20.** Problem 15   **21.** Problem 17

22. Problem 14   **23.** Problem 16   **24.** Problem 18

APPLICATIONS

25. A new computer circuit was tested and the times (in nanoseconds) required to carry out different subroutines were recorded as follows:

Difficulty Level	1	2	2	3	4	5	5	5
Time	10	11	13	8	15	18	21	19

Find r and determine whether it is statistically significant at the 1% level.

26. Ten people are given a standard IQ test. Their scores were then compared with their high school grades:

IQ	Grade (GPA)
117	3.1
105	2.8
111	2.5
96	2.8
135	3.4
81	1.9
103	2.1
99	3.2
107	2.9
109	2.3

Find r and determine whether it is statistically significant at the 1% level.

27. Find the regression line for the data in Problem 25.

28. Find the regression line for the data in Problem 26.

Problems 29–35 are based on the following table. Determine whether there is a correlation between the indicated variables, and, if so, is it at the 5% or the 1% significance level?

	Year					
	1965	**1970**	**1975**	**1979**	**1980**	**1983**
Birth Rate (Births per 1000)	19.4	18.3	14.8	15.8	16.2	15.5
Death Rate (Deaths per 1000)	9.4	9.4	8.9	8.7	8.9	8.6
Per Capita Income	$2,773	$3,893	$5,851	$8,757	$9,458	$11,675
Prime Interest Rate (6 months' commercial paper)	4.54%	7.72%	6.33%	10.91%	12.29%	12.15%

29. Birth rate and per capita income
30. Interest rate and birth rate
31. Interest rate and death rate
32. Birth rate and death rate

33. Per capita income and interest rate
34. Find the regression line for Problem 32.
35. Find the regression line for Problem 33.

7.7 Summary and Review

IMPORTANT TERMS

Average [7.2]
Bernoulli trial [7.3]
Binomial distribution [7.3]
Binomial experiment [7.3]
Binomial random variable [7.3]
Continuous distribution [7.4]
Continuous random variable [7.1]
Correlation [7.6]
Correlation coefficient [7.6]
Critical values [7.6]
Discrete random variable [7.1]
Dispersion [7.2]
Frequency distribution [7.1]
Histogram [7.1]
Least squares line [7.6]
Least squares method [7.6]
Linear correlation coefficient, r [7.6]
Mean [7.2]

Measure of central tendency [7.2]
Median [7.2]
Mode [7.2]
Normal distribution [7.4]
Population [7.2]
Probability distribution [7.1]
Random variable [7.1]
Range [7.2]
Regression analysis [7.6]
Regression line [7.6]
Relative frequency [7.1]
Scatter diagram [7.6]
Significance level [7.6]
Standard deviation [7.2]
Standard normal curve [7.4]
Variance [7.2]
Weighted mean [7.2]
z-Score [7.4]

SAMPLE TEST

For additional practice there are a large number of review problems categorized by objective in the Student Solutions Manual. The following sample test (40 minutes) is intended to review the main ideas of the chapter.

Use the following outcomes, obtained from rolling a pair of dice 50 times, in Problems 1–5.

4, 3, 6, 10, 8, 9, 2, 4, 7, 4, 6, 7, 11, 7, 8, 6, 4, 8, 3, 9, 7, 8, 7, 9, 5, 9, 6, 6, 10, 7, 3, 7, 10, 6, 11, 5, 9, 10, 6, 11, 8, 11, 7, 5, 6, 11, 12, 7, 8, 9

1. Prepare a frequency distribution.
2. Find the probability distribution.
3. Represent the data in a histogram.
4. Find the mean, median, and mode.
5. Find the range, variance, and standard deviation.
6. Use Table 3 in Appendix E to compute a binomial probability where $n = 10$, $k = 4$, and $p = .8$.
7. Find the mean, variance, and standard deviation for the distribution described in Problem 6.
8. Find the area under a normal curve with $\mu = 86$ and $\sigma = 5$ between 80 and 90.

APPLICATIONS

9. The following item once appeared in Dear Abby's column:

Dear Abby: You wrote in your column that a woman is pregnant for 266 days. Who said so? I carried my baby for ten months and five days, and there is no doubt about it because I know the exact date my baby was conceived. My husband is in the Navy and it couldn't have possibly been conceived any other time because I saw him only once for an hour, and I didn't see him again until the day before the baby was born.

I don't drink or run around, and there is no way this baby isn't his, so please print a retraction about that 266-day carrying time because otherwise I am in a lot of trouble.

San Diego Reader

Abby's answer was consoling and gracious but not very statistical:

Dear Reader: The average gestation period is 266 days. Some babies come early. Others come late. Yours was late.

If the mean pregnancy duration is 266 with a standard deviation of 16 days, what is the probability of having a pregnancy longer than 310 days?

10. Since World War II, plutonium for use in atomic weapons has been produced at an Atomic Energy Commission facility in Hanford, Washington. One of the major safety problems encountered there has been the storage of radioactive wastes. Over the years, significant quantities of these substances have leaked from their open-pit storage areas into the nearby Columbia River, which flows through parts of Oregon and eventually empties into the Pacific Ocean. To measure the health consequences of this contamination, an index of exposure was calculated for each of the nine Oregon counties having frontage on either the Columbia River or the Pacific Ocean. The cancer mortality rate for each of these counties was also determined. The data are listed in the table, where higher index values represent higher levels of contamination.*

Radioactive contamination and cancer mortality in Oregon counties

County	Index of exposure	Cancer mortality per 100,000
Clatsop	8.34	210.3
Columbia	6.41	177.9
Gilliam	3.41	129.9
Hood River	3.83	162.3
Morrow	2.57	130.1
Portland	11.64	207.5
Sherman	1.25	113.5
Umatilla	2.49	147.1
Wasco	1.62	137.5

Find the linear correlation coefficient and determine whether the variables are significantly correlated at either the 1% or 5% level.

11. Draw a scatter diagram and the regression line for the set of data in Problem 10.

---

* From Richard J. Larsen and Donna Fox Stroup, *Statistics in the Real World* (New York: Macmillan, 1976).

The following multiple-choice questions dealing with probability and random variables are from actual actuarial exams and are reprinted with permission of the Society of Actuaries, 208 South LaSalle Street, Chicago, Illinois 60604.

Actuary Exam
May 1982

12. Suppose Q and S are independent events such that the probability that at least one of them occurs is $\frac{1}{3}$ and the probability that Q occurs but S does not occur is $\frac{1}{9}$. What is $P(S)$?
 A. $\frac{4}{9}$ B. $\frac{1}{3}$ C. $\frac{2}{9}$ D. $\frac{1}{7}$ E. $\frac{1}{9}$

13. A fair coin is tossed until a head appears. Given that the first head appeared on an even-numbered toss, what is the conditional probability that the head appeared on the fourth toss?
 A. $\frac{1}{16}$ B. $\frac{1}{8}$ C. $\frac{3}{16}$ D. $\frac{1}{4}$ E. $\frac{15}{16}$

14. A card is drawn at random from an ordinary deck of 52 cards and replaced. This is done a total of 5 independent times. What is the conditional probability of drawing the ace of spades exactly 4 times, given that this ace is drawn at least 4 times?
 A. $\frac{1}{2}$ B. $\frac{12}{13}$ C. $\frac{13}{14}$ D. $\frac{60}{61}$ E. $\frac{255}{256}$

15. Suppose an experiment consists of tossing a fair coin until three heads occur. What is the probability that the experiment ends after exactly six flips of the coin with a head on the fifth toss as well as on the sixth?
 A. $\frac{1}{16}$ B. $\frac{1}{8}$ C. $\frac{5}{32}$ D. $\frac{1}{4}$ E. $\frac{10}{32}$

7.8 Cumulative Review for Chapters 5–7

1. Let $U = \{1, 2, 3, \ldots, 8, 9, 10\}$, $A = \{2, 3, 5, 7\}$, $B = \{1, 3, 5, 7, 9\}$, and $C = \{2, 4, 6, 8, 10\}$. Find:
 a. $A \cap C$ b. $B \cup \bar{C}$ c. $U \cap C$ d. $\bar{U}$ e. $\overline{A \cap (B \cup C)}$

APPLICATIONS

2. In a survey of 600 college students

> 250 had tried marijuana
> 350 had tried alcohol
> 175 had tried cocaine
> 110 had tried both cocaine and alcohol
> 140 had tried both marijuana and alcohol
> 100 had tried both marijuana and cocaine
> 70 had tried all three

 a. How many had not tried any of these drugs?
 b. How many had tried only one of these drugs?
 c. How many had tried exactly two of these drugs?

3. A psychologist has an experiment consisting of seven colored boxes. In how many ways can three boxes be selected?

4. In how many ways can a full house of three kings and two queens be drawn from a deck of cards?

5. How many subsets can be chosen from five elements?

6. In how many ways can six hats be distributed to six persons?

7. Suppose two cards are drawn from a deck of cards without replacement. Find the requested probabilities.
 a. P(both hearts)   b. P(1 heart and 1 diamond)
 c. P(heart on first draw and diamond on second draw)
 d. P(diamond on second draw given heart is drawn on first draw)
 e. P(ace of hearts drawn both times)

8. Repeat Problem 7 with replacement.

9. A sample of 100 ball bearings is drawn from a day's production and 15 are found to be defective.

   a. What is the probability of drawing one ball bearing and finding that it is defective?

   b. Is this problem an example of empirical or theoretical probability?

10. At a fast-food outlet, 15% of the hamburgers sold were cold, 5% had a missing ingredient, and 0.5% were both cold and had a missing ingredient. What is the probability that your hamburger is cold if the pickle is missing?

11. A messenger transports money between two locations and travels along one of three routes, A Street, B Street, or C Street. One day, the route is randomly selected according to the probabilities $P(A) = .3$, $P(B) = .5$, and $P(C) = .2$. On the following day, the probabilities change so that the probability of the previously chosen route is .2, with the other two routes being equally probable. Find the indicated probabilities (subscripts are used to indicate the day on which a particular route is taken; for example, A_2 means that A was chosen on the second day).

   a. $P(A_2 | A_1)$ **b.** $P(A_2 | B_1)$ **c.** $P(A_2)$ **d.** $P(C_3)$

Use the following data in Problems 12–15:

   85, 70, 75, 90, 65, 40, 70, 95, 80, 70, 55, 65, 70, 80, 95

12. Prepare a frequency distribution using five subdivisions.

13. Find the probability distribution and draw a histogram.

14. Find the mean, median, and mode.

15. Find the range, variance, and standard deviation.

16. A test for psychic ability involves having the subject pick one of five cards after a card is picked by the experimenter in another room. If the subject picks the same card, it is called a "match"; if a different card is picked, it is called a "no-match." If $p = $ P(any particular choice is a match), then $p = \frac{1}{5}$ assumes no special ability on the part of the subject. If X is the number of correctly identified cards in 10 tries, and if $p = \frac{1}{5}$, find $P(X > 5)$.

17. Find the area under the standard normal curve between -1.5 and .6 standard deviations.

18. A history teacher grades on a curve. If the mean is 65 and the standard deviation is 10, how are the grades distributed? If there are 100 students in the class, how many could expect to get As, Bs, etc.?

19. A drug has side effects for 1 person out of 5000 taking it. If the drug is administered to 500,000 people, what is the probability that more than 110 of them will experience side effects?

20. A study is conducted to test the relationship between speed (mph) and fuel consumption (mpg). The following information is obtained:

Speed	20	30	40	50	60
Fuel consumption	35	38	40	34	29

   Find the linear correlation coefficient and determine whether the variables are significantly correlated at either the 1% or the 5% level.

21. Find the regression line for the data in Problem 20.

Bonus Questions*   *The following multiple-choice questions dealing with probability and random variables are from an actuarial exam and are reprinted with permission of the Society of Actuaries, 208 South LaSalle Street, Chicago, Illinois 60604.*

---

* These problems are challenging.

Actuary Exam
November 1981

22. A family has five children. Assuming that the probability of a girl on each birth was $\frac{1}{2}$ and that the five births were independent, what is the probability that the family has at least one girl, given that they have at least one boy?

A. $\frac{31}{32}$ B. $\frac{30}{31}$ C. $\frac{15}{16}$ D. $\frac{5}{31}$ E. $\frac{5}{32}$

23. A calculator has a random number generator key that, when pushed, displays a random digit $(0, 1, \ldots, 9)$. The key is pushed four times. Assuming the numbers generated are independent, what is the probability of obtaining one 0, one 5, and two 9s in any order?

A. $\frac{10!}{2!}\left(\frac{1}{10}\right)^{10}$ B. $\frac{10!}{2!}\left(\frac{1}{10}\right)^{4}$ C. $\frac{4!}{2!}\left(\frac{1}{10}\right)^{4}$ D. $\frac{9!}{4!}\left(\frac{1}{9}\right)^{4}$ E. $\left(\frac{1}{10}\right)^{4}$

24. Mr. Flowers plants ten rose bushes in a row. Eight of the bushes are white and two are red, and he plants them in random order. What is the probability that he will consecutively plant seven or more white bushes?

A. $\frac{1}{10}$ B. $\frac{1}{9}$ C. $\frac{2}{15}$ D. $\frac{7}{15}$ E. $\frac{1}{5}$

25. Events S and T have probabilities $P(S) = P(T) = \frac{1}{3}$ and $P(S \mid T) = \frac{1}{6}$. What is $P(\overline{S} \cap \overline{T})$?

A. $\frac{1}{6}$ B. $\frac{1}{3}$ C. $\frac{7}{18}$ D. $\frac{4}{9}$ E. $\frac{1}{2}$

26. A jar has three red marbles and one white marble. A shoebox has one red marble and one white marble. Three marbles are chosen at random without replacement from the jar and placed in the shoebox. Then two marbles are chosen at random and without replacement from the shoebox. What is the probability that both marbles chosen from the shoebox are red?

A. $\frac{9}{10}\left(\frac{3}{4}\right)^3$ B. $\frac{43}{100}$ C. $\frac{3}{8}$ D. $\left(\frac{3}{4}\right)^3$ E. $\frac{9}{40}$

27. A jar contains n black balls and n white balls. Three balls are chosen at random and without replacement. What is the value of n if the probability is $\frac{1}{12}$ that all three balls are white?

A. 4   B. 5   C. 8   D. 10   E. 12

28. An automobile manufacturing company produces three different car models. The table below presents data on sales and average gasoline consumption for these three models. What is the mean miles per gallon (mpg) for the cars sold by the company, assuming that each car uses the same number of gallons of gasoline?

Model	Number of cars sold	Gas consumption (mpg)
I	2000	15
II	4000	20
III	4000	25

A. 25   B. 21   C. 20   D. 15
E. Cannot be determined from the given information

Modeling Application 7

World Running Records

World records for footraces at all distances have improved consistently ever since records have been kept. For example, if we consider the mile run, the magic 4-minute mile was broken in 1954 by Bannister of the United Kingdom. Since then, the record has decreased steadily to the present time when the world record is under 3 minutes 47 seconds!*

Write a paper that develops a mathematical model to answer the question, will a 3-minute mile ever be run? Is there an "ultimate" time for a mile run, and, if so, what should we expect that time to be? For general guidelines about writing this essay, see the commentary for Modeling Application 1 on page 36.

---

* This modeling application is from Joseph Brown, "Predicting Future Improvements in Footracing," *MATYC Journal*, Fall 1980, pp. 173–179.

CHAPTER 8
Markov Chains

APPLICATIONS

Management (*Business, Economics, Finance, and Investments*)

Stock analysis (8.1, Problem 15)
Distribution of rental cars in San Francisco (8.1, Problems 17, 19–20)
Homeowners insurance purchasing patterns in Dallas–Ft. Worth area
    (8.1, Problems 18, 21–22)
Computer "learning program" probabilities (8.1, Problems 31–38)
Purchasing patterns for buying dogfood (8.2, Problem 17)
Advertising strategies (8.2, Problem 18)
Purchasing patterns for coffee (8.2, Problems 19–22)
Probability that a defective monitor will be approved (8.3, Problem 27)
Probability of revoked credit cards (8.3, Problem 28)
Competition between greeting card companies (8.4, Problems 2–3)

Life sciences (*Biology, Ecology, Health, and Medicine*)

Distribution of a cattle herd in future generations (8.2, Problems 25–26)
Probability that a substance will be in a liquid or gaseous state in the future
    (8.4, Problems 1, 4)
Cross-pollination of plants (8.5, Problems 35–36)

Social sciences (*Demography, Political Science, Population, Psychology, Society, and Sociology*)

Living patterns in a metropolitan area (8.1, Problem 16)
Long-range voting patterns (8.2, Problems 23–24)
Campaign strategies in a presidential election (8.2, Problems 28–30)
Rat maze problem (8.3, Problems 29–30)

General interest

Financial ruin in a game (8.3, Problems 25–26; 8.4, Problem 5)

Modeling application—Genetics

CHAPTER OVERVIEW

We begin by being introduced to the notion of a Markov chain. Then we consider two types of Markov chains—absorbing and nonabsorbing. An absorbing Markov chain has the property that once a given state is reached, it is impossible to move out of that state. On the other hand, a nonabsorbing Markov chain (or, simply, a Markov chain) may move toward a limiting position that can then be used to make probabilistic predictions about the future.

PREVIEW

In this chapter we combine the ideas of probability and matrices to study experiments whose outcomes depend only on the outcome of the previous experiment. The model developed is called a *Markov chain* and is named for the Russian mathematician Andrei Markov (1856–1922) who conceived the theory of stochastic processes. A stochastic process refers to a random event that depends on previous random events. A Markov chain is a process used to analyze these sequences of experiments.

PERSPECTIVE

In this chapter we develop models or examples that use Markov chains to make stock market predictions, to study living patterns in metropolitan areas, to follow the purchases of consumers in certain markets, to study the ability of parents to pass certain physical traits to their offspring, and to monitor the outcome of some gambling situations that fall into a category called *financial ruin problems*. The study of stochastic processes is an important one in applied mathematics, and the best that can be done in an introductory course such as this is to introduce the topic and some of the terminology. However, Markov chains are an excellent way of combining two of the main topics of this course, matrices and probability.

8.1 Introduction to Markov Chains

In this chapter we build a probability model using a process called a **Markov chain**. A Markov chain process is used when a series of events or experiments consists of a finite number of trials, each with a finite number of possible outcomes having a fixed probability of occurrence. The result of each event or experiment depends only on the result of the immediately preceding experiment. For example, suppose you own 100 shares of CBS stock and are keeping a record of its progress at the close of each trading day. Each day there are three possibilities: up, down, or unchanged. After a lengthy analysis you determine that the following probabilities apply:

If the stock increases one day, the probabilities for the following day are:

If the stock remains unchanged one day, the probabilities for the following day are:

If the stock decreases one day, the probabilities for the following day are:

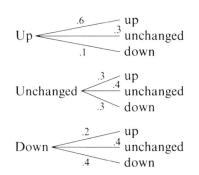

All of these probabilities can be easily summarized in matrix form:

$$
\begin{array}{c}
\\
\text{Present state}
\end{array}
\begin{array}{c}
\\
\text{Up}\\
\text{Unchanged}\\
\text{Down}
\end{array}
\begin{array}{ccc}
\text{Up} & \text{Unchanged} & \text{Down}\\
\begin{bmatrix}
.6 & .3 & .1\\
.3 & .4 & .3\\
.2 & .4 & .4
\end{bmatrix}
\end{array}
$$

The entries can each be defined as a conditional probability. For example:

P(up | up) = .6   This is the probability that the stock increases on the day following a day that it had increased

P(unchanged | up) = .3   This is the probability that the stock is unchanged on the day following a day that it had increased

P(down | up) = .1

P(up | unchanged) = .3,   P(unchanged | unchanged) = .4,

P(down | unchanged) = .3

P(up | down) = .2,   P(unchanged | down) = .4,   P(down | down) = .4

The matrix form for these probabilities is called a **transition matrix** because it gives the probability of moving from a present state to the next state. A transition matrix must have the following properties:

1. *It is square* because all possible states are used both as rows and columns.
2. *All entries are between 0 and 1 inclusive* because all entries represent probabilities.
3. *The sum of the entries in any row is 1* because the numbers in the row give the probability of changing from the state listed at the left to one of the states listed across the top.

Probability Vector

A **probability vector** is a $1 \times n$ matrix for which the sum of the n entries is 1. If $s_1, s_2, \ldots, s_n$ are the states of a Markov chain and p_{ij} is the probability that an experiment will be in state s_j if it is in state s_i now, then the row matrices

$$[p_{11} \quad p_{12} \quad \cdots \quad p_{1n}]$$
$$[p_{21} \quad p_{22} \quad \cdots \quad p_{2n}]$$
$$\vdots$$
$$[p_{n1} \quad p_{n2} \quad \cdots \quad p_{nn}]$$

Transition Matrix

are the probability vectors, and the matrix T formed by these probability vectors is called the **transition matrix**:

$$
T = \begin{bmatrix}
p_{11} & p_{12} & \cdots & p_{1n}\\
p_{21} & p_{22} & \cdots & p_{2n}\\
 & & \vdots & \\
p_{n1} & p_{n2} & \cdots & p_{nn}
\end{bmatrix}
$$

EXAMPLE 1 A mathematics instructor gives surprise quizzes. She never gives a quiz 2 days in a row, but if she does not give a quiz one day, she is just as likely to give a quiz the following day as she is not to give a quiz. This is an example of a Markov chain with two *states:* s_1 = quiz and s_2 = no quiz. The probability of a quiz the next day depends on the present state. We can summarize the probabilities in matrix form:

$$
\begin{array}{c}
 \\
\text{Quiz one day} \\
\text{No quiz one day}
\end{array}
\begin{array}{cc}
\quad\text{Quiz} & \text{No quiz} \\
\text{next day} & \text{next day} \\
\left[\begin{array}{cc} 0 & 1 \\ \frac{1}{2} & \frac{1}{2} \end{array}\right] &
\end{array}
$$

∎

Suppose the mathematics instructor in Example 1 uses the same rules for giving quizzes day after day. If today is Monday, what are the possibilities for Wednesday if the class meets daily? Let Q = quiz given and let N = no quiz given. Then we can draw a tree diagram, as shown. The probabilities for each outcome are found by multiplying along each branch of the tree; $P(NNN) = \frac{1}{4}$ means that there is a probability of $\frac{1}{4}$ that no quiz will be given on any of the 3 days.

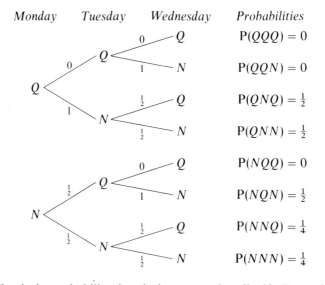

EXAMPLE 2 What is the probability that the instructor described in Example 1 will give a quiz on Wednesday, given the following information?

a. There is a quiz on Monday.
b. There is no quiz on Monday.

What is the probability that there is no quiz on Wednesday, given the following information?
c. There is a quiz on Monday.
d. There is no quiz on Monday.

Solution We use the tree diagram above and the addition principle for probability.

a. $P(Q \text{ on Wednesday} \mid Q \text{ on Monday}) = P(QQQ) + P(QNQ) = 0 + \frac{1}{2} = \frac{1}{2}$
b. $P(Q \text{ on Wednesday} \mid N \text{ on Monday}) = P(NQQ) + P(NNQ) = 0 + \frac{1}{4} = \frac{1}{4}$

c. P(N on Wednesday | Q on Monday) = P(QQN) + P(QNN) = $0 + \frac{1}{2} = \frac{1}{2}$

d. P(N on Wednesday | N on Monday) = P(NQN) + P(NNN) = $\frac{1}{2} + \frac{1}{4} = \frac{3}{4}$ ∎

EXAMPLE 3   If T is the transition matrix given in Example 1, find T^2.

Solution
$$T^2 = \begin{bmatrix} 0 & 1 \\ \frac{1}{2} & \frac{1}{2} \end{bmatrix} \begin{bmatrix} 0 & 1 \\ \frac{1}{2} & \frac{1}{2} \end{bmatrix}$$

$$= \begin{bmatrix} 0 + \frac{1}{2} & 0 + \frac{1}{2} \\ 0 + \frac{1}{4} & \frac{1}{2} + \frac{1}{4} \end{bmatrix}$$

$$= \begin{bmatrix} \frac{1}{2} & \frac{1}{2} \\ \frac{1}{4} & \frac{3}{4} \end{bmatrix}$$ ∎

Now compare the results of Examples 2 and 3:

P(Q on Wed. | Q on Mon.) = $\frac{1}{2}$ P(N on Wed. | Q on Mon.) = $\frac{1}{2}$

P(Q on Wed. | N on Mon.) = $\frac{1}{4}$ P(N on Wed. | N on Mon.) = $\frac{3}{4}$

$$\begin{array}{cc} & \begin{array}{cc} \text{Wednesday} \\ Q \quad\quad N \end{array} \\ \text{Monday} \begin{array}{c} Q \\ N \end{array} & \begin{bmatrix} \frac{1}{2} & \frac{1}{2} \\ \frac{1}{4} & \frac{3}{4} \end{bmatrix} \end{array}$$

EXAMPLE 4   What are the probabilities that the instructor described in Example 1 will or will not give a quiz on Friday (4 days after Monday)?

Solution   If the above result can be applied twice, we need to find T^4.

$$T^4 = \begin{bmatrix} 0 & 1 \\ \frac{1}{2} & \frac{1}{2} \end{bmatrix}^4 = \begin{bmatrix} 0 & 1 \\ \frac{1}{2} & \frac{1}{2} \end{bmatrix}^2 \begin{bmatrix} 0 & 1 \\ \frac{1}{2} & \frac{1}{2} \end{bmatrix}^2 = \begin{bmatrix} \frac{1}{2} & \frac{1}{2} \\ \frac{1}{4} & \frac{3}{4} \end{bmatrix} \begin{bmatrix} \frac{1}{2} & \frac{1}{2} \\ \frac{1}{4} & \frac{3}{4} \end{bmatrix}$$

$$= \begin{bmatrix} \frac{1}{4} + \frac{1}{8} & \frac{1}{4} + \frac{3}{8} \\ \frac{1}{8} + \frac{3}{16} & \frac{1}{8} + \frac{9}{16} \end{bmatrix} = \begin{bmatrix} \frac{3}{8} & \frac{5}{8} \\ \frac{5}{16} & \frac{11}{16} \end{bmatrix}$$

The row 1, column 1 entry in T^4 signifies that the probability of a quiz on Friday, given a quiz on Monday, is $\frac{3}{8}$. ∎

Eight (school) days after the initial Monday the probabilities are found by considering T^8:

$$T^8 = \begin{bmatrix} 0 & 1 \\ \frac{1}{2} & \frac{1}{2} \end{bmatrix}^4 \begin{bmatrix} 0 & 1 \\ \frac{1}{2} & \frac{1}{2} \end{bmatrix}^4 = \begin{bmatrix} \frac{3}{8} & \frac{5}{8} \\ \frac{5}{16} & \frac{11}{16} \end{bmatrix} \begin{bmatrix} \frac{3}{8} & \frac{5}{8} \\ \frac{5}{16} & \frac{11}{16} \end{bmatrix}$$

$$= \begin{bmatrix} \frac{9}{64} + \frac{25}{128} & \frac{15}{64} + \frac{55}{128} \\ \frac{15}{128} + \frac{55}{256} & \frac{25}{128} + \frac{121}{256} \end{bmatrix}$$

$$= \begin{bmatrix} \frac{43}{128} & \frac{85}{128} \\ \frac{85}{256} & \frac{171}{256} \end{bmatrix} \approx \begin{bmatrix} .3359 & .6641 \\ .3320 & .6680 \end{bmatrix}$$

This suggests that as we take higher and higher powers of T, the result gets closer and closer to

$$\begin{bmatrix} \frac{1}{3} & \frac{2}{3} \\ \frac{1}{3} & \frac{2}{3} \end{bmatrix}$$

Note that the same probability vector appears in both rows. Suppose we multiply this vector $\begin{bmatrix} \frac{1}{3} & \frac{2}{3} \end{bmatrix}$ times the original transition matrix:

$$\begin{bmatrix} \frac{1}{3} & \frac{2}{3} \end{bmatrix} \begin{bmatrix} 0 & 1 \\ \frac{1}{2} & \frac{1}{2} \end{bmatrix} = \begin{bmatrix} 0 + \frac{1}{3} & \frac{1}{3} + \frac{1}{3} \end{bmatrix} = \begin{bmatrix} \frac{1}{3} & \frac{2}{3} \end{bmatrix}$$

The answer is still $\begin{bmatrix} \frac{1}{3} & \frac{2}{3} \end{bmatrix}$! This is no coincidence, and if $T, T^2, T^3, \ldots, T^n$, approach a matrix all of whose rows are V, then $VT = V$. For this reason we call V a **fixed probability vector**. We discuss the process for finding a fixed probability vector in the next section.

> **COMPUTER APPLICATION**   Program 13 on the computer disk accompanying this book raises a transition matrix to higher and higher powers so you can "see" it moving toward a stable state.

Problem Set 8.1

Are the matrices in Problems 1–3 probability vectors?

1. a. $\begin{bmatrix} \frac{1}{2} & \frac{1}{2} & \frac{1}{2} \end{bmatrix}$ **b.** $\begin{bmatrix} \frac{1}{2} & \frac{1}{2} \end{bmatrix}$
   c. $\begin{bmatrix} \frac{2}{3} & \frac{1}{5} \end{bmatrix}$

2. a. $\begin{bmatrix} 1 & 0 & 0 \end{bmatrix}$ **b.** $\begin{bmatrix} .6 & .3 & .1 \end{bmatrix}$
   c. $\begin{bmatrix} .1 & .1 & .1 \end{bmatrix}$

3. a. $\begin{bmatrix} .5 & .6 & -.1 \end{bmatrix}$
   b. $\begin{bmatrix} .05 & .15 & .63 & .17 \end{bmatrix}$
   c. $\begin{bmatrix} 0 & 0 & 0 \end{bmatrix}$

Are the matrices in Problems 4–6 transition matrices?

4. a. $\begin{bmatrix} .3 & .2 & .5 \\ 0 & .1 & .9 \\ .8 & .1 & .1 \end{bmatrix}$ **b.** $\begin{bmatrix} .5 & 0 & .5 \\ .1 & .1 & .1 \\ 1 & 0 & 0 \end{bmatrix}$

5. a. $\begin{bmatrix} \frac{1}{3} & \frac{2}{3} \\ \frac{1}{5} & \frac{4}{5} \end{bmatrix}$ **b.** $\begin{bmatrix} \frac{2}{3} & \frac{4}{5} \\ \frac{5}{6} & \frac{1}{6} \end{bmatrix}$

6. a. $\begin{bmatrix} \frac{1}{4} & \frac{1}{2} & \frac{1}{4} \\ \frac{1}{5} & \frac{3}{5} & \frac{2}{5} \\ 0 & 0 & 0 \end{bmatrix}$ **b.** $\begin{bmatrix} 1 & 0 & 0 \\ 0 & 1 & 0 \\ 0 & 0 & 1 \end{bmatrix}$

If T is the matrix in Problems 7–12, find T^2 and T^3.

7. $\begin{bmatrix} \frac{1}{2} & \frac{1}{2} \\ \frac{1}{4} & \frac{3}{4} \end{bmatrix}$ **8.** $\begin{bmatrix} .1 & .2 & .7 \\ .1 & .8 & .1 \\ .2 & .3 & .5 \end{bmatrix}$

9. $\begin{bmatrix} .6 & .4 \\ .23 & .77 \end{bmatrix}$ **10.** $\begin{bmatrix} 1 & 0 & 0 \\ 0 & 1 & 0 \\ .4 & .3 & .3 \end{bmatrix}$

11. $\begin{bmatrix} .15 & .25 & .6 \\ .5 & .35 & .15 \\ .1 & .8 & .1 \end{bmatrix}$ **12.** $\begin{bmatrix} .29 & .53 & .18 \\ .01 & .99 & 0 \\ .6 & .3 & .1 \end{bmatrix}$

13. If a certain experiment has two states 0 and 1 with the probabilities $P(0|0) = .3$, $P(1|0) = .7$, $P(0|1) = .6$, and $P(1|1) = .4$, write the transition matrix.

14. If a certain experiment has three states 0, 1, and 2 with the probabilities $P(0|0) = .4$, $P(1|0) = .3$, $P(2|0) = .3$, $P(1|0) = .1$, $P(1|1) = .5$, $P(2|1) = .4$, $P(0|2) = .7$, $P(1|2) = .2$, $P(2|2) = .1$, write the transition matrix.

APPLICATIONS

15. A stock broker has been watching the price changes of a particular stock. Over the past year, the broker has found that on a day the stock is traded,

P(increase | increase on previous day) = .3
P(decrease | increase on previous day) = .7
P(increase | decrease on previous day) = .8
P(decrease | decrease on previous day) = .2

Write a transition matrix for this information.

16. A city planner has studied the living patterns within a certain greater metropolitan area. Over the last 10 years the study shows that

P(move to city | lived in suburbs previous year) = .1

P(stay in suburbs | lived in suburbs previous year) = .6

P(leave area | lived in suburbs previous year) = .3

P(stay in city | lived in city previous year) = .8

P(move to suburbs | lived in city previous year) = .1

P(leave area | lived in city previous year) = .1

P(move to city | lived out of the area previous year) = .4

P(move to suburbs | lived out of the area previous year) = .6

Write a transition matrix for this information.

17. San Francisco is served by three airports, San Francisco International (SFO), Oakland (OAK), and San Jose (SJC), and a car rental company plans to begin operations in San Francisco by setting up rental and car storage facilities at these three airports. The probabilities that a person will return the car to a particular location is given by the following matrix, T:

$$
\begin{array}{c}
\text{Rented from}
\end{array}
\begin{array}{cc}
 & \begin{array}{ccc} \text{Returned to} & & \\ \text{SFO} & \text{OAK} & \text{SJC} \end{array} \\
\begin{array}{c} \text{SFO} \\ \text{OAK} \\ \text{SJC} \end{array} &
\left[\begin{array}{ccc}
.8 & .1 & .1 \\
.2 & .7 & .1 \\
.3 & .2 & .6
\end{array}\right]
\end{array}
$$

a. Is this a transition matrix? Why or why not?
b. What does the entry .8 signify?
c. Which airport has the lowest percent of returns?

18. A survey of the residents in the greater Dallas–Fort Worth area shows that they purchased their homeowners insurance from one of three companies. The probabilities that a resident will purchase homeowners insurance from one of these companies in the subsequent year are summarized by the following matrix:

$$
\begin{array}{c}
\text{Present company}
\end{array}
\begin{array}{cc}
 & \begin{array}{ccc} \text{Company the} \\ \text{subsequent year} \\ \text{A} \quad \text{B} \quad \text{C} \end{array} \\
\begin{array}{c} \text{A} \\ \text{B} \\ \text{C} \end{array} &
\left[\begin{array}{ccc}
.4 & .4 & .2 \\
.1 & .8 & .1 \\
.1 & .2 & .7
\end{array}\right]
\end{array}
$$

a. Is this a transition matrix? Why or why not?
b. What does the entry .8 signify?

c. Which company looks like it has the least percentage of satisfied customers?

19. Find T^2 for the matrix in Problem 17.

20. Find T^4 for the matrix in Problem 17.

21. Find T^2 for the matrix in Problem 18.

22. Find T^4 for the matrix in Problem 18.

In Problems 23–30 suppose that the transition matrix in Example 1 is changed to

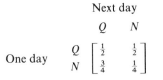

$$
\begin{array}{c}
\text{One day}
\end{array}
\begin{array}{cc}
 & \begin{array}{cc} \text{Next day} \\ Q \quad\quad N \end{array} \\
\begin{array}{c} Q \\ N \end{array} &
\left[\begin{array}{cc}
\frac{1}{2} & \frac{1}{2} \\
\frac{3}{4} & \frac{1}{4}
\end{array}\right]
\end{array}
$$

Assume that the instructor does not give a quiz on the first day of class.

23. What is the probability of a quiz on the second day of class?

24. What is the probability of a quiz on the third day of class?

25. What is the probability of a quiz on the fourth day of class?

26. What is P(Q on Wednesday | Q on Monday)?

27. What is P(Q on Wednesday | N on Monday)?

28. What is P(N on Wednesday | Q on Monday)?

29. What is P(N on Wednesday | N on Monday)?

30. Write a matrix that will answer Problems 26–29.

A computer program is called a "learning program" if it can learn by making mistakes and then correcting those mistakes. Suppose the probability that the computer will be functioning properly on a given day depends on whether it was functioning properly on the previous day. These probabilities are summarized by the following transition matrix, where C means functioning properly and E means malfunctioning:

$$
\begin{array}{c}
\text{Present state}
\end{array}
\begin{array}{cc}
 & \begin{array}{cc} \text{State next day} \\ C \quad\quad E \end{array} \\
\begin{array}{c} C \\ E \end{array} &
\left[\begin{array}{cc}
.99 & .01 \\
.9 & .1
\end{array}\right]
\end{array}
$$

In Problems 31–38 assume that the present state of the program is that it is operating properly and give answers correct to four decimal places.

31. What is the probability that the computer will be in a correct state on the second day (i.e., after two transitions)?

32. What is the probability that the computer will be in a correct state on the third day (i.e., after three transitions)?

33. What is the probability that the computer will be in a correct state on the fourth day (i.e., after four transitions)?

34. What is P(C on Wednesday | E on Monday)?

35. What is P(C on Wednesday | C on Monday)?

36. What is P(E on Wednesday | E on Monday)?

37. What is P(E on Wednesday | C on Monday)?

38. Write a matrix that will answer Problems 34–37.

8.2 Regular Markov Chains

In Section 8.1 we discussed transition matrices that summarized the probabilities for changing from one state to another. We also were introduced to the idea of a *fixed probability vector*, which is now defined:

Fixed Probability Vector

> A probability vector V such that $VT = V$ is called a **fixed probability vector** for the matrix T.

To understand this concept, consider a marketing analysis for Tide® . Suppose the following transition matrix is given:

$$
\begin{array}{cc}
 & \text{Next purchase} \\
 & \begin{array}{cc} \text{Tide} & \text{Brand X} \end{array} \\
\text{Present purchase} \quad \begin{array}{c} \text{Tide} \\ \text{Brand X} \end{array} & \begin{bmatrix} .8 & .2 \\ .4 & .6 \end{bmatrix}
\end{array}
$$

Now, suppose that there are 10 million people who purchase a detergent each month and that 25% of the consumers purchase Tide and 75% purchase Brand X (all other brands lumped together). Also, suppose that every purchaser makes one purchase per month. What share of the market will use Tide in the next month, given these assumptions?

$$
\begin{array}{cc} \text{Tide} & \text{Brand X} \\ [\ .25 & .75\ ] \end{array} \begin{bmatrix} .8 & .2 \\ .4 & .6 \end{bmatrix} = [.5 \quad .5]
$$

Second month:

$$
[.5 \quad .5] \begin{bmatrix} .8 & .2 \\ .4 & .6 \end{bmatrix} = [.6 \quad .4]
$$

Third month:

$$
[.6 \quad .4] \begin{bmatrix} .8 & .2 \\ .4 & .6 \end{bmatrix} = [.64 \quad .36]
$$

Fourth month:

$$
[.64 \quad .36] \begin{bmatrix} .8 & .2 \\ .4 & .6 \end{bmatrix} = [.656 \quad .344]
$$

This means that at the end of the fourth month, Tide could expect 65.6% of the market, or 6.56 million customers. If we continue this process, will the market share for Tide and Brand X stabilize at some specific share of the market for each? If so, this share is represented by the fixed probability vector.

Let $V = [v_1 \quad v_2]$. Now find V so that

$$[v_1 \quad v_2]\begin{bmatrix} .8 & .2 \\ .4 & .6 \end{bmatrix} = [v_1 \quad v_2]$$

$$\begin{cases} .8v_1 + .4v_2 = v_1 \\ .2v_1 + .6v_2 = v_2 \end{cases} \quad \text{or} \quad \begin{cases} .2v_1 - .4v_2 = 0 \\ .2v_1 - .4v_2 = 0 \end{cases}$$

In addition, $[v_1 \quad v_2]$ is a probability vector, so $v_1 + v_2 = 1$, giving the system

$$\begin{cases} .2v_1 - .4v_2 = 0 \\ .2v_1 - .4v_2 = 0 \\ v_1 + v_2 = 1 \end{cases}$$

$$\begin{bmatrix} .2 & -.4 & | & 0 \\ .2 & -.4 & | & 0 \\ 1 & 1 & | & 1 \end{bmatrix} \rightarrow \begin{bmatrix} 1 & 1 & | & 1 \\ 0 & -.6 & | & -.2 \\ 0 & 0 & | & 0 \end{bmatrix} \rightarrow \begin{bmatrix} 1 & 1 & | & 1 \\ 0 & 1 & | & \frac{1}{3} \\ 0 & 0 & | & 0 \end{bmatrix} \rightarrow \begin{bmatrix} 1 & 0 & | & \frac{2}{3} \\ 0 & 1 & | & \frac{1}{3} \\ 0 & 0 & | & 0 \end{bmatrix}$$

Thus $v_1 = \frac{2}{3}, v_2 = \frac{1}{3}$, and $[\frac{2}{3} \quad \frac{1}{3}]$ is the fixed probability vector. This means that, in the long run, Tide can expect to capture $\frac{2}{3}$ of the market.

The next example ties together the concepts of a fixed probability vector and the successive powers of a transition matrix.

EXAMPLE 1 Suppose a professor gives quizzes (Q) or no quizzes (N) according to the transition matrix

$$\begin{array}{cc} & \begin{array}{cc} Q & N \end{array} \\ \begin{array}{c} Q \\ N \end{array} & \begin{bmatrix} 0 & 1 \\ \frac{1}{2} & \frac{1}{2} \end{bmatrix} \end{array}$$

Find the fixed probability vector $V = [v_1 \quad v_2]$.

Solution $$[v_1 \quad v_2]\begin{bmatrix} 0 & 1 \\ \frac{1}{2} & \frac{1}{2} \end{bmatrix} = [v_1 \quad v_2]$$

We solve this equation for v_1 and v_2:

$$[v_1 \quad v_2]\begin{bmatrix} 0 & 1 \\ \frac{1}{2} & \frac{1}{2} \end{bmatrix} = [0 + \frac{1}{2}v_2 \quad v_1 + \frac{1}{2}v_2] = [\frac{1}{2}v_2 \quad v_1 + \frac{1}{2}v_2]$$

If $[\frac{1}{2}v_2 \quad v_1 + \frac{1}{2}v_2] = [v_1 \quad v_2]$, the corresponding components are equal:

$$\begin{cases} \frac{1}{2}v_2 = v_1 \\ v_1 + \frac{1}{2}v_2 = v_2 \end{cases}$$

Using this relationship, along with the fact that $v_1 + v_2 = 1$, gives the matrix system

$$\begin{bmatrix} 1 & 1 & | & 1 \\ -1 & \frac{1}{2} & | & 0 \\ 1 & -\frac{1}{2} & | & 0 \end{bmatrix} \rightarrow \begin{bmatrix} 1 & 1 & | & 1 \\ 0 & \frac{3}{2} & | & 1 \\ 0 & -\frac{3}{2} & | & -1 \end{bmatrix} \rightarrow \begin{bmatrix} 1 & 1 & | & 1 \\ 0 & 1 & | & \frac{2}{3} \\ 0 & 0 & | & 0 \end{bmatrix} \rightarrow \begin{bmatrix} 1 & 0 & | & \frac{1}{3} \\ 0 & 1 & | & \frac{2}{3} \\ 0 & 0 & | & 0 \end{bmatrix}$$

Thus $v_1 = \frac{1}{3}$ and $v_2 = \frac{2}{3}$, or $V = [\frac{1}{3} \quad \frac{2}{3}]$. ■

Now let us compare the solution of Example 1 with the powers of T found in Section 8.1:

$$T^2 = \begin{bmatrix} \frac{1}{2} & \frac{1}{2} \\ \frac{1}{4} & \frac{3}{4} \end{bmatrix} \qquad T^4 = \begin{bmatrix} \frac{3}{8} & \frac{5}{8} \\ \frac{5}{16} & \frac{11}{16} \end{bmatrix} \qquad T^8 \approx \begin{bmatrix} .3359 & .6641 \\ .3320 & .6680 \end{bmatrix}$$

It is no coincidence that the higher powers of T approach the matrix consisting of the fixed probability vector $\begin{bmatrix} \frac{1}{3} & \frac{2}{3} \end{bmatrix}$:

$$\begin{bmatrix} \frac{1}{3} & \frac{2}{3} \\ \frac{1}{3} & \frac{2}{3} \end{bmatrix}$$

This relationship between the higher powers of T and the fixed probability vector are summarized in the **transition matrix theorem**. However, before we can state this theorem, we need to consider some terminology that will help us describe when the transition matrix theorem applies.

Consider the matrix

$$\begin{bmatrix} 1 & 0 \\ \frac{1}{5} & \frac{4}{5} \end{bmatrix}$$

The powers of this matrix will always have $p_{12} = 0$. We can see this by considering

$$\begin{bmatrix} a & 0 \\ b & c \end{bmatrix} \begin{bmatrix} a & 0 \\ b & c \end{bmatrix}$$

The entry in the first row, second column of the product is

$$a \cdot 0 + 0 \cdot c = 0$$

Thus, regardless of the other entries, the powers of this matrix will always have $p_{12} = 0$. This means that the powers of this matrix will never approach a matrix whose rows are all a probability vector V. However, if for some power of T all the entries are positive, then the powers of the matrix will approach a matrix whose rows are all a probability vector. If this is the case, we call the transition matrix **regular**.

Regular Transition Matrix

A transition matrix is *regular* if there exists some power of it that contains only positive elements.

EXAMPLE 2 Is $\begin{bmatrix} \frac{1}{5} & \frac{4}{5} \\ 1 & 0 \end{bmatrix}$ regular?

Solution
$$\begin{bmatrix} \frac{1}{5} & \frac{4}{5} \\ 1 & 0 \end{bmatrix} \begin{bmatrix} \frac{1}{5} & \frac{4}{5} \\ 1 & 0 \end{bmatrix} = \begin{bmatrix} \frac{1}{25} + \frac{4}{5} & \frac{4}{25} + 0 \\ \frac{1}{5} + 0 & \frac{4}{5} + 0 \end{bmatrix}$$

$$= \begin{bmatrix} \frac{21}{25} & \frac{4}{25} \\ \frac{1}{5} & \frac{4}{5} \end{bmatrix}$$

Since the second power of the matrix has all positive elements, we see that it is regular. ∎

If a transition matrix T is regular, then the following theorem can be proved:

Transition Matrix Theorem

If T is a regular transition matrix, then:
1. T has a unique fixed probability vector, V, whose elements are all positive.
2. $T, T^2, T^3, \ldots, T^n$ approach the matrix all of whose rows are V.
3. If W is any probability vector, then $WT, WT^2, WT^3, \ldots, WT^n$ approach the fixed probability vector V.

COMPUTER APPLICATION   If you have access to a computer, you can verify this theorem using Program 13 on the computer disk accompanying this text. In this program, $a(n)$ is used as follows:

$$a(1) = WT, a(2) = WT^2, a(3) = WT^3, \ldots, a(n) = WT^n$$

EXAMPLE 3   Suppose that General Motors (GM), Ford (F), and Chrysler (C) each introduce a front-wheel-drive compact with better than 50 mpg. Assume that each company initially captures about one-third of the market. During the year:

1. General Motors keeps 85% of its customers but loses 10% to Ford and 5% to Chrysler.
2. Ford keeps 80% of its customers but loses 10% to General Motors and 10% to Chrysler.
3. Chrysler keeps 60% of its customers but loses 25% to General Motors and 15% to Ford.

If this trend continues, what share of the market is each likely to have at the end of next year, and what is the long-run prediction if this trend continues?

Solution   The transition matrix is

$$
\begin{array}{c}
 \\
\text{GM} \\
\text{F} \\
\text{C}
\end{array}
\begin{array}{ccc}
\text{GM} & \text{F} & \text{C}
\end{array}
\left[
\begin{array}{ccc}
\frac{17}{20} & \frac{1}{10} & \frac{1}{20} \\
\frac{1}{10} & \frac{4}{5} & \frac{1}{10} \\
\frac{1}{4} & \frac{3}{20} & \frac{3}{5}
\end{array}
\right]
$$

Note: $85\% = \frac{85}{100} = \frac{17}{20}$; do the same for all the percents

Since the entries are all positive, the matrix is a regular transition matrix and the transition matrix theorem applies. To determine what share of the market each company will have at the end of next year we multiply the original share vector $[\frac{1}{3} \quad \frac{1}{3} \quad \frac{1}{3}]$ by T:

$$
\begin{bmatrix} \frac{1}{3} & \frac{1}{3} & \frac{1}{3} \end{bmatrix}
\begin{bmatrix}
\frac{17}{20} & \frac{1}{10} & \frac{1}{20} \\
\frac{1}{10} & \frac{4}{5} & \frac{1}{10} \\
\frac{1}{4} & \frac{3}{20} & \frac{3}{5}
\end{bmatrix}
$$

$$
= \begin{bmatrix} \frac{17}{60} + \frac{1}{30} + \frac{1}{12} & \frac{1}{30} + \frac{4}{15} + \frac{1}{20} & \frac{1}{60} + \frac{1}{30} + \frac{1}{5} \end{bmatrix}
$$

$$
= \begin{bmatrix} \frac{2}{5} & \frac{7}{20} & \frac{1}{4} \end{bmatrix}
$$

We see that at the end of next year GM will have $\frac{2}{5}$ or 40% of the market, Ford will have 35% of the market, and Chrysler will have 25% of the market.

For the long-range prediction we find the fixed probability vector:

$$\begin{bmatrix} v_2 & v_2 & v_3 \end{bmatrix} \begin{bmatrix} \frac{17}{20} & \frac{1}{10} & \frac{1}{20} \\ \frac{1}{10} & \frac{4}{5} & \frac{1}{10} \\ \frac{1}{4} & \frac{3}{20} & \frac{3}{5} \end{bmatrix} = \begin{bmatrix} v_1 & v_2 & v_3 \end{bmatrix}$$

By the third part of the transition matrix theorem, any probability vector $\begin{bmatrix} \frac{1}{3} & \frac{1}{3} & \frac{1}{3} \end{bmatrix}$ T approaches this fixed probability vector. Multiplying, we get

$$\begin{bmatrix} \frac{17}{20}v_1 + \frac{1}{10}v_2 + \frac{1}{4}v_3 & \frac{1}{10}v_1 + \frac{4}{5}v_2 + \frac{3}{20}v_3 & \frac{1}{20}v_1 + \frac{1}{10}v_2 + \frac{3}{5}v_3 \end{bmatrix} = \begin{bmatrix} v_1 & v_2 & v_3 \end{bmatrix}$$

By the equality of vectors, we obtain the following system:

$$\begin{cases} \frac{17}{20}v_1 + \frac{1}{10}v_2 + \frac{1}{4}v_3 = v_1 \\ \frac{1}{10}v_1 + \frac{4}{5}v_2 + \frac{3}{20}v_3 = v_2 \\ \frac{1}{20}v_1 + \frac{1}{10}v_2 + \frac{3}{5}v_3 = v_3 \end{cases} \quad \text{or} \quad \begin{cases} 3v_1 - 2v_2 - 5v_3 = 0 \\ 2v_1 - 4v_2 + 3v_3 = 0 \\ v_1 + 2v_2 - 8v_3 = 0 \end{cases}$$

If you solve this system of equations, you will find that it is a dependent system. Add to this system the fact that V is a probability vector, namely, $v_1 + v_2 + v_3 = 1$, to obtain the matrix:

$$\begin{bmatrix} 3 & -2 & -5 & | & 0 \\ 2 & -4 & 3 & | & 0 \\ 1 & 2 & -8 & | & 0 \\ 1 & 1 & 1 & | & 1 \end{bmatrix} \rightarrow \begin{bmatrix} 1 & 1 & 1 & | & 1 \\ 0 & -5 & -8 & | & -3 \\ 0 & -6 & 1 & | & -2 \\ 0 & 1 & -9 & | & -1 \end{bmatrix} \rightarrow \begin{bmatrix} 1 & 1 & 1 & | & 1 \\ 0 & 1 & -9 & | & -1 \\ 0 & 0 & -53 & | & -8 \\ 0 & 0 & -53 & | & -8 \end{bmatrix}$$

$$\rightarrow \begin{bmatrix} 1 & 1 & 1 & | & 1 \\ 0 & 1 & -9 & | & -1 \\ 0 & 0 & 1 & | & \frac{8}{53} \\ 0 & 0 & 0 & | & 0 \end{bmatrix} \rightarrow \begin{bmatrix} 1 & 0 & 0 & | & \frac{26}{53} \\ 0 & 1 & 0 & | & \frac{19}{53} \\ 0 & 0 & 1 & | & \frac{8}{53} \\ 0 & 0 & 0 & | & 0 \end{bmatrix}$$

Therefore, $v_1 = \frac{26}{53}$, $v_2 = \frac{19}{53}$, and $v_3 = \frac{8}{53}$. Thus the fixed probability vector is

$$\begin{bmatrix} \frac{26}{53} & \frac{19}{53} & \frac{8}{53} \end{bmatrix} \approx \begin{bmatrix} .49 & .36 & .15 \end{bmatrix}$$

This means that the long-range prediction is that General Motors will have about 49% of the market, Ford 36%, and Chrysler 15%. Note that the transition matrix theorem tells us that this is the long-term prediction *regardless of the original market share*. Even if GM started with 10% of the market, Ford 20%, and Chrysler 70%, the long-term prediction, if the same trend continues, will still be 49% for GM, 36% for Ford, and 15% for Chrysler. ■

Problem Set 8.2

In Problems 1–12 determine whether the matrix is regular. For each matrix that is regular, find the matrix that it approaches as it is raised to higher and higher powers.

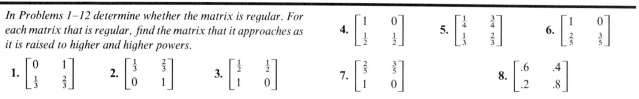

1. $\begin{bmatrix} 0 & 1 \\ \frac{1}{3} & \frac{2}{3} \end{bmatrix}$
2. $\begin{bmatrix} \frac{1}{3} & \frac{2}{3} \\ 0 & 1 \end{bmatrix}$
3. $\begin{bmatrix} \frac{1}{2} & \frac{1}{2} \\ 1 & 0 \end{bmatrix}$
4. $\begin{bmatrix} 1 & 0 \\ \frac{1}{2} & \frac{1}{2} \end{bmatrix}$
5. $\begin{bmatrix} \frac{1}{4} & \frac{3}{4} \\ \frac{1}{3} & \frac{2}{3} \end{bmatrix}$
6. $\begin{bmatrix} 1 & 0 \\ \frac{2}{5} & \frac{3}{5} \end{bmatrix}$
7. $\begin{bmatrix} \frac{2}{5} & \frac{3}{5} \\ 1 & 0 \end{bmatrix}$
8. $\begin{bmatrix} .6 & .4 \\ .2 & .8 \end{bmatrix}$

9. $\begin{bmatrix} 0 & 1 & 0 \\ \frac{1}{2} & \frac{1}{4} & \frac{1}{4} \\ 0 & \frac{1}{3} & \frac{2}{3} \end{bmatrix}$

10. $\begin{bmatrix} \frac{1}{3} & \frac{1}{3} & \frac{1}{3} \\ 0 & 0 & 1 \\ \frac{3}{10} & \frac{1}{2} & \frac{1}{5} \end{bmatrix}$

11. $\begin{bmatrix} .6 & .2 & .2 \\ .1 & .8 & .1 \\ .1 & .2 & .7 \end{bmatrix}$

12. $\begin{bmatrix} 0 & .3 & .7 \\ .4 & 0 & .6 \\ .5 & .5 & 0 \end{bmatrix}$

13. **a.** Is the following matrix regular?

   b. Does this matrix have a unique fixed probability vector?
   c. Does the transition matrix theorem apply? Why or why not?

14. **a.** Is the following matrix regular?

$$\begin{bmatrix} 1 & 0 \\ \frac{1}{2} & \frac{1}{2} \end{bmatrix}$$

   b. Does this matrix have a unique fixed probability vector?
   c. Does the transition matrix theorem apply? Why or why not?

15. Consider the following matrix:

$$\begin{bmatrix} a & 1-a \\ 1-a & a \end{bmatrix}$$

   a. What conditions on a are necessary if this is to be a transition matrix?
   b. Find the fixed probability vector (assuming that the conditions you found in part a apply).

16. Let

$$T = \begin{bmatrix} 1 & 0 & 0 \\ \frac{1}{2} & 0 & \frac{1}{2} \\ 0 & 0 & 1 \end{bmatrix}$$

   a. Find the matrix that T^n approaches.
   b. Does T have more than one fixed probability vector?

APPLICATIONS

17. A housewife buys three kinds of dogfood: canned, packaged, and dry. Her dog likes variety so she never buys the same kind on successive days. If she buys canned food one day, then the next day she buys packaged food. But if she buys packaged or dry food, then the next day she is twice as likely to buy canned food as the other types. Find the transition matrix. In the long run, how often does she buy each of the three varieties?

18. A publishing company has three books to promote. It is company policy to take out an advertisement for one of the books each month in a national journal. The author of book A is the brother-in-law of an editor and has a 60% chance of being picked for advertising each month. The other books will not be picked two months in a row, and if book A is not chosen, then books B and C have an equal chance of being picked. Find the transition matrix. In the long run, how often will each book be advertised?

19. A retailer stocks three brands of coffee. Each brand has about one-third of the market. A survey is taken and finds that each month the following occurs:

   a. Alpha Coffee Concern keeps 70% of its customers but loses 10% to Better Bean Coffee and 20% to Carolyn's Certified Coffee.
   b. Better Bean Coffee keeps 80% of its customers but loses 10% to Alpha Coffee Concern and 10% to Carolyn's Certified Coffee.
   c. Carolyn's Certified Coffee keeps 50% of its customers but loses 10% to Alpha Coffee and 40% to Better Bean Coffee.

   Write the transition matrix for this information.

20. What is the market share of each brand of coffee in Problem 19 after 1 month? After 2 months?

21. What is the market share of each brand of coffee in Problem 19 in the long run?

22. What is the market share of each brand of coffee in Problem 19 in the long run if Alpha Coffee starts with 10%, Better Bean with 30%, and Carolyn's Certified Coffee with 60% of the market? What is the market share after 1 month?

23. In an eastern state a survey finds that if a person's father is a Democrat, the person will vote Democratic 70% of the time and Independent the rest of the time. If a person's father is a Republican, the person will vote Republican 50% of the time, Democratic 40% of the time, and Independent 10% of the time. Of the children of Independents, 40% of the time they will vote Independent, 40% Democratic, and 20% Republican. What is the long-range voting pattern for each party?

24. What is the probability that the grandchild of a Democrat will vote Independent in the eastern state described in Problem 23?

25. On Niels' cattle ranch it has been shown that the probability that a Jersey cow will have a Jersey calf is .6, that it will have a Guernsey calf is .2, or that it will have a white-faced calf is .2. The probabilities that a Guernsey cow will have a Jersey, Guernsey, or white-faced calf are .1, .7, and .2, respectively; and the probabilities that a

white-faced cow will have a Jersey, Guernsey, or white-faced calf are .1, .1, and .8, respectively. Assume that the herd is now 30% Jersey, 20% Guernsey, and 50% white-faced. What will the distribution of the herd be in two generations?

26. What will the distribution of the herd of cattle described in Problem 25 be after a large number of generations?

27. Two companies compete against each other, and the transition matrix for people changing from one company to another each month is given by the following transition matrix:

$$
\begin{array}{cc}
& \text{To company} \\
& \begin{array}{cc} \text{I} & \text{II} \end{array} \\
\text{From company} \quad \begin{array}{c} \text{I} \\ \text{II} \end{array} & \begin{bmatrix} .7 & .3 \\ .2 & .8 \end{bmatrix}
\end{array}
$$

What is the long-range expectation for the companies if this trend continues?

28. In a recent close presidential race a poll finds that the incumbent would have to carry California, New York, and Texas to be reelected, so the campaign manager decides to devote the last week of the campaign to these states. She also decides that the incumbent will not campaign in the same state 2 days in a row and that if he campaigns in California, he will be twice as likely to campaign in New York the next day. Also, if he campaigns in New York, he will be twice as likely to campaign in California the next day. If he campaigns in Texas, he will be equally likely to campaign in California or New York the next day. What is the transition matrix? Is the transition matrix regular?

29. If the incumbent of Problem 28 spoke in Texas on Wednesday, what is the probability that he will be in California on Friday?

30. If the incumbent of Problem 28 spoke in California on Wednesday, what is the probability that he will be in California again on Friday?

8.3 Absorbing Markov Chains*

Under certain conditions a stochastic process will give rise to a transition matrix for a Markov chain with the property that once a given state is reached it is impossible to move out of that state. For example, suppose two people each have two quarters and decide to flip one coin each and call "match" or "no match." The winner takes both coins. If they play long enough, common sense tells us that eventually one player will go broke and the other player will end up with all four coins to put an end to the game. The states of having no or four coins are called **absorbing states**.

Absorbing States

> A state i of a Markov chain is an *absorbing state* if once the system reaches state i on same trial, the system remains in that state on all future trials. This will occur whenever $p_{ii} = 1$.

EXAMPLE 1 Let

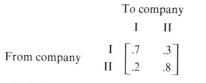

$$
\begin{array}{c}
\begin{array}{ccc} 1 & 2 & 3 \end{array} \\
T = \begin{array}{c} 1 \\ 2 \\ 3 \end{array} \begin{bmatrix} .1 & .2 & .7 \\ 0 & 1 & 0 \\ .5 & .3 & .2 \end{bmatrix}
\end{array}
$$

Then $p_{12} = .2$ is the probability of going from state 1 to state 2. Since $p_{22} = 1$, the probability of going from state 2 to state 2 (that is, remaining in state 2) is 1. Note also that if one entry in any row is 1, then the other entries must be 0s, since each row of a transition matrix is a probability vector. ∎

---

* This section is not required for the remainder of the text.

EXAMPLE 2 Suppose two people each have two coins and decide to play match–no match. Write the transition matrix for the first player. The position p_{ij} is the probability of changing from state i to state j. In this example, state i means having i coins.

$$
\begin{array}{c}
\text{Number of beginning} \\
\text{coins of player 1}
\end{array}
\begin{array}{c}
\\
\\
\\
0 \\
1 \\
2 \\
3 \\
4
\end{array}
\begin{array}{c}
\text{Number of ending coins} \\
\text{of player 1} \\
\begin{array}{ccccc}
0 & 1 & 2 & 3 & 4
\end{array} \\
\left[
\begin{array}{ccccc}
1 & 0 & 0 & 0 & 0 \\
\frac{1}{2} & 0 & \frac{1}{2} & 0 & 0 \\
0 & \frac{1}{2} & 0 & \frac{1}{2} & 0 \\
0 & 0 & \frac{1}{2} & 0 & \frac{1}{2} \\
0 & 0 & 0 & 0 & 1
\end{array}
\right]
\end{array}
$$

Note that states 0 and 4 are absorbing. Also note that at each state the player's bankroll will increase or decrease by one coin, each with a probability of $\frac{1}{2}$. ∎

Suppose we want to examine the long-term trend for a matrix with one or more absorbing states. Let

$$
T = \begin{array}{c} 1 \\ 2 \\ 3 \end{array}
\begin{array}{c}
\begin{array}{ccc} 1 & 2 & 3 \end{array} \\
\left[
\begin{array}{ccc}
1 & 0 & 0 \\
.1 & .5 & .4 \\
0 & 0 & 1
\end{array}
\right]
\end{array}
$$

Both states 1 and 3 are absorbing states—once these states are entered, the system will remain in that state. If the Markov chain begins in state 2, it too must eventually end up in either state 1 or state 3. Now consider the powers of T (correct to two decimal places):

$$
T^2 = \begin{bmatrix} 1 & 0 & 0 \\ .15 & .25 & .6 \\ 0 & 0 & 1 \end{bmatrix}
\qquad
T^4 = \begin{bmatrix} 1 & 0 & 0 \\ .19 & .06 & .75 \\ 0 & 0 & 1 \end{bmatrix}
$$

$$
T^8 = \begin{bmatrix} 1 & 0 & 0 \\ .20 & .00 & .80 \\ 0 & 0 & 1 \end{bmatrix}
$$

It appears that if the system begins in state 2, the probability that it will end in absorbing state 1 is $\frac{1}{5}$ (entry p_{21}) and in absorbing state 3, $\frac{4}{5}$ (entry p_{23}). Now remember that with regular Markov chains the final result is independent of the initial state. This is not true for absorbing Markov chains, as we can see from this example.

Absorbing Markov Chain Theorem

If T is a transition matrix for an absorbing Markov chain, then:

1. $T, T^2, T^3, \ldots$ approach some particular matrix.
2. In a finite number of steps the chain will enter an absorbing state and stay there.
3. The long-term trend depends on the initial state.

The next step is to learn a procedure for finding the probabilities that a particular initial nonabsorbing state will lead to a particular absorbing state. The following theorem outlines this procedure:

Transition Theorem for Absorbing Markov Chains

Let T be a transition matrix for an absorbing Markov chain.

Step 1 Rearrange the rows and columns of T so that the absorbing states come first:

$$T = \begin{bmatrix} I_n & \vdots & 0 \\ \hline P & \vdots & Q \end{bmatrix}$$

where I_n is an identity matrix and 0 is a matrix consisting of all 0s. Matrices P and Q are the matrices consisting of the remaining entries shown in the indicated positions.

Step 2 Let $F = [I_m - Q]^{-1}$, where I_m is an identity matrix of the same order as Q. Remember, $[I_m - Q]^{-1}$ is the inverse of the matrix $[I_m - Q]$. Call F the **fundamental matrix**.

Step 3 FP is a matrix whose entries p_{ij} represent the probability that nonabsorbing state i will lead to absorbing state j.

EXAMPLE 3   Let

$$T = \begin{array}{c} \\ 1 \\ 2 \\ 3 \end{array} \begin{array}{c} \begin{array}{ccc} 1 & 2 & 3 \end{array} \\ \begin{bmatrix} 1 & 0 & 0 \\ .1 & .5 & .4 \\ 0 & 0 & 1 \end{bmatrix} \end{array}$$

Find and interpret FP.

Solution   *Step 1*

$$\begin{array}{c} \\ 1 \\ 3 \\ 2 \end{array} \begin{array}{c} \begin{array}{ccc} 1 & 2 & 3 \end{array} \\ \begin{bmatrix} 1 & 0 & 0 \\ 0 & 0 & 1 \\ .1 & .5 & .4 \end{bmatrix} \end{array} \qquad \begin{array}{c} \\ 1 \\ 3 \\ 2 \end{array} \begin{array}{c} \begin{array}{ccc} 1 & 3 & 2 \end{array} \\ \begin{bmatrix} 1 & 0 & \vdots & 0 \\ 0 & 1 & \vdots & 0 \\ .1 & .4 & \vdots & .5 \end{bmatrix} \end{array}$$

$$I = \begin{bmatrix} 1 & 0 \\ 0 & 1 \end{bmatrix} \qquad 0 = \begin{bmatrix} 0 \\ 0 \end{bmatrix} \qquad P = [.1 \quad .4] \qquad Q = [.5]$$

Step 2 $F = [I_1 - Q]^{-1} = [1 - .5]^{-1} = [.5]^{-1} = [2]$

Step 3 $FP = [2][.1 \quad .4] = \begin{array}{c} \begin{array}{cc} 1 & 3 \end{array} \\ [.2 \quad .8] \end{array} \quad 2$

This means that if the system begins with nonabsorbing state 2, the probability that it will end up in absorbing state 1 is .2 and in absorbing state 3 is .8. This is consistent with the arithmetic we did by finding T^2, T^4, and T^8 earlier.   ∎

EXAMPLE 4 Find and interpret FP for the game of match–no match and the matrix

$$
T = \begin{array}{c} \\ 0 \\ 1 \\ 2 \\ 3 \\ 4 \end{array}
\begin{array}{ccccc}
0 & 1 & 2 & 3 & 4 \\
\begin{bmatrix}
1 & 0 & 0 & 0 & 0 \\
\frac{1}{2} & 0 & \frac{1}{2} & 0 & 0 \\
0 & \frac{1}{2} & 0 & \frac{1}{2} & 0 \\
0 & 0 & \frac{1}{2} & 0 & \frac{1}{2} \\
0 & 0 & 0 & 0 & 1
\end{bmatrix}
\end{array}
$$

Solution

$$
\begin{array}{c} \\ 0 \\ 4 \\ 1 \\ 2 \\ 3 \end{array}
\begin{array}{ccccc}
0 & 1 & 2 & 3 & 4 \\
\begin{bmatrix}
1 & 0 & 0 & 0 & 0 \\
0 & 0 & 0 & 0 & 1 \\
\frac{1}{2} & 0 & \frac{1}{2} & 0 & 0 \\
0 & \frac{1}{2} & 0 & \frac{1}{2} & 0 \\
0 & 0 & \frac{1}{2} & 0 & \frac{1}{2}
\end{bmatrix}
\end{array}
\qquad
\begin{array}{c} \\ 0 \\ 4 \\ 1 \\ 2 \\ 3 \end{array}
\begin{array}{ccccc}
0 & 4 & 1 & 2 & 3 \\
\begin{bmatrix}
1 & 0 & 0 & 0 & 0 \\
0 & 1 & 0 & 0 & 0 \\
\frac{1}{2} & 0 & 0 & \frac{1}{2} & 0 \\
0 & 0 & \frac{1}{2} & 0 & \frac{1}{2} \\
0 & \frac{1}{2} & 0 & \frac{1}{2} & 0
\end{bmatrix}
\end{array}
$$

$$
P = \begin{bmatrix} \frac{1}{2} & 0 \\ 0 & 0 \\ 0 & \frac{1}{2} \end{bmatrix}
\qquad
Q = \begin{bmatrix} 0 & \frac{1}{2} & 0 \\ \frac{1}{2} & 0 & \frac{1}{2} \\ 0 & \frac{1}{2} & 0 \end{bmatrix}
$$

$$
I - Q = \begin{bmatrix} 1 & 0 & 0 \\ 0 & 1 & 0 \\ 0 & 0 & 1 \end{bmatrix} - \begin{bmatrix} 0 & \frac{1}{2} & 0 \\ \frac{1}{2} & 0 & \frac{1}{2} \\ 0 & \frac{1}{2} & 0 \end{bmatrix} = \begin{bmatrix} 1 & -\frac{1}{2} & 0 \\ -\frac{1}{2} & 1 & -\frac{1}{2} \\ 0 & -\frac{1}{2} & 1 \end{bmatrix}
$$

$$
F = (I - Q)^{-1} = \begin{bmatrix} \frac{3}{2} & 1 & \frac{1}{2} \\ 1 & 2 & 1 \\ \frac{1}{2} & 1 & \frac{3}{2} \end{bmatrix}
$$

Note: See Section 2.3 for the method for finding the inverse of a matrix. A computer program for finding the inverse is also available (see Appendix D). Just the final result is shown here.

$$
FP = \begin{bmatrix} \frac{3}{2} & 1 & \frac{1}{2} \\ 1 & 2 & 1 \\ \frac{1}{2} & 1 & \frac{3}{2} \end{bmatrix} \begin{bmatrix} \frac{1}{2} & 0 \\ 0 & 0 \\ 0 & \frac{1}{2} \end{bmatrix}
$$

$$
= \begin{array}{c} \\ 1 \\ 2 \\ 3 \end{array}
\begin{array}{cc}
0 & 4 \\
\begin{bmatrix}
\frac{3}{4} & \frac{1}{4} \\
\frac{1}{2} & \frac{1}{2} \\
\frac{1}{4} & \frac{3}{4}
\end{bmatrix}
\end{array}
$$

This means that if the system was originally in state 1 (player 1 having one coin), then the probability of financial ruin (0 coins) is $\frac{3}{4}$ and the probability that all four coins would be obtained is $\frac{1}{4}$. If the system began in state 2 (player 1 having two coins), then there is a probability of $\frac{1}{2}$ of either winning or losing both coins. If the system began in state 3 (player 1 having three coins), then there is a probability of $\frac{1}{4}$ of losing all three coins and $\frac{3}{4}$ of winning all four coins. ∎

Problem Set 8.3

Determine whether each matrix in Problems 1–12 is a transition matrix for an absorbing Markov chain. If it is, find the absorbing states.

1. $\begin{bmatrix} .3 & .5 & .2 \\ .1 & .7 & .2 \\ .2 & .6 & .2 \end{bmatrix}$

2. $\begin{bmatrix} .1 & .6 & .3 \\ 1 & 0 & 0 \\ .4 & .3 & .3 \end{bmatrix}$

3. $\begin{bmatrix} \frac{1}{2} & \frac{1}{3} & \frac{1}{6} \\ \frac{2}{3} & \frac{1}{6} & \frac{1}{12} \\ \frac{1}{6} & \frac{1}{3} & \frac{1}{2} \end{bmatrix}$

4. $\begin{bmatrix} \frac{1}{3} & \frac{1}{3} & \frac{1}{3} \\ 0 & 1 & 0 \\ \frac{1}{6} & \frac{1}{3} & \frac{1}{2} \end{bmatrix}$

5. $\begin{bmatrix} 0 & 1 & 0 \\ .3 & .2 & .5 \\ .5 & .5 & 0 \end{bmatrix}$

6. $\begin{bmatrix} 1 & 0 & 0 \\ .3 & .2 & .5 \\ .5 & .5 & 0 \end{bmatrix}$

7. $\begin{bmatrix} 1 & 0 & 0 \\ 0 & 1 & 0 \\ 0 & 0 & 1 \end{bmatrix}$

8. $\begin{bmatrix} 0 & 1 & 0 \\ 1 & 0 & 0 \\ 0 & 0 & 0 \end{bmatrix}$

9. $\begin{bmatrix} 1 & 0 & 0 \\ 0 & 1 & 0 \\ .41 & .34 & .25 \end{bmatrix}$

10. $\begin{bmatrix} \frac{1}{2} & \frac{1}{3} & \frac{1}{4} & 0 \\ 0 & 1 & 0 & 0 \\ \frac{1}{4} & \frac{1}{4} & \frac{1}{4} & \frac{1}{4} \\ 0 & 0 & 0 & 1 \end{bmatrix}$

11. $\begin{bmatrix} \frac{1}{2} & \frac{1}{4} & \frac{1}{5} & \frac{1}{20} \\ \frac{1}{3} & 0 & \frac{1}{3} & \frac{1}{3} \\ 0 & 0 & 1 & 0 \\ \frac{1}{5} & \frac{1}{5} & \frac{1}{5} & \frac{2}{5} \end{bmatrix}$

12. $\begin{bmatrix} 0 & 1 & 0 & 0 \\ 0 & 0 & 1 & 0 \\ 0 & 0 & 0 & 1 \\ 1 & 0 & 0 & 0 \end{bmatrix}$

Find and interpret FP for the matrices in Problems 13–24. If approximate answers are necessary, give them correct to two decimal places.

13. $\begin{bmatrix} .2 & .3 & .5 \\ 0 & 1 & 0 \\ 0 & 0 & 1 \end{bmatrix}$

14. $\begin{bmatrix} 1 & 0 & 0 \\ .1 & .8 & .1 \\ 0 & 0 & 1 \end{bmatrix}$

15. $\begin{bmatrix} 1 & 0 & 0 \\ 0 & 1 & 0 \\ .15 & .35 & .5 \end{bmatrix}$

16. $\begin{bmatrix} \frac{1}{3} & \frac{1}{3} & \frac{1}{3} \\ \frac{1}{2} & \frac{1}{4} & \frac{1}{4} \\ 0 & 0 & 1 \end{bmatrix}$

17. $\begin{bmatrix} 1 & 0 & 0 \\ .1 & .9 & 0 \\ .3 & .6 & .1 \end{bmatrix}$

18. $\begin{bmatrix} \frac{1}{4} & \frac{1}{3} & \frac{5}{12} \\ 0 & 1 & 0 \\ \frac{1}{2} & 0 & \frac{1}{2} \end{bmatrix}$

19. $\begin{bmatrix} 1 & 0 & 0 & 0 \\ \frac{1}{3} & \frac{1}{3} & 0 & \frac{1}{3} \\ \frac{1}{4} & \frac{1}{4} & \frac{1}{4} & \frac{1}{4} \\ 0 & 0 & 0 & 1 \end{bmatrix}$

20. $\begin{bmatrix} \frac{1}{2} & \frac{1}{2} & 0 & 0 \\ 0 & 1 & 0 & 0 \\ 0 & 0 & 1 & 0 \\ 0 & \frac{1}{3} & \frac{1}{3} & \frac{1}{3} \end{bmatrix}$

21. $\begin{bmatrix} 1 & 0 & 0 & 0 \\ .4 & .2 & .1 & .3 \\ 0 & 0 & 1 & 0 \\ .1 & .5 & .4 & 0 \end{bmatrix}$

22. $\begin{bmatrix} \frac{1}{5} & \frac{1}{5} & \frac{1}{5} & \frac{1}{5} & \frac{1}{5} \\ 0 & 1 & 0 & 0 & 0 \\ \frac{1}{3} & \frac{1}{3} & \frac{1}{3} & 0 & 0 \\ 0 & 0 & 0 & 1 & 0 \\ 0 & 0 & 0 & 0 & 1 \end{bmatrix}$

23. $\begin{bmatrix} 1 & 0 & 0 & 0 & 0 \\ .1 & .2 & .3 & .3 & .1 \\ .2 & .2 & .2 & .2 & .2 \\ 0 & 0 & 1 & 0 & 0 \\ 0 & 0 & 0 & 0 & 1 \end{bmatrix}$

24. $\begin{bmatrix} 1 & 0 & 0 & 0 & 0 \\ \frac{1}{3} & 0 & \frac{2}{3} & 0 & 0 \\ 0 & \frac{1}{3} & 0 & \frac{2}{3} & 0 \\ 0 & 0 & \frac{1}{3} & 0 & \frac{2}{3} \\ 0 & 0 & 0 & 0 & 1 \end{bmatrix}$

APPLICATIONS

25. a. Write a transition matrix for a match–no match game for two players, each with a single coin.
  b. Find *FP*.
  c. What is the probability of financial ruin for the first player?

26. a. Write a transition matrix for a match–no match game for two players with a total of six coins between them.
  b. Find *FP*.
  c. What is the probability of financial ruin if the first player begins with three coins?

d. What is the probability of financial ruin if the first player begins with two coins?

e. What is the probability of financial ruin if the first player begins with four coins?

27. Noxin, Inc., produces color computer monitors. Each monitor is tested and categorized before it is shipped. If it passes the test, it is shipped; if it needs minor repairs, it is returned to the assembly room for repair and retesting; if it then fails the test, it is destroyed. The probabilities for these events are summarized in the following transition matrix:

		Subsequent test		
		Failed	Returned	Passed
First test	Failed	1	0	0
	Returned	.1	.2	.7
	Passed	0	0	1

What is the probability that a returned item will eventually be passed?

28. Rennolds Department Store issues credit cards and has a policy of revoking any cards with delinquent accounts for 3 months. The auditor calculates the probabilities for the following transition matrix by looking at the records of 5000 credit card customers:

		Subsequent month			
		Paid up	Delinquent 1 month	Delinquent 2 months	Revoke card
First month	Paid up	.65	.35	0	0
	Delinquent 1 month	.70	0	.30	0
	Delinquent 2 months	.3	.4	0	.3
	Revoke card	0	0	0	1

a. What is the probability that a paid-up customer will eventually have a revoked card?

b. What is the probability that a customer who is 1 month delinquent will eventually have a revoked card?

29. A rat is placed at random in one of the compartments of the maze shown here. When the rat is in some room, there are equal probabilities that it will choose any door in that room. When the rat reaches room 1, it is rewarded with food, and it no longer leaves that room. Room 3 has one-way doors that do not permit the rat to leave the room once it enters. What is the probability that the rat will end up in room 1 if it was originally placed in room 5?

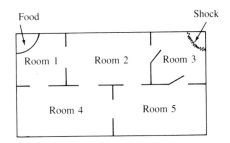

30. The experiment in Problem 29 is repeated. What is the probability that the rat will end up in room 1 if it is placed in room 2?

8.4 Summary and Review

SAMPLE TEST   *For additional practice there are a large number of review problems categorized by objective in the Student Solutions Manual. The following sample test (40 minutes) is intended to review the main ideas of the chapter.*

1. A certain chemical can exist in a solid, liquid, or gaseous state. An experiment is performed and the following information is obtained:

   If the beginning state is solid, there is a 75% probability that the chemical will go to a liquid state after 1 hour, a 5% probability that it will go to a gaseous state after 1 hour, and a 20% probability that it will remain in a solid state.

   If the beginning state is liquid, there is a 10% probability that the chemical will be in a solid state after 1 hour, a 30% probability that it will be in a gaseous state, and a 60% probability that it will remain in a liquid state.

   If the beginning state is gaseous, there is a 40% probability that the chemical will be in a liquid state in 1 hour and a 60% probability that it will stay in a gaseous state.

   Write the hourly transition matrix.

2. The yearly transition matrix for customers of Hallmark and Buzza-Cardoza cards is

$$T = \begin{array}{c} H \\ B \end{array} \begin{bmatrix} .7 & .3 \\ .5 & .5 \end{bmatrix} \begin{array}{c} \end{array}$$

with columns labeled H and B.

   If the present market is 2 million customers and Hallmark has 80% of the market while Buzza-Cardoza has the other 20%, what is the expected number of Buzza-Cardoza customers at the end of 1 year?

3. Find the fixed probability vector for the transition matrix in Problem 2.

4. What is the probable long-range composition for the chemical in Problem 1?

5. Two players start a game with $5 between them. The game has the following transition matrix for the first player:

$$\begin{array}{c} \\ 0 \\ 1 \\ 2 \\ 3 \\ 4 \\ 5 \end{array} \begin{array}{cccccc} 0 & 1 & 2 & 3 & 4 & 5 \end{array} \\ \begin{bmatrix} 1 & 0 & 0 & 0 & 0 & 0 \\ .5 & 0 & .5 & 0 & 0 & 0 \\ 0 & .5 & 0 & .5 & 0 & 0 \\ 0 & 0 & .5 & 0 & .5 & 0 \\ 0 & 0 & 0 & .5 & 0 & .5 \\ 0 & 0 & 0 & 0 & 0 & 1 \end{bmatrix}$$

   a. What are the absorbing states?
   b. Does this game favor the first player? Why or why not?
   c. If the first player begins with $4, what is the probability of financial ruin (to the nearest hundredth)?

Modeling Application 8

Genetics

A researcher interested in studying the ability of parents to pass certain physical traits to their offspring sets up a classic *brother-sister mating model* on a family of rats. In this experiment, two rats are mated, and from among their direct descendants two individuals of the opposite sex are selected at random and are mated. Then the process is repeated indefinitely and the target traits are studied. What are the expected results from such experimentation?* Assume that traits are determined by *genes* which are passed from parents to their offspring. Each parent has a pair of genes, and the basic assumption is that each offspring inherits one gene from each parent to form the offspring's own pair. The genes are selected in a random, independent way. In your model, assume that the researcher is studying a trait that is both easily identifiable (such as color of a rat's fur) and determined by a pair of genes consisting of a *dominant* gene, denoted by A, and a *recessive* gene, denoted by a. The possible pairings of genes are called *genotypes*:

> AA is *dominant*, or homozygous
>
> Aa is *hybrid*, or heterozygous; genetically, the genotype aA is the same as Aa
>
> aa is called *recessive*

Write a paper considering the possibilities for the offspring. Also, suppose one parent has genotype Aa (heterozygous) and is mated with a parent whose genotype is unknown. The offspring is mated with another heterozygous genotype (Aa) and this process is continued. Build a model to predict the genotype of the offspring over time. For general guidelines about writing this essay, see the commentary accompanying Modeling Application 1 on page 36.

---

* This modeling application is based on the work of Gregor Mendel (1822–1884), an Austrian monk, who formulated the laws of heredity and genetics. Mendel's work was later amplified and explained by a mathematician, G. H. Hardy (1877–1947), and a physician, Wilhelm Weinberg (1862–1937). For years Mendel taught science without any teaching credentials because he had failed the biology portion of the licensing examination! His work, however, laid the foundation for the very important branch of biology known today as genetic science.

CHAPTER 9
Decision Theory

APPLICATIONS

Management (*Business, Economics, Finance, and Investments*)

Life insurance expectation (9.1, Problems 11–16)
Real estate expectation (9.1, Problem 26)
Operations research (9.1, Problems 32–35)
Competition between McDonald's and Burger King (9.2, Problem 26)
Selecting a contract on which to bid (9.4, Problem 8)
Demand for cars at a rental agency (9.4, Problem 9)
Marketing a new cereal (9.4, Problem 10)
Investment strategy (9.4, Problem 12)
CPA examination questions (9.4, Problems 11–12)

Life sciences (*Biology, Ecology, Health, and Medicine*)

Feasibility for sinking a test oil well (9.1, Problems 27–28)

Social sciences (*Demography, Political Science, Population, Psychology, Society, and Sociology*)

Military strategy (9.2, Problems 21 and 24; 9.3, Problem 13)
Presidential campaign strategies (9.2, Problem 23)
Submarine strategies (9.2, Problem 24)
Sherlock Holmes' solution in "The Final Problem" (9.2, Problem 25)

General interest

Lotteries and contests (9.1, Problems 7–10; 9.4 Problem 6)
Roulette (9.1, Problems 17–25)
Game of Morra (9.2, Problem 20)
Game strategy (9.2, Problem 19)

Modeling application—Overbooking of Airline Flights

CHAPTER
OVERVIEW

This chapter presents two decision models. The first is mathematical expectation, which places numerical values on various courses of action and allows us to make decisions based on the expected return. The second is game theory, which enables us to analyze not only our own choices but also the possible alternatives open to our opponent so we can select a strategy that will maximize our return.

PREVIEW

As we move from the mechanics of mathematics to more realistic problem solving, the models we must use become more complicated because real life situations have many variables and component parts. Models must accommodate these complexities, so the methods rely to a great extent on the aid of computers. If you have not yet used a computer in this course, this is a good time to read Appendix D.

PERSPECTIVE

In business, as well as in almost all endeavors, it is frequently necessary to choose between opposing alternatives. Much of this decision making is done intuitively because the choices are not usually well defined or clear-cut. However, a relatively new branch of mathematics called *decision theory* provides a systematic way to attack problems of decision making. This problem-solving skill is an essential one, and in this chapter we develop two decision models.

9.1 Expectation

If we apply the notion of a weighted mean to the average value of a random variable, we obtain a measure called the **expected value** or **mathematical expectation**. This concept began as a means of measuring probable winnings in gambling situations, but today it has widespread applications to the decision-making process in business, economics, and operations research.

EXAMPLE 1

Suppose \$1 is bet on a single number in a game of roulette. The roulette wheel has 38 numbered slots, and a ball is rotated around the wheel until it falls into one of the slots. There are many ways to place a bet, but if a single number is chosen and the ball lands on the chosen number, the bet is returned along with 35 times the original bet. If the ball lands on any other number, the bet is lost. Let the random variable X represent the amount won by the player. Then X has either of two values, 35 or -1, and

$$P(X = +35) = \tfrac{1}{38} \quad \text{and} \quad P(X = -1) = \tfrac{37}{38}$$

Suppose now that a player is a compulsive gambler who always bets on a single number, and makes n bets. The player should win approximately $\frac{n}{38}$ times and lose approximately $\frac{37n}{38}$ times. Since each win is \$35, and each loss is \$1, we have

$$\text{Total winnings:} \quad 35 \cdot \left(\frac{n}{38}\right) = \frac{35n}{38}$$

$$\text{Total losses:} \quad 1 \cdot \left(\frac{37n}{38}\right) = \frac{37n}{38}$$

$$\text{Net winnings:} \quad \frac{35n}{38} - \frac{37n}{38} = -\frac{2n}{38}$$

Negative winnings are interpreted as a net loss. If we are concerned with the *average* winnings per game, we divide the total by the number of trials, *n*, to obtain the expected value:

$$\text{Expected value} = -\frac{2}{38} = -\frac{1}{19} \qquad \text{This is a loss of about 5.263¢ per game}$$

∎

In general:

Expected Value of a Random Variable

If a random variable X has possible values $x_1, x_2, \ldots, x_n$ occurring with probabilities $p_1, p_2, \ldots, p_n$ ($\sum p_i = 1$), we define the *expected value of* X, denoted by E(X), by

$$E(X) = x_1 p_1 + x_2 p_2 + \cdots + x_n p_n$$

EXAMPLE 2   Suppose you are going to play a game by rolling a pair of dice. You will be paid $5.00 every time you roll a pair of 6s. You will not receive anything for any other outcome. It costs $.50 to play. What is the expected value for this game?

Solution   Compare the wording of this example and Example 1. In Example 1 you would not lose the dollar you bet until after the game is played. In this example, there is an admission price of $.50, which means that this amount is paid, win or lose. In situations for which there is an "admission price" this amount must be subtracted from the proposed winnings.
Let X be the amount won. Then

$$\text{Win-}X = 5.00 - .50 = \$4.50$$
$$\text{Lose-}X = -\$.50$$

and

$$P(X = \$4.50) = \tfrac{1}{36} \qquad \text{and} \qquad P(X = -\$.50) = \tfrac{35}{36}$$

Using the notation of the above definition,

$$E(X) = x_1 p_1 + x_2 p_2 \qquad \text{where} \begin{cases} x_1 = 4.50 \\ x_2 = -0.50 \end{cases} \text{and} \begin{aligned} p_1 &= \tfrac{1}{36} \\ p_2 &= \tfrac{35}{36} \end{aligned}$$

Thus

$$E(X) = 4.50(\tfrac{1}{36}) + (-0.50)(\tfrac{35}{36})$$
$$\approx 0.125 - 0.486$$
$$\approx -0.36$$

When we say that the expectation is a loss of $.36 per game it does *not* mean that you will lose $.36 every time you play the game. (Indeed, you will *never* lose $.36; you will either win $5.00 or nothing—less the $.50 admission charge.) But, if you were to play this game a large number of times, you could expect to lose an *average* of $.36 per game. ∎

We say that a game is **fair** if $E(X)$ is zero; if $E(X)$ is positive, the game is favorable to the player; and if $E(X)$ is negative, the game is unfavorable to the player.

EXAMPLE 3 A contest offers the following prizes:

Prize	Value	Probability of winning
Grand prize trip	$1,500 = x_1$	$.000026 = p_1$
Weber Kettle	$110 = x_2$	$.000032 = p_2$
Magic Chef Range	$279 = x_3$	$.000016 = p_3$
Murray Bicycle	$191 = x_4$	$.000021 = p_4$
Lawn Boy Mower	$140 = x_5$	$.000026 = p_5$
Samsonite Luggage	$183 = x_6$	$.000016 = p_6$

What is the expected value for this contest?

Solution
$$E(X) = x_1 p_1 + x_2 p_2 + \cdots + x_6 p_6$$
$$= 1500(.000026) + 110(.000032) + 279(.000016)$$
$$+ 191(.000021) + 140(.000026) + 183(.000016)$$
$$= .057563$$

The expected value is a little less than 6¢. ■

Expectation is obviously useful in decision-making processes. For example, should you enter the contest in Example 3? What is the cost of entering? Do you need to mail in your entry? From a mathematical standpoint, you should not be willing to pay more than 5¢ to enter the contest. Let us now turn to a more practical problem in Example 4.

EXAMPLE 4 Karlin, Inc., manufactures automated vending machines and must choose between bidding on two contracts. Contract I will cost $250 in preparation costs to make the bid, but if Karlin is selected, there will be a net profit of $5,000. Contract II will cost $100 to prepare, with a net profit of $3,000 if the bid is won. The actual probability of winning contract I is 20% and of winning contract II is 25%. Should Karlin bid on contract I or contract II?

Solution The solution can be found by calculating the expectation:
$$E(\text{contract I}) = 5000(.20) + (-250)(.80)$$
$$= 1000 + (-200)$$
$$= 800$$
$$E(\text{contract II}) = 3000(.25) + (-100)(.75)$$
$$= 750 + (-75)$$
$$= 675$$

Thus, on the basis of the given information, Karlin should bid on contract I. ■

Operations research is the science of making optimal decisions, as illustrated in the next example.

EXAMPLE 5   A $150,000 house has an unstable foundation and is sliding off a hillside. The only hope for the house is to drill a horizontal well to tap a reservoir of stored water that is causing the slippage. If the cost of drilling each well is $1,000, and if the probability that a particular well will hit the water reservoir is .8, how many wells should be drilled?

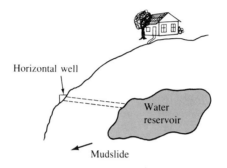

Horizontal well

Water reservoir

Mudslide

Solution   It is obvious that drilling no well is too few and drilling 150 wells is too many. Our task is to decide on the number of wells, x, to be drilled. The probability that the x wells will be unsuccessful is $.2^x$ (assume that the probability of hitting water for each well is independent). Thus the probability that at least one well will hit the water reservoir is

$$1 - .2^x$$

The expected gain is $150,000(1 - .2^x)$, while the cost of drilling the wells is $1,000x$, so

$$E(x) = \$150,000(1 - .2^x) - \$1000x$$

We want to find the maximum value of $E(x)$ for the various values of x.

$$E(0) = 0 \qquad \text{If no wells are tried, the house will slide off the hill and be lost}$$
$$E(1) = \$150,000(1 - .2^1) - \$1000 \cdot 1$$
$$= \$119,000$$
$$E(2) = \$150,000(1 - .2^2) - \$1000 \cdot 2$$
$$= \$142,000$$
$$E(3) = \$150,000(1 - .2^3) - \$1000 \cdot 3$$
$$= \$145,800$$
$$E(4) = \$150,000(1 - .2^4) - \$1000 \cdot 4$$
$$= \$145,760$$
$$E(5) = \$150,000(1 - .2^5) - \$1000 \cdot 5$$
$$= \$144,952$$
$$E(6) = \$150,000(1 - .2^6) - \$1000 \cdot 6$$
$$= \$143,990.40$$

Note that sinking additional wells does not increase the expected gain. The expected value is greatest when three wells are drilled, so operations research tells us to drill three wells.   ■

Problem Set 9.1

Given the probability functions in Problems 1–6, find the expected value of X.

1.

X	1	2	3
P(X)	$\frac{1}{4}$	$\frac{1}{2}$	$\frac{1}{4}$

2.

X	5	10	21
P(X)	.2	.5	.3

3.

X	3	6	9	12
P(X)	.4	.3	.2	.1

4.

X	4	9	12	18
P(X)	$\frac{1}{4}$	$\frac{1}{3}$	$\frac{1}{4}$	$\frac{1}{6}$

5.

X	0	1	2	3	4
P(X)	.10	.25	.30	.20	.15

6.

X	1	4	9	16	25
P(X)	.5	.25	.10	.10	.05

APPLICATIONS

7. A magazine subscription service is having a contest in which the prize is $80,000. If the company receives 1 million entries, what is the mathematical expectation of entering the contest?

8. Krinkles potato chips is having a "Lucky Seven Sweepstakes." The grand prize is $70,000; 7 second prizes each pay $7,000; 77 third prizes each pay $700; and 777 fourth prizes each pay $70. How much is the expectation of entering this contest if we assume that there are 10 million entries?

9. A company presents the following table for its contest:

Prize	Prizes available	Approximate probability of winning
$1,000.00	13	.000005
100.00	52	.00002
10.00	520	.0002
1.00	28,900	.010989
Total	29,495	.011111*

* This is $\frac{1}{90}$.

What is the expectation for playing this game one time?

10. A company holds a bingo card contest where the following chances of winning are given:

	One card 1 time	One card 7 times	One card 13 times
$25 prize	1 in 21,252	1 in 3036	1 in 1630
$ 3 prize	1 in 2125	1 in 304	1 in 163
$ 1 prize	1 in 886	1 in 127	1 in 68
Any prize	1 in 609	1 in 87	1 in 47

What is the expectation in dollars for playing one card 13 times?

Life insurance policies use mathematical expectation. The "winning" value is the value of the policy and the "winning" probability is the probability of dying during the life of the policy. The probability of dying is found from an actuarial table similar to Table 9 in Appendix E. Note that the table gives p_x (probability of living) and q_x (probability of dying). Find the expected value for the 1-year policies described in Problems 11–16.

11. $10,000 issued at age 10   **12.** $10,000 issued at age 38

13. $10,000 issued at age 65   **14.** $25,000 issued at age 19

15. $50,000 issued at age 19   **16.** $125,000 issued at age 35

A U.S. roulette wheel has 38 numbered slots (1–36, 0, and 00, as shown in the figure). Some of the more common bets and payoffs are shown in the figure. If the payoff is listed as 6 to 1, the player would receive $6 plus the $1 bet. Use the figure to find the expectation in Problems 17–25 for playing $1 one time. One play consists of the dealer spinning the wheel and a small white ball in opposite directions. As the ball slows to a stop it lands in one of the 38 numbered slots, which are also colored black, red, or green.

17. Black

18. Odd

19. Single-number bet

20. Double-number bet

21. Three-number bet

22. Four-number bet

23. Five-number bet

24. Six-number bet

25. Column bet

26. A realtor who takes the listing on a house to be sold knows that she will spend $800 trying to sell the house. If she sells it herself, she will earn 6% of the selling price. If another realtor sells a house from her list, our realtor will earn only 3% of the price. If the house is unsold in 6 months, she will lose the listing. Suppose the probabilities are as follows:

	Probability
Sell by herself	.50
Sell by another realtor	.30
Not sell in 6 months	.20

What is the expected profit for listing an $85,000 house? (Be sure to subtract the $800 from the commission because the profit is 6% of the selling price *less* selling costs.)

27. An oil drilling company knows that it will cost $25,000 to sink a test well. If oil is hit, the income for the drilling company will be $425,000. If only natural gas is hit, the income will be $125,000. If nothing is hit, the company will have no income. If the probability of hitting oil is $\frac{1}{40}$ and if the probability of hitting gas is $\frac{1}{20}$, what is the expectation for the drilling company? Should the test well be sunk? (Do not forget to subtract the cost from the income in determining the payoffs.)

28. In Problem 27, suppose the income for hitting oil is changed to $825,000 and the income for gas to $225,000. Now, what is the expectation for the drilling company? Should the test well be sunk?

29. Suppose you roll one die. You are paid $5 if you roll a 1, and you pay $1 otherwise. What is the expectation?

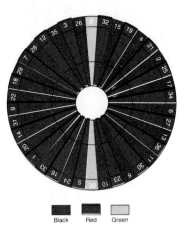

Black   Red   Green

HERE IS HOW BETS ARE PLACED ON THE ROULETTE TABLE

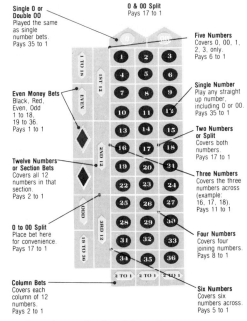

U.S. roulette wheel and board

30. A game involves drawing a single card from an ordinary deck. If an ace is drawn, the player receives 50¢; if a heart is drawn, the player receives 25¢; if the queen of spades is drawn, the player receives $1. If the cost of playing is 10¢, should you play?

31. Consider the following game where a player rolls a single die: If a prime (2, 3, or 5) is rolled, the player wins $2. If a square (1 or 4) is rolled, the player wins $1. However, if the player rolls a perfect number (6), then it costs the player $11. Is this a good deal for the player or not?

32. Suppose the probability of hitting the reservoir in Example 5 is .6 instead of .8. Now, how many wells should be tried?

33. Suppose each well in Example 5 costs $5,000 to drill instead of $1,000. Now, how many wells should be tried?

34. Kingston and Associates, a consulting firm, is hired by a national car rental agency to determine the number of cars to purchase for a new outlet. After appropriate research, the consulting firm finds that the daily demand will be from 12 to 17 cars, distributed as follows:

Number of customers	12	13	14	15	16	17
Probability	.05	1	.35	.2	.2	.1

Records also show that it costs the agency $20 per day per car whether the car is rented or not. If the rental price per car is $32, then the profit is $12 for each day the car is rented. What is the optimal number of cars that the consulting firm should recommend the agency to obtain?

35. Repeat Problem 34 for the following values:

Number of customers	10	11	12	13	14	15
Probability	.1	.2	.3	.2	.1	.1

9.2 Game Theory

Another application of decision theory is concerned with the analysis of certain types of conflict that may be either real or artificial, for example, a game of cards or chess played between friends, union and management at a bargaining table, a presidential election, opposing armies on a battlefield, an oil company drilling for oil (oil company versus nature).

The outcomes in the games of chess and poker, for example, depend on more than mere chance. A player must play according to some set of strategies and in anticipation of what the opponent will do. The analysis of these conflict situations and their corresponding strategies for decision making are developed in a relatively recent branch of mathematics called **game theory**. Game theory is primarily concerned with the logic of strategy and was first envisioned by the great German mathematician Gottfried Leibniz (1646–1716). However, the theory of games as we know it today was developed in the 1920s by John von Neumann and Emile Borel. It gained wide acceptance in 1944 in a book titled *Theory of Games and Economic Behavior* by von Neumann and Oskar Morgenstern.

Let us begin our study of game theory by considering a rather simple game. Suppose a friend of ours, Linda, likes to play games and suppose we will act as her consultant in matters relating to game theory. Her opponent, therefore, is our opponent.

The first game Linda decides to play is a variation of the game of two-finger mora. The game consists of two persons simultaneously holding out either one or two fingers from a closed fist. Linda's opponent makes her the following offer, "Let's play two-finger mora. If the sum of the fingers is two, I'll pay you 5¢. If it is four, you pay me 5¢. If the sum is odd, we break even."

This is an example of a two-person **zero-sum game**. If the payoffs for a game are such that a win for one player results in a corresponding loss for the other player, then we describe the situation as a zero-sum game. The game of two-finger mora can be summarized by a 2×2 **game matrix**:

$$
\begin{array}{c}
\text{Opponent} \\
\begin{array}{cc} 1 & 2 \end{array} \\
\text{Linda} \quad \begin{array}{c} 1 \\ 2 \end{array} \left[\begin{array}{cc} 5 & 0 \\ 0 & -5 \end{array} \right]
\end{array}
$$

In a game matrix, each element represents a payoff to the first player. The above matrix shows the payoffs to Linda. We see that if Linda and her opponent both hold out one finger, she will win 5¢; if they both hold out two fingers, she will win -5¢ (which is interpreted as a 5¢ loss for her). Linda's opponent, however, views negative entries as winning payoffs and positive entries as losing payoffs.

We would like to develop a **strategy** for this game. Each row of the game matrix represents a strategy for Linda, while each column represents a strategy for her opponent. If we examine the game matrix, we see that Linda's best strategy is to show one finger since the *worst* she can do is break even. On the other hand, the best strategy for her opponent is to show two fingers since the worst he can do is break even. The payoff when they both play their best strategies is called the **value of the game**. We see that the value of this game is 0, and whenever the value of a game is 0 we call the game **fair**.

COMPUTER APPLICATION   You might enjoy playing some of the games in this section, so, if you have access to a computer, use Program 14 on the disk accompanying this book to play them with the computer as your opponent. The advantage of playing on a computer instead of playing against a person is the ability to play a very large number of games in a short period of time so that you can "see" the effect of a particular strategy. You can tell the computer to play using an intelligent or a random strategy.

EXAMPLE 1   Determine the best strategies and value of the game of three-finger mora where player II agrees to pay player I the difference between the fingers shown.

Solution   First we write the game matrix:

$$
\begin{array}{c}
\text{Player II} \\
\begin{array}{ccc} 1 & 2 & 3 \end{array} \\
\text{Player I} \quad \begin{array}{c} 1 \\ 2 \\ 3 \end{array} \left[\begin{array}{ccc} 0 & -1 & -2 \\ 1 & 0 & -1 \\ 2 & 1 & 0 \end{array} \right]
\end{array}
$$

We see that player I can reason as follows: "If I play one finger, then I stand to lose 2; if I play two fingers, I stand to lose 1; if I play three fingers, the *worst* I can do is break even." We see that player I looks for the *minimum entry* in each row. On the other hand, player II looks for the *maximum entry* in each column (since the negative values indicate a win for player II). He reasons, "If I play column 1, I stand to lose 2; column 2, I can lose 1; but if I play column 3, the *worst* I can do is break even." In this game we see that player I would play three fingers and player II would also play three fingers. The value of the game is 0. ∎

Games such as this are said to be **strictly determined** since a knowledge of our opponent's strategy would not alter our own strategy. In a strictly determined game the smallest entry in some row is also the largest entry in some column. This point is called a **saddle point**. The reason it is called a saddle point is because it can be visualized by thinking of the surface of a saddle. The same point is a maximum for one person and at the same time a minimum for another. There is such a point on a saddle, as we can see in Figure 9.1. The **value of a strictly determined game** is the value of this saddle point.

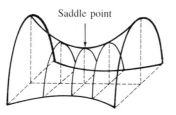

Saddle point

Figure 9.1 Saddle point

If a matrix game has a saddle point, then the row containing the saddle point is the best strategy for player I, and the column containing the saddle point is the best strategy for player II.

Let us now look at the matrices in Examples 2–4 and decide if the game determined by each matrix is strictly determined. We will use the following procedure:

Step 1 Place an asterisk (*) next to the minimum of each row.

Step 2 Check to see if each element marked by an * is the maximum in its column. If so, circle it. This is a saddle point and the game is strictly determined.

EXAMPLE 2

$$\begin{bmatrix} 1 & -2 & -3^* \\ 2 & 4 & \boxed{1^*} \\ -3 & -4^* & -2 \end{bmatrix}$$

Row minimum
-3
1
-4

Column maximum $2 \quad 4 \quad 1$

The saddle point is circled, and the value of the game is 1. The game is strictly determined. ∎

EXAMPLE 3

$$\begin{bmatrix} 2 & 1 & -3^* \\ -2^* & 0 & 2 \\ 3 & -1^* & 1 \end{bmatrix}$$

Row minimum
-3
-2
-1

Column maximum $3 \quad 1 \quad 2$

There is no saddle point. In this case, we say that the game is **nonstrictly determined**. ∎

EXAMPLE 4

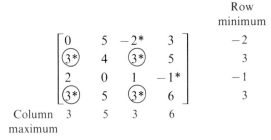

$$\begin{bmatrix} 0 & 5 & -2^* & 3 \\ \text{(3*)} & 4 & \text{(3*)} & 5 \\ 2 & 0 & 1 & -1^* \\ \text{(3*)} & 5 & \text{(3*)} & 6 \end{bmatrix}$$

Row minimum: -2, 3, -1, 3

Column maximum: 3, 5, 3, 6

The saddle points are circled, and the value of the game is 3. We see that a game can have more than one saddle point, but all saddle points must be numerically the same. The game is strictly determined. ∎

Now consider a game that is nonstrictly determined. Linda's friend says he is tired of playing two-finger mora and has thought of another game. He says it is played with a dime and a nickel. Each of them will select one of two coins and put it in a hand. On a given signal they will open their hands together. If the coins match, Linda will win both. If the coins do not match, her friend will win both coins. We advise Linda to play this silly game, provided we can first write out the game matrix.

Opponent

Nickel Dime

Linda

$$\begin{array}{cc} \text{Nickel} \\ \text{Dime} \end{array} \begin{bmatrix} 5 & -5 \\ -10 & 10 \end{bmatrix}$$

We quickly check for a saddle point and see that there is none, so this is not a strictly determined game. Game theory will not help Linda play this game if she plays only once, but if she plays over and over, we can try to develop a **mixed strategy** that will tell her how often to play a nickel and how often to play a dime.

Let p be the probability that Linda will choose row 1 and q the probability that her opponent will choose column 1 (that is, both choose nickels). Then p and q are numbers between 0 and 1 (inclusive). Also, since Linda must choose either row 1 or row 2, the probability she will choose row 2 is $1 - p$. Likewise, column 2 for her opponent will be chosen with the probability $1 - q$. Now, if we represent Linda's strategy by the row matrix

$$[p \quad 1 - p]$$

and her opponent's strategy by the column matrix

$$\begin{bmatrix} q \\ 1 - q \end{bmatrix}$$

we mean that Linda will choose row 1 of the game matrix with probability p and her opponent will choose column 1 with probability q.

Since Linda and her opponent are making their choices independently, the probability that the payoff to Linda is 5¢ is pq:

Linda picks row 1⌐ ⌐Opponent picks column 1

pq

If Linda's opponent plays column 1, then Linda's expectation is

$5p - 10(1 - p)$

If her opponent plays column 2, then her expectation is

$-5p + 10(1 - p)$

These expectations can be summarized by using matrix multiplication:

$$[p \quad 1 - p]\begin{bmatrix} 5 & -5 \\ -10 & 10 \end{bmatrix} = [5p - 10(1 - p) \quad -5p + 10(1 - p)]$$
$$= [5p - 10 + 10p \quad -5p + 10 - 10p]$$
$$= [15p - 10 \quad -15p + 10]$$

In 1928 John von Neumann proved that if a game is not strictly determined, then the best strategy for the first player (Linda) is to choose p so that the entries of this product matrix are equal. That is,

$$15p - 10 = -15p + 10$$
$$30p = 20$$
$$p = \tfrac{2}{3}$$

Thus Linda's best strategy for this game is to pick row 1 two-thirds of the time and row 2 one-third of the time. For her opponent, find

$$\begin{bmatrix} 5 & -5 \\ -10 & 10 \end{bmatrix}\begin{bmatrix} q \\ 1 - q \end{bmatrix} = [5q - 5(1 - q) \quad -10q + 10(1 - q)]$$
$$= [5q - 5 + 5q \quad -10q + 10 - 10q]$$
$$= [10q - 5 \quad -20q + 10]$$

Now we solve:

$$10q - 5 = -20q + 10$$
$$30q = 15$$
$$q = \tfrac{1}{2}$$

The best strategy for Linda's opponent is to choose each row one-half of the time.

Von Neumann also proved that if both players select the optimum strategy, then the expectation of winning for the row player is the same as the expectation of losing for the column player. This is called Von Neumann's **minimax theorem**. This common expectation is called the **value of the nonstrictly determined game**. Thus, to calculate the value of the game, we can use either player's optimum strategy:

$$[\tfrac{2}{3} \quad \tfrac{1}{3}]\begin{bmatrix} 5 & -5 \\ -10 & 10 \end{bmatrix} \quad \text{or} \quad \begin{bmatrix} 5 & -5 \\ -10 & 10 \end{bmatrix}\begin{bmatrix} \tfrac{1}{2} \\ \tfrac{1}{2} \end{bmatrix}$$

We can use the product in *any* row or column to find the value (we choose column 1 of the first product):

$$\tfrac{2}{3}(5) + \tfrac{1}{3}(-10) = \tfrac{10}{3} - \tfrac{10}{3} = 0$$

The value of this game is 0, which means that it is a fair game.

Optimal Strategy Theorem

In the zero-sum two-person game

$$\begin{bmatrix} a & b \\ c & d \end{bmatrix}$$

with strategies

$$\begin{bmatrix} p & 1-p \end{bmatrix} \quad \text{and} \quad \begin{bmatrix} q \\ 1-q \end{bmatrix}$$

the optimal strategy for the row player occurs when the entries of the product

$$\begin{bmatrix} p & 1-p \end{bmatrix} \begin{bmatrix} a & b \\ c & d \end{bmatrix}$$

are equal, provided that the game matrix has no saddle point. The optimal strategy for the column player occurs when the entries of the product

$$\begin{bmatrix} a & b \\ c & d \end{bmatrix} \begin{bmatrix} q \\ 1-q \end{bmatrix}$$

are equal, provided that the game matrix has no saddle point.

EXAMPLE 5 Find the optimal strategies and value for the game

$$\begin{bmatrix} -10 & 5 \\ 8 & -10 \end{bmatrix}$$

Solution There is no saddle point. Next, find the products. For the row player:

$$\begin{bmatrix} p & 1-p \end{bmatrix} \begin{bmatrix} -10 & 5 \\ 8 & -10 \end{bmatrix} = \begin{bmatrix} -10p + 8(1-p) & 5p - 10(1-p) \end{bmatrix}$$

Solve:

$$-10p + 8(1-p) = 5p - 10(1-p)$$
$$-10p + 8 - 8p = 5p - 10 + 10p$$
$$-18p + 8 = 15p - 10$$
$$-33p = -18$$
$$p = \tfrac{18}{33}$$

For the column player:

$$\begin{bmatrix} -10 & 5 \\ 8 & -10 \end{bmatrix} \begin{bmatrix} q \\ 1-q \end{bmatrix} = \begin{bmatrix} -10q + 5(1-q) & 8q - 10(1-q) \end{bmatrix}$$

Solve:

$$-10q + 5(1 - q) = 8q - 10(1 - q)$$
$$-10q + 5 - 5q = 8q - 10 + 10q$$
$$-15q + 5 = 18q - 10$$
$$-33q = -15$$
$$q = \tfrac{15}{33}$$

Thus the optimal strategies are $\begin{bmatrix} \tfrac{18}{33} & \tfrac{15}{33} \end{bmatrix}$ and $\begin{bmatrix} \tfrac{15}{33} \\ \tfrac{18}{33} \end{bmatrix}$. The value of the game is

$$-10\left(\tfrac{18}{33}\right) + 8\left(1 - \tfrac{18}{33}\right) = -\tfrac{60}{33}$$

The game is biased in favor of the second player since the value of the game is negative. ■

Sometimes the original game is not 2×2, but can be reduced to a 2×2 game matrix by eliminating certain rows or columns. Consider the following game matrix:

Opponent

$$\text{Linda} \begin{bmatrix} 2 & 4 & -3 \\ -1 & 3 & 2 \\ 0 & -1 & -4 \end{bmatrix}$$

In finding the optimum strategies we first look for a saddle point and see that there is none. Suppose Linda is playing this game with an opponent. Certainly she would never choose row 3 since every entry in row 3 is smaller than the corresponding entry in row 1. She can therefore eliminate row 3 from any further consideration. We say that row 1 **dominates** row 3. *We can delete* **dominated rows**. The result is a 2×3 *subgame* of the original game:

Opponent

$$\text{Linda} \begin{bmatrix} 2 & 4 & -3 \\ -1 & 3 & 2 \end{bmatrix}$$

On the other hand, Linda's opponent wants to make his choices as small as possible (negative values are good for him). He would not pick column 2 as a strategy since each entry in column 1 is smaller than the corresponding entry in column 2. We say that column 2 dominates column 1 and the opponent would *delete the* **dominating column** from further consideration. The result is a 2×2 subgame of the original game:

Opponent

$$\text{Linda} \begin{bmatrix} 2 & -3 \\ -1 & 2 \end{bmatrix}$$

The following principle can be applied to any matrix game:

Dominating Principle

> In a given game matrix, *dominated rows* can be eliminated (*eliminate the smaller row*). *Dominating columns* can be eliminated (*eliminate the larger column*).

EXAMPLE 6   Find the strategies for the game matrix:

Opponent

$$\text{Linda} \begin{bmatrix} 2 & 4 & -3 \\ -1 & 3 & 2 \\ 0 & -1 & -4 \end{bmatrix}$$

Solution   Eliminate dominated rows (row 3) and dominating columns (column 2), leaving the following subgame:

$$\begin{bmatrix} 2 & -3 \\ -1 & 2 \end{bmatrix}$$

Solve this subgame (the details are left for you) to obtain:

Linda's strategy $= [\frac{3}{8} \quad \frac{5}{8}]$ Opponent's strategy $= \begin{bmatrix} \frac{5}{8} \\ \frac{3}{8} \end{bmatrix}$

Next, it is necessary to state the strategy for the *original game*. Use a 0 to denote the eliminated row and column. Thus, since row 3 was deleted from Linda's strategy, we write

$$[\tfrac{3}{8} \quad \tfrac{5}{8} \quad 0]$$

and since column 2 was deleted, the opponent's strategy is

$$\begin{bmatrix} \frac{5}{8} \\ 0 \\ \frac{3}{8} \end{bmatrix}$$ ∎

COMPUTER APPLICATION   Program 14 on the computer disk accompanying this book will play two-person zero-sum games with you according to strategies you choose. You have the option of having the computer play an optimum strategy or a random strategy. By running this program you can gain intuitive insight into the material of this section.

Problem Set 9.2

Determine the optimal strategies and the value of the game in Problems 1–18.

1. $\begin{bmatrix} 4 & 0 \\ 3 & -2 \end{bmatrix}$ **2.** $\begin{bmatrix} 0 & -3 \\ -2 & 0 \end{bmatrix}$ **3.** $\begin{bmatrix} 1 & -1 \\ -1 & 1 \end{bmatrix}$

4. $\begin{bmatrix} 1 & -2 \\ 0 & 2 \end{bmatrix}$ **5.** $\begin{bmatrix} 3 & -2 \\ 1 & -3 \end{bmatrix}$ **6.** $\begin{bmatrix} 10 & 0 \\ -10 & 5 \end{bmatrix}$

7. $\begin{bmatrix} 5 & -1 \\ 3 & -2 \end{bmatrix}$ **8.** $\begin{bmatrix} 1.5 & -.5 \\ .5 & 2.5 \end{bmatrix}$ **9.** $\begin{bmatrix} -1 & 0 \\ \frac{1}{4} & -\frac{1}{4} \end{bmatrix}$

10. $\begin{bmatrix} 2 & 1 & 3 \\ -2 & 0 & 3 \\ 4 & -2 & -3 \end{bmatrix}$ 11. $\begin{bmatrix} 2 & 3 & 3 \\ 1 & 0 & -1 \\ 0 & 0 & 4 \end{bmatrix}$

12. $\begin{bmatrix} 2 & 1 & 3 \\ 0 & 1 & -2 \\ -1 & 0 & -3 \end{bmatrix}$ 13. $\begin{bmatrix} 4 & -2 & 3 \\ 3 & -3 & 1 \\ 2 & 0 & -1 \end{bmatrix}$

14. $\begin{bmatrix} -1 & -8 & -5 \\ 1 & 5 & -4 \\ -5 & 7 & 2 \end{bmatrix}$ 15. $\begin{bmatrix} 5 & -3 & -4 \\ 4 & -1 & 0 \\ 3 & -3 & 6 \end{bmatrix}$

16. $\begin{bmatrix} 0 & 1 & 2 & 0 \\ 1 & 2 & 2 & 1 \\ 3 & 0 & -1 & 0 \\ 2 & -1 & 0 & -2 \end{bmatrix}$ 17. $\begin{bmatrix} 2 & -1 & 4 & 3 \\ -4 & -5 & -2 & 1 \\ -1 & 0 & 3 & 0 \end{bmatrix}$

18. $\begin{bmatrix} 9 & 10 & 11 & 9 \\ 10 & 11 & 11 & 10 \\ 12 & 9 & 8 & 9 \\ 11 & 8 & 9 & 8 \end{bmatrix}$

APPLICATIONS

19. A friend suggests the following game: He will hide a quarter, dime, or half-dollar in one hand behind his back. You are to guess which coin he has. If you guess correctly, you get the coin. If you guess incorrectly, he gets the difference between your guess and the coin held. Write the game matrix and say if the game is strictly determined.

20. A friend suggests the following four-finger mora: If the sum is even, you win an amount equal to the sum of the fingers shown; if the sum is odd, your friend wins an amount equal to the sum of the fingers shown. Write the game matrix and say if the game is strictly determined.

21. A general with two regiments is trying to capture a city. He can attack from the north with both regiments, from the south with both regiments, or from the north with one regiment and from the south with one regiment. The defending forces also have two regiments that can be deployed for protection in the north, in the south, or one in the north and one in the south. The general of the attacking forces will gain 1 point if the attack succeeds, −1 if the attack fails, and 0 if the forces are held at a standoff. Whichever force has more regiments in an area wins the battle, and if one regiment is deployed against one regiment, the armies are held at a standoff. Write the game matrix and state whether this is a strictly determined game.

22. A friend proposes the following game: "We each flip an imaginary coin and call out heads or tails without knowing each other's choice. If there is a match, I win $1, and if there is no match, I lose $1." What are the strategies for the friend and her opponent, and what is the value of the game?

23. In a presidential campaign the Democratic and Republican candidates can campaign in either urban or rural areas. The units assigned to each choice are in gains or losses of thousands of votes:

		Republican	
		Urban	Rural
Democratic	Urban	−5	3
	Rural	4	2

What are the strategies, and what is the value of the game?

24. In submarine warfare, a submarine can attack enemy ships from shallow depths or can launch rockets toward enemy ships from deep water. The shallow-depth attack gives more accurate results but is also more dangerous for the submarine. The surface ships can also drop depth charges set for shallow detonation or drop charges set for deep detonation. If a submarine avoids the depth charges, we credit it with 50 points if it was deep and 80 points if it was shallow (since it is more effective if it is shallow). If it is hit with a depth charge, it is credited with −100 points. What are the strategies, and what is the value of the game?

25. In Sir Arthur Conan Doyle's story "The Final Problem," Sherlock Holmes is pursuing his archenemy Professor Moriarty. The professor is out to kill Holmes, whose only chance for escape is to flee to the Continent by taking a train from London to Dover. Just as the train is leaving London, the two men see each other and the professor is left at the station. Holmes knows that if he meets Moriarty again it means certain death. Let us assign this occurrence the value of −100 (for Holmes). Holmes can stay on the train until he reaches Dover or he can get off at Canterbury. What should he do? If he eludes Moriarty and makes it to Dover, we will assign this occurrence the value of 50 (for Holmes). If he eludes Moriarty but only reaches Canterbury, we will call this a draw and assign this occurrence the value of 0. On the other hand, Moriarty can catch Holmes by chartering a special train, but must decide whether to go to Canterbury or to Dover. Find the strategies for Holmes and Moriarty. Assume that the results are for repeated plays of the "game."

26. McDonald's is planning to build a restaurant in a large city. It can build uptown, downtown, or in the suburbs. Burger King also plans to build a restaurant in the same

city. The payoff matrix in thousands of dollars is:

$$
\begin{array}{c}
\text{Burger King} \\
\begin{array}{ccc}
\text{Uptown} & \text{Downtown} & \text{Suburbs}
\end{array}
\end{array}
$$

$$
\text{McDonald's}
\begin{array}{c}
\text{Uptown} \\
\text{Downtown} \\
\text{Suburbs}
\end{array}
\left[
\begin{array}{ccc}
4 & 1 & -3 \\
-2 & 0 & -5 \\
4 & 2 & 3
\end{array}
\right]
$$

What are the strategies, and what is the value of the game?

9.3 $m \times n$ Matrix Games*

We now discuss the solution to certain types of $m \times n$ matrix games. If the $m \times n$ matrix is strictly determined or if it can be reduced to a 2×2 game, we know how to find optimum strategies. Remember, the first steps in solving *any* matrix game are:

1. Check for a saddle point; if the matrix has one, then the game is strictly determined and the saddle point is the value of the game.
2. Eliminate dominated rows (eliminate the smaller row).
3. Eliminate dominating columns (eliminate the larger column).

The discussion that follows assumes that these steps are carried out first.

$2 \times n$ Matrix Games

Suppose we check for a saddle point and dominated rows or dominating columns and the resulting game matrix is a $2 \times n$ game, where n is some positive integer greater than 2. For a $2 \times n$ matrix game, player I has two strategies and player II has n strategies. For example, suppose Linda is playing the following game with an opponent:

$$
\text{Linda}
\begin{array}{c}
\text{Opponent} \\
\left[
\begin{array}{ccc}
5 & -1 & 1 \\
-1 & 3 & 0
\end{array}
\right]
\end{array}
$$

Recall that V is the value of the game if both Linda and her opponent use optimum strategies on repeated playing of the same game. Therefore, if Linda uses her optimal strategy and her opponent does not, Linda's winnings will be greater than or equal to V since positive entries indicate payoffs to her. On the other hand, if her opponent plays an optimum strategy and Linda does not, his winnings will be less than or equal to V since negative entries indicate payoffs to her opponent.

First consider the situation where we find Linda's optimum strategies first. We do this in the case where we have a $2 \times n$ matrix game. We denote the value of the game by V and denote her strategy by

$$
[p \quad 1 - p]
$$

---

* The material in this section is not required for the remainder of the text.

Next we calculate the payoffs if her opponent plays each of the three columns. Given

$$[p \quad 1-p]\begin{bmatrix} 5 & -1 & 1 \\ -1 & 3 & 0 \end{bmatrix}$$

for the first column,

$$[p \quad 1-p]\begin{bmatrix} 5 \\ -1 \end{bmatrix} = 5p + (-1)(1-p) = 5p - 1 + p = 6p - 1$$

For the second column,

$$[p \quad 1-p]\begin{bmatrix} -1 \\ 3 \end{bmatrix} = -p + 3(1-p) = -p + 3 - 3p = -4p + 3$$

For the third column,

$$[p \quad 1-p]\begin{bmatrix} 1 \\ 0 \end{bmatrix} = p$$

These must all be greater than or equal to V since we wish to make Linda's payoffs as great as possible:

$$6p - 1 \geq V$$
$$-4p + 3 \geq V$$
$$p \geq V$$

We wish to maximize V subject to the inequalities above (also remember that $p \geq 0$). We do this graphically as shown in Figure 9.2. The color portion shows the simultaneous solution of the system of inequalities and represents Linda's possible winnings, depending on the column her opponent plays. We see that we can *maximize* V at the intersection of the lines

$$p = V \quad \text{and} \quad -4p + 3 = V$$

This point of intersection occurs when $p = \frac{3}{5}$ and $V = \frac{3}{5}$; that is, at the point $(\frac{3}{5}, \frac{3}{5})$. Therefore Linda's optimal strategy is when $p_1 = \frac{3}{5}$ and $p_2 = 1 - p = \frac{2}{5}$ or

$$[\frac{3}{5} \quad \frac{2}{5}]$$

and the value of the game is $\frac{3}{5}$.

The optimal strategy for Linda's opponent is found by again considering the game matrix

$$\begin{bmatrix} 5 & -1 & 1 \\ -1 & 3 & 0 \end{bmatrix}$$

Her opponent can see that Linda's optimum strategy is determined by the intersection of the lines from the second and third columns, so he will concentrate his attention on these two columns:

$$\begin{bmatrix} -1 & 1 \\ 3 & 0 \end{bmatrix}$$

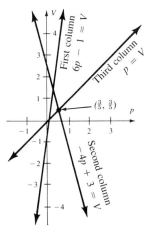

Figure 9.2

$$\begin{bmatrix} -1 & 1 \\ 3 & 0 \end{bmatrix}\begin{bmatrix} q \\ 1-q \end{bmatrix} = \begin{bmatrix} -q + 1 - q \\ 3q + 0 \end{bmatrix} = \begin{bmatrix} 1 - 2q \\ 3q \end{bmatrix}$$

Thus

$$1 - 2q = 3q$$
$$q = \tfrac{1}{5}$$

So Linda's opponent's strategy for the subgame is

$$\begin{bmatrix} \tfrac{1}{5} \\ \tfrac{4}{5} \end{bmatrix}$$

Therefore, for the original game, his strategy is

$$\begin{bmatrix} 0 \\ \tfrac{1}{5} \\ \tfrac{4}{5} \end{bmatrix}$$

$m \times 2$ Matrix Games

When Linda is playing an $m \times 2$ matrix game we use a similar method of solution except that for this game we *first* find the strategies for her opponent. For an $m \times 2$ matrix game, player I has m strategies and player II has two strategies. Consider the following game:

$$\begin{array}{c} \qquad\qquad \text{Opponent} \\ \text{Linda} \begin{bmatrix} 1 & 2 \\ 3 & -2 \\ -3 & 3 \end{bmatrix} \end{array}$$

Linda's opponent wants to play a strategy that results in a payoff that is as small as possible. If the value of the game is denoted by V and her opponent's strategy is

$$\begin{bmatrix} q \\ 1 - q \end{bmatrix}$$

where $q_1 = q$ and $q_2 = 1 - q$, we calculate her opponent's payoffs for each of the three rows that Linda might play.

$$\text{Row 1:} \quad \begin{bmatrix} 1 & 2 \end{bmatrix} \begin{bmatrix} q \\ 1 - q \end{bmatrix} = q + 2(1 - q) = -q + 2$$

$$\text{Row 2:} \quad \begin{bmatrix} 3 & -2 \end{bmatrix} \begin{bmatrix} q \\ 1 - q \end{bmatrix} = 3q - 2(1 - q) = 5q - 2$$

$$\text{Row 3:} \quad \begin{bmatrix} -3 & 3 \end{bmatrix} \begin{bmatrix} q \\ 1 - q \end{bmatrix} = -3q + 3(1 - q) = -6q + 3$$

These must be less than or equal to V since her opponent wants to make the payoffs as small as possible:

$$-q + 2 \leq V$$
$$5q - 2 \leq V$$
$$-6q + 3 \leq V$$

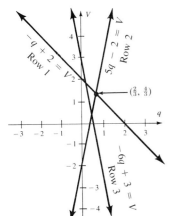

Figure 9.3

The value V is minimized subject to the inequalities above and $q \geq 0$, as shown in Figure 9.3.

Linda's opponent wishes to *minimize* V so we see that the color area is minimized at the point of intersection of the lines

$$-q + 2 = V \quad \text{and} \quad 5q - 2 = V$$

The point of intersection is $(\frac{2}{3}, \frac{4}{3})$ and the optimal strategy for her opponent is

$$\begin{bmatrix} \frac{2}{3} \\ \frac{1}{3} \end{bmatrix}$$

For Linda's strategy, we concentrate on the first and second rows (why?), and solve the matrix game

$$\begin{bmatrix} 1 & 2 \\ 3 & -2 \end{bmatrix}$$

The solution is found by multiplying

$$[p \quad 1-p]\begin{bmatrix} 1 & 2 \\ 3 & -2 \end{bmatrix} = [p + 3(1-p) \quad 2p - 2(1-p)]$$
$$= [-2p + 3 \quad 4p - 2]$$

We equate the entries and solve to obtain:

$$-2p + 3 = 4p - 2$$
$$-6p = -5$$
$$p = \tfrac{5}{6}$$

Linda's strategy for the subgame is $[\frac{5}{6} \quad \frac{1}{6}]$ and for the original game is $[\frac{5}{6} \quad \frac{1}{6} \quad 0]$. The value of the game is $\frac{8}{6} = \frac{4}{3}$.

m × *n* Matrix Games

The general solution to a 3 × 3, or even higher-order game, is a problem in linear programming and is beyond the scope of this course.

COMPUTER APPLICATION Higher-order games are not beyond the scope of this course if you have access to a computer. For this reason, they are discussed in the computer supplement *Computer-Aided Finite Mathematics* (Chapter 14). The rudimentary techniques necessary are contained in Chapters 3 and 4 on linear programming and in the earlier parts of this chapter. What you do for the first player is to multiply the strategy vector times the game matrix and then set each of these greater than or equal to the value of the game, V, and then maximize V by using the simplex method.

A word of caution is in order: The theory of games is still being developed, and at the present time we can solve only the simplest types of games. Most practical problems in game theory are too complicated and remain yet unsolved. The difficulty is in constructing a mathematical model that adequately accounts for all the interrelationships that may be present in a real game between buyer and seller, competing companies, people and nature, and so on. Nevertheless, game theory can, and does, help us begin to intelligently analyze conflict situations.

Problem Set 9.3

Find the optimum strategies and game values for the games in Problems 1–12.

1. $\begin{bmatrix} 2 & 1 \\ -3 & 2 \\ 1 & -4 \end{bmatrix}$

2. $\begin{bmatrix} 0 & -2 \\ 2 & 1 \\ -3 & -1 \end{bmatrix}$

3. $\begin{bmatrix} 1 & -1 \\ 0 & 1 \\ -1 & 0 \\ 1 & -2 \end{bmatrix}$

4. $\begin{bmatrix} 4 & 3 \\ -1 & 2 \\ -5 & 6 \\ 2 & -3 \end{bmatrix}$

5. $\begin{bmatrix} 1 & 4 & -3 \\ -2 & 1 & 2 \end{bmatrix}$

6. $\begin{bmatrix} -2 & 1 & 3 \\ 1 & -3 & 4 \end{bmatrix}$

7. $\begin{bmatrix} 1 & 0 & 3 \\ -3 & 2 & -4 \\ 2 & 3 & 1 \end{bmatrix}$

8. $\begin{bmatrix} 1 & -2 & 3 \\ 2 & -3 & 2 \\ 4 & 1 & -3 \end{bmatrix}$

9. $\begin{bmatrix} 2 & 5 & -4 \\ 1 & 0 & -1 \\ 7 & -2 & -3 \end{bmatrix}$

10. $\begin{bmatrix} 4 & -3 & 1 \\ 2 & -1 & 0 \\ 3 & -2 & 1 \end{bmatrix}$

11. $\begin{bmatrix} 5 & 4 & 2 & 1 \\ 1 & -2 & 4 & 3 \\ 4 & 3 & 2 & 2 \\ 1 & 0 & -1 & -2 \end{bmatrix}$

12. $\begin{bmatrix} -3 & 1 & 0 & -2 \\ 2 & 3 & -4 & 1 \\ 1 & -1 & -5 & 0 \\ -2 & 2 & 1 & -1 \end{bmatrix}$

APPLICATION

13. A general with two regiments is trying to capture a city. He can attack from the north with both regiments, from the south with both regiments, or from the north with one regiment and from the south with one regiment. The defending forces also have two regiments that can be deployed for protection in the north, in the south, or one in the north and one in the south. The general of the attacking forces will gain 1 point if the attack succeeds, −1 if the attack fails, and 0 if the forces are held at a standoff. Whichever force has more regiments in an area wins the battle, and if one regiment is deployed against one regiment, the armies are held at a standoff. Find the generals' strategies.

9.4 Summary and Review

IMPORTANT TERMS

Decision theory   [9.1]
Dominated rows   [9.2]
Dominating columns   [9.2]
Dominating principle   [9.2]
Expected value   [9.1]
Fair game   [9.1, 9.2]
Game matrix   [9.2]
Game theory   [9.2]
Mathematical expectation   [9.1]
Minimax theorem   [9.2]

Mixed strategy   [9.2]
Nonstrictly determined game   [9.2]
Operations research   [9.1]
Optimal strategy theorem   [9.2]
Saddle point   [9.2]
Strategy   [9.2]
Strictly determined game   [9.2]
Value of a game   [9.2]
Zero-sum game   [9.2]

SAMPLE TEST

For additional practice there are a large number of review problems categorized by objective in the Student Solutions Manual. The following sample test (40 minutes) is intended to review the main ideas of the chapter.

Find the optimal strategies for the games in Problems 1–5. Also find the value of each game, and say if the game is not strictly determined.

1. $\begin{bmatrix} 3 & 5 \\ 2 & -5 \end{bmatrix}$
2. $\begin{bmatrix} 2 & 1 \\ 0 & 1 \end{bmatrix}$
3. $\begin{bmatrix} 4 & -5 \\ 2 & 3 \end{bmatrix}$

4. $\begin{bmatrix} 50 & 20 & -20 \\ -10 & 0 & -50 \\ 40 & 20 & 30 \end{bmatrix}$
5. $\begin{bmatrix} 2 & 1 & 3 \\ 0 & 1 & -2 \\ -1 & 2 & -3 \end{bmatrix}$

6. A contest offers a $50,000 first prize, five second prizes worth $5,000 each, and ten third prizes worth $500 each. What is the expectation for this contest if there are one million entries?

7. Find the expected value of X:

X	0	1	2	3	4
$P(X)$	.4	.3	.2	.1	0

APPLICATIONS

8. A contractor must choose one of three contracts on which to bid. The costs and profits for the bids are:

Contract	Preparation costs	Profit	Probability of winning contract
I	$100	$10,000	.9
II	$5,000	$100,000	.1
III	$4,500	$135,000	.1

If the contractor only has time to submit one bid, which should it be? Base your answer on mathematical expectation and remember to subtract preparation costs from the profits when making your decision.

9. The daily demand at a car rental agency is between 20 and 25 cars as follows:

Number of cars rented	20	21	22	23	24	25
Probability	.1	.2	.3	.2	.1	.1

If it costs the agency $10 per day whether the car is rented or not, and if the net profit is $30 per day on the days the car is rented, what is the optimal number of cars that the agency should have on hand?

10. Suppose you are producing two types of cereal. One type appeals to children and the other with granola is for adults. Your competitor markets five varieties of cereals, and the resulting gain or loss in thousands of dollars of sales is shown below:

	Competitor's cereals				
	A	B	C	D	E
Children	20	10	2	0	-6
Adults	4	-2	4	10	1

If you must choose to manufacture only one type each month and your competitor can also choose only one type each month, what are your best strategies, and what is the value of this game?

The following multiple-choice questions dealing with decision theory are from actual CPA exams. They are reprinted with permission of the American Institute of Certified Public Accountants, Inc.

CPA Exam
November 1980

11. Duguid Company is considering a proposal to introduce a new product, XPL. An outside marketing consultant prepares the following table describing the relative likelihood of monthly sales volume levels and related income (loss) for XPL:

Monthly sales volume	Probability	Income (loss)
$3,000	.10	$(35,000)
$6,000	.20	$5,000
$9,000	.40	$30,000
$12,000	.20	$50,000
$15,000	.10	$70,000

If Duguid decides to market XPL, the expected value of the added monthly income will be

A. $24,000   **B.** $26,500   **C.** $30,000   **D.** $120,000

CPA Exam
May 1976

12. Your client wants your advice on which of two alternatives he should choose. One alternative is to sell an investment now for $10,000. Another alternative is to hold the investment 3 days, after which he can sell it for a certain selling price based on the following probabilities:

Selling price	Probability
$5,000	.4
$8,000	.2
$12,000	.3
$30,000	.1

Using probability theory, which of the following is the most reasonable statement?

A. Hold the investment 3 days because the expected value of holding exceeds the current selling price.

B. Hold the investment 3 days because of the chance of getting $30,000 for it.

C. Sell the investment now because the current selling price exceeds the expected value of holding.

D. Sell the investment now because there is a 60% chance that the selling price will fall in 3 days.

Modeling Application 9

Overbooking of Airline Flights

A practical problem for an airline company is how many reservations should be accepted for a particular flight. If everyone who makes a reservation shows up at the airport, the solution would be easy. The airline should simply take no more reservations than there are seats on the airplane. However, many people with reservations do not show up at the airport, so empty seats result. These could be filled if more reservations were accepted. One solution would seem to be for the airline company to accept as many reservations as possible—that is, not to impose a limit on the number of reservations accepted. However, there is a penalty if someone with a reservation shows up at the airport but is unable to board the plane because all seats are taken. Specifically, the "denied boarding compensation" rule requires that the airline company must fly the person free on another flight. For occasions when no one volunteers to fly on a later flight (that is, to be bumped), some airline companies are now offering coupons worth more than the cost of the flight as an incentive for travelers to give up their seats when necessary.*

Write a paper discussing a model that would decide how many reservations should be accepted for a particular flight. (The airline company may also have a public relations problem if too many passengers with reservations cannot fly. However, this should not be considered in the model you build for this case study.) For general guidelines about writing this essay, see the commentary for Modeling Application 1 on page 36.

---

* This modeling application is adapted from Joe Dan Austin, "Overbooking Airline Flights," *The Mathematics Teacher*, March 1982, pp. 221–223.

CHAPTER 10
Mathematics of Finance

APPLICATIONS

Management (*Business, Economics, Finance, and Investments*)

Salary increases (10.1, Problems 49–50)

Finding the effective yield of an investment (10.2, Problems 32–35)

Calculating the effect of inflation (10.2, Problems 41–52; 10.5, Problem 23; 10.6, Problem 7; 10.7, Problem 15)

Present value problem (10.2, Problems 36–40, 53–58; 10.6, Problem 11)

Setting up a retirement fund (10.3, Problems 26–27; 10.7, Problems 11–12)

Value of Social Security deposits (10.3, Problems 28–29)

Value of an insurance annuity (10.4, Problems 29–30)

Sinking funds to pay off notes (10.5, Problems 13–16, 21; 10.6, Problems 13–14)

Repayment of a bond (10.5, Problem 28; 10.6, Problem 17)

Value of a savings certificate (10.6, Problem 1)

Value of a lump-sum insurance payment (10.6, Problems 3, 4)

Management accounting examination (10.6, Problem 18)

CPA examination (10.6, Problems 19–20; 10.7, Problem 6)

Life sciences (*Biology, Ecology, Health, and Medicine*)

Cell division (10.1, Problem 53)

Price of wheat as determined by last year's production (10.1, Problem 54)

Social sciences (*Demography, Political Science, Population, Psychology, Society, and Sociology*)

Breaking news story (10.1, Problem 51)

Spread of rumors (10.1, Problem 52)

Interception of a secret courier (10.7, Problem 9)

General interest

Values of an insurance policy (10.4, Problem 30)

Calculating total interest paid for a consumer's loan (10.4, Problems 25–26)

Present value of a lottery prize (10.4, Problem 28; 10.6, Problem 15)

Saving for a purchase (10.5, Problems 18, 21, 24–26; 10.7, Problem 14)

Investing an inheritance (10.5, Problem 19)

Savings account for a child (10.5, Problem 20)

Modeling application—Investment Opportunities

CHAPTER OVERVIEW   The notion of simple interest is introduced, which leads directly to the concept of compound interest. This chapter uses a unified notation, so you should learn the meanings of the variables used early in the chapter, especially the difference between present value, P, and future value, A. Pay special attention to the relationship between annual rate, r, and the rate per period, i; and to the total time, t (in years), and the number of periods, N. One of the important skills to be learned in this chapter is the ability to look at a particular situation and decide what type of financial formula applies (that is, what model to use).

PREVIEW   One of the cornerstones upon which business is built is the payment of interest for the temporary use of another's money. In this chapter we discuss simple and compound interest, present value, annuities, and sinking funds. We will be concerned with comparing the present value, P, with its value, A, at some future time. We can make payments or deposits in two ways: deposit a lump sum or make periodic payments. Consider the following simplified chapter overview:

	Present value, P	Future value, A	Method for finding unknown value
Lump-sum payment or deposit	Known	Unknown	Compound interest
	Unknown	Known	Present value
Periodic payment	Known	Unknown	Annuity
	Unknown	Known	Sinking fund

PERSPECTIVE   Even though this is the last chapter of the book, it is one of the most important. Intelligent handling of money goes hand in hand with financial success. While understanding compound interest, annuities, amortization, and sinking funds does not guarantee success, such understanding is necessary when handling money and investments.

10.1 Difference Equations

One of the most fundamental mathematical applications for business people, as well as for consumers, is that of finances and financial formulas. In order to derive many of the financial equations we will use in this chapter, you will first need to understand the notion of a difference equation.

An infinite set of numbers $x_0, x_1, x_2, x_3, \ldots$ is called a sequence or progression. For example, the even numbers

$$2, 4, 6, 8, 10, \ldots$$

form a sequence. The individual numbers are called *terms* of the sequence. The number x_0 is the first value of the sequence and is called the initial value. In terms of investments and interest, it is usually the amount of money you have now (initially). The number x_1 is the second value. Think of the subscript as a measurement of time—the second value of the sequence, x_1, is the value at the end of the first year. Thus, x_5 is the value at the end of the fifth year (which is the sixth term of the sequence).

Sequences are usually defined by giving what is called the *general term*, which is a formula such as

$$x_n = 3n + 2$$

where $n = 0, 1, 2, 3, 4, \ldots$. You can then find any term by evaluation:

$$x_0 = 3(0) + 2 = 2$$
$$x_1 = 3(1) + 2 = 5$$
$$x_2 = 3(2) + 2 = 8$$
$$x_3 = 3(3) + 2 = 11$$

Thus, the sequence is 2, 5, 8, 11, ...

Although sequences are used throughout mathematics in a variety of contexts and applications, we will use them in connection with financial formulas by considering general terms of the type

$$x_{n+1} = ax_n + b$$

where a and b are constants. An equation of this type is called a **first-order linear difference equation**.

EXAMPLE 1   Find the first four terms of the sequence that satisfies the difference equation

$$x_{n+1} = 6x_n + 50$$

and has an initial value of 100.

Solution   Given $x_0 = 100$; then

$$x_1 = 6x_0 + 50 = 6(100) + 50 = 650$$
$$x_2 = 6x_1 + 50 = 6(650) + 50 = 3950$$
$$x_3 = 6x_2 + 50 = 6(3950) + 50 = 23{,}750$$

The sequence is 100, 650, 3950, 23,750, ...   ∎

A **solution** of a linear difference equation is a sequence in which the successive terms will satisfy the equation. That is, a solution of $x_{n+1} = ax_n + b$ is completely determined when a, b, and x_0 are all known. The difference between finding the solution and finding the general term is that the solution allows us to calculate any term without first calculating all the preceding terms. Our goal, therefore, is to find this solution for the first-order linear difference equation. First, however, we will solve two special cases, called *arithmetic* and *geometric* sequences.

Arithmetic Sequence

If $a = 1$, then the general difference equation becomes

$$x_{n+1} = x_n + b$$

EXAMPLE 2 Let $b = 2$ in the difference equation

$$x_{n+1} = x_n + 2$$

with initial value 1. Find the first four terms of this sequence.

Solution This difference equation is $x_{n+1} = x_n + 2$ with $x_0 = 1$:

$$x_0 = 1$$
$$x_1 = x_0 + 2 = 1 + 2 = 3$$
$$x_2 = x_1 + 2 = 3 + 2 = 5$$
$$x_3 = x_2 + 2 = 5 + 2 = 7$$

The sequence is 1, 3, 5, 7, ... ∎

A difference equation with $a = 1$, namely

$$x_{n+1} = x_n + b$$

is called an **arithmetic sequence** or **progression**. It is characterized by the fact that there is a *common difference*, b. Notice other terms of the sequence are

$$x_0$$
$$x_1 = x_0 + b$$
$$x_2 = x_1 + b = x_0 + b + b = x_0 + 2b$$
$$x_3 = x_2 + b = x_0 + 2b + b = x_0 + 3b$$
$$x_4 = x_3 + b = x_0 + 3b + b = x_0 + 4b$$
$$\vdots$$
$$x_n = x_0 + nb$$

Arithmetic Sequence

> The solution of the arithmetic sequence with first term x_0 and common difference b is
> $$x_n = x_0 + nb$$

EXAMPLE 3 Find the x_{10} of the arithmetic sequence whose first term is -3 and whose common difference is 4.

Solution $x_0 = -3; b = 4$, so

$$x_n = x_0 + nb$$
$$x_{10} = -3 + 10(4) = 37$$ ∎

Geometric Sequence

If $b = 0$, then the general difference equation becomes

$$x_{n+1} = ax_n$$

EXAMPLE 4 Find the first four terms of the sequence $x_{n+1} = ax_n$ for the initial value of 1 and $a = 2$.

Solution The difference equation is $x_{n+1} = 2x_n$:

$$x_0 = 1$$
$$x_1 = 2x_0 = 2(1) = 2$$
$$x_2 = 2x_1 = 2(2) = 4$$
$$x_3 = 2x_2 = 2(4) = 8$$
$$\vdots$$

The sequence is 1, 2, 4, 8, ... ∎

A difference equation with $b = 0$, namely

$$x_{n+1} = ax_n$$

is called a **geometric sequence** or **progression**. It is characterized by the fact that there is a *common ratio, a*. Notice the terms of the sequence are

$$x_0$$
$$x_1 = ax_0$$
$$x_2 = ax_1 = a(ax_0) = a^2x_0$$
$$x_3 = ax_2 = a(a^2x_0) = a^3x_0$$
$$x_4 = ax_3 = a(a^3x_0) = a^4x_0$$
$$\vdots$$
$$x_n = a^nx_0$$

Geometric Sequence

> The solution of the geometric sequence with first term x_0 and common ratio a is
> $$x_n = a^nx_0$$

EXAMPLE 5 Find the x_5 of the geometric sequence whose first term is 100 and whose common ratio is $\frac{1}{2}$.

Solution $x_0 = 100$, $a = \frac{1}{2}$, so

$$x_n = a^nx_0$$
$$x_5 = (\tfrac{1}{2})^5(100) = 3.125$$ ∎

General Solution of a Difference Equation

We can now find the general solution of the difference equation

$$x_{n+1} = ax_n + b$$

Begin by looking for a pattern. Let x_0 be the initial value.

$$x_1 = ax_0 + b$$
$$x_2 = ax_1 + b = a(ax_0 + b) + b = a^2x_0 + ab + b$$
$$x_3 = ax_2 + b = a(a^2x_0 + ab + b) = a^3x_0 + a^2b + ab + b$$
$$x_4 = ax_3 + b = a(a^3x_0 + a^2b + ab + b) + b = a^4x_0 + a^3b + a^2b + ab + b$$
$$\vdots$$
$$x_n = a^nx_0 + a^{n-1}b + a^{n-2}b + \cdots + a^2b + ab + b$$

Even though this is the general solution, it is not particularly useful because of its algebraic form, so we now change the algebraic form so that it is easier to use.

$$x_n = a^nx_0 + a^{n-1}b + a^{n-2}b + \cdots + a^2b + ab + b$$
$$= a^nx_0 + b(a^{n-1} + a^{n-2} + \cdots + a^2 + a + 1)$$

Now, consider the sum

$$S_n = 1 + a + a^2 + \cdots + a^n$$

and

$$aS_n = a + a^2 + a^3 + \cdots + a^{n+1} \qquad \text{Multiply both sides by } a$$
$$S_n - aS_n = 1 - a^{n+1} \qquad \text{Subtract (notice the arrows in the lines above; all of those terms are zero when you subtract)}$$

$$(1 - a)S_n = 1 - a^{n+1} \qquad \text{Factor}$$
$$S_n = \frac{1 - a^{n+1}}{1 - a} \qquad a \neq 1$$

Thus,

$$a^{n-1} + a^{n-2} + \cdots + a^2 + a + 1 = \frac{1 - a^{(n-1)+1}}{1 - a}$$
$$= \frac{1 - a^n}{1 - a}$$

The general solution is now stated (by substitution in the second step)

$$x_n = a^nx_0 + b(a^{n-1} + a^{n-2} + \cdots + a^2 + a + 1)$$
$$x_n = a^nx_0 + b\left(\frac{1 - a^n}{1 - a}\right)$$
$$= a^nx_0 + \frac{b}{1 - a}(1 - a^n)$$
$$= a^nx_0 + \frac{b}{1 - a} - \frac{b}{1 - a}a^n$$
$$= \frac{b}{1 - a} + a^nx_0 - \frac{b}{1 - a}a^n$$
$$= \frac{b}{1 - a} + \left(x_0 - \frac{b}{1 - a}\right)a^n$$

General Solution of a Difference Equation

> The difference equation $x_{n+1} = ax_n + b$ with $a \neq 1$ has solution
> $$x_n = \frac{b}{1-a} + \left(x_0 - \frac{b}{1-a}\right)a^n$$

EXAMPLE 6 Solve each difference equation and find x_4 for each.

a. $x_{n+1} = 5x_n + 8$, $x_0 = 1$ **b.** $x_{n+1} = 3x_n$, $x_0 = 4$
c. $x_{n+1} = x_n + 4$, $x_0 = 3$

Solution **a.** $a = 5$, $b = 8$, $x_0 = 1$, so using the general solution,

$$\frac{b}{1-a} = \frac{8}{1-5} = \frac{8}{-4} = -2$$

$$x_n = -2 + (1+2)5^n$$
$$= -2 + 3 \cdot 5^n \qquad \text{This is the solution}$$

Also, $x_4 = -2 + 3 \cdot 5^4 = 1873$

b. $a = 3$, $b = 0$, $x_0 = 4$, so using the general solution,

$$\frac{b}{1-a} = \frac{0}{1-3} = 0$$

$$x_n = 0 + (4+0)3 = 4 \cdot 3^n \qquad \text{This is the solution}$$

Notice, however, since $b = 0$, this is a geometric sequence, so using the theorem for geometric sequences we find (directly)

$$x_n = 4 \cdot 3^n$$

Also, $x_4 = 4 \cdot 3^4 = 324$

c. $a = 1$, $b = 4$, $x_0 = 3$. This is an arithmetic sequence. From the theorem for arithmetic sequences, we find (directly)

$$x_n = 3 + 4n$$

Also, $x_4 = 3 + 4 \cdot 4 = 19$ ∎

EXAMPLE 7 Suppose the average beginning salary for a recent college graduate is $25,000 and that this graduate can expect annual salary increases of $1,000 a year plus a 5% cost-of-living increase. What is the graduate's salary for the fifth year?

Solution Let x_n be the salary in the nth year. Now,

$$x_n = \text{last year's salary} + .05\,(\text{last year's salary}) + 1000$$
$$= x_{n-1} + .05x_{n-1} + 1000$$
$$= 1.05x_{n-1} + 1000$$

Note that the general solution is given for a difference equation of the form $x_{n+1} = ax_n + b$ and we have modeled this example in terms of x_n (instead of x_{n+1}):

$$x_n = 1.05x_{n-1} + 1000$$

These equations are the same since the relationship between x_n and x_{n+1} is the same as that between x_{n+1} and x_n (that is, one year change). Therefore, we can use the general solution formula for a difference equation where $a = 1.05$, $b = 1000$, and $x_0 = 25,000$.

$$x_n = \frac{1000}{1 - 1.05} + \left(25,000 - \frac{1000}{1 - 1.05}\right)(1.05)^n$$

In particular, we want $n = 5$ (fifth year):

$$x_5 = \frac{1000}{-.05} + \left(25,000 - \frac{1000}{-.05}\right)(1.05)^5 \approx 37,432.67$$

The graduate's salary in five years should be $37,433.   ∎

Sometimes models using difference equations are developed using the idea of **proportionality**. We say that two quantities are proportional if one is equal to a constant times the other. For example, if x and y are proportional then $x = ky$ for some constant k, which is called the *constant of proportionality*.

EXAMPLE 8 Suppose that a major news story breaks on the four national television networks at the same time. The number of people learning the news each hour after it broke is proportional to the number who have not heard it by the end of the preceding hour. If we assume that the population is 220 million, how many people would hear of the news 24 hours after it broke if the constant of proportionality is 0.2?

Solution Let x_n = number of people who have heard the news after n hours. If P is the population (220 million in this example), then the number of people who have not heard the news after n hours is $P - x_n$. Thus,

$$\begin{pmatrix} \text{number who have} \\ \text{heard the news} \\ \text{after } n + 1 \text{ hours} \end{pmatrix} = \begin{pmatrix} \text{number who know} \\ \text{after } n \text{ hours} \end{pmatrix} + \begin{pmatrix} \text{number who learn} \\ \text{news during the} \\ (n + 1)\text{st hour} \end{pmatrix}$$

$$x_{n+1} \qquad = \qquad x_n \qquad + \qquad k(P - x_n)$$

For this example, $k = .2$ and $P = 220$:

$$x_{n+1} = x_n + 0.2(220 - x_n) = x_n + 44 - 0.2x_n = 0.8x_n + 44$$

If we assume that no one heard the news initially ($x_0 = 0$) we can find the solution where $a = 0.8$, $b = 44$, and $x_0 = 0$:

$$x_n = \frac{b}{1-a} + \left(x_0 - \frac{b}{1-a}\right)a^n = \frac{44}{1 - 0.8} + \left(0 - \frac{44}{1 - 0.8}\right)(0.8)^n$$

$$= 220 - 220(0.8)^n$$

For $n = 24$,

$$x_{24} = 220 - 220(0.8)^{24} \approx 219$$

Thus, after one day (24 hours) approximately 1 million people would not have heard the news.   ∎

Sigma Notation

Sometimes we want to write a sum of terms of a sequence using a shorthand notation. This notation uses the uppercase Greek letter sigma, so is often referred to as **sigma notation**. For example, we wrote

$$S_n = 1 + a + a^2 + a^3 + \cdots + a^n$$

and using sigma notation this sum could be written as

$$S_n = \sum_{k=0}^{n} a^k \quad \text{since} \quad \sum_{k=0}^{n} a^k = 1 + a + a^2 + a^3 + \cdots + a^n$$

In other words, the sigma notation evaluates the expression immediately following the sigma (a^k in this example) first for $k = 0$, then for $k = 1$, next for $k = 2$, and so on, where k counts up (one unit at a time) until it reaches the final value of n. The value 0 in this example is called the *lower limit of summation* and n the *upper limit of summation*, while the variable k is called the *index of summation*.

EXAMPLE 9   Find $\sum_{i=1}^{4} i(i + 3)$.

Solution   We have $\sum_{i=1}^{4} i(i + 3) = \overbrace{1(1 + 3)}^{i = 1} + \overbrace{2(2 + 3)}^{i = 2} + \overbrace{3(3 + 3)}^{i = 3} + \overbrace{4(4 + 3)}^{i = 4}$

$$= 1 \cdot 4 + 2 \cdot 5 + 3 \cdot 6 + 4 \cdot 7$$
$$= 4 + 10 + 18 + 28$$
$$= 60 \qquad \blacksquare$$

Problem Set 10.1

Compute the first four terms of the solution to the linear difference equations given in Problems 1–12.

1. $x_{n+1} = x_n + 10; x_0 = 6$

2. $x_{n+1} = x_n - 5; x_0 = 3$

3. $x_{n+1} = 4x_n; x_0 = 1$

4. $x_{n+1} = 10x_n; x_0 = 0$

5. $x_{n+1} = -5x_n; x_0 = 10$

6. $x_{n+1} = 3x_n; x_0 = 4$

7. $x_{n+1} = 2x_n + 3; x_0 = 4$

8. $x_{n+1} = -3x_n - 4; x_0 = 5$

9. $y_{n+1} = 2y_n + 3; y_0 = 1$

10. $y_{n+1} = 3y_n - 2; y_0 = 4$

11. $y_{n+1} = 5y_n - 2; y_0 = 3$

12. $y_{n+1} = 4y_n + 3; y_0 = 2$

Find x_4 for each difference equation given in Problems 13–24.

13. $x_{n+1} = x_n + 8; x_0 = 0$

14. $x_{n+1} = x_n + 100; x_0 = 100$

15. $x_{n+1} = 3x_n; x_0 = 1$

16. $x_{n+1} = 2x_n; x_0 = 1$

17. $x_n = (\frac{1}{2})x_{n-1} + 2; x_0 = 100$

18. $x_n = (\frac{1}{10})x_{n-1} + 2; x_0 = 1000$

19. $y_{n+1} = 5y_n + 2; y_0 = 0$

20. $y_{n+1} = 2y_n - 3; y_0 = 10$

21. $y_{n+1} = 2y_n + 1; y_0 = 8$

22. $y_{n+1} = 1 - 2y_n; y_0 = 0$

23. $y_{n+1} = 1 - (\frac{1}{2})y_n; y_0 = 0$

24. $y_{n+1} = 10 - (\frac{1}{10})y_n; y_0 = 0$

Solve the difference equations in Problems 25–40.

25. $x_{n+1} = x_n + 25, x_0 = 5$

26. $x_{n+1} = x_n - 20, x_0 = 20$

27. $x_{n+1} = x_n - 2, x_0 = 4$

28. $x_{n+1} = x_n + 1, x_0 = 0$

29. $x_n = x_{n-1} + 4, x_0 = 0$

30. $x_n = x_{n-1} - 5, x_0 = 5$

31. $x_n = 3x_{n-1}, x_0 = 1$

32. $x_n = (\frac{1}{5})x_{n-1}, x_0 = 5$

33. $x_n = 2x_{n-1}, x_0 = 1$

34. $x_n = 4x_{n-1}, x_0 = 1$

35. $x_n = 2x_{n-1} - 3, x_0 = 0$

36. $x_n = -2x_{n-1} + 15, x_0 = 0$

37. $x_n = 5x_{n-1} + 9, x_0 = 2$

38. $x_n = -3x_{n-1} + 5, x_0 = 1$

39. $x_n = 4x_{n-1} + 2, x_0 = 1$

40. $x_n = 3x_{n-1} - 4, x_0 = 10$

In Problems 41–48 evaluate the given expression.

41. $\sum_{k=2}^{6} k$ **42.** $\sum_{m=1}^{4} m^2$ **43.** $\sum_{n=0}^{6} (2n + 1)$

44. $\sum_{k=2}^{5} (10 - 2k)$ **45.** $\sum_{k=1}^{5} (-2)^{k-1}$ **46.** $\sum_{k=0}^{4} 3(-2)^k$

47. $\sum_{k=0}^{3} 2(3^k)$ **48.** $\sum_{k=2}^{5} (100 - 5k)$

APPLICATIONS

49. Suppose a job pays a starting salary of $20,000 with a $500 per year increase and a 3% cost-of-living increase. What will the salary be in the tenth year?

50. Suppose a job pays a starting salary of $38,000 with a $2,500 per year increase and a 5% cost-of-living increase. What will the salary be in the fifth year?

51. Suppose a news story breaks in a town of 200,000 and the number of people learning of the news after the story broke is proportional to the number who have not heard it by the end of the preceding hour. How many will have heard the story in 8 hours if the constant of proportionality is 0.3?

52. Suppose the number of persons in a company of 3000 employees who have heard a particular piece of gossip is proportional to the number of persons who have not heard it by the end of the previous day. How many will have heard the gossip one week after the rumor started if the constant of proportionality is 0.4?

53. Suppose a cell divides every 20 minutes. If there is initially one cell, how many cells will there be in 8 hours?

54. According to the Department of Agriculture (Statistical Reporting Service), the U.S. wheat production was approximately 2.6 billion bushels. It is known that the current wheat crop affects next year's level of production and next year's price. Let p_n denote the price of wheat (in dollars per bushel) and q_n denote the quantity of wheat produced (in billions of bushels), and suppose that p_n and q_n are related by the equations

$$p_n = 10 - .01q_n$$
$$q_{n+1} = .75p_n - 5$$

Solve the difference equation for production.

10.2 Interest

One of the most fundamental mathematical concepts for business people and consumers is the idea of interest. Simply stated, **interest** is rent paid for the use of another's money. That is, we receive interest when we let others use our money (when we deposit money in a savings account, for example), and we pay interest when we use the money of others (for example, when we borrow from a bank).

The amount of the deposit or loan is called the **principal**, or **present value**, and the **interest rate** is stated as a percentage of the principal over a given period of **time**.

Simple Interest Formula

The *simple interest formula* is

$$I = Prt$$ where $I =$ Amount of interest

$P =$ Principal or Present value

$r =$ Annual interest rate

$t =$ Time (in years)

EXAMPLE 1 How much interest does a $73 deposit earn in 3 years if the interest rate is 8%?

Solution
$$I = Prt$$
$$= 73(0.08)(3) \quad \text{Notice that interest rates are written as decimals when}$$
$$= 17.52 \quad \text{substituted into formulas}$$

The amount of interest is $17.52. ∎

The **future value**, A, of a deposit is the amount of money on deposit after a given amount of time. For Example 1, the future value of the account in 3 years is $A = \$73 + \$17.52 = \$90.52$.

Future Value
(simple interest)

> The *future value*, A, is
> $$A = P + I \quad \text{where} \quad P = \text{Principal}$$
> $$I = \text{Amount of interest}$$

Example 1 is an example of simple interest, but banks and businesses pay **compound interest**. For compound interest, after some designated period of time, the earned interest is added to the account so that future calculations include this earned interest as part of the principal. For Example 1, suppose that interest is *compounded annually*. This means that at the end of the first year the value of the account is found as follows:

$$I = Prt$$
$$= 73(0.08)(1) \quad t = 1 \text{ at the end of the first year}$$
$$= 5.84$$

Therefore
$$A = P + I$$
$$= 73 + 5.84$$
$$= 78.84$$

This amount then becomes the principal for the second year:

$$I = Prt$$
$$= 78.84(0.08)(1)$$
$$= 6.3072 \quad \text{This is \$6.31 interest for the second year}$$

At the end of the second year,
$$A = P + I$$
$$= 78.84 + 6.31$$
$$= 85.15$$

For the third year,
$$I = 85.15(0.08)(1) \quad \text{and} \quad A = 85.15 + 6.81$$
$$= 6.812 \quad\quad\quad\quad\quad = 91.96$$

Note that with simple interest (Example 1) the future value in 3 years is $90.52 and with compound interest it is $91.96.

The process just described with numbers is easy to follow, but to find the general formula we need to repeat the steps algebraically.

First Year $A = P + I$

$\qquad = P + Pr$ $\qquad I = Prt = Pr$ when $t = 1$

$\qquad = P(1 + r)$ $\qquad$ Factor out the common factor P; this number becomes the principal for the second year

Second Year $A = P(1 + r) + I$ $\qquad$ The second year principal is

$\qquad = P(1 + r) + P(1 + r)r$ $\qquad P(1 + r)$. Also,

$\qquad = P(1 + r)[1 + r]$ $\qquad I = $ principal $\times$ rate $\times$ time

$\qquad = P(1 + r)^2$ $\qquad = \underbrace{P(1 + r)}r(1)$

$\qquad\qquad\qquad\qquad\qquad\qquad\qquad$ Second-year principal

This becomes the principal for the third year.

Third Year $A = P(1 + r)^2 + I$

$\qquad = P(1 + r)^2 + P(1 + r)^2 r$

$\qquad = P(1 + r)^2[1 + r]$

$\qquad = P(1 + r)^3$

If you continue in the same fashion for t years at a rate of r compounded annually, you can see the formula is

$\qquad A = P(1 + r)^t$

We can also derive this formula using a difference equation.

$$\begin{matrix} \text{BALANCE AT BEGINNING} \\ \text{OF NEXT YEAR} \end{matrix} = \begin{matrix} \text{BALANCE AT BEGINNING} \\ \text{OF THIS YEAR} \end{matrix} + \text{INTEREST}$$

$\qquad A_{n+1} = A_n + I$

$\qquad A_{n+1} = A_n + A_n r$ $\qquad$ Since $I = Prt = A_n r(1)$

$\qquad A_{n+1} = A_n(1 + r)$

This is a geometric sequence whose first term is P and whose common ratio is $(1 + r)$, so after t years the formula is

$\qquad A_t = (1 + r)^t P$ or $A = P(1 + r)^t$

EXAMPLE 2 Find the compound interest for a $73 deposit at 8% for 3 years compounded annually.

Solution $\qquad A = 73(1 + 0.08)^3$

$\qquad\qquad = 73(1.08)^3$

On a calculator,

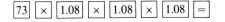

or, if your calculator has an exponent key,

$$\boxed{73}\ \boxed{\times}\ \boxed{1.08}\ \boxed{y^x}\ \boxed{3}\ \boxed{=}$$

The rest of this chapter assumes that calculators have an exponent key; if yours does not, you will need to use repeated multiplication as shown on the first line. The result is **91.958976** or $91.96; the interest is $91.96 − $73 = $18.96. ■

If you wish, you may use a table for compound interest problems. Table 5 in Appendix E shows the compound interest for $1 for N periods. For Example 2, find the column headed 8% and look in the row $N = 3$. The entry is 1.259712. Multiply this number by the principal (73) to obtain **91.958976** or $91.96.

EXAMPLE 3 You are considering a 6-year $1,000 certificate of deposit paying 12% compounded annually. How much money will you have at the end of 5 years?

Solution

By calculator

$A = 1000(1.12)^5$

$$\boxed{1.12}\ \boxed{y^x}\ \boxed{5}\ \boxed{\times}\ \boxed{1000}\ \boxed{=}$$

The result **1762.341683** rounded to the nearest cent is $1,762.34.

By Table 5, Appendix E

Find the column headed 12% and the row labeled $N = 5$ to find the entry 1.762342. Multiply:

$$1000(1.762342) \approx \$1,762.34$$

■

The disadvantages of working with tables are (1) a different table is needed for each different rate; (2) tables are not as readily available as calculators; and (3) even when using a table, a final multiplication by the principal is necessary.

If you have deposited money or seen savings and loan advertisements lately, you know that most financial institutions compound interest more frequently than annually. This is to your advantage because the shorter the compounding period, the sooner you earn interest on your interest.

***Future Value
(compound interest)***

$A = P(1 + i)^N$ where A = Future value

P = Present value

r = Annual interest rate

t = Number of years

n = Number of times compounded per year

i = Rate per period $= \dfrac{r}{n}$

N = Number of periods $= nt$

EXAMPLE 4 Reconsider Example 3 for 12% compounded monthly.

Solution By calculator:

$$A = 1000\left(1 + \frac{0.12}{12}\right)^{12(5)}$$

$$= 1000(1.01)^{60}$$

$$= 1816.696699 \qquad \text{PRESS:} \quad \boxed{1.01}\ \boxed{y^x}\ \boxed{60}\ \boxed{\times}\ \boxed{1000}\ \boxed{=}$$

Answer rounded to the nearest cent: $1,816.70 ∎

We can also use Table 5 in Appendix E to work problems like Example 4. This is particularly important if your calculator does not have an exponent key. To find the rate per period (this is i in Table 5), divide the annual rate by

    2   if semiannually
    4   if quarterly
  12   if monthly
360   if daily

Also, to find the number of periods (this is N in Table 5), multiply the time by

    2   if semiannually
    4   if quarterly
  12   if monthly
360   if daily

To find the amount of Example 4 using Table 5, look up a rate of 1% (12% ÷ 12) and $N = 60$. The entry is 1.816697, so

$$A = 1000(1.816697)$$

$$\approx \$1,816.70 \qquad \text{Answer is rounded to the nearest cent}$$

Note that when paying interest, banks use 360 for the number of days in a year; this is called **ordinary interest**. Unless instructed otherwise, use 360 for the number of days in a year. If 365 days are used, it is called **exact interest**.

EXAMPLE 5 Use Table 5 in Appendix E to find the amount you would have if you invested $1,000 for 10 years at 8% interest compounded:

  a. Annually   **b.** Semiannually   **c.** Quarterly

Solution **a.** $N = 10$, and the rate is 8%: $A = 1000 \times 2.158925$
                                                    $= \$2,158.93$

  b. $N = 20$, and the rate per period is 4%: $A = 1000 \times 2.191123$
                                                    $= \$2,191.12$

  c. $N = 40$, and the rate per period is 2%: $A = 1000 \times 2.208040$
                                                    $= \$2,208.04$ ∎

Table 5 is not extensive enough to find monthly or daily compounding of $1,000 at an 8% annual interest; these problems require a calculator. Also, keep in mind the difference between interest and future value for both the simple and compound interest formulas. With simple interest, you find the interest first and then the future value. For compound interest, you first find the future value and then find the interest. Table 10.1 summarizes this comparison.

TABLE 10.1
Comparison of simple and compound interest

	Interest, I	Future value, A
Simple interest formula	$I = Prt$ This is found first for simple interest	$A = P + I$ This is found after using the simple interest formula
Compound interest	$I = A - P$ This is found after using Table 5	$A = P\left(1 + \dfrac{r}{n}\right)^{nt}$ or $A = P(\text{Table 5 number})$ This is found first for compound interest

You have no doubt, seen advertisements that say

$$8.00\% = 8.33\%*$$

* Effective annual yield when principal and interest are left in the account.

In order to understand the phrase **effective annual yield**, look at the earnings of $1 at 8% compounded annually and quarterly:

Annually	Quarterly
$1(1 + 0.08) = 1.08$	$1\left(1 + \dfrac{0.08}{4}\right)^4 = (1.02)^4$
	$= 1.08243216$

The $1 compounded quarterly at 8% is the same as $1 compounded annually at a rate of 8.243216%. Therefore 8.24% is called the *effective annual yield* of 8% compounded quarterly.

Effective Annual Yield

The *effective annual yield*, or *effective rate*, for an account paying r percent compounded n times per year is the simple annual interest rate that would pay an equivalent amount. It is found by the following formula:

$$\left(1 + \frac{r}{n}\right)^n - 1$$

EXAMPLE 6 Verify the claims of the bank advertisement shown. (You need a calculator with an exponent key for this example.)

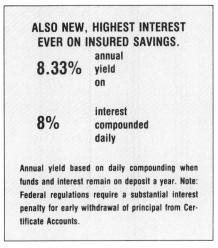

Solution   The compounding is on a daily basis ($n = 360$), so:

$$\left(1 + \frac{0.08}{360}\right)^{360} \approx (1.00022222)^{360}$$ This calculation was done on a calculator with an exponent key

$$\approx 1.08327735$$

Effective rate $\approx 1.08327735 - 1$

$$\approx 0.0833 \quad \text{or} \quad 8.33\%$$ ∎

Businesses sometimes need to have a given amount of money on hand at some time in the future. That is, suppose a business has a \$1,000 note payable due in 5 years. The business would like to know the amount of money that must be deposited *today* so that in 5 years the total amount of principal *and* interest will be \$1,000. This sum to be deposited is called the **present value** and is the same variable P that appeared in the future value formulas. That is, for present value, the interest formulas are solved for P since the variable A is known.

Present Value

SIMPLE INTEREST: $P = A - I$

COMPOUND INTEREST: By formula, $P = A(1 + i)^{-N}$
By table, $P = A \div$ (Table 5, Appendix E, entry)

The variables in this formula have the same meaning they had in the compound interest formula:

P = Present value (principal)
A = Future value
n = Number of times compounded per year
r = Annual rate
t = Number of years
$i = \frac{r}{n}$ (rate per period)
$N = tn$ (number of periods)

It is easy to see where the simple interest formula comes from (subtract I from both sides of the future value formula). The compound interest formula is also easy to derive when you remember the meaning of a negative exponent:

$$A = P(1 + i)^N \qquad \text{Future value formula}$$

$$P = \frac{A}{(1 + i)^N} \qquad \text{Divide both sides by } (1 + i)^N$$

$$= A(1 + i)^{-N} \qquad (1 + i)^{-N} = \frac{1}{(1 + i)^N}$$

EXAMPLE 7   Don's Shoe Repair needs \$1,000 for a note payable in 5 years, and Don wants to make a deposit now to pay for that note. What is the present value if the money is deposited in a savings account paying 8% compounded quarterly?

Solution $i = \dfrac{r}{n} = \dfrac{0.08}{4} = 0.02$

$N = nt = 4(5) = 20$

By calculator

$A = 1000(1.02)^{-20}$

$\boxed{1.02}\ \boxed{y^x}\ \boxed{20}\ \boxed{+/-}\ \boxed{\times}\ \boxed{1000}\ \boxed{=}$

The display shows: 672.9713331

Don should deposit \$672.97.

By Table 5, Appendix E

Find the column headed 2% and the row labeled $N = 20$ to find the entry 1.485947.

$A = 1000 \div 1.485947$

$\approx \$672.97$ ∎

EXAMPLE 8   Tom Mikalson has built a good reputation for his sporting goods store, and the business is growing at an annual rate of 21%. Tom plans on selling out and retiring in 5 years. Yesterday a developer offered Tom \$240,000 for the business, and Tom is trying to decide whether he should sell now.

The business was recently appraised for \$198,000, so Tom uses this as the present value of the business.

If he sells, he will invest the money with his stockbroker in an account paying 12% compounded quarterly.

He will sell if the price today is better than the one he can expect in 5 years. What should Tom do?

Solution   First, find the expected value of the business in 5 years, then find the present value of that amount.

Future value

$A = P(1 + i)^N$

$\quad = 198,000(1 + 0.21)^5$

$\quad = 513,561.0071$

$n = 1$ (compounded annually)

$r = $ Growth rate (21%)

$N = nt = 1(5) = 5$

$i = r/n = \frac{0.21}{1} = 0.21$

Present value

$$P = A(1 + i)^{-N}$$
$$= 513,561.0071(1 + 0.03)^{-20}$$
$$= 284,346.2779$$

$n = 4$ (compounded quarterly)
$r = $ Interest rate (12%)
$N = nt = 4(5) = 20$
$i = \frac{0.12}{4} = 0.03$

This means that Tom should not accept less than $284,346.28 for his business (even though it is presently appraised at $198,000). ■

Inflation

Example 8 involves a very important application of present and future value—inflation. **Inflation** indicates an increase in the amount of currency in circulation, which leads to a fall in the currency's value and a consequent rise in prices. Inflation is usually specified as a percent, and for our purposes we will use the future and present value formulas with $n = 1$ so that $N = $ the number of years and $i = $ the annual rate.

EXAMPLE 9 The inflation rate in 1988 is about 2%. A person earning a $25,000 salary wants to know what salary to expect in 1994 if this rate of inflation continues and if her salary keeps pace with inflation.

Solution This is a future value problem; use Table 5 in Appendix E or a calculator.

$$A = 25,000(1 + 0.02)^6$$
$$\approx 28,154.06$$

Future value formula
$P = 25,000$
$i = 0.02$ (inflation rate)
$N = 6$ (number of years)

By table, $A = 25,000(1.126162)$ In Table 5, $N = 6, i = 2\% = 1.126162$
$$= 28,154.05$$

The expected wage will be about $28,000. ■

With inflation problems, the best figure we can arrive at is an estimation since in reality the inflation rate is not a constant. Therefore be careful not to try to project exact or precise answers.

EXAMPLE 10 An insurance agent wants to sell you a policy that will pay you $20,000 in 30 years. If you assume an average rate of inflation of 9% over the next 30 years, what is the value of that insurance payment in terms of today's dollars?

Solution Use Table 5 to find the present value ($N = 30, i = 9\%$):

$$P = 20,000 \div 13.267678$$
$$= 1507.42$$

The present value of the $20,000 is about $1,500. ■

The effects of compound interest can sometimes be rather dramatic. It is not uncommon to find some investments that pay 16% (although such investments are much riskier than insured savings accounts).

EXAMPLE 11   Your first child has just been born. You want to give her a million dollars when she retires at age 65. If you invest your money at 16% compounded quarterly, how much do you need to invest today so your child will have a million dollars in 65 years?

Solution   Find the present value of 1,000,000 where $n = 4$, $r = 16\%$, $t = 65$, $i = \frac{0.16}{4} = 0.04$, and $N = 4(65) = 260$. Use the present value formula:

$$P = 1,000,000(1 + 0.04)^{-260}$$
$$= 37.26763062$$

This means that if you invest $37.28 at 16% compounded quarterly in your child's name, she will have $1,000,000 at age 65.   ∎

Problem Set 10.2

APPLICATIONS

In Problems 1–6 compare the amount of simple interest and the interest if the investment is compounded annually.

1. $1,000 at 8% for 5 years

2. $5,000 at 10% for 3 years

3. $2,000 at 12% for 3 years

4. $2,000 at 12% for 5 years

5. $5,000 at 12% for 20 years

6. $1,000 at 14% for 30 years

In Problems 7–12 compare the future amount you would have if the money were invested at simple interest or invested with interest compounded annually.

7. $1,000 at 8% for 5 years

8. $5,000 at 10% for 3 years

9. $2,000 at 12% for 3 years

10. $2,000 at 12% for 5 years

11. $5,000 at 12% for 20 years

12. $1,000 at 14% for 30 years

Fill in the blanks for Problems 13–26.

Compounding period, n	Principal, P	Yearly rate, r	Time, t	Period rate, $i = r/n$	Number of periods, $N = nt$	Entry in Table 5	Total amount	Total interest
13. Annually, 1	$1,000	9%	5 years	_____	_____	_____	_____	_____
14. Semiannually, 2	$1,000	9%	5 years	_____	_____	_____	_____	_____
15. Annually, 1	$500	8%	3 years	_____	_____	_____	_____	_____
16. Semiannually, 2	$500	8%	3 years	_____	_____	_____	_____	_____
17. Quarterly, 4	$500	8%	3 years	_____	_____	_____	_____	_____
18. Semiannually, 2	$3,000	18%	3 years	_____	_____	_____	_____	_____
19. Quarterly, 4	$5,000	18%	10 years	_____	_____	_____	_____	_____
20. Quarterly, 4	$624	16%	5 years	_____	_____	_____	_____	_____
21. Quarterly, 4	$5,000	20%	10 years	_____	_____	_____	_____	_____
22. Monthly, 12	$350	12%	5 years	_____	_____	_____	_____	_____
23. Monthly, 12	$4,000	24%	5 years	_____	_____	_____	_____	_____
24. Quarterly, 4	$800	12%	90 days	_____	_____	_____	_____	_____
25. Quarterly, 4	$1,250	16%	450 days	_____	_____	_____	_____	_____
26. Quarterly, 4	$1,000	12%	900 days	_____	_____	_____	_____	_____

Answer the questions in Problems 27–35. A calculator with an exponent key is required for these exercises.

27. What is the future amount of $12,000 invested for 5 years at 14% compounded monthly?

28. What is the future amount of $800 invested for 1 year at 10% compounded daily?

29. What is the future amount of $9,000 invested for 4 years at 20% compounded monthly?

30. If $5,000 is compounded annually at $5\frac{1}{2}\%$ for 12 years, what is the total interest received at the end of that time?

31. If $10,000 is compounded annually at 8% for 18 years, what is the total interest received at the end of that time?

32. What is the effective yield of 6% compounded quarterly?

33. What is the effective yield of 6% compounded monthly?

34. What is the effective yield of 8% compounded monthly?

35. What is the effective yield of 12% compounded daily?

Find the present value in Problems 36–40.

	Amount needed	Time	Interest	Compounded
36.	$7,000	5 years	8%	Annually
37.	$25,000	5 years	8%	Semiannually
38.	$165,000	10 years	12%	Semiannually
39.	$500,000	10 years	12%	Quarterly
40.	$3,000,000	20 years	12%	Semiannually

In Problems 41–52 calculate the expected price in the year 2008 if you assume a 10% inflation rate and use the given 1988 price. Answers should be rounded to the nearest dollar.

41. Cup of coffee, $0.75

42. Small car, $6,000

43. Big Mac, $1.85

44. Movie admission, $5.00

45. Monthly rent, $400

46. Sunday paper, $1.00

47. Textbook, $28.00

48. Tuition at a private college, $14,000

49. Yearly salary, $25,000

50. Condominium, $65,000

51. House, $115,000

52. Business, $675,000

53. You owe $5,000 due in 3 years, but you would like to pay the debt today. If the present interest rate is compounded annually 11%, how much should you pay today so that the present value is equivalent to the $5,000 payment in 3 years?

54. Ricon Bowling Alley will need $20,000 in 5 years to resurface the lanes. What deposit should be made today in an account that pays 9% compounded semiannually?

55. An accounting firm agrees to purchase a computer for $150,000. It will be delivered in 270 days. How much do the owners need to deposit in an account paying 18% compounded quarterly?

56. The Fair View Market must be remodeled in 3 years. Remodeling will cost $200,000. How much should be deposited now (to the nearest dollar) in order to pay for this remodeling if the account pays 10% compounded monthly?

57. A laundromat will need seven new washing machines in $2\frac{1}{2}$ years for a total cost of $2,900. How much money (to the nearest dollar) should be deposited at the present time to pay for these machines? The interest rate is 11% compounded semiannually.

58. A computerized checkout system is planned for Able's Grocery Store. The system will be delivered in 18 months at a cost of $560,000. How much should be deposited today into an account paying 7.5% compounded daily?

59. A contest offers the winner $50,000 now or $10,000 now and $45,000 in 1 year. Which is the best choice if the current interest rate is 10% compounded monthly and the winner does not intend to use any of the money for 1 year?

10.3 Annuities

In Section 10.2 we discussed the present and future values of a lump-sum deposit bearing compound interest. However, it is far more typical to make monthly or other periodic payments into an account. Suppose you decide to give up smoking and save the $1 per day you spend on cigarettes. How much will you save in 5 years? If you save the money without earning any interest, you will have

$$\$1 \times 365 \times 5 = \$1,825$$

However, let us assume that you save $1 per day and at the end of each month you deposit the $30 (assume all months are 30 days) into a savings account earning 12% interest compounded monthly. Now, how much will you have in 5 years?

TABLE 10.2

Time	Amount saved	Interest earned	Total amount
Start	$0	$0	$0
1 month	$30	$0	$30
2 months	$60	$.30	$60.30
3 months	$90	$.60	$90.90
4 months	$120	$.90	$121.80

If we continue to calculate the entries in Table 10.2 month by month, we will tire long before we reach 5 years (that would be 60 entries to calculate). Instead, let us shorten the problem to 6 months and try to notice a pattern so we can use a formula. At the same time, we will use the variables of the last section to mean the same thing: P = present value; A = future value; r = annual rate; t = number of years; n = number of times compounded per year; $i = \frac{r}{n}$; and $N = nt$. We will also introduce a new variable, m, as the periodic payment (monthly payment in this example).

Detail of Table 10.2 *For $n = 6$, $i = \frac{0.12}{12} = 0.01$, and $m = \$30$:*

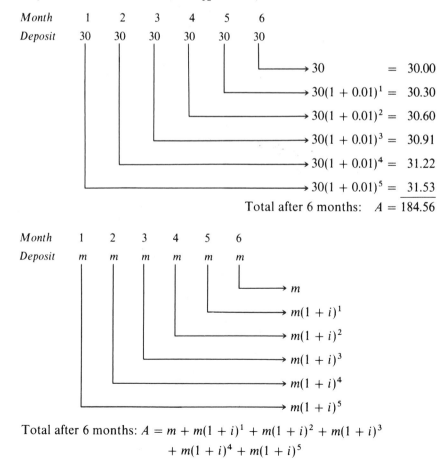

$$
\begin{array}{llll}
& 30 & = & 30.00 \\
& 30(1 + 0.01)^1 & = & 30.30 \\
& 30(1 + 0.01)^2 & = & 30.60 \\
& 30(1 + 0.01)^3 & = & 30.91 \\
& 30(1 + 0.01)^4 & = & 31.22 \\
& 30(1 + 0.01)^5 & = & 31.53 \\
\end{array}
$$

Total after 6 months: $A = 184.56$

Month 1 2 3 4 5 6

Deposit m m m m m m

$$
\begin{array}{l}
m \\
m(1 + i)^1 \\
m(1 + i)^2 \\
m(1 + i)^3 \\
m(1 + i)^4 \\
m(1 + i)^5 \\
\end{array}
$$

Total after 6 months: $A = m + m(1 + i)^1 + m(1 + i)^2 + m(1 + i)^3$
$$+ m(1 + i)^4 + m(1 + i)^5$$

A sequence of payments into an interest-bearing account is called an **annuity**. If the payments are made at the end of the time period, and if the frequency of payments is the same as the frequency of compounding, the annuity is called an **ordinary annuity**.

To find a general equation for an annuity, let

A_N = The amount of money in the annuity after N periods

i = Interest rate per period of time

m = Monthly deposit

Also assume that $A_0 = 0$ (that is, there is nothing in the account when the account is opened). Now,

CURRENT AMOUNT = PREVIOUS AMOUNT + INTEREST + DEPOSIT

$$A_N \quad = \quad A_{N-1} \quad + \quad iA_{N-1} \quad + \quad m$$

$$A_N = (1 + i)A_{N-1} + m$$

The solution of this difference equation is found by noticing that $a = 1 + i$, $b = m$, and $A_0 = 0$, so

$$A_N = \frac{m}{1 - (1 + i)} + \left[0 - \frac{m}{1 - (1 + i)}\right](1 + i)^N$$

$$= \frac{m}{-i} + \frac{m}{i}(1 + i)^N$$

$$= m\left[\frac{(1 + i)^N - 1}{i}\right]$$

Ordinary Annuity

By formula	By table
Let $i = \dfrac{r}{n}$ and $N = nt$ with monthly payment m; then	Use Table 6, Appendix E:
	Find this table entry as usual using i and N
$A = m\left[\dfrac{(1 + i)^N - 1}{i}\right]$	$A = m(\text{Table 6 entry})$

EXAMPLE 1 How much do you save in 5 years if you deposit $30 at the end of each month into an account paying 12% compounded monthly?

Solution The rate i is $\frac{0.12}{12} = 0.01$ and $N = 5(12) = 60$. From Table 6 in Appendix E,

$$A = 30(81.669670)$$

$$\approx \$2,450.09$$

∎

EXAMPLE 2 Repeat Example 1 for 2 years.

Solution This time $N = 2(12) = 24$. Table 6 does not have an entry for $N = 24$, so you must use the formula:

$$i = \frac{0.12}{12} = 0.01 \qquad A = 30\left[\frac{(1 + 0.01)^{24} - 1}{0.01}\right]$$

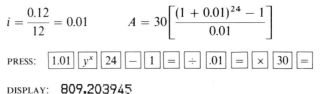

PRESS: $\boxed{1.01}\;\boxed{y^x}\;\boxed{24}\;\boxed{-}\;\boxed{1}\;\boxed{=}\;\boxed{\div}\;\boxed{.01}\;\boxed{=}\;\boxed{\times}\;\boxed{30}\;\boxed{=}$

DISPLAY: 809.203945

You would have $809.20. ∎

The annuities described thus far, in which the deposit is made at the end of each period, are ordinary annuities. Another type of annuity is one in which the payments are made at the *beginning* of each period. These are called **annuities due**. To derive a formula for an annuity due, look at the detail for Table 10.2 and note that if the periodic deposit is put into the account at the beginning of the period instead of at the end, the only difference will be that each exponent is increased by 1 (for one more period). This leads to the formula shown below:

Annuity Due

By formula	By table
Let $i = \dfrac{r}{n}$ and $N = nt$ with monthly payment m; then $$A = m\left[\frac{(1 + i)^{N+1} - 1}{i} - 1\right]$$	Use Table 7, Appendix E: $A = m(\text{Table 7 entry})$

EXAMPLE 3 What is the value of an annuity due for which $100 per month is deposited into an account paying 18% compounded monthly for $7\frac{1}{2}$ years?

Solution Here, $i = \frac{0.18}{12} = 0.015$ and $N = 7\frac{1}{2}(12) = 90$. Use Table 7 in Appendix E to find: 190.748849. Thus

$$A = \$100(190.748849) = \$19{,}074.88$$ ∎

Problem Set 10.3

Find the value of each annuity in Problems 1–23 at the end of the indicated number of years. Assume that the interest is compounded with the same frequency as the deposits. Give your answers to the nearest dollar.

	Amount of deposit	Frequency compounded	Rate	Number of years	Type of annuity
1.	$500	Annually	8%	30	Ordinary
2.	$500	Annually	6%	30	Ordinary
3.	$250	Semiannually	8%	30	Ordinary

	Amount of deposit	Frequency compounded	Rate	Number of years	Type of annuity
4.	$600	Semiannually	12%	10	Ordinary
5.	$300	Quarterly	12%	10	Ordinary
6.	$100	Monthly	12%	5	Ordinary
7.	$500	Annually	8%	30	Due
8.	$500	Annually	6%	30	Due
9.	$250	Semiannually	8%	30	Due

Amount of deposit	Frequency compounded	Rate	Number of years	Type of annuity
10.	$600	Semiannually	12%	10	Due
11.	$300	Quarterly	12%	10	Due
12.	$100	Monthly	12%	5	Due
13.	$50	Monthly	18%	5	Due
14.	$200	Quarterly	18%	20	Ordinary
15.	$400	Quarterly	16%	20	Ordinary
16.	$2,500	Semiannually	18%	30	Due
17.	$30	Monthly	18%	5	Due
18.	$30	Monthly	18%	5	Ordinary
19.	$5,000	Annually	8%	10	Ordinary
20.	$5,000	Annually	8%	10	Due
21.	$100	Monthly	18%	$7\frac{1}{2}$	Due
22.	$2,500	Semiannually	12%	20	Ordinary
23.	$1,250	Quarterly	12%	20	Due

APPLICATIONS

24. The owner of Sebastopol Tree Farm deposits $650 at the beginning of each quarter for 5 years into an account paying 8% compounded quarterly. What is the value of the account at the end of 5 years?

25. The owner of Oak Hill Squirrel Farm deposits $1,000 at the end of each quarter for 5 years into an account paying 8% compounded quarterly. What is the value at the end of 5 years?

26. John and Rosamond want to retire in 5 years and can save $150 every three months. They plan to deposit the money at the beginning of each quarter into an account paying 8% compounded quarterly. How much will they have at the end of 5 years?

27. You want to retire at age 65. You decide to make a deposit to yourself at the beginning of each year into an account paying 13%, compounded annually. Assuming you are now 25 and can spare $1,200 per year, how much will you have (to the nearest dollar) when you retire? (Alternate question: use your present age.)

28. In 1988 the maximum Social Security deposit was $3,818.10. Suppose you are 25 and make a deposit of this amount into an account at the end of each year. How much would you have (to the nearest dollar) when you retire if the account pays 8% compounded annually and you retire at age 65?

29. Repeat Problem 28 using your own age.

10.4   Amortization

Present Value of an Annuity

Suppose that instead of making monthly payments at periodic intervals, we want to deposit a lump sum today that will have the same value as an annuity at the end of some time period.

EXAMPLE 1   Chen and Mai are partners who have decided to set aside a fund for their secretary who will retire in 10 years. Chen wants to make a $50 monthly deposit into this account while Mai wants to make a lump-sum deposit today, but both want to deposit equal amounts. How much will be in the account in 10 years if the interest rate is 9% compounded monthly, and how much should Mai deposit in order to equal Chen's contribution?

Solution   Chen's deposits are an ordinary annuity with $m = \$50$, $r = 9\%$, $t = 10$, $i = \frac{0.09}{12}$, and $N = 10(12) = 120$:

$$A = m\left[\frac{(1 + i)^N - 1}{i}\right]$$

$$= \$50\left[\frac{(1 + \frac{0.09}{12})^{120} - 1}{\frac{0.09}{12}}\right]$$

$$= \$9,675.71$$

Mai's contribution is a present value problem. When we speak of the **present value of an annuity**, we mean the lump-sum deposit today that will equal the future value of a given annuity. Using the present value formula from Section 10.2, we find

$$P = A(1 + i)^{-N} = 9675.71(1 + \tfrac{0.09}{12})^{-120}$$
$$= \$3,947.08$$

For Chen and Mai to contribute equal amounts, Chen will deposit \$50 per month for 10 years and Mai will make a lump-sum contribution of \$3,947.08. The total value in the account when the secretary retires will be \$9,675.71. ∎

The goal of this section is to develop a formula that will allow us to calculate a lump-sum deposit (see Example 1) *without first finding the value of an annuity.* Recall that \$P deposited today will amount to $A = P(1 + i)^N$ as before, $i = \tfrac{r}{n}$ and $N = nt$. Substitute this into the ordinary annuity formula with monthly payment m to obtain

$$P(1 + i)^N = m\left[\frac{(1 + i)^N - 1}{i}\right]$$

Solve this formula for P and simplify to obtain a formula for the present value of an annuity:

$$P = \frac{m}{i}\left[\frac{(1 + i)^N - 1}{(1 + i)^N}\right]$$
$$= \frac{m}{i}\left[1 - \frac{1}{(1 + i)^N}\right]$$
$$= \frac{m}{i}[1 - (1 + i)^{-N}]$$

Present Value of an Annuity

By formula

$$P = m\left[\frac{1 - (1 + i)^{-N}}{i}\right]$$

By table

Use Table 8, Appendix E:

$$P = m(\text{Table 8 entry})$$

EXAMPLE 2 Use the formula for the present value of an annuity to calculate Mai's deposit as described in Example 1. That is, find the present value of an annuity where $m = \$50$, $r = 9\%$, and $t = 10$.

Solution
$$P = m\left[\frac{1 - (1 + i)^{-N}}{i}\right]$$
$$= \$50\left[\frac{1 - (1 + \tfrac{0.09}{12})^{-120}}{\tfrac{0.09}{12}}\right]$$
$$\approx \$3,947.08$$

PRESS: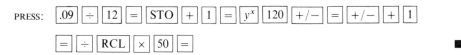

The calculator steps are rather complicated, so you might want to use Table 8, Appendix E. This procedure is illustrated in Example 3.

EXAMPLE 3   What is the present value of an annuity of $30 per month for 5 years deposited into an account paying 12% compounded monthly?

Solution   We have $i = \frac{0.12}{12} = 0.01$ and $N = 5(12) = 60$. From Table 8 in Appendix E,

$$P = 30(44.955038)$$
$$= \$1{,}348.65114$$

It would require a deposit of $1,348.65.

EXAMPLE 4   Teresa's Social Security benefit is $450 per month if she retires at age 62 instead of age 65. What is the present value of an annuity that would pay $450 per month for 3 years if the interest rate is 10% compounded monthly?

Solution   Here, $i = \frac{0.10}{12} = 0.008\overline{3}$ and $N = 12(3) = 36$. These values are not included in Table 8, so you must use the formula:

$$P = 450\left[\frac{1 - (1 + i)^{-36}}{i}\right]$$

To use this formula you need a calculator with an exponent key. First, store i in memory:

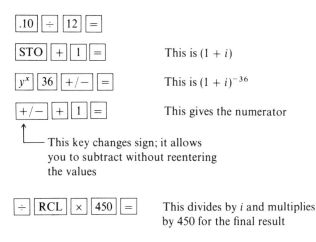

DISPLAY: 13946.056

The present value is $13,946.06.

Amortization

Installment loans are one of the most common examples of the present value of an annuity. The process of paying off a debt by systematically making partial payments until the debt (principal) and the interest are repaid is called **amortization**. If the loan is paid off in regular equal installments, then we can use the formula for the present value of an annuity to find the monthly payments by algebraically solving for m:

$$P = m\left[\frac{1 - (1 + i)^{-N}}{i}\right]$$

Multiply both sides by i:

$$Pi = m[1 - (1 + i)^{-N}]$$

Divide by $[1 - (1 - i)^{-N}]$ to solve for m:

$$\frac{Pi}{1 - (1 + i)^{-N}} = m$$

Installment Payments

The monthly payment, m, for a fully amortized installment loan with annual percentage rate r is found as follows:

By formula

$$m = \frac{Pi}{1 - (1 + i)^{-N}}$$

where $i = \dfrac{r}{12}$

By table

Use the present value of an annuity table, Table 8 in Appendix E:

$$m = P \div (\text{Table 8 entry})$$

EXAMPLE 5 In 1986 the average price of a new home was $112,000 and the interest rate was 11%. If this amount is financed for 40 years at 11% interest, what is the monthly payment?

Solution We have $i = \frac{0.11}{12}$, $N = 12(40) = 480$, and $P = 112{,}000$. Thus

$$m = \frac{112{,}000i}{1 - (1 + i)^{-480}}$$

PRESS:		
$\boxed{11}\ \boxed{\div}\ \boxed{12}\ \boxed{=}\ \boxed{\text{STO}}$	Store i for future use
$\boxed{+}\ \boxed{1}\ \boxed{=}$	Continue to find $(1 + i)$
$\boxed{y^x}\ \boxed{480}\ \boxed{+/-}\ \boxed{=}$	This gives $(1 + i)^{-480}$
$\boxed{+/-}\ \boxed{+}\ \boxed{1}\ \boxed{=}$	This gives the denominator

Subtracts without reentering the number

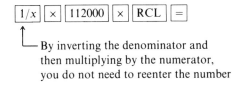

By inverting the denominator and
then multiplying by the numerator,
you do not need to reenter the number

DISPLAY: **1039.68973**

The monthly payment is $1,039.69. This is the payment for interest and principal to pay off a 40-year 11% loan of $112,000. ∎

The amount found in Example 5 is the payment for interest and principal. It does not include taxes and insurance. In passing, it might be interesting to see how much interest is paid for the home loan of Example 5. There are 480 payments of $1,039.69, so the total repaid is

$$480(1039.69) = \$499,051.20$$

Since the loan was for $112,000, the interest is

$$\$499,051.20 - \$112,000 = \$387,051.20$$

The amount of interest paid can be reduced by making a large down payment. For Example 5, if a 20% down payment was made, then $112,000(0.80) = \$89,600$ is the amount to be financed.

Problem Set 10.4

Find the present value of the ordinary annuities in Problems 1–12.

	Amount of deposit	Frequency compounded	Rate	Number of years
1.	$500	Annually	8%	30
2.	$500	Annually	6%	30
3.	$250	Semiannually	8%	30
4.	$600	Semiannually	12%	10
5.	$300	Quarterly	12%	10
6.	$100	Monthly	12%	5
7.	$200	Quarterly	18%	20
8.	$400	Quarterly	16%	20
9.	$30	Monthly	18%	5
10.	$75	Monthly	10%	10
11.	$50	Monthly	11%	30
12.	$100	Monthly	13%	40

Find the monthly payment for the loans in Problems 13–24.

13. $500 loan for 12 months at 12%

14. $100 loan for 18 months at 18%

15. $4,560 loan for 20 months at 21%

16. $3,520 loan for 30 months at 19%

17. Used-car financing of $2,300 for 24 months at 15%

18. New car financing of 2.9% on a 30-month $12,450 loan

19. Furniture financed at $3,456 for 36 months at 23%

20. A refrigerator financed for $985 at 17% for 15 months

21. A $112,000 home bought with a 20% down payment and the balance financed for 30 years at 11.5%

22. A $108,000 home bought with a 30% down payment and the balance financed for 30 years at 12.05%

23. Finance $450,000 for a warehouse with a 12.5% 30-year loan

24. Finance $859,000 for an apartment complex with a 13.2% 20-year loan

25. How much interest would be saved in Problem 21 if the home were financed for 15 rather than for 30 years?

26. How much interest would be saved in Problem 22 if the home were financed for 15 rather than for 30 years?

APPLICATIONS

27. Pat agrees to contribute $500 to the alumni fund at the end of each year for the next 5 years. Karl wants to match Pat's gift, but he wants to make a lump-sum contribution. If the current interest rate is 12.5% compounded annually, how much should Karl contribute to equal Pat's gift?

28. A $1,000,000 lottery prize pays $50,000 per year for the next 20 years. If the current rate of return is 12.25%, what is the present value of this prize?

29. Suppose you have an annuity from an insurance policy and you have the option of being paid $250 per month for 20 years or having a lump-sum payment of $25,000. Which has the most value if the current rate of return is 10% compounded monthly?

30. An insurance policy offers you the option of being paid $750 per month for 20 years or a lump sum of $50,000. Which has the most value if the current rate of return is 9% compounded monthly and you expect to live for 20 years?

10.5 Sinking Funds*

The last financial application we consider is the situation in which we need to have a lump sum of money in a certain period of time. The present value formula will tell us how much we need to have today, but we frequently do not have that amount available. Suppose your goal is $10,000 in 5 years. You can obtain 8% compounded quarterly, so the present value is $6,729.71. However, this is more than you can afford to put into the bank now. The next choice is to make a series of small equal investments to accumulate at 8% compounded quarterly, so that the end result is the same, namely, $10,000 in 5 years. The account you set up to receive those investments is called a **sinking fund**.

To find a formula for a sinking fund, we begin with the formula for an ordinary annuity and solve for m:

$$A = m\left[\frac{(1 + i)^N - 1}{i}\right]$$

$$m[(1 + i)^N - 1] = Ai$$

$$m = \frac{Ai}{(1 + i)^N - 1}$$

Sinking Fund	*By formula*	*By table*
		Use the ordinary annuity table, Table 6, Appendix E:
	$m = \dfrac{Ai}{(1 + i)^N - 1}$	$m = A \div$ (Table 6 entry)

EXAMPLE 1 A business needs to raise $1,000,000 in 20 years by making equal quarterly payments into an account paying 12% interest compounded quarterly. What is the required amount of each deposit?

---

* This section is not required for subsequent textual development.

Solution Here, $i = \frac{0.12}{4} = 0.03$ and $N = 4(20) = 80$. From Table 6 in Appendix E, we obtain 321.363019. Thus

$$m = 1,000,000 \div 321.363019$$
$$= \$3,111.745723$$

The business needs to make quarterly deposits of $3,112 (rounded to the nearest dollar). ∎

The primary reason that businesses set up a sinking fund is to pay off bonds. A **bond** is a certificate (a written promise) of a business to repay a certain amount at some future time. It is often how a business, corporation, or government agency borrows to buy new equipment or raise money for construction. Usually, each bond has a face value of $1,000 and a specified interest rate, called the *contract rate*. This interest is paid to the bondholders at specified intervals (usually twice a year). The bonds are usually sold to an underwriter, who sells them to investors. Since the contract rate is fixed but prevailing interest rates are not, bonds are bought and sold for either more or less than the face value. This, in effect, changes the interest rate actually received (because of the changes in the principal). This interest rate is called the *market rate* or *effective rate*. Bond prices are quoted as a percent of their face amounts. A quote of 103 is 103% of the face amount, and a quote of $85\frac{1}{2}$ is $85\frac{1}{2}\%$ of the face amount.

EXAMPLE 2 The Packard-Hue Corporation issues $50,000,000 worth of bonds for a new plant. These bonds are 20-year bonds with interest payable semiannually at a contract rate of 6%. In addition to paying interest on the bonds, the company sets up a sinking fund into which they will make semiannual payments and receive 8% interest compounded semiannually. What is the amount of each semiannual payment necessary to pay the interest on the bonds and for the sinking fund?

Solution *Bond interest*: This is simple interest because interest is paid to bondholders and does not accumulate.

$$I = Prt$$

⎰──── Semiannual payment

$$= \$50,000,000(0.06)(\tfrac{1}{2})$$
$$= \$1,500,000$$

Sinking fund payment: $i = \frac{0.08}{2} = 0.04$ and $N = 2(20) = 40$; thus

$$m = \$50,000,000 \div 95.025516 \longleftarrow \text{Table 6 entry}$$
$$\approx \$526,174 \qquad \text{Rounded to the nearest dollar}$$

The total semiannual cost is $1,500,000 + $526,174 = $2,026,174 ∎

Summary

For many students, the most difficult part of working with business formulas is determining *which* formula or *which* table to use. To make the processes easier for you, consistent notation and consistently constructed tables are used here. Your main task, therefore, is one of classification, summarized below:

TABLE 10.3   Classification and formulas for financial problems

DEFINITION OF VARIABLES:

P = Present value (sometimes called principal)

A = Future value

I = Amount of interest

r = Annual interest rate

t = Number of years

n = Periods; that is, the number of times per year that the interest is compounded

m = Periodic payment

i = Rate per period = $\dfrac{r}{n}$

N = Number of periods = nt

FINANCIAL TABLES (Appendix E):   *For all tables use i and N to find entry.*

Future value (Table 5):   Lump sum known, to find future lump sum

Present value (Table 5):   Future lump sum known, to find present value

Ordinary annuity (Table 6):   Periodic payments at the end of each period, to find the future value

Annuity due (Table 7):   Periodic payments at the beginning of each period, to find the future value

Present value of an annuity (Table 8):   Find the present lump sum necessary to deposit to equal the future value of periodic payments for a given period of time

Installment payments (Table 8):   Find the periodic payments to pay off both principal and interest in a given period of time

Sinking fund (Table 6):   Find the amount of periodic payment in order to have a given amount at some future date

Problem Set 10.5

Find the amount of payment necessary for each deposit to a sinking fund in Problems 1–12. Assume that the deposit is made at the end of each compounding period.

	Amount needed	Time	Interest rate	Compounded
1.	$7,000	5 years	8%	Annually
2.	$25,000	5 years	11%	Annually
3.	$25,000	5 years	12%	Semiannually
4.	$50,000	10 years	14%	Semiannually
5.	$165,000	10 years	12%	Semiannually
6.	$3,000,000	20 years	12%	Semiannually
7.	$500,000	10 years	12%	Quarterly
8.	$55,000	5 years	16%	Quarterly
9.	$100,000	8 years	10%	Quarterly
10.	$35,000	12 years	18%	Quarterly
11.	$45,000	6 years	18%	Monthly
12.	$120,000	30 years	14%	Monthly

APPLICATIONS

13. Clearlake Optical has a $50,000 note that comes due in 4 years. The owners wish to create a sinking fund to pay this note. If the fund earns 8% compounded semiannually, how much should each semiannual deposit be?

Classification of Types:

LUMP SUM				
Present value, P	**Future value,** A	**Classification**	**Formula**	**Table**
Known	Unknown	Future value	$A = P(1 + i)^N$	$A = P$(Table 5 entry)
Unknown	Known	Present value	$P = \dfrac{A}{(1 + i)^N}$	$P = A \div$ (Table 5 entry)

PERIODIC PAYMENTS					
Periodic payment, m	**Present value,** P	**Future value,** A	**Classification**	**Formula**	**Table**
Known		Unknown	Ordinary annuity (end of each period)	$A = m\left[\dfrac{(1 + i)^N - 1}{i}\right]$	$A = m$(Table 6 entry)
Known		Unknown	Annuity due (start of each period)	$A = m\left[\dfrac{(1 + i)^{N+1} - 1}{i} - 1\right]$	$A = m$(Table 7 entry)
Known	Unknown*		Present value of an annuity	$P = m\left[\dfrac{1 - (1 + i)^{-N}}{i}\right]$	$P = m$(Table 8 entry)
Unknown	Known		Installment payment	$m = \dfrac{Pi}{1 - (1 + i)^{-N}}$	$m = P \div$ (Table 8 entry)
Unknown		Known	Sinking fund	$m = \dfrac{Ai}{(1 + i)^N - 1}$	$m = A \div$ (Table 6 entry)

* Amount needed to deposit today to equal the future value of an annuity; this is the amount you can borrow with a given monthly payment on an installment loan.

14. A business must raise $70,000 in 5 years. What should be the size of the owners' quarterly payment to a sinking fund paying 8% compounded quarterly?

15. Clearlake Optical has developed a new lens. The owners plan on issuing a $4,000,000 30-year bond with a contract rate of $5\frac{1}{2}$% paid annually to raise capital to market this new lens. To pay off the debt, they will also set up a sinking fund paying 8% interest compounded annually. What size annual payment is necessary for interest and sinking fund combined?

16. The owners of Bardoza Greeting Cards wish to introduce a new line of cards but need to raise $200,000 to do it. They decide to issue 10-year bonds with a contract rate of 6%

paid semiannually. They also set up a sinking fund paying 8% interest compounded semiannually. How much money will they need to make the semiannual interest payments as well as payments to the sinking fund?

Problems 17–30 provide a mixture of financial problems. For each problem: a. Classify the type. b. Answer the questions. (Round to the nearest dollar.)

17. Rincon Bowling Alley will need $80,000 in 4 years to resurface the lanes. What lump sum would be necessary today if the owner of the business can deposit it in an account that pays 9% compounded semiannually?

18. Rita wants to save for a trip to Tahiti, so she puts $2.00 per day into a jar. After 1 year she has saved $730 and puts the money into a bank account paying 10% compounded annually. She continues to save in this manner and makes her annual $730 deposit for 15 years. How much does she have at the end of that time period?

19. Karen receives a $12,500 inheritance that she wants to save until she retires in 22 years. If she deposits the money in a fixed 11% account, compounded daily, how much will she have when she retires?

20. You want to give your child $1,000,000 when he retires at age 65. How much money do you need to deposit to an account paying 9% compounded monthly if your child is now 10 years old?

21. An accounting firm agrees to purchase a computer for $150,000 (cash on delivery) and the delivery date is in 270 days. How much do the owners need to deposit in an account paying 18% compounded quarterly?

22. For 15 years, Thompson Cleaners deposits $900 at the beginning of each quarter into an account paying 8% compounded quarterly. What is the value of the account to the nearest dollar at the end of 5 years?

23. In 1980 the inflation rate hit 16%. Suppose that the average cost of a textbook in 1980 was $15. What is the expected cost in the year 2000 if we project this rate of inflation on the cost?

24. What is the necessary amount of monthly payments to an account paying 18% compounded monthly in order to have $100,000 in $8\frac{1}{3}$ years if the deposits are made at the end of the month?

25. Thomas' Grocery Store is going to be remodeled in 5 years, and the remodeling will cost $300,000. How much should be deposited now (to the nearest dollar) in order to pay for this remodeling if the account pays 12% compounded monthly?

26. If an apartment complex will need painting in $3\frac{1}{2}$ years and the job will cost $45,000, what amount needs to be deposited into an account now in order to have the necessary funds? The account pays 12% interest compounded semiannually.

27. Teal and Associates needs to borrow $45,000. The best loan they can find is one at 12% that must be repaid in monthly installments over the next $3\frac{1}{2}$ years. How much are the monthly payments?

28. A city issues $20 million in tax-exempt 25-year bonds with 8% interest payable quarterly. In addition to paying interest on these bonds, the city sets up an account into which quarterly payments are made and 12% interest compounded quarterly is received. How much needs to be paid into this account to pay off the $20 million in 25 years?

29. Certain Concrete Company deposits $4,000 at the end of each quarter into an account paying 10% interest compounded quarterly. What is the value of the account at the end of $7\frac{1}{2}$ years?

30. Major Magic Corporation deposits $1,000 at the beginning of each month into an account paying 18% interest compounded monthly. What is the value of the account at the end of $8\frac{1}{3}$ years?

10.6 Summary and Review

IMPORTANT TERMS

Amortization   [10.4]
Annuity   [10.3]
Annuity due   [10.3]
Arithmetic sequence   [10.1]
Bond   [10.5]
Compound interest formula   [10.2]
Effective annual yield   [10.2]
Effective rate   [10.2]
Exact interest   [10.2]
First-order difference equation   [10.1]
Future value   [10.2]
Geometric sequence   [10.1]

Inflation   [10.2]
Installment loan   [10.4]
Interest   [10.2]
Ordinary annuity   [10.3]
Ordinary interest   [10.2]
Present value   [10.2]
Present value of an annuity   [10.4]
Principal   [10.2]
Rate of interest   [10.2]
Sigma notation   [10.1]
Simple interest formula   [10.2]
Sinking fund   [10.5]

SAMPLE TEST   *For additional practice there are a large number of review problems categorized by objective in the Student Solutions Manual. The following Sample Test (40 minutes) is intended to review the main ideas of this chapter.*

In Problems 1–17 classify the type of financial problem, state the appropriate formula, identify the variables and constants, and then answer the question.

1. Find the value of a $1,000 certificate in $2\frac{1}{2}$ years if the interest rate is 12% compounded monthly.

2. You deposit $300 at the end of each year into an account paying 12% compounded annually. How much is in the account in 10 years?

3. An insurance policy pays $10,000 in 5 years. What lump-sum deposit today will yield $10,000 in 5 years if it is deposited at 12% compounded quarterly?

4. A 5-year term policy has an annual premium of $300, and at the end of 5 years, all payments and interest are refunded. What lump-sum deposit is necessary to equal this amount if you assume an interest rate of 10% compounded annually?

5. What annual deposit is necessary to give $10,000 in 5 years if the money is deposited at 9% interest compounded annually?

6. A $5,000,000 apartment complex loan is to be paid off in 10 years by making 10 equal annual payments. How much is each payment if the interest rate is 14% compounded annually?

7. The prices of automobiles have increased at 6.25% per year. How much would you expect a $10,000 automobile to cost in 5 years?

8. The amount to be financed on a new car is $9,500. The terms are 17% for 4 years. What is the monthly payment?

9. At the beginning of each year, you deposit $450 into an account paying 11% compounded annually. How much is in the account in 25 years?

10. At the end of each year, you deposit $825 into an account paying 7.5% compounded annually. How much is in the account in 23 years?

11. What lump-sum deposit today will yield $1,000,000 in 37 years if it is deposited at 12% compounded annually?

12. What deposit today is equal to 33 annual deposits of $500 into an account paying 8% compounded annually?

13. What annual deposit is necessary into an account paying 11.5% to give $5,000 in 15 years?

14. You want to have $10,000 in 6 years by making a deposit at the end of each year into an account paying 12% interest compounded annually. How much should your annual deposit be?

15. A lottery offers you a choice of $1,000,000 per year for 5 years or a lump-sum payment. What lump-sum payment would equal the annual payments if the current interest rate is 14% compounded annually?

16. What is the monthly payment for a home costing $125,000 with a 20% down payment and the balance financed for 30 years at 12%?

17. Western Electric decides to raise $150,000,000 in order to develop a new geothermal energy source. They issue a 50-year bond with a contract rate of 9% paid annually. They also set up a fund to repay the debt. This fund pays 10% compounded annually. What is the size of the annual payment for both the interest and the bond repayment?

The following multiple-choice questions have appeared on recent Management accounting and CPA examinations. The questions are reprinted with permission of the National Association of Accountants and the American Institute of Certified Public Accountants, Inc.

Management Accounting Exam June 1983

18. A firm has daily cash receipts of $200,000. A commercial bank has offered to reduce the collection time by 3 days. The bank requires a monthly fee of $4,000 for providing this service. If money market rates will average 12% during the year, the additional annual income (loss) of having the service is
A. $(24,000)   B. $24,000   C. $66,240   D. $68,000
E. Some amount other than those given above

CPA Exam November 1976

19. A businesswoman wants to withdraw $3,000 (including principal) from an investment fund at the end of each year for 5 years. How should she compute her required initial investment at the beginning of the first year if the fund earns 6% compounded annually?

A. $3,000 times the amount of an annuity of $1 at 6% at the end of each year for 5 years
B. $3,000 divided by the amount of an annuity of $1 at 6% at the end of each year for 5 years
C. $3,000 times the present value of an annuity of $1 at 6% at the end of each year for 5 years
D. $3,000 divided by the present value of an annuity of $1 at 6% at the end of each year for 5 years

CPA Exam November 1976

20. A businessman wants to invest a certain sum of money at the end of each year for 5 years. The investment will earn 6% compounded annually. At the end of 5 years, he will need a total of $30,000 accumulated. How should he compute his required annual investment?

A. $30,000 times the amount of an annuity of $1 at 6% at the end of each year for 5 years
B. $30,000 divided by the amount of an annuity of $1 at 6% at the end of each year for 5 years
C. $30,000 times the present value of an annuity of $1 at 6% at the end of each year for 5 years
D. $30,000 divided by the present value of an annuity of $1 at 6% at the end of each year for 5 years

10.7 Cumulative Review for Chapters 8–10

1. Suppose the probability that a manufacturer will buy raw materials from three companies each month is summarized by the transition matrix below.

$$
\begin{array}{c c c c}
 & A & B & C \\
A & \begin{bmatrix} .50 \\ .30 \\ .30 \end{bmatrix} & \begin{matrix} .30 \\ .50 \\ .30 \end{matrix} & \begin{matrix} .20 \\ .20 \\ .40 \end{matrix} \\
B & & & \\
C & & &
\end{array}
$$

If the manufacturer buys from a company one time, it is more probable that the manufacturer will buy from that company the next time. If the manufacturer is currently giving one-third of the business to each of the three companies, how much business will the manufacturer do with each of the companies after 1 month?

2. Repeat Problem 1 for 2 months.

3. What are the long-term prospects for doing business with companies A, B, and C in Problem 1?

4. Find the fixed probability vector for

$$\begin{bmatrix} \frac{2}{3} & \frac{1}{3} \\ \frac{2}{5} & \frac{3}{5} \end{bmatrix}$$

5. **a.** What is an absorbing Markov chain?
   b. Find and interpret FP for

$$\begin{bmatrix} .1 & .5 & .4 \\ .5 & .3 & .2 \\ 0 & 0 & 1 \end{bmatrix}$$

The following multiple-choice question is reprinted with permission of the American Institute of Certified Public Accountants, Inc.

CPA Exam
May 1975

6. The Stat Company wants more information on the demand for its products. The following data are relevant:

Units demanded	Probability of unit demand	Total cost of units demanded
0	.10	$0
1	.15	1.00
2	.20	2.00
3	.40	3.00
4	.10	4.00
5	.05	5.00

What is the total expected value or payoff with perfect information?
A. $2.40   **B.** $7.40   **C.** $9.00   **D.** $9.15

7. Martha, a realtor, knows that the expenses for selling a house are $1,500 per listing. If she sells the house, she will receive 6% of the selling price. If another realtor sells the house, Martha will receive 3% of the selling price. If the house is unsold in 3 months, she will lose the listing and receive nothing. Now she must decide on only one of the following listings. Which should she take if she wants to maximize her mathematical expectation?

House	Value	Probability of selling the house		
		Listing agent	Another agent	Not selling
A	$125,000	.5	.4	.1
B	$168,000	.4	.5	.1
C	$225,000	.2	.3	.5

8. Determine the value of the following game, as well as both strategies.

$$\begin{bmatrix} -3 & 1 & 2 \\ -5 & 0 & 2 \\ -1 & 5 & 6 \end{bmatrix}$$

9. A secret courier for the SSO must carry a coded message from London to Paris. This courier can take a train or can fly. Her opponent from Scotland Yard must prevent her (the courier) from delivering the message and must guess which route she will take so that he (the Scotland Yard inspector) can intercept the courier. If he intercepts her on the train, he will dispose of her and her message, resulting in a "score" of 100 for Scotland Yard. On the other hand, if he intercepts her on the plane, he will only be able to intercept the message, resulting in a "score" of 25 for Scotland Yard. If the courier delivers the message, her "score" is -10. Find the strategies for the courier and the Scotland Yard inspector, and also find the value of the game.

10. What is the future value $5,680 deposited at 5.85% interest compounded daily for $2\frac{1}{2}$ years?

11. Donna will retire in 20 years and is depositing $2,000 into an IRA account at the end of each year. If the IRA is paying 9.5% interest compounded annually, how much will she have in 20 years?

12. Dale wants to retire in 20 years with the same amount of money in his retirement fund that Donna (Problem 9) has in her account, but he wants to make a one-time deposit now (because he just received an inheritance and wants to invest the money). He will place the money into a bond paying a guaranteed 9.5% interest compounded annually for the next 20 years. How much is his inheritance?

13. Carlos buys a $10,000 car and agrees to finance it for 3 years at 12% interest (APR). What is his monthly payment?

14. Richard wants to have $10,000 to purchase a car in 3 years. How much money should he deposit each month into a savings account paying 12% compounded monthly?

15. An insurance policy pays $25,000 in 20 years. What is the present value of this policy if inflation is a constant 6% per year?

Modeling Application 10

Investment Opportunities

You have just inherited $30,000 and need to decide what to do with the money. You check around and find that the current interest rate for deposits is 13.58%, while for loans it ranges from 14% to 21%. Inflation has been around 4%, but long-range estimates are that it will average about 10%. For this problem, assume that you have no unusual debts, are 30 years old, and have a home worth $60,000 with a $30,000 mortgage at an 8% interest rate with payments of $225 per month for the next 25 years.

Write a paper discussing your investment alternatives. Omit any discussion of tax considerations. For general guidelines about writing this essay, see the commentary for Modeling Application 1 on page 36.

APPENDIXES

APPENDIX A

Logic

Logic is a method of reasoning that accepts only inescapable conclusions. This is possible because of the strict way every concept is *defined*, or accepted without definition. So, we begin by defining a crucial term: a **statement**:

Statement

> A *statement* is a meaningful sentence that is either true or false, but not both true and false.

Statements can be either **simple** or **compound**. Simple statements are those without **connectives**—words such as *not, and, or, neither...nor, if ...then, either...or, unless, because,* and so on; while compound statements have at least one connective. A compound statement is formed by using connectives to combine simple statements. Because of the definition of a statement, the **truth value** of *any* statement labels it as either true (T) or false (F). The true value of a *compound* statement depends on the truth values of its component parts.

Letters such as $p, q, r, s, \ldots$ are used to denote simple statements, and then certain connectives are defined. There are three **fundamental connectives: conjunction** (*and*), **disjunction** (*or*), and **negation** (*not*). These connectives are defined by a **truth table**, which lists all possibilities. The truth values for each connective are chosen to correspond to everyday usage.

Fundamental Connectives

Conjunction:	*p and q*	symbolized by	$p \wedge q$
Disjunction:	*p or q*	symbolized by	$p \vee q$
Negation:	*not p*	symbolized by	$\sim p$

Truth table definition

p	q	$p \wedge q$	$p \vee q$	$\sim p$
T	T	T	T	F
T	F	F	T	F
F	T	F	T	T
F	F	F	F	T

The definition of the fundamental connectives, along with a truth table, are used to determine the truth or falsity of a compound statement.

EXAMPLE 1 Construct a truth table for the compound statement:

> *Alfie did **not** come last night **and** he did **not** pick up his money.*

Solution Let

> p: Alfie came last night
> q: Alfie picked up his money

Then the statement can be written $\sim p \wedge \sim q$. To begin, list all the possible combinations of truth values for the simple statements p and q:

p	q	
T	T	
T	F	
F	T	
F	F	

Insert the truth values for $\sim p$ and $\sim q$:

p	q	$\sim p$	$\sim q$	
T	T	F	F	
T	F	F	T	
F	T	T	F	
F	F	T	T	

Finally, insert the truth values for $\sim p \wedge \sim q$:

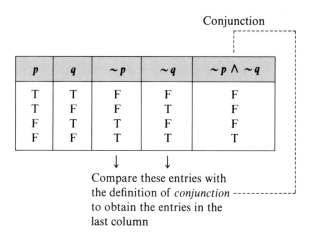

Conjunction

p	q	$\sim p$	$\sim q$	$\sim p \wedge \sim q$
T	T	F	F	F
T	F	F	T	F
F	T	T	F	F
F	F	T	T	T

Compare these entries with the definition of *conjunction* to obtain the entries in the last column

The only time the compound statement is true is when *both* p and q are false. ■

EXAMPLE 2   Construct a truth table for $\sim(\sim p)$.

Solution

Negation

p	$\sim p$	$\sim(\sim p)$
T	F	T
F	T	F

↓

Look at this column. Use
the definition of *negation*
for all the truth values in
this column

Note that $\sim(\sim p)$ and p have the same truth values. If two statements have the same truth values, one can replace the other in any logical expression. This means that the double negative of a statement is the same as the original statement.

Law of Double Negation

> $\sim(\sim p)$ may be replaced by p in any logical expression.

Truth tables can be used to prove certain useful results and to introduce some additional operators. The first one we will consider is called the **conditional**. The statement "if p, then q" is called a *conditional statement*. It is symbolized by $p \rightarrow q$; p is called the **antecedent**, and q is called the **consequent**.

Definition of the Conditional

> The *conditional* is defined by the following truth table:
>
> | p | q | $p \rightarrow q$ |
> |-----|-----|-------------------|
> | T | T | T |
> | T | F | F |
> | F | T | T |
> | F | F | T |

The *if* part of an implication need not be stated first. All the following statements have the same meaning:

Conditional translation	*Example*
If p, then q	If you are 18, then you can vote.
q, if p	You can vote, if you are 18.
p, only if q	You are 18 only if you can vote.
$p \subseteq q$	p is a subset of q.

Some statements, although not originally written as a conditional, can be put into if–then form. For example, "All ducks are birds" can be rewritten as "If it is a duck, then it is a bird." Thus we add one more form to the list:

All p are q All 18-year-olds can vote.

EXAMPLE 3   Translate the following sentence (which appeared on the 1986 federal income tax form) into symbolic form:

If you received capital gains distributions for the year and you do not need Schedule D to report any other gains or losses, do not file that schedule.

Solution   *Step 1:*   Isolate the simple statements and assign them variables. Let

c: You received capital gains distributions for the year

g: You need Schedule D to report other gains for the year

l: You need Schedule D to report other losses for the year

f: You need to file Schedule D

Your choice of variables is, of course, arbitrary, but you must be careful not to let variables represent compound statements

Step 2:   Rewrite the sentence, making substitutions for the variables.

If c and ($\sim g$ or $\sim l$), then $\sim f$

Step 3:   Complete the translation to symbols:

$$[c \wedge (\sim g \vee \sim l)] \to \sim f$$ ∎

Related to the conditional $p \to q$ are other statements, which are defined below.

Converse, Inverse, and Contrapositive

Given the conditional $p \to q$, we define:

Converse: $q \to p$
Inverse: $\sim p \to \sim q$
Contrapositive: $\sim q \to \sim p$

Not all these statements are equivalent, as shown by Table A.1.

TABLE A.1

p	q	$\sim p$	$\sim q$	Statement $p \to q$	Converse $q \to p$	Inverse $\sim p \to \sim q$	Contrapositive $\sim q \to \sim p$
T	T	F	F	T	T	T	T
T	F	F	T	F	T	T	F
F	T	T	F	T	F	F	T
F	F	T	T	T	T	T	T

Note that, if the conditional is true, the converse and inverse are not necessarily true. However, the contrapositive is always true if the original conditional is true.

Law of Contraposition

> A conditional may always be replaced by its contrapositive without having its truth value affected.

EXAMPLE 4   Consider the following:

> p: You obey the law.
> q: You will go to jail.

Given $p \to \sim q$, write the converse, inverse, and contrapositive.

Solution

Statement:	$p \to \sim q$	*Converse*:	$\sim q \to p$
Inverse:	$\sim p \to q$	*Contrapositive*:	$q \to \sim p$

The double negative $\sim(\sim q)$ is replaced by q in the inverse and contrapositive. Make this simplification whenever possible

Statement:	$p \to \sim q$	If you obey the law, then you will not go to jail.
Contrapositive:	$q \to \sim p$	If you go to jail, then you did not obey the law.
Converse:	$\sim q \to p$	If you do not go to jail, then you obey the law.
Inverse:	$\sim p \to q$	If you do not obey the law, then you will go to jail.

∎

The power of logic is in drawing conclusions not explicitly stated or known. There are three main types of reasoning that allow us to draw such conclusions. These are *direct reasoning*, *indirect reasoning*, and *transitivity*.

Direct reasoning is the drawing of a conclusion from two premises. The following example, called a **syllogism**, is an illustration:

$p \to q$	If you receive an A on the final, then you will pass the course.
p	You receive an A on the final.
q	You pass the course.

This argument consists of two *premises*, or *hypotheses*, and a *conclusion*; the argument is valid if

$$[(p \to q) \land p] \to q$$

is always true, as shown in Table A.2.

TABLE A.2
Truth table for direct reasoning

p	q	$p \to q$	$(p \to q) \land p$	$[(p \to q) \land p] \to q$
T	T	T	T	T
T	F	F	F	T
F	T	T	F	T
F	F	T	F	T

The pattern of argument is illustrated below:

Direct Reasoning

Major premise:	$p \to q$
Minor premise:	p
Conclusion:	$\therefore q$ Three dots ($\therefore$) are used to mean *therefore*

This reasoning is also sometimes called *modus ponens, law of detachment,* or *assuming the antecedent.*

EXAMPLE 5   If you play chess, then you are intelligent.
You play chess.

$\therefore$ You are intelligent. ■

EXAMPLE 6   If you are a logical person, then you will understand this example.
You are a logical person.

$\therefore$ You understand this example. ■

We say that these arguments are *valid* since we recognize them as direct reasoning. The following syllogism illustrates what we call **indirect reasoning**:

$p \to q$ If you receive an A on the final, then you will pass the course.

$\sim q$ You did not pass the course.

$\therefore \sim p$ $\therefore$ You did not receive an A on the final.

We can prove this result is valid by direct reasoning as follows:

$$[(\sim q \to \sim p) \land \sim q] \to \sim p$$

Also, since $\sim q \to \sim p$ is logically equivalent to the contrapositive $p \to q$, we have

$$[(p \to q) \land \sim q] \to \sim p$$

We can also prove that indirect reasoning is valid by using a truth table, and you are asked to do this in the problem set. The pattern of argument is illustrated below:

Indirect Reasoning

Major premise:	$p \to q$
Minor premise:	$\sim q$
Conclusion:	$\therefore \sim p$

This method is also known as reasoning by *denying the consequent* or as *modus tollens.*

EXAMPLE 7   If the cat takes the rat, then the rat will take the cheese.
The rat does not take the cheese.

$\therefore$ The cat does not take the rat. ■

EXAMPLE 8   If you received an A on the test, then I am Napoleon.

I am not Napoleon.

∴ You did not receive an A on the test.   ∎

Sometimes we must consider some extended arguments. **Transitivity** allows us to reason through several premises to some conclusion. The argument form is as follows:

Transitivity

Premise:	$p \rightarrow q$
Premise:	$q \rightarrow r$
Conclusion:	$\therefore p \rightarrow r$

Transitivity is proved by a truth table, as Table A.3 shows.

TABLE A.3
Truth table for transitivity

p	q	r	$p \rightarrow q$	$q \rightarrow r$	$p \rightarrow r$	$(p \rightarrow q) \wedge (q \rightarrow r)$	$[(p \rightarrow q) \wedge (q \rightarrow r)] \rightarrow (p \rightarrow r)$
T	T	T	T	T	T	T	T
T	T	F	T	F	F	F	T
T	F	T	F	T	T	F	T
T	F	F	F	T	F	F	T
F	T	T	T	T	T	T	T
F	T	F	T	F	T	F	T
F	F	T	T	T	T	T	T
F	F	F	T	T	T	T	T

EXAMPLE 9   If you attend class, then you will pass the course. $p \rightarrow q$

If you pass the course, then you will graduate. $q \rightarrow r$

∴ If you attend class, then you will graduate. $\therefore p \rightarrow r$

Transitivity can be extended so that a chain of several if–then sentences is connected together. For example, we could continue:

If you graduate, then you will get a good job.

If you get a good job, then you will meet the right people.

If you meet the right people, then you will become well-known.

∴ If you attend class, then you will become well-known.   ∎

Several of these argument forms may be combined into one argument. Remember:

Procedure for Drawing
Logical Conclusions

1. Translate into symbols.
2. Simplify the symbolic argument. Replace a statement by its contrapositive, use direct or indirect reasoning, or use transitivity.
3. Translate the conclusion back into words.

EXAMPLE 10 Form a valid conclusion using all these statements:

1. If I receive a check for $500, then we will go on vacation.
2. If the car breaks down, then we will not go on vacation.
3. The car breaks down.

Solution First, change into symbolic form:

1. $c \rightarrow v$	where	c = I receive a $500 check
2. $b \rightarrow \sim v$		v = We will go on vacation
3. b		b = Car breaks down

Next, simplify the symbolic argument. In this example the premises must be rearranged:

2. $b \rightarrow \sim v$ Second premise
1. $c \rightarrow v$ First premise
3. b

Now $c \rightarrow v$ is the same as $\sim v \rightarrow \sim c$ (replace the statement by its contrapositive). That is, if 1 is replaced by 1' $\sim v \rightarrow \sim c$, the following argument is obtained:

2. $\quad b \rightarrow \sim v$ Second premise
1. $\quad \underline{\sim v \rightarrow \sim c}$ Contrapositive of first premise
$\quad\ \therefore b \rightarrow \sim c$ Transitive
3. $\quad \underline{b}$ Third premise
$\quad\ \therefore \sim c$ Direct reasoning

Finally, translate the conclusion back into words: "I did not receive a check for $500." ∎

Problem Set A

According to the definition, which of the examples in Problems 1–4 are statements?

1. **a.** Hickory, Dickory, Dock, the mouse ran up the clock.
   b. $3 + 5 = 9$ **c.** Is John ugly?
   d. John has a wart on the end of his nose.

2. **a.** March 13, 1984, is Monday.
   b. Division by zero is impossible.
   c. Logic is not as difficult as I had anticipated.
   d. $4 - 6 = 2$

3. **a.** $6 + 9 \neq 7 + 8$
   b. Thomas Jefferson was the twenty-third president.
   c. Sit down and be quiet!
   d. If wages continue to rise, then prices will also rise.

4. **a.** John and Mary were married on August 3, 1979.
   b. $6 + 12 \neq 10 + 8$
   c. Do not read this sentence.
   d. Do you have a cold?

Translate the statements in Problems 5–15 into symbols. For each simple statement, be sure to indicate the meanings of the symbols you use. Answers are not unique.

5. W. C. Fields is eating, drinking, and having a good time.

6. Sam will not seek and will not accept the nomination.

7. Jack will not go tonight and Rosamond will not go tomorrow.

8. Fat Albert lives to eat and does not eat to live.

9. The decision will depend on judgment or intuition, and not on who paid the most.

10. Everything happens to everybody sooner or later if there is time enough. (G. B. Shaw)

11. We are not weak if we make a proper use of those means which the God of Nature has placed in our power. (Patrick Henry)

12. A useless life is an early death. (Goethe)

13. All work is noble. (Thomas Carlyle)

14. Everything's got a moral if only you can find it. (Lewis Carroll)

15. You don't have to itemize deductions on Schedule A or complete the worksheet if you have earned income of $2,300 or more. (1983 tax form)

Construct a truth table for the statements in Problems 16–39.

16. $\sim p \vee q$ **17.** $\sim p \wedge \sim q$ **18.** $\sim(p \wedge q)$

19. $\sim r \vee \sim s$ **20.** $\sim(\sim r)$ **21.** $(r \wedge s) \vee \sim s$

22. $p \wedge \sim q$ **23.** $\sim p \vee \sim q$

24. $(\sim p \wedge q) \vee \sim q$ **25.** $(p \wedge \sim q) \wedge p$

26. $(\sim p \vee q) \wedge (q \wedge p)$ **27.** $(p \vee q) \vee (p \wedge \sim q)$

28. $p \vee (p \rightarrow q)$ **29.** $p \rightarrow (\sim p \rightarrow q)$

30. $(p \wedge q) \rightarrow p$ **31.** $\sim p \rightarrow \sim (p \wedge q)$

32. $(p \wedge q) \wedge (p \rightarrow \sim q)$ **33.** $(p \rightarrow p) \rightarrow (q \rightarrow \sim q)$

34. $(p \rightarrow \sim q) \rightarrow (q \rightarrow \sim p)$ **35.** $(p \rightarrow q) \rightarrow (\sim q \rightarrow \sim p)$

36. $[p \wedge (p \vee q)] \rightarrow p$ **37.** $(p \wedge q) \wedge \sim r$

38. $[(p \vee q) \wedge \sim r] \wedge r$ **39.** $[p \wedge (q \vee \sim p)] \vee r$

40. Prove indirect reasoning by using a truth table.

Name the type of reasoning illustrated by the arguments in Problems 41–48.

41. If I inherit $1,000, I will buy you a cookie.
I inherit $1,000.
Therefore, I will buy you a cookie.

42. All snarks are fribbles.
All fribbles are ugly.
Therefore, all snarks are ugly.

43. If you understand a problem, it is easy.
The problem is not easy.
Therefore, you do not understand the problem.

44. If I do not get a raise in pay, I will quit.
I do not get a raise.
Therefore, I quit.

45. If Fermat's last theorem is ever proved, then my life is complete.
My life is not complete.
Therefore, Fermat's last theorem is not proved.

46. All mathematicians are eccentrics.
All eccentrics are rich.
Therefore, all mathematicians are rich.

47. No students are enthusiastic.
You are enthusiastic.
Therefore, you are not a student.

48. If you like beer, you will like Bud.
You do not like Bud.
Therefore, you do not like beer.

In Problems 49–65 form a valid conclusion using all the statements for each argument.

49. If you can learn mathematics, then you are intelligent.
If you are intelligent, then you understand human nature.

50. If I am idle, then I become lazy.
I am idle.

51. All trebbles are frebbles.
All frebbles are expensive.

52. If we interfere with the publication of false information, we are guilty of suppressing the freedom of others.
We are not guilty of suppressing the freedom of others.

53. If a nail is lost, then a shoe is lost.
If a shoe is lost, then a horse is lost.
If a horse is lost, then a rider is lost.
If a rider is lost, then a battle is lost.
If a battle is lost, then a kingdom is lost.

54. If you climb the highest mountain, you will feel great.
If you feel great, then you are happy.

55. If $b = 0$, then $a = 0$.
$a \neq 0$.

56. If $a \cdot b = 0$, then $a = 0$ or $b = 0$.
$a \cdot b = 0$.

57. If I eat that piece of pie, I will get fat.
I will not get fat.

58. If we win first prize, we will go to Europe.
If we are ingenious, we will win first prize.
We are ingenious.

59. If I can earn enough money this summer, I will attend college in the fall.
If I do not participate in student demonstrations, then I will not attend college.
I earned enough money this summer.

60. If I am tired, then I cannot finish my homework.
If I understand the material, then I can finish my homework.

61. If you go to college, then you will get a good job.
If you get a good job, then you will make a lot of money.
If you do not obey the law, then you will not make a lot of money.
You go to college.

62. All puppies are nice.
This animal is a puppy.
No nice creatures are dangerous.

63. Babies are illogical.
Nobody is despised who can manage a crocodile.
Illogical persons are despised.

64. Everyone who is sane can do logic.
No lunatics are fit to serve on a jury.
None of your sons can do logic.

65. No ducks waltz.
No officers ever decline to waltz.
All my poultry are ducks.

APPENDIX B

Mathematical Induction

Mathematical induction is an important method of proof in mathematics, allowing us to prove results involving the set of positive integers. Mathematical induction is not the scientific method or the inductive logic used in the experimental sciences; it is a form of deductive logic in which conclusions are inescapable.

The first step in establishing a result by mathematical induction is often the observation of a pattern. Let us begin with a simple example. Suppose we wish to know the sum of the first n odd integers. We could begin by looking for a pattern:

$$
\begin{aligned}
1 &= 1 \\
1 + 3 &= 4 \\
1 + 3 + 5 &= 9 \\
1 + 3 + 5 + 7 &= 16 \\
1 + 3 + 5 + 7 + 9 &= 25
\end{aligned}
$$

Do you see a pattern? It appears that the sum of the first n odd numbers is n^2 since the sum of the first three odd numbers is 3^2, of the first four odd numbers is 4^2, and so on. We now wish to *prove* deductively that

$$
1 + 3 + 5 + \cdots + \underbrace{(2n - 1)}_{\uparrow} = n^2
$$

nth odd number

is true for all positive integers n. How can we proceed? We prove certain propositions about the positive integers. The proposition is denoted by $P(n)$. For example, in the above problem we let

$$
P(n) = 1 + 3 + 5 + \cdots + (2n - 1) = n^2
$$

This means that

$$
\begin{aligned}
&P(1): &&1 = 1^2 \\
&P(2): &&1 + 3 = 2^2 \\
&P(3): &&1 + 3 + 5 = 3^2 \\
&P(4): &&1 + 3 + 5 + 7 = 4^2 \\
&&&\vdots \\
&P(100): &&1 + 3 + 5 + \cdots + 199 = 100^2 \\
&&&\vdots \\
&P(x - 1): &&1 + 3 + 5 + \cdots + (2x - 3) = (x - 1)^2 \\
&P(x): &&1 + 3 + 5 + \cdots + (2x - 1) = x^2 \\
&P(x + 1): &&1 + 3 + 5 + \cdots + (2x + 1) = (x + 1)^2
\end{aligned}
$$

Now we need to show that $P(n)$ is true for all n (n a positive integer).

Principle of Mathematical Induction (PMI)

If a given proposition $P(n)$ is true for $P(1)$ and if the truth of $P(k)$ implies the truth of the proposition for $P(k + 1)$, then $P(n)$ is true for all positive integers.

Thus, for proof by mathematical induction, we need to

1. Prove $P(1)$ is true.
2. Assume $P(k)$ is true.
3. Prove $P(k + 1)$ is true.
4. Conclude that $P(n)$ is true for all positive integers n.

Students often have a certain uneasiness when they first use the principle of mathematical induction as a method of proof. Suppose we use this principle with a stack of dominoes, as shown in the cartoon below.

Reproduced by special permission of PLAYBOY magazine; copyright ©1961 by Playboy.

How can the man in the cartoon be certain of knocking over all the dominoes? He would have to be able to knock over the first one. He would have to have the dominoes arranged so that *if* the kth domino falls, then the next one, the $(k + 1)$st, will also fall. That is, each domino is set up so that if it falls, it causes the next one to fall. This is a kind of "chain reaction." The first domino falls; this knocks over the next one (the second domino); the second one knocks over the next one (the third domino); the third one knocks over the next one; this continues until all the dominoes are knocked over.

Let us now return to the example to prove (for all positive integers n)

$$1 + 3 + 5 + \cdots + (2n - 1) = n^2$$

Step 1:   Prove $P(1)$ true: $1 = 1^2$ is true.

Step 2:   Assume $P(k)$ true: $1 + 3 + 5 + \cdots + (2k - 1) = k^2$

Step 3: Prove $P(k + 1)$ true.

TO PROVE: $1 + 3 + 5 + \cdots + [2(k + 1) - 1] = (k + 1)^2$

This is found by substituting $(k + 1)$ for n in the original statement we want to prove. Next, we simplify. This is so we will know when we are finished with step 3 of the proof.

$$1 + 3 + 5 + \cdots + [2(k + 1) - 1] = (k + 1)^2$$
$$1 + 3 + 5 + \cdots + (2k + 2 - 1) \quad = (k + 1)^2$$
$$1 + 3 + 5 + \cdots + (2k + 1) \quad\quad = (k + 1)^2$$

The procedure for step 3 is to begin with the hypotheses (from step 2) and *prove*

$$1 + 3 + 5 + \cdots + (2k + 1) = (k + 1)^2$$

Statements	Reasons
1. $1 + 3 + 5 + \cdots + (2k - 1) \qquad\qquad = k^2$	1. By hypothesis (step 2)
2. $1 + 3 + 5 + \cdots + (2k - 1) + (2k + 1) = k^2 + (2k + 1)$	2. Add $(2k + 1)$ to both sides
3. $1 + 3 + 5 + \cdots + (2k - 1) + (2k + 1) = k^2 + 2k + 1$	3. Associative
4. $1 + 3 + 5 + \cdots + (2k - 1) + (2k + 1) = (k + 1)^2$	4. Factoring (distributive)

Step 4: The proposition is true for all positive integers by the principle of mathematical induction (PMI).

EXAMPLE 1 Prove or disprove: $2 + 4 + 6 + \cdots + 2n = n(n + 1)$.

Proof *Step 1:* Prove $P(1)$ true:

$2 \overset{?}{=} 1(1 + 1)$

$2 = 2$

True.

Step 2: Assume $P(k)$. HYPOTHESIS: $2 + 4 + 6 + \cdots + 2k = k(k + 1)$

Step 3: Prove $P(k + 1)$. TO PROVE: $2 + 4 + 6 + \cdots + 2(k + 1) = (k + 1)(k + 2)$

Statements	Reasons
1. $2 + 4 + 6 + \cdots + 2k \qquad\qquad = k(k + 1)$	1. Hypothesis (step 2)
2. $2 + 4 + 6 + \cdots + 2k + 2(k + 1) = k(k + 1) + 2(k + 1)$	2. Add $2(k + 1)$ to both sides
3. $\qquad\qquad\qquad\qquad\qquad\qquad = (k + 1)(k + 2)$	3. Factor

Step 4: The proposition is true for all positive integers by PMI. ∎

EXAMPLE 2 Prove or disprove: $n^3 + 2n$ is divisible by 3.

Proof *Step 1:* Prove $P(1)$: $1^3 + 2 \cdot 1 = 3$, which is divisible by 3.

Step 2: Assume $P(k)$. HYPOTHESIS: $k^3 + 2k$ is divisible by 3.

Step 3:   Prove $P(k + 1)$. TO PROVE: $(k + 1)^3 + 2(k + 1)$ is divisible by 3.

Statements	Reasons
1. $(k + 1)^3 + 2(k + 1) = k^3 + 3k^2 + 3k + 1 + 2k + 2$	1. Distributive, associative, and commutative axioms
2. $\qquad\qquad\qquad = (3k^2 + 3k + 3) + (k^3 + 2k)$	2. Commutative and associative axioms
3. $\qquad\qquad\qquad = 3(k^2 + k + 1) + (k^3 + 2k)$	3. Distributive axiom
4. $3(k^2 + k + 1)$ is divisible by 3	4. Definition of divisibility by 3
5. $k^3 + 2k$ is divisible by 3	5. Hypothesis
6. $(k + 1)^3 + 2(k + 1)$ is divisible by 3	6. Both terms are divisible by 3 and therefore the sum is divisible by 3

Step 4:   The proposition is true for all positive integers n by PMI.   ■

Example 3 shows that *even though* we make an assumption in step 2, it is not going to help if the proposition is not true.

EXAMPLE 3   Prove or disprove: $n + 1$ is prime.

Proof   *Step 1:*   Prove $P(1)$: $1 + 1 = 2$ is a prime.

Step 2:   Assume $P(k)$. HYPOTHESIS: $k + 1$ is a prime.

Step 3:   Prove $P(k + 1)$. TO PROVE: $(k + 1) + 1$ is a prime. This is not possible since $(k + 1) + 1 = k + 2$, which is not prime whenever k is an even positive integer.

Step 4:   Any conclusions? We cannot conclude that the statement is false, only that induction does not work. But this statement is, in fact, false, and a counterexample is $n = 3$ since $n + 1 = 4$ is not prime.   ■

EXAMPLE 4   Prove or disprove: $1 \cdot 2 \cdot 3 \cdot 4 \cdot\cdots\cdot n < 0$.

Proof   Students often slip into the habit of skipping either the first or second step in a proof by mathematical induction. This is dangerous, and it is important to check every step. Suppose you do not verify the first step.

Step 2:   Assume $P(k)$. HYPOTHESIS: $1 \cdot 2 \cdot 3 \cdot\cdots\cdot k < 0$

Step 3:   Prove $P(k + 1)$. TO PROVE: $1 \cdot 2 \cdot 3 \cdot\cdots\cdot k \cdot (k + 1) < 0$
$1 \cdot 2 \cdot 3 \cdot\cdots\cdot k < 0$ by hypothesis.
$k + 1$ is positive since k is a positive integer. Then, since we know that the product of a negative and a positive is negative,

$$\underbrace{1 \cdot 2 \cdot 3 \cdot\cdots\cdot k}_{\substack{\uparrow \\ \text{Negative}}} \cdot \underbrace{(k + 1)}_{\substack{\uparrow \\ \text{Positive}}} < 0$$

Step 3 is proved.

Step 4:   The proposition is not true for all positive integers since the first step, $1 < 0$, does not hold.   ■

Problem Set B

1. Prove: $1 + 2 + 3 + \cdots + n = \dfrac{n(n+1)}{2}$

for all positive integers n.

2. Prove: $1^2 + 2^2 + 3^2 + \cdots + n^2 = \dfrac{n(n+1)(2n+1)}{6}$

for all positive integers n.

3. Prove:

$1^2 + 3^2 + 5^2 + \cdots + (2n-1)^2 = \dfrac{n(2n-1)(2n+1)}{3}$

for all positive integers n.

4. Prove: $1^3 + 2^3 + 3^3 + \cdots + n^3 = \dfrac{n^2(n+1)^2}{4}$

for all positive integers n.

5. Prove:

$2^2 + 4^2 + 6^2 + \cdots + (2n)^2 = \dfrac{2n(n+1)(2n+1)}{3}$

for all positive integers n.

6. Prove:

$1 \cdot 2 + 2 \cdot 3 + 3 \cdot 4 + \cdots + n(n+1) = \dfrac{n(n+1)(n+2)}{3}$

for all positive integers n.

7. Prove:

$1 \cdot 3 + 2 \cdot 4 + 3 \cdot 5 + \cdots + n(n+2) = \dfrac{n(n+1)(2n+7)}{6}$

for all positive integers n.

8. Prove: $1 + r + r^2 + \cdots + r^n = \dfrac{r^{n+1} - 1}{r - 1}$

for all positive integers n.

9. Prove: $n^5 - n$ is divisible by 5 for all positive integers n.

10. Prove: $n(n+1)(n+2)$ is divisible by 6 for all positive integers n.

11. Prove: $(1 + n)^2 \geq 1 + n^2$ for all positive integers n.

12.
$$1^3 = 1^2$$
$$1^3 + 2^3 = 3^2$$
$$1^3 + 2^3 + 3^3 = 6^2$$
$$1^3 + 2^3 + 3^3 + 4^3 = 10^2$$

Make a conjecture based on the above pattern and then prove or disprove your conjecture.

13.
$$1 = 1$$
$$1 + 4 = 5$$
$$1 + 4 + 7 = 12$$
$$1 + 4 + 7 + 10 = 22$$

Make a conjecture based on the above pattern and then prove or disprove your conjecture.

14. Prove: $\dbinom{k}{r} + \dbinom{k}{r-1} = \dbinom{k+1}{r}$.

Use the formula $\dbinom{n}{m} = \dfrac{n!}{m!(n-m)!}$ and not induction.
You will need this result in Problem 15.

15. The binomial theorem can be proved for any positive integer n by using mathematical induction. Fill in the missing steps and reasons.

TO PROVE:

$$(a + b)^n = \sum_{j=0}^{n} \binom{n}{j} a^{n-j} b^j$$

$$= \binom{n}{0} a^n + \binom{n}{1} a^{n-1} b + \binom{n}{2} a^{n-2} b^2 + \cdots$$

$$+ \binom{n}{r} a^{n-r} b^r + \cdots + \binom{n}{n-1} ab^{n-1} + \binom{n}{n} b^n$$

Step 1: Prove the binomial theorem is true for $n = 1$.
    a. Fill in these details.

Step 2: Assume the theorem is true for $n = k$.
    b. Fill in the statement of the hypothesis.

Step 3: Prove the theorem is true for $n = k + 1$.

TO PROVE: $(a + b)^{k+1} = \sum_{j=0}^{k+1} \binom{k+1}{j} a^{k+1-j} b^j$

BY HYPOTHESIS:
    c. Fill in the statement of the hypothesis.
    d. Fill in the details; the final simplified form is

$(a + b)^{k+1} = a^{k+1} + \cdots$

$$+ \left[\binom{k}{r} + \binom{k}{r-1} \right] a^{k-r+1} b^r + \cdots + b^{k+1}$$

    e. Use Problem 14 to complete the proof.

APPENDIX C

Calculators

This appendix provides a brief introduction to calculators and their use. In the last few years, pocket calculators have been one of the fastest selling items in the United States. This is probably because most people (including mathematicians!) do not like to do arithmetic and a good calculator can be purchased for less than $20.

This book was written with the assumption that you have or will have a calculator. This appendix is included to help you choose a calculator and understand the calculator comments in this book.

Calculators are classified by the types of problems they are equipped to handle, as well as by the type of logic for which they are programmed. The problem of selecting a calculator is compounded by the multiplicity of brands from which to choose.

The different types of calculators are distinguished primarily by their price.

1. *Four-function calculators* (under $10). These calculators have a keyboard consisting of the numerals and the four arithmetic operations, or functions: addition $\boxed{+}$, subtraction $\boxed{-}$, multiplication $\boxed{\times}$, and division $\boxed{\div}$.

2. *Four-function calculators with memory* ($10–$20). Usually these are no more expensive than four-function calculators, and offer a memory register: $\boxed{M}$, $\boxed{STO}$, or $\boxed{M^+}$. The more expensive models may have more than one memory register. Memory registers allow you to store partial calculations for later recall. Some models will even remember the total when they are turned off.

3. *Scientific calculators* ($20–$50). These calculators add additional mathematical functions, such as square root $\boxed{\sqrt{}}$, trigonometric $\boxed{\sin}$, $\boxed{\cos}$, and $\boxed{\tan}$, and logarithmic $\boxed{\log}$ and $\boxed{\exp}$. Depending on the particular brand, a scientific model may have other keys as well.

4. *Special-purpose calculators* ($40–$400). Special-use calculators for business, statistics, surveying, medicine, or even gambling and chess are available.

5. *Programmable calculators* ($50–$600). With these calculators you can enter a *sequence* of steps for the calculator to repeat on your command. Some of these calculators allow the insertion of different cards that "remember" the sequence of steps for complex calculations.

For most nonscientific purposes, a four-function calculator with memory is sufficient for everyday usage. **For this book you need a scientific calculator**. Three types of logic are used by scientific calculators: *arithmetic*, *algebraic*, and *RPN*. You need to know the type of logic used by your calculator. To determine the type of logic used by a particular calculator, try this test problem:

$$\boxed{2}\ \boxed{+}\ \boxed{3}\ \boxed{\times}\ \boxed{4}\ \boxed{=}$$

If the answer shown is 20, it is an arithmetic-logic calculator. If the answer is 14 (the correct answer), then it is an algebraic-logic calculator. If the calculator has no equal key $\boxed{=}$ but has an $\boxed{\text{ENTER}}$ or $\boxed{\text{SAVE}}$ key, then it is an **RPN**-logic calculator. An RPN-logic calculator will give the answer as 14. In algebra you learn to perform multiplication before addition, so that the correct value for

$$2 + 3 \times 4$$

is 14 (multiply first). An algebraic calculator will "know" this fact and will give the correct answer, whereas an arithmetic calculator will simply work from left to right to obtain the incorrect answer, 20. Therefore, if you have an arithmetic-logic calculator, you will need to be careful about the order of operations. Some arithmetic-logic calculators provide parentheses $\boxed{(}\boxed{)}$ so that operations can be grouped as in

$$\boxed{2}\boxed{+}\boxed{(}\boxed{3}\boxed{\times}\boxed{4}\boxed{)}\boxed{=}$$

but then you must remember to insert the parentheses.

With an RPN calculator, the operation symbol is entered after the numbers have been entered. These three types of logic can be illustrated by the problem $2 + 3 \times 4$:

Arithmetic logic	RPN logic	Algebraic logic
$\boxed{3}$	$\boxed{2}$	$\boxed{2}$
$\boxed{\times}$	$\boxed{\text{ENTER}}$	$\boxed{+}$
$\boxed{4}$	$\boxed{3}$	$\boxed{3}$
$\boxed{=}$	$\boxed{\text{ENTER}}$	$\boxed{\times}$
$\boxed{+}$	$\boxed{4}$	$\boxed{4}$
$\boxed{2}$	$\boxed{\times}$	$\boxed{=}$
$\boxed{=}$	$\boxed{+}$	

Figure C.1
a. The Sharp calculator is an example of a calculator with arithmetic logic.
b. The Hewlett-Packard calculator uses RPN logic.
c. This Texas Instruments SR-51-II is an example of a calculator with algebraic logic.

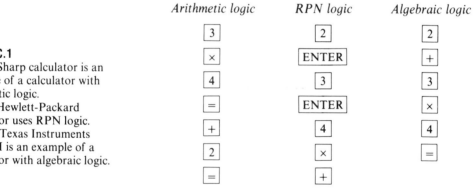

In this book the examples use algebraic- and RPN-logic calculators. The keys to be pressed are indicated by boxes drawn around the numbers and operational signs, as shown above. Numerals for calculator display are illustrated as

$$0 \; 1 \; 2 \; 3 \; 4 \; 5 \; 6 \; 7 \; 8 \; 9$$

Regardless of the type of logic your calculator uses, it is a good idea to check your owner's manual for each type of problem illustrated in the text because there are many different brands of calculators on the market, and many have slight variations in keyboards.

There is also a limit to the accuracy of your calculator. You may have a calculator with a 6-, 8-, or 10-digit display. Test its accuracy with the following example.

EXAMPLE 1   Find $2 \div 3$.

Solution   *Algebraic*: $\boxed{2}$ $\boxed{\div}$ $\boxed{3}$ $\boxed{=}$

RPN: $\boxed{2}$ $\boxed{\text{ENTER}}$ $\boxed{3}$ $\boxed{\div}$

Display: 2   2.   3   .6666666667   ∎

There may be a discrepancy between this answer and the one you obtain on your calculator. Some machines will not round the answer as shown here but will show the display

.6666666666

Others will show a display such as

6.6666 -01

This is a number in scientific notation and should be interpreted as

6.6666×10^{-1} or   .66666

Most calculators will also use scientific notation when the numbers become larger than that allowed by their display register. Example 2 shows how your calculator handles large numbers. Some calculators simply show an overflow and will not accept larger numbers. You need a calculator that will accept and handle large and small numbers.

EXAMPLE 2   Find 50^6.

Solution   Check your owner's manual to find out how to use an exponent key. Most calculators will work as shown below.

Algebraic: $\boxed{50}$ $\boxed{y^x}$ $\boxed{6}$

RPN: $\boxed{50}$ $\boxed{\text{ENTER}}$ $\boxed{6}$ $\boxed{y^x}$

When the maximum size of the display has been reached, the calculator should automatically switch to scientific notation. The point at which a calculator will do this varies from one type or brand to another. The answer for this example is **1.5625 10**, which means

$$1.5625 \times 10^{10} = 15{,}625{,}000{,}000 \qquad ∎$$

Problem Set C

Use a calculator to evaluate each of the expressions in Problems 1–31.

1. (14)(351)

2. (218)(263)

3. (4158)(0.00456)

4. $(3.00)^3(182)$

5. $(2.00)^4(1245)(277)$

6. $(6.00)^5(1456)(288)$

7. $\dfrac{(1979)(1356)}{452}$

8. $\dfrac{(515)(20,600)}{200}$

9. $\dfrac{(618)(460)(125)}{650}$

10. $[0.14 + (197)(25.08)]19$

11. $(990)(1117)(342) - 89$

12. $\dfrac{1.00}{0.005 + 0.020}$

13. $\dfrac{1.500 \times 10^4 + (7.000)(67.00)}{20,000}$

14. $(6.28)^{1/2}(4.85)$

15. $(8.23)^{1/2}(6.14)$

16. $\dfrac{1.00}{\sqrt{8.48} - \sqrt{21.3}}$

17. $\dfrac{1.00}{\sqrt{4.83} + \sqrt{2.51}}$

18. $[(4.083)^2(4.283)^3]^{-2/3}$

19. $[(6.128)^4(3.412)^2]^{-1/2}$

20. $\dfrac{16^2 + 25^2 - 9^2}{(2.0)(16)(25)}$

21. $\dfrac{216^2 + 418^2 - 315^2}{(2.00)(216)(418)}$

22. $\dfrac{4.82^2 + 6.14^2 - 9.13^2}{(2.00)(4.82)(6.14)}$

23. $\sqrt{\dfrac{(51)(36)}{212}}$

24. $\sqrt{\dfrac{25 + 49}{1 + 4(51)}}$

25. $\sqrt{\dfrac{45 + 156}{2 + 51(19)}}$

26. $\dfrac{18.361^2 + 15.215^2 - 13.815^2}{(2.0000)(18.361)(15.215)}$

27. $\dfrac{17.813^2 + 13.451^2 - 19.435^2}{2(17.813)(13.451)}$

28. $\dfrac{(2.51)^2 + (6.48)^2 - (2.51)(6.48)(.3462)}{(2.51)(6.48)}$

29. $\dfrac{241^2 + 568^2 - (241)(568)(0.5213)}{(241)(568)}$

30. $1500\left(1 + \dfrac{0.105}{12}\right)^{8(12)}$

APPENDIX D

Computers

Microcomputers are changing the curriculum today as much as calculators changed the curriculum in the last decade. The personal computer is accessible to a great many students, and more and more schools and colleges have computer labs in which programs tailored for specific classes can be used.

For this reason, *Computer Aided Finite Mathematics* has been written as a companion book to this text and is available at your book store. It explains how to use a microcomputer to help you with computationally difficult problems. The programs do not simply "do calculations," but, rather, help you through a process so that you understand each and every step. Additional examples and practice problems are also provided. The programs themselves are available on a computer disk for Apple II, IIe, IIc, IBM PC, or TRS-80 formats to schools adopting this textbook.

Computer Aided Finite Mathematics will be particularly useful in the sections of the book listed in Table D.1.

TABLE D.1
Programs on disk accompanying this book

Text Section	Options	Program number/name
1.3, 3.1 1.1, 1.4, 3.1	Graph a line Print out a table of values Graph a set of lines	1. Straight Line Plotter
2.1	Manual product (instructional option) Automatic product	2. Matrix Product

2.2	Manual reduction
	   (instructional option)
	   1. Interchange rows
	   2. Multiply a row by a constant
	   3. Divide a row by a constant
	   4. Add multiple of one row to
	      another
	Automatic reduction
		3. Row Reduction of a System

2.3	Manual inverse (instructional option) Automatic reduction	4. Matrix Inverse by Reduction
2.3, 2.4, 2.5	Solve a system of equations by finding the inverse	5. Matrix Equations

4.2, 4.3, 4.4	Manual reduction
	   (instructional option)
	   1. Interchange rows
		6. Simplex Method

Text Section	Options	Program number/name
	2. Multiply a row by a constant 3. Divide a row by a constant 4. Add multiple of one row to another Automatic reduction	
Chapter 5 Review	Program generates a random set of questions involving the fundamental counting principle, permutations, distinguishable permutations, and combinations.	7. Counting Problems
Chapter 6 Review	Program generates a random set of questions involving probability problems.	8. Probability Problems
Chapter 6 Review	Program generates a random set of questions involving union, intersection, conditional probability, and independent events.	9. Probability Properties
7.2	Mean, median, standard deviation, frequency distribution, and histogram for random or input data	10. Statistics
5.4, 7.3	Binomial coefficients Single binomial coefficient Binomial probabilities Binomial probability histogram	11. Binomial Coefficients and Probabilities
7.6	Coefficient of correlation Least squares line Random or input data	12. Correlation and Regression
	Raise a transition matrix to a power Probability vector Market share after n transitions (instructional option) Automatic reduction	13. Markov Chains
	Plays a two-person zero-sum game a large number of times when operator enters a strategy. The computer then plays either a random or intelligent strategy.	14. Two-Person Zero-Sum Game
10.1, 10.3, 10.4, 10.5	Lump sum; future value or present value Ordinary annuity or sinking fund Annuity due Present value of an annuity or installment payment	15. Formulas for Finance
Chapter 10 Review	Program generates a random set of questions involving finance problems.	16. Finance Problems
10.4, General Interest	Prints out an amortization table for paying off a loan.	17. Loan Amortization

Additional Available Software

Calculus

Calculus Blackboard. Less Hill (Wadsworth Advanced Books and Software. 511 Forest Lodge Road, Pacific Grove, CA 93950. $40; Apple). Can be used for instructor demonstrations and for graphics.

Calculus I and II (PLATO; Control Data Publishing, Higher Education Marketing Division, MNB 04 A. 3601 West 77th Street, Bloomington, MN 55435. $1,150 each; IBM PC). Nice calculus standalone, but cost is prohibitive.

Introductory Calculus Series (Microphys. 1737 West 2nd Street, Brooklyn, NY 11223. $200; Apple, IBM PC, TRS-80). Includes drill, practice, and a problem generator.

Graphing

Cactusplot. John Losse (Cactusplot. 1442 North McAllister, Tempe, AZ 85281. $60; Apple, IBM PC). Includes demonstrations and graphics.

Electronic Blackboard: Function Plotter. Richard O'Farrell et al. (Brooks/Cole. 511 Forest Lodge Road, Pacific Grove, CA 93950. $50; Apple). Includes demonstrations.

SURF. Mark Bridger (Bridge Software. 31 Champa Street, Newton Upper Falls, MA 02164. $35; IBM PC). Does three-dimensional graphing.

Surfaces for Multivariate Calculus. Roy E. Myers (CONDUIT. University of Iowa, Oakdale Campus, Iowa City, IA 52242. $65; Apple). Includes demonstrations and graphics.

Finite Mathematics

Computer-Aided Finite Mathematics. Chris Avery and Charles Barker (Brooks/Cole. 511 Forest Lodge Road, Pacific Grove, CA 93950. $16.95; Apple). Specifically designed for the finite mathematics portion of this book, with a very attractive price.

Logic

Reasoning: The Logical Process. Connie Ouding (MCE. 157 South Kalamazoo Mall, Suite 250, Kalamazoo, MI 49007. $54.95; Apple). A tutorial that includes problem solving.

Statistics

Advanced Simulation and Statistics Package. P. Lewis, E. Orav, and L. Uribe (Wadsworth Advanced Books and Software. 511 Forest Lodge Road, Pacific Grove, CA 93950. $59.95; IBM PC). Includes graphics and simulation.

KEYSTAT. H. R. Strang and A. H. Innes (Brooks/Cole. 511 Forest Lodge Road, Pacific Grove, CA 93950. $75.75; Apple, CP/M). Computational.

MacSpin (Wadsworth Advanced Books and Software. 511 Forest Lodge Road, Pacific Grove, CA 93950. $79.95; Macintosh). Graphical data analysis package.

Micro Statistics Package, The. E. A. Keller and P. Marsh (Keller, Marsh Associates. 1414 Smith Court, San Luis Obispo, CA 93401. $49.50; IBM PC, HP 110/150).

Symbolic Mathematics

MacMaple: Maple for the Macintosh. Symbolic Computation Group, University of Waterloo (Brooks/Cole Publishing Company. 511 Forest Lodge Road, Pacific Grove, CA 93950. $350; Macintosh).

muMATH, muSIMP. The Soft Warehouse (Microsoft. 10700 Northrup Way, Box 97200, Bellevue, WA 98009. $225; Apple, MS-DOS, CP/M).

PowerMath (Brainpower. 24009 Ventura Boulevard, Suite 250, Calabasas, CA 91302. $99.95; Macintosh).

TK! Solver (Lotus. 55 Cambridge Parkway, Cambridge, MA 02142. $399; Apple, Macintosh, IBM PC, TRS-80, DEC).

APPENDIX E

Tables

TABLE 1
Pascal's triangle—combinatorics

n	$\binom{n}{0}$	$\binom{n}{1}$	$\binom{n}{2}$	$\binom{n}{3}$	$\binom{n}{4}$	$\binom{n}{5}$	$\binom{n}{6}$	$\binom{n}{7}$	$\binom{n}{8}$	$\binom{n}{9}$	$\binom{n}{10}$
0	1										
1	1	1									
2	1	2	1								
3	1	3	3	1							
4	1	4	6	4	1						
5	1	5	10	10	5	1					
6	1	6	15	20	15	6	1				
7	1	7	21	35	35	21	7	1			
8	1	8	28	56	70	56	28	8	1		
9	1	9	36	84	126	126	84	36	9	1	
10	1	10	45	120	210	252	210	120	45	10	1
11	1	11	55	165	330	462	462	330	165	55	11
12	1	12	66	220	495	792	924	792	495	220	66
13	1	13	78	286	715	1287	1716	1716	1287	715	286
14	1	14	91	364	1001	2002	3003	3432	3003	2002	1001
15	1	15	105	455	1365	3003	5005	6435	6435	5005	3003
16	1	16	120	560	1820	4368	8008	11440	12870	11440	8008
17	1	17	136	680	2380	6188	12376	19448	24310	24310	19448
18	1	18	153	816	3060	8568	18564	31824	43758	48620	43758
19	1	19	171	969	3876	11628	27132	50388	75582	92378	92378
20	1	20	190	1140	4845	15504	38760	77520	125970	167960	184756

Note: $\binom{n}{m} = \dfrac{n(n-1)(n-2)\cdots\cdots(n-m+1)}{m(m-1)(m-2)\cdots\cdots 3\cdot 2\cdot 1}$; $\binom{n}{0} = 1$; $\binom{n}{1} = n$

For coefficients missing from the above table, use the relation

$$\binom{n}{m} = \binom{n}{n-m}$$

For example,

$$\binom{20}{11} = \binom{20}{9} = 167{,}960$$

TABLE 2
Squares and square roots

n	n^2	$\sqrt{n}$	$\sqrt{10n}$	n	n^2	$\sqrt{n}$	$\sqrt{10n}$
1	1	1.000	3.162	51	2601	7.141	22.583
2	4	1.414	4.472	52	2704	7.211	22.804
3	9	1.732	5.477	53	2809	7.280	23.022
4	16	2.000	6.325	54	2916	7.348	23.238
5	25	2.236	7.071	55	3025	7.416	23.452
6	36	2.449	7.746	56	3136	7.483	23.664
7	49	2.646	8.367	57	3249	7.550	23.875
8	64	2.828	8.944	58	3364	7.616	24.083
9	81	3.000	9.487	59	3481	7.681	24.290
10	100	3.162	10.000	60	3600	7.746	24.495
11	121	3.317	10.488	61	3721	7.810	24.698
12	144	3.464	10.954	62	3844	7.874	24.900
13	169	3.606	11.402	63	3969	7.937	25.100
14	196	3.742	11.832	64	4096	8.000	25.298
15	225	3.873	12.247	65	4225	8.062	25.495
16	256	4.000	12.649	66	4356	8.124	25.690
17	289	4.123	13.038	67	4489	8.185	25.884
18	324	4.243	13.416	68	4624	8.246	26.077
19	361	4.359	13.784	69	4761	8.307	26.268
20	400	4.472	14.142	70	4900	8.367	26.458
21	441	4.583	14.491	71	5041	8.426	26.646
22	484	4.690	14.832	72	5184	8.485	26.833
23	529	4.796	15.166	73	5329	8.544	27.019
24	576	4.899	15.492	74	5476	8.602	27.203
25	625	5.000	15.811	75	5625	8.660	27.386
26	676	5.099	16.125	76	5776	8.718	27.568
27	729	5.196	16.432	77	5929	8.775	27.749
28	784	5.292	16.733	78	6084	8.832	27.928
29	841	5.385	17.029	79	6241	8.888	28.107
30	900	5.477	17.321	80	6400	8.944	28.284
31	961	5.568	17.607	81	6561	9.000	28.460
32	1024	5.657	17.889	82	6724	9.055	28.636
33	1089	5.745	18.166	83	6889	9.110	28.810
34	1156	5.831	18.439	84	7056	9.165	28.983
35	1225	5.916	18.708	85	7225	9.220	29.155
36	1296	6.000	18.974	86	7396	9.274	29.326
37	1369	6.083	19.235	87	7569	9.327	29.496
38	1444	6.164	19.494	88	7744	9.381	29.665
39	1521	6.245	19.748	89	7921	9.434	29.833
40	1600	6.325	20.000	90	8100	9.487	30.000
41	1681	6.403	20.248	91	8281	9.539	30.166
42	1764	6.481	20.494	92	8464	9.592	30.332
43	1849	6.557	20.736	93	8649	9.644	30.496
44	1936	6.633	20.976	94	8836	9.695	30.659
45	2025	6.708	21.213	95	9025	9.747	30.822
46	2116	6.782	21.448	96	9216	9.798	30.984
47	2209	6.856	21.679	97	9409	9.849	31.145
48	2304	6.928	21.909	98	9604	9.899	31.305
49	2401	7.000	22.136	99	9801	9.950	31.464
50	2500	7.071	22.361	100	10000	10.000	31.623

TABLE 3
Binomial probabilities

This table computes the probability of exactly k successes in n independent binomial trials with the probability of success on a single trial equal to p. It thus contains the individual terms for specified choices of k, n, and p.* For entries $0+$, the probability is less than .0005, but greater than 0.

$$\binom{n}{k}p^k(1-p)^{n-k}$$

n	k	.01	.05	.10	.15	.20	.25	.30	.35	.40	.45	.50	k
2	0	.980	.903	.810	.723	.640	.563	.490	.423	.360	.303	.250	0
	1	.020	.095	.180	.255	.320	.375	.420	.455	.480	.495	.500	1
	2	0+	.003	.010	.023	.040	.063	.090	.122	.160	.202	.250	2
3	0	.970	.857	.729	.614	.512	.422	.343	.275	.216	.166	.125	0
	1	.029	.135	.243	.325	.384	.422	.441	.444	.432	.408	.375	1
	2	0+	.007	.027	.057	.096	.141	.189	.239	.288	.334	.375	2
	3	0+	0+	.001	.003	.008	.016	.027	.043	.064	.091	.125	3
4	0	.961	.815	.656	.522	.410	.316	.240	.179	.130	.092	.063	0
	1	.039	.171	.292	.368	.410	.422	.412	.384	.346	.299	.250	1
	2	.001	.014	.049	.098	.154	.211	.265	.311	.346	.368	.375	2
	3	0+	0+	.004	.011	.026	.047	.076	.111	.154	.200	.250	3
	4	0+	0+	0+	.001	.002	.004	.008	.015	.026	.041	.062	4
5	0	.951	.774	.590	.444	.328	.237	.168	.116	.078	.050	.031	0
	1	.048	.204	.328	.392	.410	.396	.360	.312	.259	.206	.156	1
	2	.001	.021	.073	.138	.205	.264	.309	.336	.346	.337	.313	2
	3	0+	.001	.008	.024	.051	.088	.132	.181	.230	.276	.312	3
	4	0+	0+	0+	.002	.006	.015	.028	.049	.077	.113	.156	4
	5	0+	0+	0+	0+	0+	.001	.002	.005	.010	.018	.031	5
6	0	.941	.735	.531	.377	.262	.178	.118	.075	.047	.028	.016	0
	1	.057	.232	.354	.399	.393	.356	.303	.244	.187	.136	.094	1
	2	.001	.031	.098	.176	.246	.297	.324	.328	.311	.278	.234	2
	3	0+	.002	.015	.041	.082	.132	.185	.235	.276	.303	.313	3
	4	0+	0+	.001	.005	.015	.033	.060	.095	.138	.186	.234	4
	5	0+	0+	0+	0+	.002	.004	.010	.020	.037	.061	.094	5
	6	0+	0+	0+	0+	0+	0+	.001	.002	.004	.008	.016	6
7	0	.932	.698	.478	.321	.210	.133	.082	.049	.028	.015	.008	0
	1	.066	.257	.372	.396	.367	.311	.247	.185	.131	.087	.055	1
	2	.002	.041	.124	.210	.275	.311	.318	.298	.261	.214	.164	2
	3	0+	.004	.023	.062	.115	.173	.227	.268	.290	.292	.273	3
	4	0+	0+	.003	.011	.029	.058	.097	.144	.194	.239	.273	4
	5	0+	0+	0+	.001	.004	.012	.025	.047	.077	.117	.164	5
	6	0+	0+	0+	0+	0+	.001	.004	.008	.017	.032	.055	6
	7	0+	0+	0+	0+	0+	0+	0+	.001	.002	.004	.008	7
8	0	.923	.663	.430	.272	.168	.100	.058	.032	.017	.008	.004	0
	1	.075	.279	.383	.385	.336	.267	.198	.137	.090	.055	.031	1
	2	.003	.051	.149	.238	.294	.311	.296	.259	.209	.157	.109	2
	3	0+	.005	.033	.084	.147	.208	.254	.279	.279	.257	.219	3
	4	0+	0+	.005	.018	.046	.087	.136	.188	.232	.263	.273	4
	5	0+	0+	0+	.003	.009	.023	.047	.081	.124	.172	.219	5
	6	0+	0+	0+	0+	.001	.004	.010	.022	.041	.070	.109	6
	7	0+	0+	0+	0+	0+	0+	.001	.003	.008	.016	.031	7
	8	0+	0+	0+	0+	0+	0+	0+	0+	.001	.002	.004	8

* For $p > .50$, the value of $\binom{n}{k}p^k(1-p)^{n-k}$ is found by using the table entry for $\binom{n}{n-k}(1-p)^{n-k}p^k$.

TABLE 3 (Continued)

n	k	.01	.05	.10	.15	.20	.25	.30	.35	.40	.45	.50	k
9	0	.914	.630	.387	.232	.134	.075	.040	.021	.010	.005	.002	0
	1	.083	.299	.387	.368	.302	.225	.156	.100	.060	.034	.018	1
	2	.003	.063	.172	.260	.302	.300	.267	.216	.161	.111	.070	2
	3	0+	.008	.045	.107	.176	.234	.267	.272	.251	.212	.164	3
	4	0+	.001	.007	.028	.066	.117	.172	.219	.251	.260	.246	4
	5	0+	0+	.001	.005	.017	.039	.074	.118	.167	.213	.246	5
	6	0+	0+	0+	.001	.003	.009	.021	.042	.074	.116	.164	6
	7	0+	0+	0+	0+	0+	.001	.004	.010	.021	.041	.070	7
	8	0+	0+	0+	0+	0+	0+	0+	.001	.004	.008	.018	8
	9	0+	0+	0+	0+	0+	0+	0+	0+	0+	.001	.002	9
10	0	.904	.599	.349	.197	.107	.056	.028	.013	.006	.003	.001	0
	1	.091	.315	.387	.347	.268	.188	.121	.072	.040	.021	.010	1
	2	.004	.075	.194	.276	.302	.282	.233	.176	.121	.076	.044	2
	3	0+	.010	.057	.130	.201	.250	.267	.252	.215	.166	.117	3
	4	0+	.001	.011	.040	.088	.146	.200	.238	.251	.238	.205	4
	5	0+	0+	.001	.008	.026	.058	.103	.154	.201	.234	.246	5
	6	0+	0+	0+	.001	.006	.016	.037	.069	.111	.160	.205	6
	7	0+	0+	0+	0+	.001	.003	.009	.021	.042	.075	.117	7
	8	0+	0+	0+	0+	0+	0+	.001	.004	.011	.023	.044	8
	9	0+	0+	0+	0+	0+	0+	0+	.001	.002	.004	.010	9
	10	0+	0+	0+	0+	0+	0+	0+	0+	0+	0+	.001	10
11	0	.895	.569	.314	.167	.086	.042	.020	.009	.004	.001	0+	0
	1	.099	.329	.384	.325	.236	.155	.093	.052	.027	.013	.005	1
	2	.005	.087	.213	.287	.295	.258	.200	.140	.089	.051	.027	2
	3	0+	.014	.071	.152	.221	.258	.257	.225	.177	.126	.081	3
	4	0+	.001	.016	.054	.111	.172	.220	.243	.236	.206	.161	4
	5	0+	0+	.002	.013	.039	.080	.132	.183	.221	.236	.226	5
	6	0+	0+	0+	.002	.010	.027	.057	.099	.147	.193	.226	6
	7	0+	0+	0+	0+	.002	.006	.017	.038	.070	.113	.161	7
	8	0+	0+	0+	0+	0+	.001	.004	.010	.023	.046	.081	8
	9	0+	0+	0+	0+	0+	0+	.001	.002	.005	.013	.027	9
	10	0+	0+	0+	0+	0+	0+	0+	0+	.001	.002	.005	10
	11	0+	0+	0+	0+	0+	0+	0+	0+	0+	0+	0+	11
12	0	.886	.540	.282	.142	.069	.032	.014	.006	.002	.001	0+	0
	1	.107	.341	.377	.301	.206	.127	.071	.037	.017	.008	.003	1
	2	.006	.099	.230	.292	.283	.232	.168	.109	.064	.034	.016	2
	3	0+	.017	.085	.172	.236	.258	.240	.195	.142	.092	.054	3
	4	0+	.002	.021	.068	.133	.194	.231	.237	.213	.170	.121	4
	5	0+	0+	.004	.019	.053	.103	.158	.204	.227	.222	.193	5
	6	0+	0+	0+	.004	.016	.040	.079	.128	.177	.212	.226	6
	7	0+	0+	0+	.001	.003	.011	.029	.059	.101	.149	.193	7
	8	0+	0+	0+	0+	.001	.002	.008	.020	.042	.076	.121	8
	9	0+	0+	0+	0+	0+	0+	.001	.005	.012	.028	.054	9
	10	0+	0+	0+	0+	0+	0+	0+	.001	.002	.007	.016	10
	11	0+	0+	0+	0+	0+	0+	0+	0+	0+	.001	.003	11
	12	0+	0+	0+	0+	0+	0+	0+	0+	0+	0+	0+	12
13	0	.878	.513	.254	.121	.055	.024	.010	.004	.001	0+	0+	0
	1	.115	.351	.367	.277	.179	.103	.054	.026	.011	.004	.002	1
	2	.007	.111	.245	.294	.268	.206	.139	.084	.045	.022	.010	2
	3	0+	.021	.100	.190	.246	.252	.218	.165	.111	.066	.035	3
	4	0+	.003	.028	.084	.154	.210	.234	.222	.184	.135	.087	4
	5	0+	0+	.006	.027	.069	.126	.180	.215	.221	.199	.157	5
	6	0+	0+	.001	.006	.023	.056	.103	.155	.197	.217	.209	6
	7	0+	0+	0+	.001	.006	.019	.044	.083	.131	.177	.209	7
	8	0+	0+	0+	0+	.001	.005	.014	.034	.066	.109	.157	8
	9	0+	0+	0+	0+	0+	.001	.003	.010	.024	.050	.087	9
	10	0+	0+	0+	0+	0+	0+	.001	.002	.006	.016	.035	10
	11	0+	0+	0+	0+	0+	0+	0+	0+	.001	.004	.010	11
	12	0+	0+	0+	0+	0+	0+	0+	0+	0+	0+	.002	12
	13	0+	0+	0+	0+	0+	0+	0+	0+	0+	0+	0+	13

TABLE 3 (Continued)

n	k	.01	.05	.10	.15	.20	.25	.30	.35	.40	.45	.50	k
14	0	.869	.488	.229	.103	.044	.018	.007	.002	.001	0+	0+	0
	1	.123	.359	.356	.254	.154	.083	.041	.018	.007	.003	.001	1
	2	.008	.123	.257	.291	.250	.180	.113	.063	.032	.014	.006	2
	3	0+	.026	.114	.206	.250	.240	.194	.137	.085	.046	.022	3
	4	0+	.004	.035	.100	.172	.220	.229	.202	.155	.104	.061	4
	5	0+	0+	.008	.035	.086	.147	.196	.218	.207	.170	.122	5
	6	0+	0+	.001	.009	.032	.073	.126	.176	.207	.209	.183	6
	7	0+	0+	0+	.002	.009	.028	.062	.108	.157	.195	.209	7
	8	0+	0+	0+	0+	.002	.008	.023	.051	.092	.140	.183	8
	9	0+	0+	0+	0+	0+	.002	.007	.018	.041	.076	.122	9
	10	0+	0+	0+	0+	0+	0+	.001	.005	.014	.031	.061	10
	11	0+	0+	0+	0+	0+	0+	0+	.001	.003	.009	.022	11
	12	0+	0+	0+	0+	0+	0+	0+	0+	.001	.002	.006	12
	13	0+	0+	0+	0+	0+	0+	0+	0+	0+	0+	.001	13
	14	0+	0+	0+	0+	0+	0+	0+	0+	0+	0+	0+	14
15	0	.860	.463	.206	.087	.035	.013	.005	.002	0+	0+	0+	0
	1	.130	.366	.343	.231	.132	.067	.031	.013	.005	.002	0+	1
	2	.009	.135	.267	.286	.231	.156	.092	.048	.022	.009	.003	2
	3	0+	.031	.129	.218	.250	.225	.170	.111	.063	.032	.014	3
	4	0+	.005	.043	.116	.188	.225	.219	.179	.127	.078	.042	4
	5	0+	.001	.010	.045	.103	.165	.206	.212	.186	.140	.092	5
	6	0+	0+	.002	.013	.043	.092	.147	.191	.207	.191	.153	6
	7	0+	0+	0+	.003	.014	.039	.081	.132	.177	.201	.196	7
	8	0+	0+	0+	.001	.003	.013	.035	.071	.118	.165	.196	8
	9	0+	0+	0+	0+	.001	.003	.012	.030	.061	.105	.153	9
	10	0+	0+	0+	0+	0+	.001	.003	.010	.024	.051	.092	10
	11	0+	0+	0+	0+	0+	0+	.001	.002	.007	.019	.042	11
	12	0+	0+	0+	0+	0+	0+	0+	0+	.002	.005	.014	12
	13	0+	0+	0+	0+	0+	0+	0+	0+	0+	.001	.003	13
	14	0+	0+	0+	0+	0+	0+	0+	0+	0+	0+	0+	14
	15	0+	0+	0+	0+	0+	0+	0+	0+	0+	0+	0+	15
16	0	.851	.440	.185	.074	.028	.010	.003	.001	0+	0+	0+	0
	1	.138	.371	.329	.210	.113	.053	.023	.009	.003	.001	0+	1
	2	.010	.146	.275	.277	.211	.134	.073	.035	.015	.006	.002	2
	3	0+	.036	.142	.229	.246	.208	.146	.089	.047	.022	.009	3
	4	0+	.006	.051	.131	.200	.225	.204	.155	.101	.057	.028	4
	5	0+	.001	.014	.056	.120	.180	.210	.201	.162	.112	.067	5
	6	0+	0+	.003	.018	.055	.110	.165	.198	.198	.168	.122	6
	7	0+	0+	0+	.005	.020	.052	.101	.152	.189	.197	.175	7
	8	0+	0+	0+	.001	.006	.020	.049	.092	.142	.181	.196	8
	9	0+	0+	0+	0+	.001	.006	.019	.044	.084	.132	.175	9
	10	0+	0+	0+	0+	0+	.001	.006	.017	.039	.075	.122	10
	11	0+	0+	0+	0+	0+	0+	.001	.005	.014	.034	.067	11
	12	0+	0+	0+	0+	0+	0+	0+	.001	.004	.011	.028	12
	13	0+	0+	0+	0+	0+	0+	0+	0+	.001	.003	.009	13
	14	0+	0+	0+	0+	0+	0+	0+	0+	0+	.001	.002	14
	15	0+	0+	0+	0+	0+	0+	0+	0+	0+	0+	0+	15
	16	0+	0+	0+	0+	0+	0+	0+	0+	0+	0+	0+	16
17	0	.843	.418	.167	.063	.023	.008	.002	.001	0+	0+	0+	0
	1	.145	.374	.315	.189	.096	.043	.017	.006	.002	.001	0+	1
	2	.012	.158	.280	.267	.191	.114	.058	.026	.010	.004	.001	2
	3	.001	.041	.156	.236	.239	.189	.125	.070	.034	.014	.005	3
	4	0+	.008	.060	.146	.209	.221	.187	.132	.080	.041	.018	4
	5	0+	.001	.017	.067	.136	.191	.208	.185	.138	.087	.047	5
	6	0+	0+	.004	.024	.068	.128	.178	.199	.184	.143	.094	6
	7	0+	0+	.001	.007	.027	.067	.120	.168	.193	.184	.148	7
	8	0+	0+	0+	.001	.008	.028	.064	.113	.161	.188	.185	8
	9	0+	0+	0+	0+	.002	.009	.028	.061	.107	.154	.185	9
	10	0+	0+	0+	0+	0+	.002	.009	.026	.057	.101	.148	10
	11	0+	0+	0+	0+	0+	.001	.003	.009	.024	.052	.094	11
	12	0+	0+	0+	0+	0+	0+	.001	.002	.008	.021	.047	12
	13	0+	0+	0+	0+	0+	0+	0+	.001	.002	.007	.018	13
	14	0+	0+	0+	0+	0+	0+	0+	0+	0+	.002	.005	14
	15	0+	0+	0+	0+	0+	0+	0+	0+	0+	0+	.001	15
	16	0+	0+	0+	0+	0+	0+	0+	0+	0+	0+	0+	16
	17	0+	0+	0+	0+	0+	0+	0+	0+	0+	0+	0+	17

TABLE 3 (Continued)

p

n	k	.01	.05	.10	.15	.20	.25	.30	.35	.40	.45	.50	k
18	0	.835	.397	.150	.054	.018	.006	.002	0+	0+	0+	0+	0
	1	.152	.376	.300	.170	.081	.034	.013	.004	.001	0+	0+	1
	2	.013	.168	.284	.256	.172	.096	.046	.019	.007	.002	.001	2
	3	.001	.047	.168	.241	.230	.170	.105	.055	.025	.009	.003	3
	4	0+	.009	.070	.159	.215	.213	.168	.110	.061	.029	.012	4
	5	0+	.001	.022	.079	.151	.199	.202	.166	.115	.067	.033	5
	6	0+	0+	.005	.030	.082	.144	.187	.194	.166	.118	.071	6
	7	0+	0+	.001	.009	.035	.082	.138	.179	.189	.166	.121	7
	8	0+	0+	0+	.002	.012	.038	.081	.133	.173	.186	.167	8
	9	0+	0+	0+	0+	.003	.014	.039	.079	.128	.169	.185	9
	10	0+	0+	0+	0+	.001	.004	.015	.038	.077	.125	.167	10
	11	0+	0+	0+	0+	0+	.001	.005	.015	.037	.074	.121	11
	12	0+	0+	0+	0+	0+	0+	.001	.005	.015	.035	.071	12
	13	0+	0+	0+	0+	0+	0+	0+	.001	.004	.013	.033	13
	14	0+	0+	0+	0+	0+	0+	0+	0+	.001	.004	.012	14
	15	0+	0+	0+	0+	0+	0+	0+	0+	0+	.001	.003	15
	16	0+	0+	0+	0+	0+	0+	0+	0+	0+	0+	.001	16
	17	0+	0+	0+	0+	0+	0+	0+	0+	0+	0+	0+	17
	18	0+	0+	0+	0+	0+	0+	0+	0+	0+	0+	0+	18
19	0	.826	.377	.135	.046	.014	.004	.001	0+	0+	0+	0+	0
	1	.159	.377	.285	.153	.068	.027	.009	.003	.001	0+	0+	1
	2	.014	.179	.285	.243	.154	.080	.036	.014	.005	.001	0+	2
	3	.001	.053	.180	.243	.218	.152	.087	.042	.017	.006	.002	3
	4	0+	.011	.080	.171	.218	.202	.149	.091	.047	.020	.007	4
	5	0+	.002	.027	.091	.164	.202	.192	.147	.093	.050	.022	5
	6	0+	0+	.007	.037	.095	.157	.192	.184	.145	.095	.052	6
	7	0+	0+	.001	.012	.044	.097	.153	.184	.180	.144	.096	7
	8	0+	0+	0+	.003	.017	.049	.098	.149	.180	.177	.144	8
	9	0+	0+	0+	.001	.005	.020	.051	.098	.146	.177	.176	9
	10	0+	0+	0+	0+	.001	.007	.022	.053	.098	.145	.176	10
	11	0+	0+	0+	0+	0+	.002	.008	.023	.053	.097	.144	11
	12	0+	0+	0+	0+	0+	0+	.002	.008	.024	.053	.096	12
	13	0+	0+	0+	0+	0+	0+	.001	.002	.008	.023	.052	13
	14	0+	0+	0+	0+	0+	0+	0+	.001	.002	.008	.022	14
	15	0+	0+	0+	0+	0+	0+	0+	0+	.001	.002	.007	15
	16	0+	0+	0+	0+	0+	0+	0+	0+	0+	0+	.002	16
	17	0+	0+	0+	0+	0+	0+	0+	0+	0+	0+	0+	17
	18	0+	0+	0+	0+	0+	0+	0+	0+	0+	0+	0+	18
	19	0+	0+	0+	0+	0+	0+	0+	0+	0+	0+	0+	19
20	0	.818	.358	.122	.039	.012	.003	.001	0+	0+	0+	0+	0
	1	.165	.377	.270	.137	.058	.021	.007	.002	0+	0+	0+	1
	2	.016	.189	.285	.229	.137	.067	.028	.010	.003	.001	0+	2
	3	.001	.060	.190	.243	.205	.134	.072	.032	.012	.004	.001	3
	4	0+	.013	.090	.182	.218	.190	.130	.074	.035	.014	.005	4
	5	0+	.002	.032	.103	.175	.202	.179	.127	.075	.036	.015	5
	6	0+	0+	.009	.045	.109	.169	.192	.171	.124	.075	.037	6
	7	0+	0+	.002	.016	.055	.112	.164	.184	.166	.122	.074	7
	8	0+	0+	0+	.005	.022	.061	.114	.161	.180	.162	.120	8
	9	0+	0+	0+	.001	.007	.027	.065	.116	.160	.177	.160	9
	10	0+	0+	0+	0+	.002	.010	.031	.069	.117	.159	.176	10
	11	0+	0+	0+	0+	0+	.003	.012	.034	.071	.119	.160	11
	12	0+	0+	0+	0+	0+	.001	.004	.014	.035	.073	.120	12
	13	0+	0+	0+	0+	0+	0+	.001	.004	.015	.037	.074	13
	14	0+	0+	0+	0+	0+	0+	0+	.001	.005	.015	.037	14
	15	0+	0+	0+	0+	0+	0+	0+	0+	.001	.005	.015	15
	16	0+	0+	0+	0+	0+	0+	0+	0+	0+	.001	.005	16
	17	0+	0+	0+	0+	0+	0+	0+	0+	0+	0+	.001	17
	18	0+	0+	0+	0+	0+	0+	0+	0+	0+	0+	0+	18
	19	0+	0+	0+	0+	0+	0+	0+	0+	0+	0+	0+	19
	20	0+	0+	0+	0+	0+	0+	0+	0+	0+	0+	0+	20

TABLE 4
Standard normal cumulative distribution

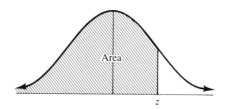

z	0.00	0.01	0.02	0.03	0.04	0.05	0.06	0.07	0.08	0.09
−3.4	0.0003	0.0003	0.0003	0.0003	0.0003	0.0003	0.0003	0.0003	0.0003	0.0002
−3.3	0.0005	0.0005	0.0005	0.0004	0.0004	0.0004	0.0004	0.0004	0.0004	0.0003
−3.2	0.0007	0.0007	0.0006	0.0006	0.0006	0.0006	0.0006	0.0005	0.0005	0.0005
−3.1	0.0010	0.0009	0.0009	0.0009	0.0008	0.0008	0.0008	0.0008	0.0007	0.0007
−3.0	0.0013	0.0013	0.0013	0.0012	0.0012	0.0011	0.0011	0.0011	0.0010	0.0010
−2.9	0.0019	0.0018	0.0017	0.0017	0.0016	0.0016	0.0015	0.0015	0.0014	0.0014
−2.8	0.0026	0.0025	0.0024	0.0023	0.0023	0.0022	0.0021	0.0021	0.0020	0.0019
−2.7	0.0035	0.0034	0.0033	0.0032	0.0031	0.0030	0.0029	0.0028	0.0027	0.0026
−2.6	0.0047	0.0045	0.0044	0.0043	0.0041	0.0040	0.0039	0.0038	0.0037	0.0036
−2.5	0.0062	0.0060	0.0059	0.0057	0.0055	0.0054	0.0052	0.0051	0.0049	0.0048
−2.4	0.0082	0.0080	0.0078	0.0075	0.0073	0.0071	0.0069	0.0068	0.0066	0.0064
−2.3	0.0107	0.0104	0.0102	0.0099	0.0096	0.0094	0.0091	0.0089	0.0087	0.0084
−2.2	0.0139	0.0136	0.0132	0.0129	0.0125	0.0122	0.0119	0.0116	0.0113	0.0110
−2.1	0.0179	0.0174	0.0170	0.0166	0.0162	0.0158	0.0154	0.0150	0.0146	0.0143
−2.0	0.0228	0.0222	0.0217	0.0212	0.0207	0.0202	0.0197	0.0192	0.0188	0.0183
−1.9	0.0287	0.0281	0.0274	0.0268	0.0262	0.0256	0.0250	0.0244	0.0239	0.0233
−1.8	0.0359	0.0352	0.0344	0.0336	0.0329	0.0322	0.0314	0.0307	0.0301	0.0294
−1.7	0.0446	0.0436	0.0427	0.0418	0.0409	0.0401	0.0392	0.0384	0.0375	0.0367
−1.6	0.0548	0.0537	0.0526	0.0516	0.0505	0.0495	0.0485	0.0475	0.0465	0.0455
−1.5	0.0668	0.0655	0.0643	0.0630	0.0618	0.0606	0.0594	0.0582	0.0571	0.0559
−1.4	0.0808	0.0793	0.0778	0.0764	0.0749	0.0735	0.0722	0.0708	0.0694	0.0681
−1.3	0.0968	0.0951	0.0934	0.0918	0.0901	0.0885	0.0869	0.0853	0.0838	0.0823
−1.2	0.1151	0.1131	0.1112	0.1093	0.1075	0.1056	0.1038	0.1020	0.1003	0.0985
−1.1	0.1357	0.1335	0.1314	0.1292	0.1271	0.1251	0.1230	0.1210	0.1190	0.1170
−1.0	0.1587	0.1562	0.1539	0.1515	0.1492	0.1469	0.1446	0.1423	0.1401	0.1379
−0.9	0.1841	0.1814	0.1788	0.1762	0.1736	0.1711	0.1685	0.1660	0.1635	0.1611
−0.8	0.2119	0.2090	0.2061	0.2033	0.2005	0.1977	0.1949	0.1922	0.1894	0.1867
−0.7	0.2420	0.2389	0.2358	0.2327	0.2296	0.2266	0.2236	0.2206	0.2177	0.2148
−0.6	0.2743	0.2709	0.2676	0.2643	0.2611	0.2578	0.2546	0.2514	0.2483	0.2451
−0.5	0.3085	0.3050	0.3015	0.2981	0.2946	0.2912	0.2877	0.2843	0.2810	0.2776
−0.4	0.3446	0.3409	0.3372	0.3336	0.3300	0.3264	0.3228	0.3192	0.3156	0.3121
−0.3	0.3821	0.3783	0.3745	0.3707	0.3669	0.3632	0.3594	0.3557	0.3520	0.3483
−0.2	0.4207	0.4168	0.4129	0.4090	0.4052	0.4013	0.3974	0.3936	0.3897	0.3859
−0.1	0.4602	0.4562	0.4522	0.4483	0.4443	0.4404	0.4364	0.4325	0.4286	0.4247
−0.0	0.5000	0.4960	0.4920	0.4880	0.4840	0.4801	0.4761	0.4721	0.4681	0.4641

TABLE 4 (Continued)

z	0.00	0.01	0.02	0.03	0.04	0.05	0.06	0.07	0.08	0.09
0.0	0.5000	0.5040	0.5080	0.5120	0.5160	0.5199	0.5239	0.5279	0.5319	0.5359
0.1	0.5398	0.5438	0.5478	0.5517	0.5557	0.5596	0.5636	0.5675	0.5714	0.5753
0.2	0.5793	0.5832	0.5871	0.5910	0.5948	0.5987	0.6026	0.6064	0.6103	0.6141
0.3	0.6179	0.6217	0.6255	0.6293	0.6331	0.6368	0.6406	0.6443	0.6480	0.6517
0.4	0.6554	0.6591	0.6628	0.6664	0.6700	0.6736	0.6772	0.6808	0.6844	0.6879
0.5	0.6915	0.6950	0.6985	0.7019	0.7054	0.7088	0.7123	0.7157	0.7190	0.7224
0.6	0.7257	0.7291	0.7324	0.7357	0.7389	0.7422	0.7454	0.7486	0.7517	0.7549
0.7	0.7580	0.7611	0.7642	0.7673	0.7704	0.7734	0.7764	0.7794	0.7823	0.7852
0.8	0.7881	0.7910	0.7939	0.7967	0.7995	0.8023	0.8051	0.8078	0.8106	0.8133
0.9	0.8159	0.8186	0.8212	0.8238	0.8264	0.8289	0.8315	0.8340	0.8365	0.8389
1.0	0.8413	0.8438	0.8461	0.8485	0.8508	0.8531	0.8554	0.8577	0.8599	0.8621
1.1	0.8643	0.8665	0.8686	0.8708	0.8729	0.8749	0.8770	0.8790	0.8810	0.8830
1.2	0.8849	0.8869	0.8888	0.8907	0.8925	0.8944	0.8962	0.8980	0.8997	0.9015
1.3	0.9032	0.9049	0.9066	0.9082	0.9099	0.9115	0.9131	0.9147	0.9162	0.9177
1.4	0.9192	0.9207	0.9222	0.9236	0.9251	0.9265	0.9278	0.9292	0.9306	0.9319
1.5	0.9332	0.9345	0.9357	0.9370	0.9382	0.9394	0.9406	0.9418	0.9429	0.9441
1.6	0.9452	0.9463	0.9474	0.9484	0.9495	0.9505	0.9515	0.9525	0.9535	0.9545
1.7	0.9554	0.9564	0.9573	0.9582	0.9591	0.9599	0.9608	0.9616	0.9625	0.9633
1.8	0.9641	0.9649	0.9656	0.9664	0.9671	0.9678	0.9686	0.9693	0.9699	0.9706
1.9	0.9713	0.9719	0.9726	0.9732	0.9738	0.9744	0.9750	0.9756	0.9761	0.9767
2.0	0.9772	0.9778	0.9783	0.9788	0.9793	0.9798	0.9803	0.9808	0.9812	0.9817
2.1	0.9821	0.9826	0.9830	0.9834	0.9838	0.9842	0.9846	0.9850	0.9854	0.9857
2.2	0.9861	0.9864	0.9868	0.9871	0.9875	0.9878	0.9881	0.9884	0.9887	0.9890
2.3	0.9893	0.9896	0.9898	0.9901	0.9904	0.9906	0.9909	0.9911	0.9913	0.9916
2.4	0.9918	0.9920	0.9922	0.9925	0.9927	0.9929	0.9931	0.9932	0.9934	0.9936
2.5	0.9938	0.9940	0.9941	0.9943	0.9945	0.9946	0.9948	0.9949	0.9951	0.9952
2.6	0.9953	0.9955	0.9956	0.9957	0.9959	0.9960	0.9961	0.9962	0.9963	0.9964
2.7	0.9965	0.9966	0.9967	0.9968	0.9969	0.9970	0.9971	0.9972	0.9973	0.9974
2.8	0.9974	0.9975	0.9976	0.9977	0.9977	0.9978	0.9979	0.9979	0.9980	0.9981
2.9	0.9981	0.9982	0.9982	0.9983	0.9984	0.9984	0.9985	0.9985	0.9986	0.9986
3.0	0.9987	0.9987	0.9987	0.9988	0.9988	0.9989	0.9989	0.9989	0.9990	0.9990
3.1	0.9990	0.9991	0.9991	0.9991	0.9992	0.9992	0.9992	0.9992	0.9993	0.9993
3.2	0.9993	0.9993	0.9994	0.9994	0.9994	0.9994	0.9994	0.9995	0.9995	0.9995
3.3	0.9995	0.9995	0.9995	0.9996	0.9996	0.9996	0.9996	0.9996	0.9996	0.9997
3.4	0.9997	0.9997	0.9997	0.9997	0.9997	0.9997	0.9997	0.9997	0.9997	0.9998

TABLE 5
Compound interest

Compounded amount of $1 for N periods at i% per period.

N	1%	1½%	2%	2½%	3%	3½%	4%	N
1	1.010000	1.015000	1.020000	1.025000	1.030000	1.035000	1.040000	1
2	1.020100	1.030225	1.040400	1.050625	1.060900	1.071225	1.081600	2
3	1.030301	1.045678	1.061208	1.076891	1.092727	1.108718	1.124864	3
4	1.040604	1.061364	1.082432	1.103813	1.125509	1.147523	1.169859	4
5	1.051010	1.077284	1.104081	1.131408	1.159274	1.187686	1.216653	5
6	1.061520	1.093443	1.126162	1.159693	1.194052	1.229255	1.265319	6
7	1.072135	1.109845	1.148686	1.188686	1.229874	1.272279	1.315932	7
8	1.082857	1.126493	1.171659	1.218403	1.266770	1.316809	1.368569	8
9	1.093685	1.143390	1.195093	1.248863	1.304773	1.362897	1.423312	9
10	1.104622	1.160541	1.218994	1.280085	1.343916	1.410599	1.480244	10
11	1.115668	1.177949	1.243374	1.312087	1.384234	1.459970	1.539454	11
12	1.126825	1.195618	1.268242	1.344889	1.425761	1.511069	1.601032	12
13	1.138093	1.213552	1.293607	1.378511	1.468534	1.563956	1.665074	13
14	1.149474	1.231756	1.319479	1.412974	1.512590	1.618695	1.731676	14
15	1.169069	1.250232	1.345868	1.448298	1.557967	1.675349	1.800944	15
16	1.172579	1.268986	1.372786	1.484506	1.604706	1.733986	1.872981	16
17	1.184304	1.288020	1.400241	1.521618	1.652848	1.794676	1.947900	17
18	1.196147	1.307341	1.428246	1.559659	1.702433	1.857489	2.025817	18
19	1.208109	1.326951	1.456811	1.598650	1.753506	1.922501	2.106849	19
20	1.220190	1.346855	1.485947	1.638616	1.806111	1.989789	2.191123	20
25	1.282432	1.450945	1.640606	1.853944	2.093778	2.363245	2.665836	25
30	1.347849	1.563080	1.811362	2.097568	2.427262	2.806794	3.243398	30
35	1.416603	1.683881	1.999890	2.373205	2.813862	3.333590	3.946089	35
40	1.488864	1.814018	2.208040	2.685064	3.262038	3.959260	4.801021	40
45	1.564811	1.954213	2.437854	3.037903	3.781596	4.702359	5.841176	45
50	1.644632	2.105242	2.691588	3.437109	4.383906	5.584927	7.106683	50
55	1.728525	2.267944	2.971731	3.888773	5.082149	6.633141	8.646367	55
60	1.816697	2.443220	3.281031	4.399790	5.891603	7.878091	10.519627	60
65	1.909366	2.632042	3.622523	4.977958	6.829983	9.356701	12.798735	65
70	2.006763	2.835456	3.999558	5.632103	7.917822	11.112825	15.571618	70
75	2.109128	3.054592	4.415835	6.372207	9.178926	13.198550	18.945255	75
80	2.216715	3.290663	4.875439	7.209568	10.640891	15.675738	23.049799	80
85	2.329790	3.544978	5.382879	8.156964	12.335709	18.617859	28.043605	85
90	2.448633	3.818949	5.943133	9.228856	14.300467	22.112176	34.119333	90
95	2.573538	4.114092	6.561699	10.441604	16.578161	26.262329	41.511386	95
100	2.704814	4.432046	7.244646	11.813716	19.218632	31.191408	50.504948	100

N	4½%	5%	5½%	6%	6½%	7%	7½%	N
1	1.045000	1.050000	1.055000	1.060000	1.065000	1.070000	1.075000	1
2	1.092025	1.102500	1.113025	1.123600	1.134225	1.144900	1.155625	2
3	1.141166	1.157625	1.174241	1.191016	1.207950	1.225043	1.242297	3
4	1.192519	1.215506	1.238825	1.262477	1.286466	1.310796	1.335469	4
5	1.246182	1.276282	1.306960	1.338226	1.370087	1.402552	1.435629	5
6	1.302260	1.340096	1.378843	1.418519	1.459142	1.500730	1.543302	6
7	1.360862	1.407100	1.454679	1.503630	1.553987	1.605781	1.659049	7
8	1.422101	1.477455	1.534687	1.593848	1.654996	1.718186	1.783478	8
9	1.486147	1.551328	1.619094	1.689479	1.762570	1.838459	1.917239	9
10	1.552969	1.628895	1.708144	1.790848	1.877137	1.967151	2.061032	10
11	1.622853	1.710339	1.802092	1.898299	1.999151	2.104852	2.215609	11
12	1.695881	1.795856	1.901207	2.012196	2.129096	2.252192	2.381780	12
13	1.772196	1.885649	2.005774	2.132928	2.267487	2.409845	2.560413	13
14	1.851945	1.979932	2.116091	2.260904	2.414874	2.578534	2.752444	14
15	1.935282	2.078928	2.232476	2.396558	2.571841	2.759032	2.958877	15
16	2.022370	2.182875	2.355263	2.540352	2.739011	2.952164	3.180793	16
17	2.113377	2.292018	2.484802	2.692773	2.917046	3.158815	3.419353	17
18	2.208479	2.406619	2.621466	2.854339	3.106654	3.379932	3.675804	18
19	2.307860	2.526950	2.765647	3.025600	3.308587	3.616528	3.951489	19
20	2.411714	2.653298	2.917757	3.207135	3.523645	3.869684	4.247851	20
25	3.005434	3.386355	3.813392	4.291871	4.827699	5.427433	6.098340	25
30	3.745318	4.321942	4.983951	5.743491	6.614366	7.612255	8.754955	30
35	4.667348	5.516015	6.513825	7.686087	9.062255	10.676581	12.568870	35
40	5.816365	7.039989	8.513309	10.285718	12.416075	14.974458	18.044239	40
45	7.248248	8.985008	11.126554	13.764611	17.011098	21.002452	25.904839	45
50	9.032636	11.467400	14.541961	18.420154	23.306679	29.457025	37.189746	50
55	11.256308	14.635631	19.005762	24.650322	31.932170	41.315001	53.390690	55
60	14.027408	18.679186	24.839770	32.987691	43.749840	57.946427	76.649240	60
65	17.480702	23.839701	32.464587	44.144972	59.941072	81.272861	110.039897	65
70	21.784136	30.426426	42.429916	59.075930	82.124463	113.989392	157.976504	70
75	27.146996	38.832686	55.454204	79.056921	112.517632	159.876019	226.795701	75
80	33.830096	49.561441	72.476426	105.795993	154.158907	224.234388	325.594560	80
85	42.158455	63.254353	94.723791	141.578904	211.211062	314.500328	467.433099	85
90	52.537105	80.730365	123.800206	189.464511	289.377460	441.102980	671.060665	90
95	65.470792	103.034676	161.801918	253.546255	396.472198	618.669748	963.394370	95
100	81.588518	131.501258	211.468636	339.302084	543.201271	867.716326	1383.077210	100

TABLE 5 (Continued)
Future and present value

Use this table to find the future value (multiply principle by table number) and to find present value (divide future value by table number).

N	8%	9%	10%	11%	12%	13%	14%	N
1	1.080000	1.090000	1.100000	1.110000	1.120000	1.130000	1.140000	1
2	1.166400	1.188100	1.210000	1.232100	1.254400	1.276900	1.299600	2
3	1.259712	1.295029	1.331000	1.367631	1.404928	1.442897	1.481544	3
4	1.360489	1.411582	1.464100	1.518070	1.573519	1.630474	1.688960	4
5	1.469328	1.538624	1.610510	1.685058	1.762342	1.842435	1.925415	5
6	1.586874	1.677100	1.771561	1.870415	1.973823	2.081952	2.194973	6
7	1.713824	1.828039	1.948717	2.076160	2.210681	2.352605	2.502269	7
8	1.850930	1.992563	2.143589	2.304538	2.475963	2.658444	2.852586	8
9	1.999005	2.171893	2.357948	2.558037	2.773079	3.004042	3.251949	9
10	2.158925	2.367364	2.593742	2.839421	3.105848	3.394567	3.707221	10
11	2.331639	2.580426	2.853117	3.151757	3.478550	3.835861	4.226232	11
12	2.518170	2.812665	3.138428	3.498451	3.895976	4.334523	4.817905	12
13	2.719624	3.065805	3.452271	3.883280	4.363493	4.898011	5.492411	13
14	2.937194	3.341727	3.797498	4.310441	4.887112	5.534753	6.261349	14
15	3.172169	3.642482	4.177248	4.784589	5.473566	6.254270	7.137938	15
16	3.425943	3.970306	4.594973	5.310894	6.130394	7.067326	8.137249	16
17	3.700018	4.327633	5.054470	5.895093	6.866041	7.986078	9.276464	17
18	3.996019	4.717120	5.559917	6.543553	7.689966	9.024268	10.575169	18
19	4.315701	5.141661	6.115909	7.263344	8.612762	10.197423	12.055693	19
20	4.660957	5.604411	6.727500	8.062312	9.646293	11.523088	13.743490	20
25	6.848475	8.623081	10.834706	13.585464	17.000064	21.230542	26.461916	25
30	10.062657	13.267678	17.449402	22.892297	29.959922	39.115898	50.950159	30
35	14.785344	20.413968	28.102437	38.574851	52.799620	72.068506	98.100178	35
40	21.724521	31.409420	45.259256	65.000867	93.050970	132.781552	188.883514	40
45	31.920449	48.327286	72.890484	109.530242	163.987604	244.641402	363.679072	45
50	46.901613	74.357520	117.390853	184.564827	289.002190	450.735925	700.232988	50
55	68.913856	114.408262	189.059142	311.002466	509.320606	830.451725	1348.238807	55
60	101.257064	176.031292	304.481640	524.057242	897.596933	1530.053473	2595.918660	60
65	148.779847	270.845963	490.370725	883.066930	1581.872491	2819.024345	4998.219642	65
70	218.606406	416.730086	789.746957	1488.019132	2787.799828	5193.869624	9623.644985	70
75	321.204530	641.190893	1271.895371	2507.398773	4913.055841	9569.368113	18529.506390	75
80	471.954834	986.551668	2048.400215	4225.112750	8658.483100	17630.940454	35676.981807	80
85	693.456489	1517.932029	3298.969030	7119.560696	15259.205681	32483.864937	68692.981028	85
90	1018.915089	2335.526582	5313.022612	11996.873812	26891.934223	59849.415520	132262.467379	90
95	1497.120549	3593.497147	8556.676047	20215.430053	47392.776624	110268.668614	254660.083396	95
100	2199.761256	5529.040792	13780.612340	34064.175270	83522.265727	203162.874228	490326.238126	100

N	15%	16%	17%	18%	19%	20%	N
1	1.150000	1.160000	1.170000	1.180000	1.190000	1.200000	1
2	1.322500	1.345600	1.368900	1.392400	1.416100	1.440000	2
3	1.520875	1.560896	1.601613	1.643032	1.685159	1.728000	3
4	1.749006	1.810639	1.873887	1.938778	2.005339	2.073600	4
5	2.011357	2.100342	2.192448	2.287758	2.386354	2.488320	5
6	2.313061	2.436396	2.565164	2.699554	2.839761	2.985984	6
7	2.660020	2.826220	3.001242	3.185474	3.379315	3.583181	7
8	3.059023	3.278415	3.511453	3.758859	4.021385	4.299817	8
9	3.517876	3.802961	4.108400	4.435454	4.785449	5.159780	9
10	4.045558	4.411435	4.806828	5.233836	5.694684	6.191736	10
11	4.652391	5.117265	5.623989	6.175926	6.776674	7.430084	11
12	5.350250	5.936027	6.580067	7.287593	8.064242	8.916100	12
13	6.152788	6.885791	7.698679	8.599319	9.596448	10.699321	13
14	7.075706	7.987518	9.007454	10.147244	11.419773	12.839185	14
15	8.137062	9.265521	10.538721	11.973748	13.589530	15.407022	15
16	9.357621	10.748004	12.330304	14.129023	16.171540	18.488426	16
17	10.761264	12.467685	14.426456	16.672247	19.244133	22.186111	17
18	12.375454	14.462514	16.878953	19.673251	22.900518	26.623333	18
19	14.231772	16.776517	19.748375	23.214436	27.251616	31.948000	19
20	16.366537	19.460759	23.105599	27.393035	32.429423	38.337600	20
25	32.918953	40.874244	50.657826	62.668627	77.388073	95.396217	25
30	66.211772	85.849877	111.064650	143.370638	184.675312	237.376314	30
35	133.175523	180.314073	243.503474	327.997290	440.700607	590.668229	35
40	267.863546	378.721158	533.868713	750.378345	1051.667507	1469.771568	40
45	538.769269	795.443826	1170.479411	1716.683879	2509.650603	3657.261988	45
50	1083.657442	1670.703804	2566.215284	3927.356860	5988.913902	9100.438150	50
55	2179.622184	3509.048796	5626.293659	8984.841120	14291.666609	22644.802257	55
60	4383.998746	7370.201365	12335.356482	20555.139966	34104.970919	56347.514353	60
65	8817.787387	15479.940952	27044.628088	47025.180900	81386.522174	140210.646915	65
70	17735.720039	32513.164839	59293.941729	107582.222368	194217.025056	348888.956932	70
75	35672.867976	68288.754533	129998.886072	246122.063716	463470.508558	868147.369314	75
80	71750.879401	143429.715890	285015.802412	563067.660386	1106004.544354	2160228.462010	80
85	144316.646994	301251.407222	624882.336142	1288162.407650	2639317.992285	5375339.686589	85
90	290272.325206	632730.879999	1370022.050417	2947003.540121	6298346.150529	13375565.248934	90
95	583841.327636	1328551.025313	3003702.153303	6742030.208228	15030081.387632	33282686.520228	95
100	1174313.450700	2791651.199375	6585460.885837	15424131.905453	35867089.727971	82817974.522015	100

TABLE 6
Ordinary annuity

Amount of ordinary annuity of $1 at compound interest. Remember, an ordinary annuity requires payment at the end of the period.

N	1%	1½%	2%	2½%	3%	3½%	4%	N
1	1.000000	1.000000	1.000000	1.000000	1.000000	1.000000	1.000000	1
2	2.010000	2.015000	2.020000	2.025000	2.030000	2.035000	2.040000	2
3	3.030100	3.045225	3.060400	3.075625	3.090900	3.106225	3.121600	3
4	4.060401	4.090903	4.121608	4.152516	4.183627	4.214943	4.246464	4
5	5.101005	5.152267	5.204040	5.256329	5.309136	5.362466	5.416323	5
6	6.152015	6.229551	6.308121	6.387737	6.468410	6.550152	6.632975	6
7	7.213535	7.322994	7.434283	7.547430	7.662462	7.779408	7.898294	7
8	8.285671	8.432839	8.582969	8.736116	8.892336	9.051687	9.214226	8
9	9.368527	9.559332	9.754628	9.954519	10.159106	10.368496	10.582795	9
10	10.462213	10.702722	10.949721	11.203382	11.463879	11.731393	12.006107	10
11	11.566835	11.863262	12.168715	12.483466	12.807796	13.141992	13.486351	11
12	12.682503	13.041211	13.412090	13.795553	14.192030	14.601962	15.025805	12
13	13.809328	14.236830	14.680332	15.140442	15.617790	16.113030	16.626838	13
14	14.947421	15.450382	15.973938	16.518953	17.086324	17.676986	18.291911	14
15	16.096896	16.682138	17.293417	17.931927	18.598914	19.295681	20.023588	15
16	17.257864	17.932370	18.639285	19.380225	20.156881	20.971030	21.824531	16
17	18.430443	19.201355	20.012071	20.864730	21.761588	22.705016	23.697512	17
18	19.614748	20.489376	21.412312	22.386349	23.414435	24.499691	25.645413	18
19	20.810895	21.796716	22.840559	23.946007	25.116868	26.357180	27.671229	19
20	22.019004	23.123667	24.297370	25.544658	26.870374	28.279682	29.778079	20
25	28.243200	30.063024	32.030300	34.157764	36.459264	38.949857	41.645908	25
30	34.784892	37.538681	40.568079	43.902703	47.575416	51.622677	56.084938	30
35	41.660276	45.592088	49.994478	54.928207	60.462082	66.674013	73.652225	35
40	48.886373	54.267894	60.401983	67.402554	75.401260	84.550278	95.025516	40
45	56.481075	63.614201	71.892710	81.516131	92.719861	105.781673	121.029392	45
50	64.463182	73.682828	84.579401	97.484349	112.796867	130.997910	152.667084	50
55	72.852457	84.529599	98.586534	115.550921	136.071620	160.946890	191.159173	55
60	81.669670	96.214652	114.051539	135.991590	163.053437	196.516883	237.990685	60
65	90.936649	108.802772	131.126155	159.118330	194.332758	238.762876	294.968380	65
70	100.676337	122.363753	149.977911	185.284114	230.594064	288.937865	364.290459	70
75	110.912847	136.972781	170.791773	214.888297	272.630856	348.530011	448.631367	75
80	121.671522	152.710852	193.771958	248.382713	321.363019	419.306787	551.244977	80
85	132.978997	169.665226	219.143939	286.278570	377.856952	503.367394	676.000123	85
90	144.863267	187.929900	247.156656	329.154253	443.348904	603.205027	827.983334	90
95	157.353755	207.606142	278.084960	377.664154	519.272026	721.780816	1012.784648	95
100	170.481383	228.803043	312.232306	432.548654	607.287733	862.611657	1237.623705	100

N	4½%	5%	5½%	6%	6½%	7%	7½%	N
1	1.000000	1.000000	1.000000	1.000000	1.000000	1.000000	1.000000	1
2	2.045000	2.050000	2.055000	2.060000	2.065000	2.070000	2.075000	2
3	3.137025	3.152500	3.168025	3.183600	3.199225	3.214900	3.230625	3
4	4.278191	4.310125	4.342266	4.374616	4.407175	4.439943	4.472922	4
5	5.470710	5.525631	5.581091	5.637093	5.693641	5.750739	5.808391	5
6	6.716892	6.801913	6.888051	6.975319	7.063728	7.153291	7.244020	6
7	8.019152	8.142008	8.266894	8.393838	8.522870	8.654021	8.787322	7
8	9.380014	9.549109	9.721573	9.897468	10.076856	10.259803	10.446371	8
9	10.802114	11.026564	11.256260	11.491316	11.731852	11.977989	12.229849	9
10	12.288209	12.577893	12.875354	13.180795	13.494423	13.816448	14.147087	10
11	13.841179	14.206787	14.583498	14.971643	15.371560	15.783599	16.208119	11
12	15.464032	15.917127	16.385591	16.869941	17.370711	17.888451	18.423728	12
13	17.159913	17.712983	18.286798	18.882138	19.499808	20.140643	20.805508	13
14	18.932109	19.598632	20.292572	21.015066	21.767295	22.550488	23.365921	14
15	20.784054	21.578564	22.408663	23.275970	24.182169	25.129022	26.118365	15
16	22.719337	23.657492	24.641140	25.672528	26.754010	27.888054	29.077242	16
17	24.741707	25.840366	26.996403	28.212880	29.493021	30.840217	32.258035	17
18	26.855084	28.132385	29.481205	30.905653	32.410067	33.999033	35.677388	18
19	29.063562	30.539004	32.102671	33.759992	35.516722	37.378965	39.353192	19
20	31.371423	33.065954	34.868318	36.785591	38.825309	40.995492	43.304681	20
25	44.565210	47.727099	51.152588	54.864512	58.887679	63.249038	67.977862	25
30	61.007070	66.438848	72.435478	79.058186	86.374864	94.460786	103.399403	30
35	81.496618	90.320307	100.251364	111.434780	124.034690	138.236878	154.251606	35
40	107.030323	120.799774	136.605614	154.761966	175.631916	199.635112	227.256520	40
45	138.849965	159.700156	184.119165	212.743514	246.324587	285.749311	332.064515	45
50	178.503028	209.347996	246.217476	290.335905	343.179672	406.528929	482.529947	50
55	227.917959	272.712618	327.377486	394.172027	475.879533	575.928593	698.542534	55
60	289.497954	353.583718	433.450372	533.128181	657.689842	813.520383	1008.656538	60
65	366.237831	456.798011	572.083392	719.082861	906.785722	1146.755161	1453.865297	65
70	461.869680	588.528511	753.271204	967.932170	1248.068666	1614.134174	2093.020048	70
75	581.044362	756.653718	990.076429	1300.948680	1715.655875	2269.657419	3010.609352	75
80	729.557699	971.228821	1299.571387	1746.599891	2356.290874	3189.062680	4327.927467	80
85	914.632336	1245.087069	1704.068919	2342.981741	3234.016343	4478.576120	6219.107984	85
90	1145.269007	1594.607301	2232.731017	3141.075187	4436.576302	6287.185427	8934.142195	90
95	1432.684259	2040.693529	2923.671235	4209.104250	6084.187663	8823.853541	12831.924930	95
100	1790.855956	2610.025157	3826.702467	5638.368059	8341.558016	12381.661794	18427.696132	100

TABLE 6 (Continued)

N	8%	9%	10%	11%	12%	13%	14%	N
1	1.000000	1.000000	1.000000	1.000000	1.000000	1.000000	1.000000	1
2	2.080000	2.090000	2.100000	2.110000	2.120000	2.130000	2.140000	2
3	3.246400	3.278100	3.310000	3.342100	3.374400	3.406900	3.439600	3
4	4.506112	4.573129	4.641000	4.709731	4.779328	4.849797	4.921144	4
5	5.866601	5.984711	6.105100	6.227801	6.352847	6.480271	6.610104	5
6	7.335929	7.523335	7.715610	7.912860	8.115189	8.322706	8.535519	6
7	8.922803	9.200435	9.487171	9.783274	10.089012	10.404658	10.730491	7
8	10.636628	11.028474	11.435888	11.859434	12.299693	12.757263	13.232760	8
9	12.487558	13.021036	13.579477	14.163972	14.775656	15.415707	16.085347	9
10	14.486562	15.192930	15.937425	16.722009	17.548735	18.419749	19.337295	10
11	16.645487	17.560293	18.531167	19.561430	20.654583	21.814317	23.044516	11
12	18.977126	20.140720	21.384284	22.713187	24.133133	25.650178	27.270749	12
13	21.495297	22.953385	24.522712	26.211638	28.029109	29.984701	32.088654	13
14	24.214920	26.019189	27.974983	30.094918	32.392602	34.882712	37.581065	14
15	27.152114	29.360916	31.772482	34.405359	37.279715	40.417464	43.842414	15
16	30.324283	33.003399	35.949730	39.189948	42.753280	46.671735	50.980352	16
17	33.750226	36.973705	40.544703	44.500843	48.883674	53.739060	59.117601	17
18	37.450244	41.301338	45.599173	50.395936	55.749715	61.725138	68.394066	18
19	41.446263	46.018458	51.159090	56.939488	63.439681	70.749406	78.969235	19
20	45.761964	51.160120	57.274999	64.202832	72.052442	80.946829	91.024928	20
25	73.105940	84.700896	98.347059	114.413307	133.333870	155.619556	181.870827	25
30	113.283211	136.307539	164.494023	199.020878	241.332684	293.199215	356.786847	30
35	172.316804	215.710755	271.024368	341.589555	431.663496	546.680819	693.572702	35
40	259.056519	337.882445	442.592556	581.826066	767.091420	1013.704243	1342.025099	40
45	386.505617	525.858734	718.904837	986.638559	1358.230032	1874.164630	2590.564800	45
50	573.770156	815.083556	1163.908529	1668.771152	2400.018249	3459.507117	4994.521346	50
55	848.923201	1260.091796	1880.591425	2818.204240	4236.005047	6380.397885	9623.134336	55
60	1253.213296	1944.792133	3034.816395	4755.065839	7471.641112	11761.949792	18535.133283	60
65	1847.248083	2998.288474	4893.707253	8018.790212	13173.937422	21677.110345	35694.426015	65
70	2720.080074	4619.223180	7687.469566	13518.355744	23223.331897	39945.150956	68733.178463	70
75	4002.556624	7113.232148	12708.953714	22785.443391	40933.798673	73602.831635	132346.474212	75
80	5886.935428	10950.574090	20474.002146	38401.025004	72145.692501	135614.926571	254828.441480	80
85	8655.706112	16854.800326	32979.690296	64714.188149	127151.714005	249868.191823	490657.007341	85
90	12723.938616	25939.184247	53120.226118	109053.398293	224091.118528	460372.427073	944724.766995	90
95	18701.506857	39916.634964	85556.760466	183767.545936	394931.471864	848212.835490	1818993.452831	95
100	27484.515704	61422.675465	137796.123398	309665.229724	696010.547721	1562783.647911	3502323.129475	100

N	15%	16%	17%	18%	19%	20%	N
1	1.000000	1.000000	1.000000	1.000000	1.000000	1.000000	1
2	2.150000	2.160000	2.170000	2.180000	2.190000	2.200000	2
3	3.472500	3.505600	3.538900	3.572400	3.606100	3.640000	3
4	4.993375	5.066496	5.140513	5.215432	5.291259	5.368000	4
5	6.742381	6.877135	7.014400	7.154210	7.296598	7.441600	5
6	8.753738	8.977477	9.206848	9.441968	9.682952	9.929920	6
7	11.066799	11.413873	11.772012	12.141522	12.522713	12.915904	7
8	13.726819	14.240093	14.773255	15.326996	15.902028	16.499085	8
9	16.785842	17.518508	18.284708	19.085855	19.923413	20.798902	9
10	20.303718	21.321469	22.393108	23.521309	24.708862	25.958682	10
11	24.349276	25.732904	27.199937	28.755144	30.403546	32.150419	11
12	29.001667	30.850169	32.823926	34.931070	37.180220	39.580562	12
13	34.351917	36.786196	39.403993	42.218663	45.244461	48.496603	13
14	40.504705	43.671987	47.102672	50.818022	54.840909	59.195923	14
15	47.580411	51.659505	56.110126	60.965266	66.260682	72.035108	15
16	55.717472	60.925026	66.648848	72.939014	79.850211	87.442129	16
17	65.075093	71.673030	78.979152	87.068036	96.021751	105.930555	17
18	75.836357	84.140715	93.405608	103.740283	115.265884	128.116666	18
19	88.211811	98.603230	110.284561	123.413534	138.166402	154.740000	19
20	102.443583	115.379747	130.032936	146.627970	165.418018	186.688000	20
25	212.793017	249.214024	292.104856	342.603486	402.042491	471.981083	25
30	434.745146	530.311731	647.439118	790.947991	966.712169	1181.881569	30
35	881.170156	1120.712955	1426.491022	1816.651612	2314.213721	2948.341146	35
40	1779.090308	2360.757241	3134.521839	4163.213027	5529.828982	7343.857840	40
45	3585.128460	4965.273911	6879.290650	9531.577105	13203.424228	18281.309940	45
50	7217.716277	10435.648773	15089.501673	21813.093666	31515.336327	45497.190750	50
55	14524.147893	21925.304976	33089.962703	49910.228445	75214.034786	113219.011287	55
60	29219.991638	46057.508533	72555.038129	114189.666478	179494.583786	281732.571766	60
65	58778.582580	96743.380952	159080.165226	261245.449442	428344.853547	701048.234576	65
70	118231.466926	203201.030246	348782.010169	597673.457599	1022189.605560	1744439.784661	70
75	237812.453171	426798.465828	764693.447483	1367339.242866	2439313.202939	4340731.846568	75
80	478332.529343	896429.474315	1676557.661247	3128148.113254	5821071.286073	10801137.310052	80
85	962104.313290	1882815.045139	3675772.565539	7156452.264725	13891142.064658	26876693.432947	85
90	1935142.168042	3954561.749997	8058947.355395	16372236.334003	33149185.002785	66877821.244672	90
95	3892268.850907	8305937.658205	17668830.313546	37455717.823488	79105686.250695	166413427.601142	95
100	7828749.671335	17445313.746092	38737999.328452	85689616.141407	188774151.199846	414089867.610073	100

TABLE 6 (Continued)

TABLE 7
Annuity due

Amount of annuity due for $1 at compound interest. Remember, annuity due requires payment at the beginning of the period.

N	1%	1½%	2%	2½%	3%	3½%	4%	N
1	1.010000	1.015000	1.020000	1.025000	1.030000	1.035000	1.040000	1
2	2.030100	2.045225	2.060400	2.075625	2.090900	2.106225	2.121600	2
3	3.060401	3.090903	3.121608	3.152516	3.183627	3.214943	3.246464	3
4	4.101005	4.152267	4.204040	4.256329	4.309136	4.362466	4.416323	4
5	5.152015	5.229551	5.308121	5.387737	5.468410	5.550152	5.632975	5
6	6.213535	6.322994	6.434283	6.547430	6.662462	6.779408	6.898294	6
7	7.285671	7.432839	7.582969	7.736116	7.892336	8.051687	8.214226	7
8	8.368527	8.559332	8.754628	8.954519	9.159106	9.368496	9.582795	8
9	9.462213	9.702722	9.949742	10.203382	10.463879	10.731393	11.006107	9
10	10.566835	10.863262	11.168715	11.483466	11.807796	12.141992	12.486351	10
11	11.682503	12.041211	12.412090	12.795553	13.192030	13.601962	14.025805	11
12	12.809328	13.236830	13.680332	14.140442	14.617790	15.113030	15.626838	12
13	13.947421	14.450382	14.973938	15.518953	16.086324	16.676986	17.291911	13
14	15.096896	15.682138	16.293417	16.931927	17.598914	18.295681	19.023588	14
15	16.257864	16.932370	17.639285	18.380225	19.156881	19.971030	20.824531	15
16	17.430443	18.201355	19.012071	19.864730	20.761588	21.705016	22.697512	16
17	18.614748	19.489376	20.412312	21.386349	22.414435	23.499691	24.645413	17
18	19.810895	20.796716	21.840559	22.946007	24.116868	25.357180	26.671229	18
19	21.019004	22.123667	23.297370	24.544658	25.870374	27.279682	28.778079	19
20	22.239194	23.470522	24.783317	26.183274	27.676486	29.269471	30.969202	20
25	28.525631	30.513969	32.670906	35.011708	37.553042	40.313102	43.311745	25
30	35.132740	38.101762	41.379441	45.000271	49.002678	53.429471	58.328335	30
35	42.076878	46.275969	50.994367	56.301413	62.275944	69.007603	76.598314	35
40	49.375237	55.081912	61.610023	69.087617	77.663298	87.509537	98.826536	40
45	57.045885	64.568414	73.330564	83.554034	95.501457	109.484031	125.870568	45
50	65.107814	74.788070	86.270989	99.921458	116.180773	135.582837	158.773767	50
55	73.580982	85.797543	100.558240	118.439694	140.153768	166.580031	198.805540	55
60	82.486367	97.657871	116.332570	139.391380	167.945040	203.394974	247.510313	60
65	91.846015	110.434814	133.748679	163.096289	200.162741	247.119577	306.767116	65
70	101.683100	124.199209	152.977469	189.916217	237.511886	299.050690	378.862077	70
75	112.021975	139.027372	174.207608	220.260504	280.809781	360.728561	466.576621	75
80	122.888237	155.001515	197.647397	254.592280	331.003909	433.982524	573.294776	80
85	134.308787	172.210204	223.526818	293.435534	389.192660	520.985253	703.133728	85
90	146.311900	190.748849	252.099789	337.383110	456.649371	624.317203	861.102667	90
95	158.927293	210.720235	283.646659	387.105758	534.850186	747.043145	1053.296034	95
100	172.186197	232.235089	318.476952	443.362370	625.506365	892.803065	1287.128653	100

N	4½%	5%	5½%	6%	6½%	7%	7½%	N
1	1.045000	1.050000	1.055000	1.060000	1.065000	1.070000	1.075000	1
2	2.137025	2.152500	2.168025	2.183600	2.199225	2.214900	2.230625	2
3	3.278191	3.310125	3.342266	3.374616	3.407175	3.439943	3.472922	3
4	4.470710	4.525631	4.581091	4.637093	4.693641	4.750739	4.808391	4
5	5.716892	5.801913	5.888051	5.975319	6.063728	6.153291	6.244020	5
6	7.019152	7.142008	7.266894	7.393838	7.522870	7.654021	7.787322	6
7	8.380014	8.549109	8.721573	8.897468	9.076856	9.259803	9.446371	7
8	9.802114	10.026564	10.256260	10.491316	10.731852	10.977989	11.229849	8
9	11.288209	11.577893	11.875354	12.180795	12.494423	12.816448	13.147087	9
10	12.841179	13.206787	13.583498	13.971643	14.371560	14.783599	15.208119	10
11	14.464032	14.917127	15.385591	15.869941	16.370711	16.888451	17.423728	11
12	16.159913	16.712983	17.286798	17.882138	18.499808	19.140643	19.805508	12
13	17.932109	18.598632	19.292572	20.015066	20.767295	21.550488	22.365921	13
14	19.784054	20.578564	21.408663	22.275970	23.182169	24.129022	25.118365	14
15	21.719337	22.657492	23.641140	24.672528	25.754010	26.888054	28.077242	15
16	23.741707	24.840366	25.996403	27.212880	28.493021	29.840217	31.258035	16
17	25.855084	27.132385	28.481205	29.905653	31.410067	32.999033	34.677388	17
18	28.063562	29.539004	31.102671	32.759992	34.516722	36.378965	38.353192	18
19	30.371423	32.065954	33.868318	35.785591	37.825309	39.995492	42.304681	19
20	32.783137	34.719252	36.786076	38.992727	41.348954	43.865177	46.552532	20
25	46.570645	50.113454	53.965981	58.156383	62.715378	67.676470	73.076201	25
30	63.752388	69.760790	76.419429	83.801677	91.989230	101.073041	111.154358	30
35	85.163966	94.836323	105.765189	118.120867	132.096945	147.913460	165.820476	35
40	111.846688	126.839763	144.118923	164.047684	187.047990	213.609570	244.300759	40
45	145.098214	167.685164	194.245719	225.508125	262.335685	305.751763	356.969354	45
50	186.535665	219.815396	259.759438	307.756059	365.486351	434.985955	518.719693	50
55	238.174268	286.348249	345.383247	417.822348	506.811702	616.243594	750.933224	55
60	302.525362	371.262904	457.290142	565.115872	700.439682	870.466810	1084.305779	60
65	382.718533	479.637912	603.547978	762.227832	965.726794	1227.028022	1562.905195	65
70	482.653815	617.954936	794.701120	1026.008100	1329.193129	1727.123566	2249.996552	70
75	607.191358	794.486404	1044.530633	1379.005601	1827.173507	2428.533438	3236.405054	75
80	762.387795	1019.790262	1371.047813	1851.395885	2509.449781	3412.297067	4652.522027	80
85	955.790791	1307.341422	1797.792710	2483.560646	3444.227405	4792.076448	6685.541082	85
90	1196.806112	1674.337666	2355.531223	3329.539698	4724.953761	6727.288407	9604.202860	90
95	1497.155051	2142.728205	3084.473153	4461.650505	6479.659861	9441.523288	13794.319300	95
100	1871.444474	2740.526415	4037.171102	5976.670142	8883.759287	13248.378119	19809.773342	100

TABLE 7 (Continued)

N	8%	9%	10%	11%	12%	13%	14%	N
1	1.080000	1.090000	1.100000	1.110000	1.120000	1.130000	1.140000	1
2	2.246400	2.278100	2.310000	2.342100	2.374400	2.406900	2.439600	2
3	3.506112	3.573129	3.641000	3.709731	3.779328	3.849797	3.921144	3
4	4.866601	4.984511	5.105100	5.227801	5.352847	5.480271	5.610104	4
5	6.335929	6.523335	6.715610	6.912860	7.115189	7.322706	7.535519	5
6	7.922803	8.200435	8.487171	8.783274	9.089012	9.404658	9.730491	6
7	9.636628	10.028474	10.435888	10.859434	11.299693	11.757263	12.232760	7
8	11.487558	12.021036	12.579477	13.163972	13.775656	14.415707	15.085347	8
9	13.486562	14.192930	14.937425	15.722009	16.548735	17.419749	18.337295	9
10	15.645487	16.560293	17.531167	18.561430	19.654583	20.814317	22.044516	10
11	17.977126	19.140720	20.384284	21.713187	23.133133	24.650178	26.270749	11
12	20.495287	21.953385	23.522712	25.211638	27.029109	28.984701	31.088654	12
13	23.214920	25.019189	26.974983	29.094918	31.392602	33.882712	36.581065	13
14	26.152114	28.360916	30.772482	33.405359	36.279715	39.417464	42.842414	14
15	29.324283	32.003399	34.949730	38.189948	41.753280	45.671735	49.980352	15
16	32.750226	35.973705	39.544703	43.500843	47.883674	52.739060	58.117601	16
17	36.450244	40.301338	44.599173	49.395936	54.749715	60.725138	67.394066	17
18	40.446263	45.018458	50.159090	55.939488	62.439681	69.749406	77.969235	18
19	44.761964	50.160120	56.274999	63.202832	71.052442	79.946829	90.024928	19
20	49.422921	55.764530	63.002499	71.265144	80.698736	91.469917	103.768418	20
25	78.954415	92.323977	108.181765	126.998771	149.333934	175.850098	207.332743	25
30	122.345868	148.575217	180.943425	220.913174	270.292606	331.315113	406.737006	30
35	186.102148	235.124723	298.126805	379.164406	483.463116	617.749325	790.672881	35
40	279.781040	368.291865	486.851811	645.826934	859.142391	1145.485795	1529.908613	40
45	417.426067	573.186021	790.795321	1095.168801	1521.217636	2117.806032	2953.243872	45
50	619.671769	888.441076	1280.299382	1852.335979	2688.020438	3909.243042	5693.754335	50
55	916.837058	1373.500057	2068.650567	3128.206707	4744.325653	7209.849610	10970.373143	55
60	1353.470360	2119.823425	3338.298035	5278.123082	8368.238046	13291.003265	21130.051943	60
65	1995.027929	3268.134436	5383.077978	8900.857202	14754.809912	24495.134690	40691.645657	65
70	2937.686480	5034.953266	8676.216525	15005.374875	26010.131725	45138.020581	78355.823448	70
75	4322.761154	7753.423041	13979.849085	25291.842164	45845.854514	83171.199747	150874.980601	75
80	6357.890263	11936.125758	22521.402360	42625.137755	80803.175601	153244.867025	290504.423288	80
85	9348.162601	18371.732355	36277.659326	71832.748846	142409.919685	282351.056760	559348.988369	85
90	13741.853705	28273.710829	58432.248730	121049.272105	250982.052751	520220.842593	1076986.234374	90
95	20197.627405	43509.132110	94112.436513	203981.975989	442323.248488	958480.504103	2073652.536227	95
100	29683.276961	66950.716257	151575.735738	343728.404994	779531.813448	1765945.522139	3992648.367601	100

N	15%	16%	17%	18%	19%	20%	N
1	1.150000	1.160000	1.170000	1.180000	1.190000	1.200000	1
2	2.472500	2.505600	2.538900	2.572400	2.606100	2.640000	2
3	3.993375	4.066496	4.140513	4.215432	4.291259	4.368000	3
4	5.742381	5.877135	6.014400	6.154210	6.296598	6.441600	4
5	7.753738	7.977477	8.206848	8.441968	8.682952	8.929920	5
6	10.066799	10.413873	10.772012	11.141522	11.522713	11.915904	6
7	12.726819	13.240093	13.773255	14.326996	14.902028	15.499685	7
8	15.785842	16.518508	17.284708	18.085855	18.923413	19.798902	8
9	19.303718	20.321469	21.393108	22.521309	23.708862	24.958682	9
10	23.349276	24.732904	26.199937	27.755144	29.403544	31.150419	10
11	28.001667	29.850169	31.823926	33.931070	36.180220	38.580502	11
12	33.351917	35.786196	38.403993	41.218663	44.244461	47.496603	12
13	39.504705	42.671987	46.102672	49.818022	53.840909	58.195923	13
14	46.580411	50.659505	55.110126	59.965266	65.260682	71.035108	14
15	54.717472	59.925026	65.648848	71.939014	78.850211	86.442125	15
16	64.075093	70.673030	77.979152	86.068036	95.021751	104.930555	16
17	74.836357	83.140715	92.405608	102.740283	114.265884	127.116666	17
18	87.211811	97.603230	109.284561	122.413534	137.166402	153.740000	18
19	101.443583	114.379747	129.032936	145.627970	164.418018	185.688000	19
20	117.810120	133.840506	152.138535	173.021005	196.847442	224.025600	20
25	244.711970	289.088267	341.762681	404.272113	478.430565	566.377300	25
30	499.956918	615.161608	757.503768	933.318630	1150.387481	1418.257883	30
35	1013.345680	1300.027028	1668.994496	2143.648902	2753.914328	3538.009375	35
40	2045.953854	2738.478399	3667.390552	4912.591372	6580.496488	8812.629408	40
45	4122.897729	5759.717737	8048.770061	11247.260984	15712.074831	21937.571928	45
50	8300.373719	12105.352576	17654.716957	25739.450526	37503.250230	54596.628900	50
55	16702.770077	25433.353773	38715.256362	58894.069565	89504.701395	135862.813544	55
60	33602.990383	53426.709898	84889.394611	134743.806444	213598.554705	338079.086119	60
65	67595.369968	112222.321904	186123.793315	308269.630342	509730.375721	841257.881492	65
70	135966.186964	235713.195085	408074.951898	705254.679967	1216405.630617	2093327.741593	70
75	273484.321146	495086.220361	894691.333555	1613460.306582	2902782.711497	5208878.215881	75
80	550082.408744	1039858.190205	1961572.463659	3691214.773639	6927074.830429	12961364.772062	80
85	1106419.960284	2184065.452361	4300653.901680	8444613.672375	16530459.056942	32252032.119537	85
90	2225413.493248	4587291.629996	9428968.405813	19319238.874124	39447530.153314	80253385.493606	90
95	4476109.178543	9634887.683518	20672531.466849	44197747.031716	94135766.638327	199696113.121370	95
100	9003062.122036	20236563.945467	45323459.214289	101113747.046860	224641239.927817	496690841.132087	100

TABLE 8
Present value of an annuity

Amount needed to deposit today to equal the future value of the annuity; it is the amount you can borrow with a given monthly payment on an installment loan.

N	1%	1½%	2%	2½%	3%	3½%	4%	N
1	.990099	.985222	.980392	.975610	.970874	.966184	.961538	1
2	1.970395	1.955883	1.941561	1.927424	1.913470	1.899694	1.886095	2
3	2.940985	2.912200	2.883883	2.856024	2.828611	2.801637	2.775091	3
4	3.901966	3.854385	3.807729	3.761974	3.717098	3.673079	3.629895	4
5	4.853431	4.782645	4.713460	4.645828	4.579707	4.515052	4.451822	5
6	5.795476	5.697187	5.601431	5.508125	5.417191	5.328553	5.242137	6
7	6.728195	6.598214	6.471991	6.349391	6.230283	6.114544	6.002055	7
8	7.651678	7.485925	7.325481	7.170137	7.019692	6.873956	6.732745	8
9	8.566018	8.360517	8.162237	7.970866	7.786109	7.607687	7.435332	9
10	9.471305	9.222185	8.982585	8.752064	8.530203	8.316605	8.110896	10
11	10.367628	10.071118	9.786848	9.514209	9.252624	9.001551	8.760477	11
12	11.255077	10.907505	10.575341	10.257765	9.954004	9.663334	9.385074	12
13	12.133740	11.731532	11.348374	10.983185	10.634955	10.302738	9.985648	13
14	13.003703	12.543382	12.106249	11.690912	11.296973	10.920520	10.563123	14
15	13.865053	13.343233	12.849264	12.381378	11.937935	11.517411	11.118387	15
16	14.717874	14.131264	13.577709	13.055003	12.561102	12.094117	11.652296	16
17	15.562251	14.907649	14.291872	13.712198	13.166118	12.651321	12.165669	17
18	16.398269	15.672561	14.992031	14.353364	13.753513	13.189682	12.659297	18
19	17.226008	16.426168	15.678462	14.978891	14.323799	13.709837	13.133939	19
20	18.045553	17.168639	16.351433	15.589162	14.877475	14.212403	13.590326	20
25	22.023156	20.719611	19.523456	18.424376	17.413148	16.481515	15.622080	25
30	25.807708	24.015838	22.396456	20.930293	19.600441	18.392045	17.292033	30
35	29.408580	27.075595	24.998619	23.145157	21.487220	20.000661	18.664613	35
40	32.834686	29.915845	27.355479	25.102775	23.114772	21.355072	19.792774	40
45	36.094508	32.552337	29.490160	26.833024	24.518713	22.495450	20.720040	45
50	39.196118	34.999688	31.423606	28.362312	25.729764	23.455618	21.482185	50
55	42.147192	37.271467	33.174788	29.713979	26.774428	24.264053	22.108612	55
60	44.955038	39.380269	34.760887	30.908656	27.675564	24.944734	22.623490	60
65	47.626608	41.337786	36.197466	31.964577	28.452892	25.517849	23.046682	65
70	50.168514	43.154872	37.498619	32.897857	29.123421	26.000397	23.394515	70
75	52.587051	44.841600	38.677114	33.722740	29.701826	26.406689	23.680408	75
80	54.888206	46.407323	39.744514	34.451817	30.200763	26.748776	23.915392	80
85	57.077676	47.860722	40.711290	35.096215	30.631151	27.036804	24.108531	85
90	59.160881	49.209855	41.586929	35.665768	31.002407	27.279316	24.267278	90
95	61.142980	50.462201	42.380023	36.169171	31.322656	27.483504	24.397756	95
100	63.028879	51.624704	43.098352	36.614105	31.598905	27.655425	24.504999	100

N	4½%	5%	5½%	6%	6½%	7%	7½%	N
1	.956938	.952381	.947867	.943396	.938967	.934579	.930233	1
2	1.872668	1.859410	1.846320	1.833393	1.820626	1.808018	1.795565	2
3	2.748964	2.723248	2.697933	2.673012	2.648476	2.624316	2.600526	3
4	3.587526	3.545951	3.505150	3.465106	3.425799	3.387211	3.349326	4
5	4.389977	4.329477	4.270284	4.212364	4.155679	4.100197	4.045885	5
6	5.157872	5.075692	4.995530	4.917324	4.841014	4.766540	4.693846	6
7	5.892701	5.786373	5.682967	5.582381	5.484520	5.389289	5.296601	7
8	6.595886	6.463213	6.334566	6.209794	6.088751	5.971299	5.857304	8
9	7.268790	7.107822	6.952195	6.801692	6.656104	6.515232	6.378887	9
10	7.912718	7.721735	7.537626	7.360087	7.188830	7.023582	6.864081	10
11	8.528917	8.306414	8.092536	7.886875	7.689042	7.498674	7.315424	11
12	9.118581	8.863252	8.618518	8.383844	8.158725	7.942686	7.735278	12
13	9.682852	9.393573	9.117070	8.852683	8.599742	8.357651	8.125640	13
14	10.222825	9.898641	9.589648	9.294984	9.013842	8.745468	8.489154	14
15	10.739546	10.379658	10.037581	9.712249	9.402669	9.107914	8.827120	15
16	11.234015	10.837770	10.462162	10.105895	9.767764	9.446649	9.141507	16
17	11.707191	11.274066	10.864609	10.477260	10.110577	9.763223	9.433960	17
18	12.159992	11.689587	11.246074	10.827603	10.432466	10.059087	9.706009	18
19	12.593294	12.085321	11.607654	11.158116	10.734710	10.335595	9.959078	19
20	13.007936	12.462210	11.950382	11.469921	11.018507	10.594014	10.194491	20
25	14.828209	14.093945	13.413933	12.783356	12.197877	11.653583	11.146946	25
30	16.288889	15.372451	14.533745	13.764831	13.058676	12.409041	11.810386	30
35	17.461012	16.374194	15.390552	14.498246	13.686957	12.947672	12.272511	35
40	18.401584	17.159086	16.046125	15.046297	14.145527	13.331709	12.594409	40
45	19.156347	17.774070	16.547726	15.455832	14.480228	13.605522	12.818629	45
50	19.762008	18.255925	16.931518	15.761861	14.724521	13.800746	12.974812	50
55	20.248021	18.633472	17.225170	15.990543	14.902825	13.939939	13.083602	55
60	20.638022	18.929290	17.449854	16.161428	15.032966	14.039181	13.159381	60
65	20.950979	19.161070	17.621767	16.289123	15.127953	14.109940	13.212165	65
70	21.202112	19.342677	17.753304	16.384544	15.197282	14.160389	13.248933	70
75	21.403634	19.484970	17.853947	16.455848	15.247885	14.196359	13.274543	75
80	21.565345	19.596460	17.930953	16.509131	15.284818	14.222005	13.292383	80
85	21.695110	19.683816	17.989873	16.548947	15.311775	14.240291	13.304809	85
90	21.799241	19.752262	18.034954	16.578699	15.331451	14.253328	13.313464	90
95	21.882800	19.805891	18.069447	16.600932	15.345812	14.262623	13.319493	95
100	21.949853	19.847910	18.095839	16.617546	15.356293	14.269251	13.323693	100

TABLE 8 (Continued)

N	8%	9%	10%	11%	12%	13%	14%	N
1	.925926	.917431	.909091	.900901	.892857	.884956	.677193	1
2	1.783265	1.759111	1.735537	1.712523	1.690051	1.668102	1.646661	2
3	2.577097	2.531295	2.486852	2.443715	2.401831	2.361153	2.321632	3
4	3.312127	3.239720	3.169865	3.102446	3.037349	2.974471	2.913712	4
5	3.992710	3.889651	3.790787	3.695897	3.604776	3.517231	3.433081	5
6	4.622880	4.485919	4.355261	4.230538	4.111407	3.997550	3.888668	6
7	5.206370	5.032953	4.868419	4.712196	4.563757	4.422610	4.288305	7
8	5.746639	5.534819	5.334926	5.146123	4.967640	4.798770	4.638864	8
9	6.246888	5.995247	5.759024	5.537048	5.328250	5.131655	4.946372	9
10	6.710081	6.417658	6.144567	5.889232	5.650223	5.426243	5.216116	10
11	7.138964	6.805191	6.495061	6.206515	5.937699	5.686941	5.452733	11
12	7.536078	7.160725	6.813692	6.492356	6.194374	5.917647	5.660292	12
13	7.903776	7.486904	7.103356	6.749870	6.423548	6.121812	5.842362	13
14	8.244237	7.786150	7.366687	6.981865	6.628168	6.302488	6.002072	14
15	8.559479	8.060688	7.606080	7.190870	6.810864	6.462379	6.142168	15
16	8.851369	8.312558	7.823709	7.379162	6.973986	6.603875	6.265060	16
17	9.121638	8.543631	8.021553	7.548794	7.119630	6.729093	6.372859	17
18	9.371887	8.755625	8.201412	7.701617	7.249670	6.839905	6.467420	18
19	9.603599	8.950115	8.364920	7.839294	7.365777	6.937969	6.550369	19
20	9.818147	9.128546	8.513564	7.963328	7.469444	7.024752	6.623131	20
25	10.674776	9.822580	9.077040	8.421745	7.843139	7.329985	6.872927	25
30	11.257783	10.273654	9.426914	8.693793	8.055184	7.495653	7.002664	30
35	11.654568	10.566821	9.644159	8.855240	8.175504	7.585572	7.070045	35
40	11.924613	10.757360	9.779051	8.951051	8.243777	7.634376	7.105041	40
45	12.108402	10.881197	9.862808	9.007910	8.282516	7.660864	7.123217	45
50	12.233485	10.961683	9.914814	9.041653	8.304498	7.675242	7.132656	50
55	12.318614	11.013993	9.947106	9.061678	8.316972	7.683045	7.137559	55
60	12.376552	11.047991	9.967157	9.073562	8.324049	7.687280	7.140106	60
65	12.415983	11.070087	9.979607	9.080614	8.328065	7.689579	7.141428	65
70	12.442820	11.084449	9.987338	9.084800	8.330344	7.690827	7.142115	70
75	12.461084	11.093782	9.992138	9.087283	8.331637	7.691504	7.142472	75
80	12.473514	11.099849	9.995118	9.088757	8.332371	7.691871	7.142657	80
85	12.481974	11.103791	9.996969	9.089632	8.332787	7.692071	7.142753	85
90	12.487732	11.106354	9.998118	9.090151	8.333023	7.692179	7.142803	90
95	12.491651	11.108019	9.998831	9.090459	8.333157	7.692238	7.142829	95
100	12.494318	11.109102	9.999274	9.090642	8.333234	7.692270	7.142843	100

N	15%	16%	17%	18%	19%	20%	N
1	.869565	.862069	.854701	.847458	.840336	.833333	1
2	1.625709	1.605232	1.585214	1.565642	1.546501	1.527778	2
3	2.283225	2.245890	2.209585	2.174273	2.139917	2.106481	3
4	2.854978	2.798181	2.743235	2.690062	2.638586	2.588735	4
5	3.352155	3.274294	3.199346	3.127171	3.057635	2.990612	5
6	3.784483	3.684736	3.589185	3.497603	3.409777	3.325510	6
7	4.160420	4.038565	3.922380	3.811528	3.705695	3.604592	7
8	4.487322	4.343591	4.207163	4.077566	3.954366	3.837160	8
9	4.771584	4.606544	4.450566	4.303022	4.163332	4.030967	9
10	5.018769	4.833227	4.658604	4.494086	4.338935	4.192472	10
11	5.233712	5.028644	4.836413	4.656005	4.486500	4.327060	11
12	5.420619	5.197107	4.988387	4.793225	4.610504	4.439217	12
13	5.583147	5.342334	5.118280	4.909513	4.714709	4.532681	13
14	5.724476	5.467529	5.229299	5.008062	4.802277	4.610567	14
15	5.847370	5.575456	5.324187	5.091578	4.875863	4.675473	15
16	5.954235	5.668497	5.405288	5.162354	4.937700	4.729561	16
17	6.047161	5.748704	5.474605	5.222334	4.989664	4.774634	17
18	6.127966	5.817848	5.533851	5.273846	5.033331	4.812195	18
19	6.198231	5.877455	5.584488	5.316241	5.070026	4.843496	19
20	6.259331	5.928841	5.627767	5.352746	5.100862	4.869580	20
25	6.464149	6.097092	5.766234	5.466906	5.195148	4.947585	25
30	6.565980	6.177198	5.829390	5.516806	5.234658	4.978936	30
35	6.616607	6.215338	5.858196	5.538618	5.251215	4.991535	35
40	6.641778	6.233497	5.871335	5.548152	5.258153	4.996598	40
45	6.654293	6.242143	5.877327	5.552319	5.261061	4.998633	45
50	6.660515	6.246259	5.880061	5.554141	5.262279	4.999451	50
55	6.663608	6.248219	5.881307	5.554937	5.262790	4.999779	55
60	6.665146	6.249152	5.881876	5.555285	5.263004	4.999911	60
65	6.665911	6.249596	5.882135	5.555437	5.263093	4.999964	65
70	6.666291	6.249808	5.882254	5.555504	5.263131	4.999986	70
75	6.666480	6.249908	5.882308	5.555533	5.263147	4.999994	75
80	6.666574	6.249956	5.882332	5.555546	5.263153	4.999998	80
85	6.666620	6.249979	5.882344	5.555551	5.263156	4.999999	85
90	6.666644	6.249990	5.882349	5.555554	5.263157	5.000000	90
95	6.666655	6.249995	5.882351	5.555555	5.263158	5.000000	95
100	6.666661	6.249998	5.882352	5.555555	5.263158	5.000000	100

TABLE 9
Mortality table based on
100,000 persons living at age 0
l_x = number of living;
d_x = number of deaths;
p_x = probability of living;
q_x = probability of dying, for
age x from 0 to 99

x	l_x	d_x	p_x	q_x	x	l_x	d_x	p_x	q_x
0	100,000	708	.9929	.0071	50	87,624	729	.9917	.0083
1	99,292	175	.9982	.0018	51	86,895	792	.9909	.0091
2	99,117	151	.9985	.0015	52	86,103	858	.9900	.0100
3	98,966	144	.9986	.0015	53	85,245	928	.9891	.0109
4	98,822	138	.9986	.0014	54	84,317	1,003	.9881	.0119
5	98,684	133	.9987	.0014	55	83,314	1,083	.9870	.0130
6	98,551	128	.9987	.0013	56	82,231	1,168	.9858	.0142
7	98,423	124	.9987	.0013	57	81,063	1,260	.9845	.0156
8	98,299	121	.9988	.0012	58	79,803	1,357	.9830	.0170
9	98,178	119	.9988	.0012	59	78,446	1,458	.9814	.0186
10	98,059	119	.9988	.0012	60	76,988	1,566	.9797	.0204
11	97,940	120	.9988	.0012	61	75,422	1,677	.9778	.0222
12	97,820	123	.9988	.0013	62	73,745	1,793	.9757	.0243
13	97,697	129	.9987	.0013	63	71,952	1,912	.9734	.0266
14	97,568	136	.9986	.0014	64	70,040	2,034	.9710	.0291
15	97,432	142	.9986	.0015	65	68,006	2,159	.9683	.0318
16	97,290	150	.9985	.0016	66	65,847	2,287	.9653	.0347
17	97,140	157	.9984	.0016	67	63,560	2,418	.9620	.0381
18	96,983	164	.9983	.0017	68	61,142	2,548	.9583	.0417
19	96,819	168	.9983	.0017	69	58,594	2,672	.9544	.0456
20	96,651	173	.9982	.0018	70	55,922	2,784	.9502	.0498
21	96,478	177	.9982	.0018	71	53,138	2,877	.9459	.0542
22	96,301	179	.9982	.0019	72	50,261	2,948	.9414	.0587
23	96,122	182	.9981	.0019	73	47,313	2,993	.9368	.0633
24	95,940	183	.9981	.0019	74	44,320	3,019	.9319	.0681
25	95,757	185	.9981	.0019	75	41,301	3,030	.9266	.0734
26	95,572	187	.9981	.0020	76	38,271	3,030	.9208	.0792
27	95,385	190	.9980	.0020	77	35,241	3,020	.9143	.0857
28	95,195	193	.9980	.0020	78	32,221	2,998	.9070	.0931
29	95,002	198	.9979	.0021	79	29,223	2,957	.8988	.1012
30	94,804	202	.9979	.0021	80	26,266	2,888	.8901	.1100
31	94,602	207	.9978	.0022	81	23,378	2,790	.8807	.1194
32	94,395	212	.9978	.0023	82	20,588	2,659	.8709	.1292
33	94,183	218	.9977	.0023	83	17,929	2,499	.8606	.1394
34	93,965	226	.9976	.0024	84	15,430	2,314	.8500	.1500
35	93,739	235	.9975	.0025	85	13,116	2,113	.8389	.1611
36	93,504	247	.9974	.0027	86	11,003	1,901	.8272	.1728
37	93,257	261	.9972	.0028	87	9,102	1,685	.8149	.1851
38	92,996	280	.9970	.0030	88	7,417	1,470	.8018	.1982
39	92,716	301	.9968	.0033	89	5,947	1,263	.7876	.2124
40	92,415	326	.9965	.0035	90	4,684	1,068	.7720	.2280
41	92,089	354	.9962	.0039	91	3,616	888	.7544	.2456
42	91,735	383	.9958	.0042	92	2,728	725	.7342	.2658
43	91,352	414	.9955	.0045	93	2,003	579	.7109	.2891
44	90,938	447	.9951	.0049	94	1,424	450	.6840	.3160
45	90,491	484	.9947	.0054	95	974	341	.6499	.3501
46	90,007	525	.9942	.0058	96	633	253	.6003	.3997
47	89,482	569	.9937	.0064	97	380	185	.5132	.4869
48	88,913	618	.9931	.0070	98	195	129	.3385	.6615
49	88,295	671	.9924	.0076	99	66	66	.0000	1.0000

Based on the 1958 CSO Mortality Table prepared in cooperation with the National Association of Insurance Commissioner. Courtesy of the Society of Actuaries, Chicago, Illinois.

APPENDIX F

Answers to Odd-Numbered Problems

CHAPTER 1
PROBLEM SET 1.1, PAGES 9–10

1. Answers vary. **3.** 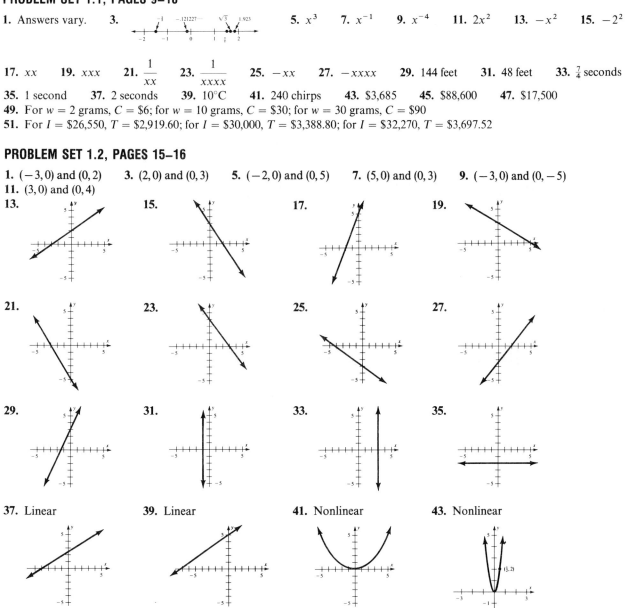 **5.** x^3 **7.** x^{-1} **9.** x^{-4} **11.** $2x^2$ **13.** $-x^2$ **15.** -2^2

17. xx **19.** xxx **21.** $\dfrac{1}{xx}$ **23.** $\dfrac{1}{xxxx}$ **25.** $-xx$ **27.** $-xxxx$ **29.** 144 feet **31.** 48 feet **33.** $\frac{7}{4}$ seconds

35. 1 second **37.** 2 seconds **39.** $10°C$ **41.** 240 chirps **43.** \$3,685 **45.** \$88,600 **47.** \$17,500

49. For $w = 2$ grams, $C = \$6$; for $w = 10$ grams, $C = \$30$; for $w = 30$ grams, $C = \$90$

51. For $I = \$26,550$, $T = \$2,919.60$; for $I = \$30,000$, $T = \$3,388.80$; for $I = \$32,270$, $T = \$3,697.52$

PROBLEM SET 1.2, PAGES 15–16

1. $(-3, 0)$ and $(0, 2)$ **3.** $(2, 0)$ and $(0, 3)$ **5.** $(-2, 0)$ and $(0, 5)$ **7.** $(5, 0)$ and $(0, 3)$ **9.** $(-3, 0)$ and $(0, -5)$
11. $(3, 0)$ and $(0, 4)$

13. **15.** **17.** **19.**

21. **23.** **25.** **27.**

29. **31.** **33.** **35.**

37. Linear **39.** Linear **41.** Nonlinear **43.** Nonlinear

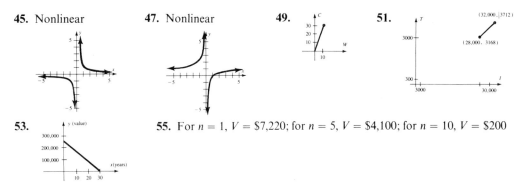

45. Nonlinear   **47.** Nonlinear   **49.**   **51.**

53.

55. For $n = 1$, $V = \$7,220$; for $n = 5$, $V = \$4,100$; for $n = 10$, $V = \$200$

PROBLEM SET 1.3, PAGES 24–25

1. Slope $= 2$, y-intercept is $(0, 4)$ **3.** Slope $= 9$, y-intercept is $(0, 1)$ **5.** Slope $= -3$, y-intercept is $(0, 4)$
7. Slope $= -\frac{1}{2}$, y-intercept is $(0, 5)$ **9.** Slope $= -\frac{4}{3}$, y-intercept is $(0, \frac{2}{3})$ **11.** Slope $= 1$, y-intercept is $(0, -5)$
13. Slope $= -2$, y-intercept is $(0, 5)$ **15.** Slope $= 2$, y-intercept is $(0, -5)$ **17.** Slope $= -\frac{3}{2}$, y-intercept is $(0, \frac{7}{2})$
19. Slope $= 0$, y-intercept is $(0, -2)$ **21.** Slope is not defined; there is no y-intercept   **23.** Slope $= \frac{2}{5}$, y-intercept is $(0, -240)$

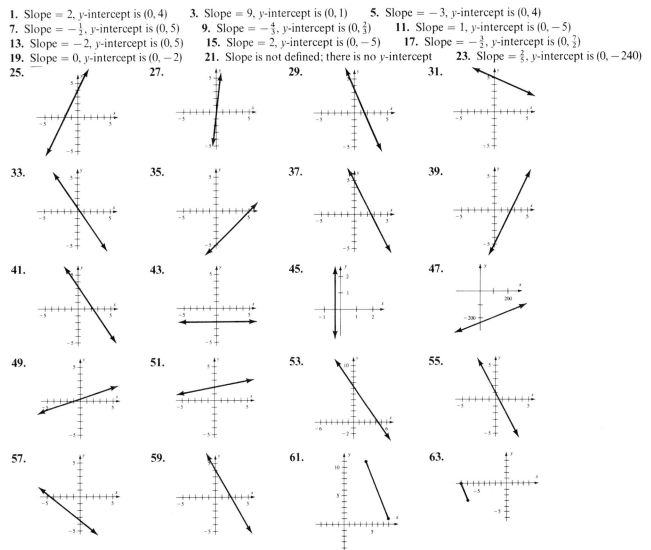

65.   **67.**   **69.**   **71.**

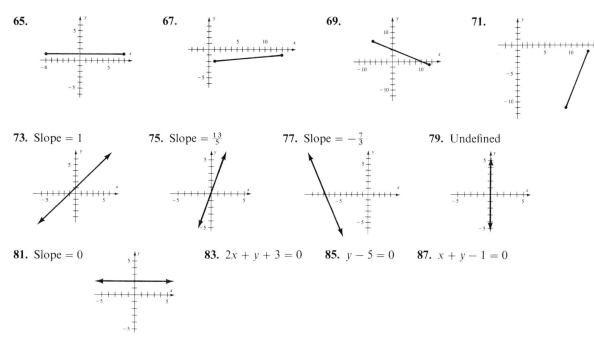

73. Slope $= 1$ **75.** Slope $= \frac{13}{5}$ **77.** Slope $= -\frac{7}{3}$ **79.** Undefined

81. Slope $= 0$ **83.** $2x + y + 3 = 0$ **85.** $y - 5 = 0$ **87.** $x + y - 1 = 0$

89. $2x - 5y - 20 = 0$ **91.** $2x - y - 4 = 0$ **93.** $y - 6 = 0$ **95.** $10x - y = 0$
97. $3x - 10y + 52 = 0$, 17.2 million in 1990   **99.** $y - 60 = 0$ **101.** $40x - y + 25{,}000 = 0$

PROBLEM SET 1.4, PAGES 31–33

1.   **3.**   **5.** Dependent system

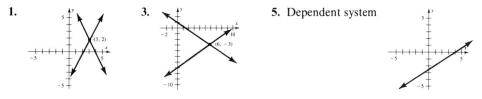

The answers for variables and constants in Problems 7–17 may vary, but the solutions to the problems will not vary.
7. $a_{11} = 1, a_{12} = -3, b_1 = -7$ **9.** $a_{11} = 3, a_{12} = -2, b_1 = -21$ **11.** $a_{11} = 2, a_{12} = -3, b_1 = 9$
   $a_{21} = 1, a_{22} = 2, b_2 = 8$ $a_{21} = 3, a_{22} = 2, b_2 = 3$ $a_{21} = -2, a_{22} = 3, b_2 = -9$
   Solution: $(a, b) = (2, 3)$ Solution: $(m, n) = (-3, 6)$ Dependent system
13. $a_{11} = 1, a_{12} = 1, b_1 = 2$ **15.** $a_{11} = 3, a_{12} = -4, b_1 = 3$ **17.** $a_{11} = 7, a_{12} = 1, b_1 = 5$
   $a_{21} = 2, a_{22} = -1, b_2 = 1$ $a_{21} = 5, a_{22} = 3, b_2 = 5$ $a_{21} = 14, a_{22} = -2, b_2 = -2$
   Solution: $(c, d) = (1, 1)$ Solution: $(q_1, q_2) = (1, 0)$ Solution: $(\frac{2}{7}, 3)$

19. $(\frac{3}{5}, \frac{1}{2})$ **21.** $(-6, 2)$ **23.** $(-\frac{8}{5}, -\frac{21}{5})$ **25.** $(31, 19)$ **27.** $\left(\dfrac{a}{a^2 - b^2}, \dfrac{-b}{a^2 - b^2} \right)$ **29.** $(1800, 200)$

31. $A(0, 0), B(\frac{7}{2}, 0), C(\frac{5}{2}, 2), D(0, 3)$ **33.** $A(0, 0), B(\frac{8}{3}, 0), C(\frac{5}{3}, \frac{5}{3}), D(0, \frac{8}{3})$ **35.** $A(0, 0), B(2, 0), C(\frac{7}{4}, \frac{1}{4}), D(1, \frac{5}{4}), E(0, \frac{3}{2})$
37. 50 miles   **39.** 6 hours   **41.** \$15   **43. a.** 125 supplied; 75 demanded   **b.** \$200   **c.** \$400
d. \$233.33   **e.** 83
45. The line through (1, 800) and (7, 0) intersects the line through (1, 200) and (7, 600) at (\$4, 400 items)
47. If 1000 pairs are produced, both cost and revenue will be \$2,000
49. If 10,000 disks are produced, both cost and revenue will be \$30,000
51. If 3733 cards are produced, both cost and revenue will be \$5,600
53. If 2240 cards are produced, both cost and revenue will be \$4,480
55. 500 shares of Standard Oil, 200 shares of Xerox   **57.** 60 units grade A, 20 units grade B

PROBLEM SET 1.5, PAGES 34–35

1. 5000 items will cost $680,000. Average cost is $136   **3.**   **5.**   **7.**

9. $-\frac{5}{3}$ **11.** $3x + 5y - 12 = 0$ **13.** $6x + 2y - 1 = 0$ **15.** $x = 1, y = 3$ **17.** $A(0,0), B(9,0), C(\frac{78}{11}, \frac{42}{11}), D(0,5)$
19. 100 miles

CHAPTER 2
PROBLEM SET 2.1, PAGES 49–52

1. a. 3×3 (square matrix)   **b.** 3×1 (column matrix)   **c.** 3×2 **d.** 4×1 (column matrix)   **e.** 2×3
f. 1×4 (row matrix)   **g.** 1×2 (row matrix)   **3. a.** $x = 2, y = 4$ **b.** $x = 4, y = -9, z = 6$
c. $a = 1, b = 0, c = 0, d = 1$ **5.**

	Part 1	Part 2	Part 3	Part 4
Factory A	25	42	193	0
Factory B	16	39	150	0
Factory C	50	50	50	50
Factory D	0	0	0	320

7. a. $3 \times 3, c_{12}$ **b.** $3 \times 3, c_{32}$ **c.** $3 \times 3, c_{21}$ **d.** $2 \times 2, c_{11}$ **e.** $2 \times 2, c_{21}$ **f.** $2 \times 2, c_{12}$
9. a. $[3w + 5x + 8y + 9z], 1 \times 1$ **b.** $[2x - 3x + 4y + 5z], 1 \times 1$ **c.** $[2a + 3b + 5c \quad 2d + 3e + 5f], 1 \times 2$

d. $[3a - 2c + e \quad 3b - 2d + f], 1 \times 2$ **11. a.** $\begin{bmatrix} 8w & -w & 3w & -2w \\ 8x & -x & 3x & -2x \\ 8y & -y & 3y & -2y \\ 8z & -z & 3z & -2z \end{bmatrix}, 4 \times 4$ **b.** $\begin{bmatrix} 6w & -w & 3w & 2w \\ 6x & -x & 3x & 2x \\ 6y & -y & 3y & 2y \\ 6z & -z & 3z & 2z \end{bmatrix}, 4 \times 4$

c. $\begin{bmatrix} 12 & -24 & 18 \\ 6 & -12 & 9 \\ -2 & 4 & -3 \end{bmatrix}, 3 \times 3$ **d.** $\begin{bmatrix} 10 & 30 & 20 \\ -4 & -12 & -8 \\ -2 & -6 & -4 \end{bmatrix}, 3 \times 3$ **13.** $\begin{bmatrix} 2 & 4 & 2 \\ 6 & -2 & 4 \\ 2 & 2 & 5 \end{bmatrix}$ **15.** $\begin{bmatrix} 0 & 4 & -2 \\ 0 & 0 & 0 \\ -6 & 0 & 5 \end{bmatrix}$

17. Not conformable   **19.** Not conformable   **21.** Not conformable   **23.** Not conformable

25. $\begin{bmatrix} -1 & -16 & 6 \\ -3 & 1 & -2 \\ 20 & -1 & -20 \end{bmatrix}$ **27.** Not conformable   **29.** Not conformable   **31.** $\begin{bmatrix} 19 & 14 & 12 \\ 15 & 7 & 20 \\ 2 & 21 & 19 \end{bmatrix}$

33. $\begin{bmatrix} 10 & 5 & 8 \\ 9 & 3 & 11 \\ 6 & 11 & 7 \end{bmatrix}$ **35.** Not conformable   **37.** $\begin{bmatrix} 13 & -4 & 10 \\ 8 & 3 & 4 \\ 21 & 4 & -2 \end{bmatrix}$ **39.** $\begin{bmatrix} 35 & 5 & 18 \\ 46 & 2 & 16 \\ 39 & -7 & 26 \end{bmatrix}$

41. $\begin{bmatrix} -1 & 37 & 32 \\ 4 & 14 & 45 \\ 27 & 9 & 28 \end{bmatrix}$ **43.** $\begin{bmatrix} 34 & 117 & 44 \\ 45 & 143 & 97 \\ 109 & 100 & 151 \end{bmatrix}$ **45.** $\begin{bmatrix} 132 & 83 & 70 \\ 89 & 59 & 77 \\ 172 & 23 & 150 \end{bmatrix}$ **47.** $\begin{bmatrix} 14 & 153 & 100 \\ 20 & 198 & 84 \\ -34 & 189 & 116 \end{bmatrix}$

49. $\begin{bmatrix} 13 & 25 & 20 \\ 25 & 31 & 23 \\ 42 & 24 & 33 \end{bmatrix}$ **51.** $\begin{bmatrix} 13 & 25 & 20 \\ 25 & 31 & 23 \\ 42 & 24 & 33 \end{bmatrix}$ **53.** Not conformable   **55.** $\begin{bmatrix} 1 & 2 \\ 4 & 0 \\ -1 & 3 \\ 2 & 1 \end{bmatrix}$ **57.** Not conformable

59. Not conformable (for multiplication)   **61.** Not conformable   **63.** $\begin{bmatrix} 42 & 70 \\ -35 & 7 \end{bmatrix}$

65. a.

	SF	D	A	KC
SF	0	1	0	1
D	1	0	1	0
A	0	1	0	1
KC	1	0	1	0

b. There is no way to go from Kansas City to San Francisco making exactly one stop

c.

Origin	First Stop	Second Stop	Destination
SF	Dallas	SF	KC
SF	Dallas	Atlanta	KC
SF	KC	Atlanta	KC
SF	KC	SF	KC

4 ways

d.

	SF	D	A	KC
SF	2	0	2	0
D	0	2	0	2
A	2	0	2	0
KC	0	2	0	2

e.

	SF	D	A	KC
SF	0	4	0	4
D	4	0	4	0
A	0	4	0	4
KC	4	0	4	0

67.

US	USSR	Cuba	Mexico	
2	4	1	3	US
4	2	3	4	USSR
1	3	0	1	Cuba
3	4	1	2	Mexico

69. False
71. True
73. True

So, using *two* intermediaries, the US can communicate with:
the USSR in four ways, Cuba in only one way, and Mexico in three ways.

PROBLEM SET 2.2, PAGES 65–67

1. a. $\begin{bmatrix} 4 & 5 & | & -16 \\ 3 & 2 & | & 5 \end{bmatrix}$ **b.** $\begin{bmatrix} 1 & 1 & 1 & | & 4 \\ 3 & 2 & 1 & | & 7 \\ 1 & -3 & 2 & | & 0 \end{bmatrix}$ **c.** $\begin{bmatrix} 1 & 3 & 1 & 1 & | & 3 \\ 1 & 0 & -2 & 2 & | & 0 \\ 0 & 0 & 1 & 5 & | & -14 \\ 0 & 1 & -3 & -1 & | & 2 \end{bmatrix}$

3. a. $\begin{cases} 2x_1 + x_2 + 4x_3 = 3 \\ 6x_1 + 2x_2 - x_3 = -4 \\ -3x_1 - x_2 = 1 \end{cases}$ **b.** $\begin{cases} x_1 = 5 \\ x_2 = -3 \\ x_3 = 4 \end{cases}$ **c.** $\begin{cases} x_1 = 3 \\ x_2 = 2 \\ x_3 = -8 \\ 0 = 1 \end{cases}$ **5. a.** $\begin{bmatrix} 1 & 5 & 3 & | & -2 \\ -1 & 3 & -2 & | & 3 \\ -2 & 2 & 4 & | & 8 \end{bmatrix}$

b. $\begin{bmatrix} -2 & 2 & 4 & | & 8 \\ -2 & 6 & -4 & | & 6 \\ 1 & 5 & 3 & | & -2 \end{bmatrix}$ **c.** $\begin{bmatrix} -1 & 1 & 2 & | & 4 \\ -1 & 3 & -2 & | & 3 \\ 1 & 5 & 3 & | & -2 \end{bmatrix}$ **d.** $\begin{bmatrix} 0 & 12 & 10 & | & 4 \\ -1 & 3 & -2 & | & 3 \\ 1 & 5 & 3 & | & -2 \end{bmatrix}$ **e.** $\begin{bmatrix} -2 & 2 & 4 & | & 8 \\ -2 & -2 & -5 & | & 5 \\ 1 & 5 & 3 & | & -2 \end{bmatrix}$

7. $(1, 2)$ **9.** $(-6, -4)$ **11.** $(\frac{10}{3} + 2t, 3t)$ **13.** Inconsistent system   **15.** Inconsistent system
17. Inconsistent system   **19.** $(1, 2, 3)$ **21.** $(3, -1, 2)$ **23.** $(3, 2, 5)$ **25.** $(2, -1, 0)$ **27.** $(6, 2, -9)$ **29.** $(1, 1, 2)$
31. $(\frac{8}{5} - 7t, -\frac{19}{5} + t, 5t)$ **33.** Inconsistent system   **35.** $(-7 + 11t, -2 + 7t, 3t)$
37. $(w, x, y, z) = (\frac{18}{5} + 3s - t, \frac{7}{5} + 7s - 14t, 5s, 5t)$ **39.** $(w, x, y, z) = (\frac{11}{2} - 6s, 2 - 4s + t, 3s, 3t)$
41. $(-\frac{3}{2} - 2t, 1, -\frac{1}{4} + 3t, 12t)$ **43.** $(4, -1, 1, -3)$ **45.** $(1, -1, 1, 2)$ **47.** $(\frac{3}{2} + 51t, \frac{1}{3} - 94t, -\frac{1}{3} - 62t, -\frac{3}{2} + 33t, 6t)$
49. Three containers of spray I and four containers of spray II are needed
51. Four units of candy I, five units of candy II, and six units of candy III   **53.** No. The large scoop is the best buy

PROBLEM SET 2.3, PAGES 74–75

1. Yes   **3.** Yes   **5.** Yes   **7.** No   **9.** Yes   **11.** No   **13.** $\begin{bmatrix} 2 & 7 \\ 1 & 4 \end{bmatrix}$ **15.** $\begin{bmatrix} 2 & -5 \\ -1 & 3 \end{bmatrix}$

17. $\dfrac{1}{6}\begin{bmatrix} 0 & 3 \\ 2 & -1 \end{bmatrix}$ **19.** $\dfrac{1}{22}\begin{bmatrix} 2 & -3 \\ 1 & 4 \end{bmatrix}$ **21.** $\begin{bmatrix} 4 & 3 \\ 2 & 2 \end{bmatrix}$ **23.** $\begin{bmatrix} 3 & -17 & -20 \\ 3 & -18 & -20 \\ -1 & 6 & 7 \end{bmatrix}$ **25.** $\begin{bmatrix} 3 & 3 & -1 \\ -2 & -2 & 1 \\ -4 & -5 & 2 \end{bmatrix}$

27. $\dfrac{1}{12}\begin{bmatrix} -2 & 2 & 2 \\ 8 & -8 & 4 \\ 7 & -1 & -1 \end{bmatrix}$ **29.** $\dfrac{1}{10}\begin{bmatrix} -12 & 0 & 8 & 1 \\ 6 & 0 & -4 & 2 \\ 2 & 0 & 2 & -1 \\ 0 & 10 & 0 & 0 \end{bmatrix}$ **31.** $(3, 2)$ **33.** $(5, -4)$ **35.** $(-1, 4)$ **37.** $(3, 1)$

39. $(-2, 2)$ **41.** $(23, -30)$ **43.** $(5, 6, 1)$ **45.** $(1, -2, 3)$ **47.** $(5, 4, -1)$ **49.** $(1, 2, 3)$ **51.** $(-3, 2, 7)$
53. $(w, x, y, z) = (3, -2, 1, -4)$ **55.** Six bags of grain I, three bags of grain II, and one bag of grain III
57. Twenty-two bags of grain I, eight bags of grain II, and eleven bags of grain III   **59.** Ten oz of each food

PROBLEM SET 2.4, PAGES 80–81

1. $x = t,\ y = 3t,\ z = 2t$ **3.** $x = -6t,\ y = 2t,\ z = 3t$ **5.** $x_1 = 154t,\ x_2 = 89t,\ x_3 = 30t$
7. Answers vary. One possible solution is $x_1 = \$50{,}000,\ x_2 = \$47{,}500,\ x_3 = \$30{,}000$
9. Answers vary. One possible solution is $x_1(\text{farmer}) = \$1{,}000,\ x_2(\text{builder}) = \$1{,}250,\ x_3(\text{tailor}) = \$1{,}250,\ x_4(\text{merchant}) = \$1{,}000$
11. The required output in 3 years is about $136 for farming, $139 for construction, and $155 for clothing
13. Farming output = $400, construction output = $658, clothing output = $382

15. $\begin{bmatrix} 1.41 & 0.14 & 0.34 \\ 0.33 & 1.41 & 0.39 \\ 0.17 & 0.06 & 1.16 \end{bmatrix}$ **17.** 347, 251, 214   **19.** $\begin{bmatrix} 1.41 & 0.00 & 0.00 \\ 0.04 & 1.26 & 0.03 \\ 0.08 & 0.02 & 1.28 \end{bmatrix}$

PROBLEM SET 2.5, PAGE 82

1. a. $\left(\dfrac{-25}{7}, \dfrac{64}{21}\right)$ **b.** $(3, 2)$ **c.** $(5, 8)$ **d.** -2 **3. a.** $\begin{bmatrix} 8 & 3 & 7 \\ -2 & 16 & 17 \end{bmatrix}$ **b.** Not conformable

5. a. $\begin{bmatrix} 11 & 8 & -3 \\ 3 & 8 & -6 \\ 11 & -6 & 14 \end{bmatrix}$ **b.** $\begin{bmatrix} 5 & 2 & 0 \\ 3 & 0 & 0 \\ 2 & 3 & 5 \end{bmatrix}$ **7. a.** $(3, -2)$ **b.** $(2, -5)$ **c.** $(\frac{7}{5} + 2t, -\frac{24}{5} + 11t, 5t)$

9. a. $A^{-1} = \begin{bmatrix} 1 & -1 & 1 \\ 0 & 2 & -1 \\ 2 & 3 & 0 \end{bmatrix}$ **b.** $\begin{bmatrix} 1 & -1 & 1 \\ 0 & 2 & -1 \\ 2 & 3 & 0 \end{bmatrix}\begin{bmatrix} 8 \\ -1 \\ 3 \end{bmatrix} = \begin{bmatrix} 12 \\ -5 \\ 13 \end{bmatrix}$ $x = 12,\ y = -5,\ z = 13$

CHAPTER 3
PROBLEM SET 3.1, PAGES 90–91

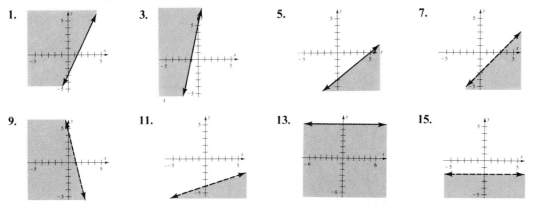

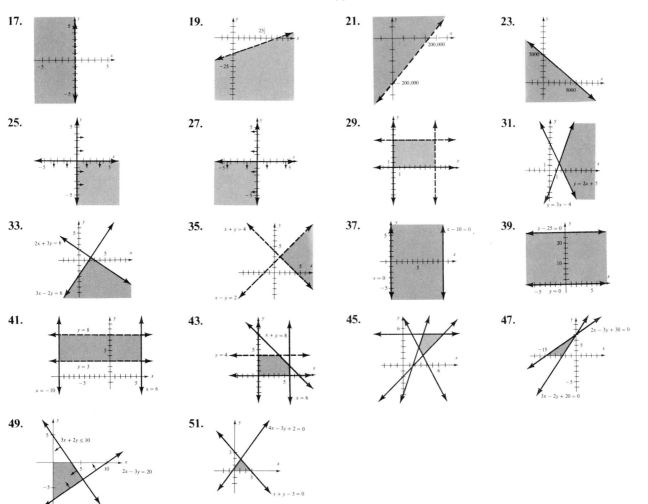

PROBLEM SET 3.2, PAGES 98–101

1. Let x = Number of acres of corn
y = Number of acres of wheat
Maximize: $P = (1.20)(100x) + (2.50)(40y)$
Subject to: $\begin{cases} 120x + 60y \le 24{,}000 \\ 100x + 40y \ge 20{,}000 \\ x + y \le 500 \\ x \ge 0, \quad y \ge 0 \end{cases}$

3. Let x = Number of standard items produced
y = Number of economy items produced
P = Profit
Maximize: $P = 45x + 30y$
Subject to: $\begin{cases} 3x + 3y \le 1500 \\ 2x + 0y \le 800 \\ x \ge 0, \quad y \ge 0 \end{cases}$

5. Let x = Ounces of Corn Flakes
y = Ounces of Honeycombs
Minimize: $C = 0.07x + 0.19y$
Subject to: $\begin{cases} 23x + 14y \ge 322 \\ 7x + 17y \ge 119 \\ x \ge 0, \quad y \ge 0 \end{cases}$

7. Let x = Amount invested in Pertec stock
y = Amount invested in Campbell Municipal Bonds
Maximize: $R = 0.20x + 0.10y$
Subject to: $\begin{cases} x + y \le 100{,}000 \\ x \le 70{,}000 \\ y \ge 20{,}000 \\ y \le 3x \\ x \ge 0, \quad y \ge 0 \end{cases}$

9. Let x = Number of Glassbelt tires produced
  y = Number of Rainbelt tires produced
  Maximize: $R = 52x + 36y$
  Subject to: $\begin{cases} 3x + y \le 200 \\ 2x + 2y \le 400 \\ x \ge 0, \quad y \ge 0 \end{cases}$

11. Let x = Number of commercial guests
  y = Number of other guests
  Maximize: $P = 4.50x + 3.50y$
  Subject to: $\begin{cases} 0.40x + 0.20y \le 50 \\ x + y \le 200 \\ x \ge 0, \quad y \ge 0 \end{cases}$

13. Let x = Number of Alpha units produced
  y = Number of Beta units produced
  Maximize: $P = 5x + 8y$
  Subject to: $\begin{cases} x \le 700 \\ x + 3y \le 1200 \\ x + 2y \le 1000 \\ x \ge 0, \quad y \ge 0 \end{cases}$

15. Let x = Number of A1 animals
  y = Number of A2 animals
  z = Number of A3 animals
  Maximize: $T = x + y + z$
  Subject to: $\begin{cases} 12x + 15y + 20z \le 8000 \\ 6x + 20y + 10z \le 10{,}000 \\ 16x + 5y + 10z \le 6000 \\ x \ge 0, \quad y \ge 0, \quad z \ge 0 \end{cases}$

17. Let x = Number of tables manufactured
  y = Number of shelves manufactured
  P = Profit
  Maximize: $P = 4x + 2y$
  Subject to: $\begin{cases} x \ge 30, \quad y \ge 25 \\ 2x + y \le 200 \\ 2x + 2y \le 240 \\ 2x + 3y \le 300 \\ x \ge 0, \quad y \ge 0 \end{cases}$

19. Let x = Number of hours of jogging
  y = Number of hours of handball
  z = Number of hours of dancing
  E = Total exercise time
  Minimize: $E = x + y + z$
  Subject to: $\begin{cases} x \ge 3 \\ y \ge 2 \\ z \ge 5 \\ x \ge y \\ 900x + 600y + 800z \ge 9000 \\ x \ge 0, \quad y \ge 0, \quad z \ge 0 \end{cases}$

21. Let x = Number of units on day shift
  y = Number of units on swing shift
  z = Number of units on graveyard shift
  C = Labor costs
  Minimize: $C = 100x + 150y + 180z$
  Subject to: $\begin{cases} x \ge 20, \quad x \le 50, \quad y \le 50, \quad z \le 50 \\ y + z \ge 50 \\ 60x \le 1800 \\ 60y \le 1500 \\ 60z \le 3000 \\ x \ge 0, \quad y \ge 0, \quad z \ge 0 \end{cases}$

23. Let x = Number of cases shipped from LA to Chicago
  y = Number of cases shipped from LA to Dallas
  z = Number of cases shipped from Seattle to Chicago
  w = Number of cases shipped from Seattle to Dallas
  C = Shipping cost
  Minimize: $C = 9x + 7y + 7z + 8w$
  Subject to: $\begin{cases} x + y \le 90 \\ z + w \le 130 \\ x + z \ge 80 \\ y + w \ge 110 \\ x \ge 0, \quad y \ge 0, \quad z \ge 0, \quad w \ge 0 \end{cases}$

PROBLEM SET 3.3, PAGES 110–112

1. Maximum at D, minimum at A **3.** No maximum, minimum at C

5. $(0, \frac{9}{2}), (6, 0), (5, 2), (0, 0)$ **7.** $(0, 0), (0, 4), (4, 0), (2, 3)$ **9.** $(6, 0), (6, 2), (4, 4), (0, 4), (0, 0)$

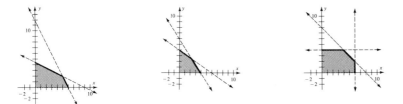

11. (*Note: This is Example 2,*
   Section 3.2.) $(0, \frac{8}{5}), (0, 4), (\frac{24}{13}, \frac{16}{13})$

13. (*Note: This is Example 3, Section 3.2.*)
   $(0, 40), (8, 24), (\frac{200}{7}, \frac{60}{7}), (50, 0)$

15. $(3, 2), (5, 5), (7, 5), (\frac{10}{3}, \frac{4}{3})$

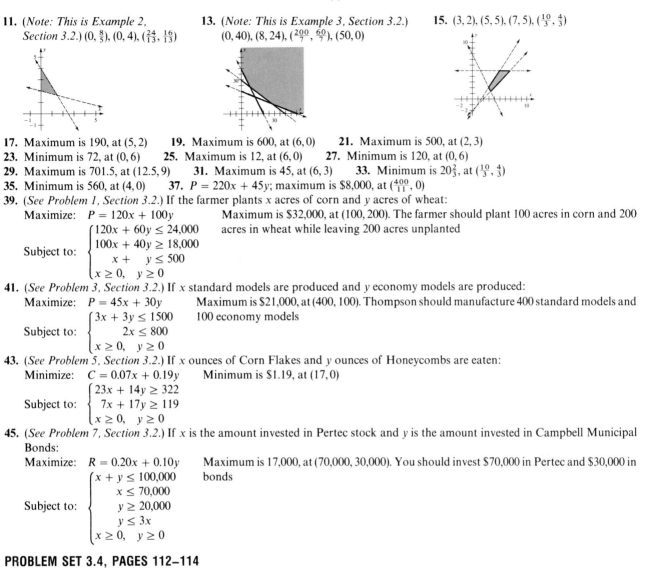

17. Maximum is 190, at $(5, 2)$ **19.** Maximum is 600, at $(6, 0)$ **21.** Maximum is 500, at $(2, 3)$
23. Minimum is 72, at $(0, 6)$ **25.** Maximum is 12, at $(6, 0)$ **27.** Minimum is 120, at $(0, 6)$
29. Maximum is 701.5, at $(12.5, 9)$ **31.** Maximum is 45, at $(6, 3)$ **33.** Minimum is $20\frac{2}{3}$, at $(\frac{10}{3}, \frac{4}{3})$
35. Minimum is 560, at $(4, 0)$ **37.** $P = 220x + 45y$; maximum is \$8,000, at $(\frac{400}{11}, 0)$
39. (*See Problem 1, Section 3.2.*) If the farmer plants x acres of corn and y acres of wheat:

Maximize: $P = 120x + 100y$

Subject to: $\begin{cases} 120x + 60y \leq 24,000 \\ 100x + 40y \geq 18,000 \\ x + y \leq 500 \\ x \geq 0, \quad y \geq 0 \end{cases}$

Maximum is \$32,000, at $(100, 200)$. The farmer should plant 100 acres in corn and 200 acres in wheat while leaving 200 acres unplanted

41. (*See Problem 3, Section 3.2.*) If x standard models are produced and y economy models are produced:

Maximize: $P = 45x + 30y$

Subject to: $\begin{cases} 3x + 3y \leq 1500 \\ 2x \leq 800 \\ x \geq 0, \quad y \geq 0 \end{cases}$

Maximum is \$21,000, at $(400, 100)$. Thompson should manufacture 400 standard models and 100 economy models

43. (*See Problem 5, Section 3.2.*) If x ounces of Corn Flakes and y ounces of Honeycombs are eaten:

Minimize: $C = 0.07x + 0.19y$

Subject to: $\begin{cases} 23x + 14y \geq 322 \\ 7x + 17y \geq 119 \\ x \geq 0, \quad y \geq 0 \end{cases}$

Minimum is \$1.19, at $(17, 0)$

45. (*See Problem 7, Section 3.2.*) If x is the amount invested in Pertec stock and y is the amount invested in Campbell Municipal Bonds:

Maximize: $R = 0.20x + 0.10y$

Subject to: $\begin{cases} x + y \leq 100,000 \\ x \leq 70,000 \\ y \geq 20,000 \\ y \leq 3x \\ x \geq 0, \quad y \geq 0 \end{cases}$

Maximum is 17,000, at $(70,000, 30,000)$. You should invest \$70,000 in Pertec and \$30,000 in bonds

PROBLEM SET 3.4, PAGES 112–114

1.

3. Maximum is $P = 3(210) - 4(0) = 630$

5. C   **7.** A   **9.** C

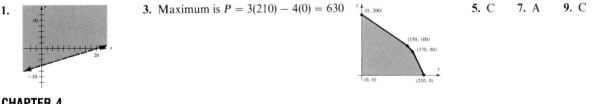

CHAPTER 4
PROBLEM SET 4.1, PAGES 122–123

1. Maximize: $z = 30x_1 + 20x_2$

Subject to: $\begin{cases} 2x_1 + x_2 \leq 12 \\ 5x_1 + 8x_2 \leq 40 \\ x_1 \geq 0, \quad x_2 \geq 0 \end{cases}$

3. Maximize: $z = 100x_1 + 100x_2$

Subject to: $\begin{cases} 3x_1 + 2x_2 \leq 12 \\ x_1 + 2x_2 \leq 8 \\ x_1 \geq 0, \quad x_2 \geq 0 \end{cases}$

5. Conditions 1 and 3 are violated

7. Conditions 2 and 3 are violated   **9.** Condition 3 is violated   **11.** Condition 2 is violated

13.

x_1	x_2	y_1	y_2	z	
3	1	1	0	0	300
2	2	0	1	0	400
−2	−3	0	0	1	0

15.

x_1	x_2	y_1	y_2	z	
1	1	1	0	0	200
4	2	0	1	0	500
−45	−35	0	0	1	0

17.

x_1	x_2	y_1	y_2	y_3	z	
12	150	1	0	0	0	1200
6	200	0	1	0	0	1200
16	50	0	0	1	0	800
−1	−1	0	0	0	1	0

19.

x_1	x_2	x_3	y_1	y_2	z	
8	5	3	1	0	0	1000
5	1	3	0	1	0	800
−12	−7	−5	0	0	1	0

21.

x_1	x_2	x_3	x_4	y_1	y_2	y_3	y_4	z	
1	1	0	0	1	0	0	0	0	90
0	0	1	1	0	1	0	0	0	130
1	0	1	0	0	0	1	0	0	80
0	1	0	1	0	0	0	1	0	110
−9	−7	−7	−8	0	0	0	0	1	0

23.

x_1	x_2	y_1	y_2	z	
1	0	$\frac{1}{3}$	0	0	30
0	2	−2	1	1	−162
0	−6	4	0	1	360

25.

x_1	x_2	y_1	y_2	z	
1	0	$\frac{1}{3}$	0	0	30
0	1	−1	$\frac{1}{2}$	0	−81
0	0	−2	3	1	−126

27.

x_1	x_2	y_1	y_2	z	
2	1	$\frac{1}{4}$	0	0	10
0	0	$-\frac{3}{2}$	1	0	540
12	0	$\frac{5}{2}$	0	1	100

29.

x_1	x_2	y_1	y_2	y_3	z	
0	−3	1	−1	0	0	100
1	3	0	$\frac{1}{2}$	0	0	50
0	−7	0	$-\frac{3}{2}$	1	0	150
0	25	0	5	0	1	500

31.

x_1	x_2	y_1	y_2	y_3	z	
$\frac{5}{2}$	0	1	$\frac{5}{2}$	0	0	350
$\frac{1}{2}$	1	0	$\frac{1}{2}$	0	0	10
−4	0	0	−6	1	0	100
2	0	0	8	0	1	340

33.

x_1	x_2	x_3	y_1	y_2	y_3	z	
−15	−6	0	1	0	−2	0	60
−11	−2	0	0	1	$-\frac{3}{2}$	0	45
4	2	1	0	0	$\frac{1}{2}$	0	5
14	5	0	0	0	2	1	20

PROBLEM SET 4.2, PAGES 132–134

1. a. $x_1 = 0, x_2 = 0, y_1 = 30, y_2 = 50, z = 0$ **b.** $x_1 = 0, x_2 = 0, y_1 = 120, y_2 = 180, z = 0$

3. a. $x_1 = 0, x_2 = 0, x_3 = 0, y_1 = 60, y_2 = 30, y_3 = 40, z = 0$ **b.** $x_1 = 0, x_2 = 0, x_3 = 0, y_1 = 80, y_2 = 50, y_3 = 60, y_4 = 90, z = 0$

5. a. $x_1 = 10, y_1 = 12, y_3 = 20, x_2 = y_2 = 0, z = 32$. Final tableau

   b. $x_1 = 120, x_3 = 80, y_1 = 20, x_2 = y_2 = y_3 = 0, z = 360$. Not the final tableau

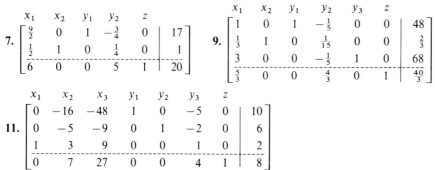

7.

x_1	x_2	y_1	y_2	z	
$\frac{9}{2}$	0	1	$-\frac{3}{4}$	0	17
$\frac{1}{2}$	1	0	$\frac{1}{4}$	0	1
6	0	0	5	1	20

9.

x_1	x_2	y_1	y_2	y_3	z	
1	0	1	$-\frac{1}{5}$	0	0	48
$\frac{1}{3}$	1	0	$\frac{1}{15}$	0	0	$\frac{2}{3}$
3	0	0	$-\frac{1}{5}$	1	0	68
$\frac{5}{3}$	0	0	$\frac{4}{3}$	0	1	$\frac{40}{3}$

11.

x_1	x_2	x_3	y_1	y_2	y_3	z	
0	−16	−48	1	0	−5	0	10
0	−5	−9	0	1	−2	0	6
1	3	9	0	0	1	0	2
0	7	27	0	0	4	1	8

13. Maximum is $z = 600$, when $x_1 = 0$, $x_2 = 200$ **15.** Maximum is 480, when $x_1 = 0$ and $x_2 = 30$
17. Maximum is $z = 7500$, when $x_1 = 50$, $x_2 = 150$ **19.** Maximum is 4700, when $x_1 = 700$, $x_2 = 150$
21. Maximum is $z = 50$, when $x_1 = 50$, $x_2 = 0$ **23.** Maximum is 400, when $x_1 = 100$, $x_2 = 0$
25. Maximum is $z = 1600$, when $x_1 = 0$, $x_2 = 50$, $x_3 = 250$ **27.** Maximum is $z = 1600$, when $x_1 = 80$, $x_4 = 110$
29. The company should manufacture 700 Alpha products and 150 Beta products. Then the (maximum) profit will be $4,700
31. Maximum profit is $1,600, when $x_1 = 0$, $x_2 = 50$, $x_3 = 250$

PROBLEM SET 4.3, PAGES 141–142

1. Maximum is 400, when $x_1 = 0$ and $x_2 = 10$ **3.** Maximum is 600, when $x_1 = 20$ and $x_2 = 0$
5. Maximum is $\frac{405}{2}$, when $x_1 = \frac{15}{4}$ and $x_2 = -\frac{9}{2}$ **7.** Maximum is $\frac{1840}{13}$, when $x_1 = \frac{24}{13}$ and $x_2 = \frac{16}{13}$
9. Maximum is 12, when $x_1 = 0$ and $x_2 = 4$ **11.** Minimum is -40, when $x_1 = 2$ and $x_2 = 0$
13. Minimum is 125, when $x_1 = 3$ and $x_2 = 2$ **15.** Maximum is 80, when $x_1 = 15$, $x_2 = 0$, and $x_3 = 10$
17. Minimum cost is $\frac{115}{2}$ or $57.50, when no units of A, $\frac{115}{2}$ units of B, and no units of C are mixed
19. Any ordered pair on the line segment with endpoints $(80, 40)$ and $(87.5, 25)$ will yield the maximum, $400, when substituted into the objective function. Since we desire integral solutions, we could start at $(80, 40)$ and use the idea of the slope of the line segment, which is $-\frac{2}{1}$, to generate other integral solutions: $(81, 38)$, $(82, 36)$, $(83, 34)$, $(84, 32)$, $(85, 30)$, $(86, 28)$, $(87, 26)$
21. Answers vary, but the key idea is that the simplex method always gives the maximum, *but only one* set of values where it occurs. The occurrence of a maximum will not be unique if and only if the objective function has the same slope as one of the sides of the (polygonal) feasibility region. In this case we might have noticed that the objective function and the first constraint both have slopes of -2. Perhaps in such a situation, one should resort to the graphing method when possible

PROBLEM SET 4.4, PAGES 149–151

1. a. $A^{\mathsf{T}} = \begin{bmatrix} 6 & 4 \\ 9 & 8 \end{bmatrix}$ **b.** $B^{\mathsf{T}} = \begin{bmatrix} 5 & 3 \\ 6 & 8 \end{bmatrix}$ **c.** $C^{\mathsf{T}} = \begin{bmatrix} 4 & 6 \\ 9 & 1 \\ 1 & 4 \end{bmatrix}$ **3. a.** $G^{\mathsf{T}} = \begin{bmatrix} 1 \\ 3 \\ 5 \end{bmatrix}$ **b.** $H^{\mathsf{T}} = \begin{bmatrix} 4 & 9 & 6 \end{bmatrix}$

c. $J^{\mathsf{T}} = \begin{bmatrix} 1 & 0 & 3 & 2 \end{bmatrix}$ **5.** Maximize: $z' = 10y_1 + 30y_2$ **7.** Maximize: $z' = 10y_1 + 2y_2 + y_3 + 15y_4$

Subject to: $\begin{cases} 2y_1 + 3y_2 \le 3 \\ 8y_1 + 5y_2 \le 4 \\ y_1 \ge 0, \quad y_2 \ge 0 \end{cases}$ Subject to: $\begin{cases} y_1 + 2y_2 + 3y_4 \le 3 \\ y_1 + y_3 \le 2 \\ y_1 + 3y_2 + 2y_3 + 5y_4 \le 5 \\ y_1 \ge 0, \quad y_2 \ge 0, \quad y_3 \ge 0, \quad y_4 \ge 0 \end{cases}$

9. Minimum value is 24, when $x_1 = 0$ and $x_2 = 6$ **11.** Minimum value is 25, when $x_1 = 5$, $x_2 = 5$, $x_3 = 0$
13. Minimum is $\frac{2000}{3}$, when $x_1 = 0$, $x_2 = 0$, $x_3 = \frac{100}{3}$ **15.** Minimum is 72, when $x_1 = 0$, $x_2 = 6$
17. Minimum is 120, when $x_1 = 0$, $x_2 = 6$ **19.** Minimum is $\frac{62}{3}$, when $x_1 = \frac{10}{3}$, $x_2 = \frac{4}{3}$
21. Minimum is 560, when $x_1 = 4$, $x_2 = 0$
23. The Gainesville plant should operate 30 days and the Sacramento plant should operate 20 days. The minimum cost is $900,000
25. She should spend $\frac{38}{9}$ hours (about 4 hours, 13 minutes) jogging, 2 hours playing handball, and 5 hours dancing per week
27. The day shift should produce 30 units, the swing shift should produce 25 units, and the graveyard shift should produce 50 units
29. The dealer should ship 30 sets from Burlingame to Hillsborough and 35 sets from San Jose to Palo Alto
31. The dealer should ship 15 refrigerators from Dallas to Fort Worth, 10 from Dallas to Houston, and 25 from San Antonio to Houston

PROBLEM SET 4.5, PAGES 152–153

1.

	x_1	x_2	x_3	y_1	y_2	y_3	z	
	2	0	1	1	0	0	0	50
	0	④	1	0	1	0	0	90
	3	4	0	0	0	1	0	100
	-6	-25	-3	0	0	0	1	0

First pivot is circled. For extra practice, you can solve this problem; the maximum is 582.5, which occurs when $x_1 = \frac{10}{3}$, $x_2 = \frac{45}{2}$, and $x_3 = 0$.

3.
$$\left[\begin{array}{cccccc|c}
 & y_1 & y_2 & y_3 & x_1 & x_2 & z^1 \\
9 & 3 & 2 & 1 & 0 & 0 & 50 \\
1 & \textcircled{12} & 3 & 0 & 1 & 0 & 80 \\
\hline
-18 & -36 & -30 & 0 & 0 & 1 & 0
\end{array}\right]$$

First pivot is circled. For extra practice, you can solve this problem; the minimum is 750, which occurs when $x_1 = 15$ and $x_2 = 0$.

5. C   **7.** C   **9.** C

PROBLEM SET 4.6, PAGES 153–156

1. $\left[\begin{array}{ccc} 14 & -1 & 12 \\ 20 & -8 & 5 \\ -13 & 9 & 6 \end{array}\right]$ **3.** $\left[\begin{array}{ccc} 12 & -1 & 1 \\ 3 & -3 & 4 \\ 2 & 3 & -4 \end{array}\right]$ **5.** $(4, -1)$ **7.** $(-t, 1 - t, t)$ **9.**

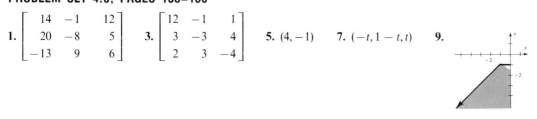

11. Maximum is 12,000, when $x_1 = \frac{2400}{7}$, $x_2 = \frac{1200}{7}$

13. The maximum number of animals is 400 (all species C)

15. The combination should consist of 2 pounds of peanuts   **17.** C

19.
$$\left[\begin{array}{cccccc|c}
y_1 & y_2 & y_3 & A & B & z & \\
4 & \frac{2}{3} & \frac{5}{4} & 1 & 0 & 0 & 23 \\
2 & 1 & 1 & 0 & 1 & 0 & 24 \\
\hline
-1800 & -400 & -600 & 0 & 0 & 1 & 0
\end{array}\right]$$

CHAPTER 5
PROBLEM SET 5.1, PAGES 169–171

1. (Answers vary.) An *element* of a set is a member of the set, and not a subset of the set. A *subset* of a set is not a member of the set, and is a set. For example, 4 is an element (not a subset) of the set of even integers, $\{4\}$ is a subset (not an element) of the set of even integers

3. (Answers vary.) The universal set is the set of all things (or elements) under consideration and is usually nonempty. The empty set contains no elements

5. $\{m\}, \{y\}, \{m, y\}, \{\ \}$ **7.** $\{y\}, \{o\}, \{u\}, \{y, o\}, \{y, u\}, \{o, u\}, \{y, o, u\}, \{\ \}$

9. $\{\ \}, \{3\}, \{6\}, \{9\}, \{3, 6\}, \{3, 9\}, \{6, 9\}, \{3, 6, 9\}$

11. $\{\ \}, \{1\}, \{2\}, \{3\}, \{4\}, \{1, 2\}, \{1, 3\}, \{1, 4\}, \{2, 3\}, \{2, 4\}, \{3, 4\}, \{1, 2, 3\}, \{1, 2, 4\}, \{2, 3, 4\}, \{1, 3, 4\}, \{1, 2, 3, 4\}$

13. $\{2, 6, 8, 10\}$ **15.** $\{3, 4, 5\}$ **17.** $\{2, 3, 5, 6, 8, 9\}$ **19.** $\{1, 3, 4, 5, 6, 7, 10\}$ **21.** $\{1, 2, 3, 4, 5, 6, 7, 9, 10\}$

23. $\{1, 2, 3, 4, 5, 6\}$ **25.** $\{1, 2, 3, 5, 6, 7\}$ **27.** $\{3\}$ **29.** $\{5, 6, 7\}$ **31.** $\{1, 2, 4, 6\}$ **33.** $\{1, 2, 3, 4, 5, 6, 7\}$

35. $\{1, 2, 3, 4, 5\}$ **37.** $\{5, 6, 7\}$ **39. a.** False   **b.** True   **c.** True   **d.** True   **e.** False

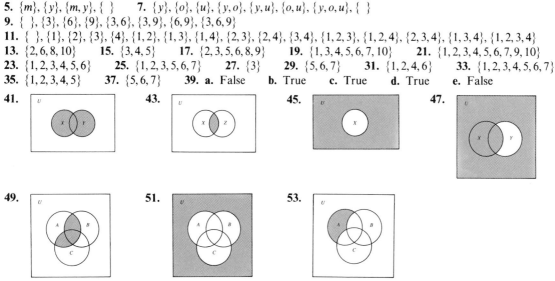

41.   **43.**   **45.**   **47.**

49.   **51.**   **53.**

55. a. $A \cup B = \{$Bob Wisner, Joan Marsh, Craig Barth, Phyllis Niklas, Shannon Smith, Christy Anton$\}$; union
b. $A \cap B = \{$Phyllis Niklas, Craig Barth$\}$; intersection

57. a. Answers vary. 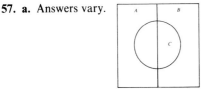 **b.** False **c.** True **d.** True
e. False **f.** True **g.** True **h.** False **i.** False

59. Let M be the set of tires with defective materials and W be the set of tires with defective workmanship.
Then: $n(M \cup W) = n(M) + n(W) - n(M \cap W) = 72 + 89 - 17 = 144$ **61.** 70

63. **65. a.** 12 **b.** 7 **c.**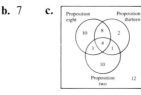

67. a. B **b.** A **c.** U **d.** $\varnothing$ **69.** 1,001,101 **71.** Answers vary

PROBLEM SET 5.2, PAGES 178–179

1. 696 **3.** 40,200 **5.** 24 **7.** 144 **9.** 3,628,800 **11.** 72 **13.** 11,880 **15.** 126 **17.** 330 **19. a.** 9

b. 72 **c.** 504 **d.** 3024 **e.** 1 **21. a.** 95,040 **b.** 60 **c.** 1680 **d.** 1 **e.** $\dfrac{g!}{(g-h)!}$ **23. a.** 210 **b.** 120

c. $\dfrac{50!}{2}$ **d.** 25 **e.** $\dfrac{m!}{(m-3)!}$ or $m(m-1)(m-2)$ **25.** 7! or 5040 **27.** $\dfrac{6!}{2!} = 360$ **29.** $\dfrac{11!}{4!4!2!} = 34,650$

31. $\dfrac{11!}{2!2!2!} = 4,989,600$ **33.** $\dfrac{11!}{3!2!2!2!} = 831,600$ **35.** 210 **37.** 15 **39.** 40,320 **41.** 47,045,520 **43.** 60

45. 175,760,000 **47.** 2,300,000 **49.** 17,331,000 **51.** 1,976,000 **53.** Answers vary

PROBLEM SET 5.3, PAGES 186–187

1. a. 9 **b.** 36 **c.** 84 **d.** 126 **e.** 1 **3. a.** 35 **b.** 1 **c.** 1225 **d.** 25 **e.** $\dfrac{g!}{h!(g-h)!}$ **5. a.** 2520

b. 1 **c.** 45 **d.** $\dfrac{n!}{(n-4)!}$ **e.** $\dfrac{n!}{4!(n-4)!}$ **7. a.** 1 **b.** 23,426 **c.** 1 **d.** 5040 **e.** $\dfrac{n!}{(n-5)!}$ **9. a.** 20

b. 420 **11. a.** 2520 **b.** 1680 **13.** $\dbinom{100}{5} = 75,287,520$ ways **15.** $\dbinom{4}{2} = 6$ ways **17.** $\dbinom{13}{5} = 1287$ ways

19. Permutation; $_8P_5 = 6720$ **21.** Permutation; $_5P_5 = 120$ **23.** Combination; $\dbinom{100}{6} = 1,192,052,400$

25. Permutation; $_6P_6 = 720$ **27.** None; $3 \cdot 3 \cdot 2 \cdot 2 \cdot 1 \cdot 1 = 36$ ways **29.** Combination; $\dbinom{5}{3} = 10$

31. Permutation; $_6P_5 = 720$ **33.** $\dbinom{5}{2}\dbinom{25}{5}\dbinom{18}{5} = 4,552,178,400$ **35.** $\dfrac{50!}{15!10!25!}$

37. $\dbinom{50}{7,5,38} \approx 96,149,000,000,000$ **39.** $\dbinom{10}{4,5,1} = 1260$ **41. a.** $\dbinom{10}{3} = 120$ **b.** $\dbinom{11}{2,8,1} = 495$

c. $\dbinom{10}{1}\dbinom{11}{2} = 550$ **d.** $\dbinom{10}{1}\dbinom{6}{1}\dbinom{5}{1} = 300$ **e.** $\dbinom{6}{2}\dbinom{5}{1} = 75$ **43.** $\dbinom{10}{3}\dbinom{12}{3} = 26,400$

45. $\dbinom{n}{k-1} + \dbinom{n}{k} = \dfrac{n!}{(k-1)!(n-k+1)!} + \dfrac{n!}{k!(n-k)!} = \dfrac{k \cdot n!}{k!(n-k+1)!} + \dfrac{(n-k+1) \cdot n!}{k!(n-k+1)!} = \dfrac{(n+1)n!}{k!(n-k+1)!} = \dbinom{n+1}{k}$

PROBLEM SET 5.4, PAGES 192–193

1. 8   **3.** 28   **5.** 56   **7.** 12   **9.** 1   **11.** 153   **13.** $x^5 + 5x^4y + 10x^3y^2 + 10x^2y^3 + 5xy^4 + y^5$
15. $x^4 + 4x^3y + 6x^2y^2 + 4xy^3 + y^4$ **17.** $x^5 - 5x^4y + 10x^3y^2 - 10x^2y^3 + 5xy^4 - y^5$
19. $x^5 + 10x^4 + 40x^3 + 80x^2 + 80x + 32$ **21.** $16x^4 + 96x^3y + 216x^2y^2 + 216xy^3 + 81y^4$
23. $1 - 10x + 45x^2 - 120x^3 + 210x^4 - 252x^5 + 210x^6 - 120x^7 + 45x^8 - 10x^9 + x^{10}$
25. $\binom{15}{0}x^{15} - \binom{15}{1}x^{14}y + \cdots + (-1)^r\binom{15}{r}x^{15-r}y^r + \cdots + \binom{15}{14}xy^{14} - \binom{15}{15}y^{15}$
27. $\binom{20}{0}x^{20} + \binom{20}{1}x^{19}y + \cdots + \binom{20}{r}x^{20-r}y^r + \cdots + \binom{20}{19}xy^{19} + \binom{20}{20}y^{20}$ **29.** -330 **31.** 81,081
33. -16 **35.** 2^{100} **37.** 210   **39.** Look at the expansion of $(x + y)^n$ with $x = 1, y = 1$
41. $\binom{n}{r} = \dfrac{n!}{r!(n-r)!}, \binom{n}{n-r} = \dfrac{n!}{(n-r)!r!}$

PROBLEM SET 5.5, PAGES 193–194

1. $\{2, 3, 4, 8, 9, 10, 12\}$ **3.** $\{5, 6\}$ **5. a.** 5040   **b.** 5016   **c.** 6   **d.** 9900   **7. a.** 28   **b.** 168   **9.** 42
11. $\binom{4}{2} 19 \cdot 18 \cdot 17 \cdot 16 \cdot \left(\dfrac{32!}{6!26!}\right) \approx 5.06 \times 10^{11}$ **13.** $2^{10} = 1024$ **15. a.** 125   **b.** 5   **c.** 25   **d.** 10   **e.** 60

CHAPTER 6
PROBLEM SET 6.1, PAGES 201–202

1. $S = \{2, 3, 4, 5, 6, 7, 8, 9, 10, 11, 12\}$ **3.** $\{0, 1, 2, 3, 4, 5, 6, 7, 8, 9, 10\}$ **5.** $\{4, 5, 6, 7\}$
7. {Nonnegative integers less than or equal to 500}   **9.** {bbb, bbg, bgb, bgg, gbb, gbg, ggb, ggg}
11. a. $E = \{2, 4, 6, 8, 10, 12\}$ **b.** $F = \{2, 3, 5, 7, 11\}$; not mutually exclusive   **13. a.** $G = \{2\}$
b. $H = \{3, 4, 5, 6, 7, 8, 9, 10, 11, 12\}$; mutually exclusive   **15. a.** $G = \{HHH, HHT, HTH, HTT\}$
b. $H = \{HHH, HHT, THH, THT\}$; not mutually exclusive   **17. a.** $F = \{fd, fr, fi, fn\}$
b. $D = \{md, fd\}$; not mutually exclusive   **19. a.** $E = \{fd, fr, fi, fn, mr\}$ **b.** $R = \{md, mr, mi, fd, fr, fi\}$; not mutually exclusive
21. $L \cup E = \{$Roll a number less than 3 or roll an even number$\} = \{1, 2, 4, 6\}$
23. $E \cap L = \{$Roll a number less than 3 and even$\} = \{2\}$ **25.** $\bar{L} = \{$Roll a 3 or larger$\} = \{3, 4, 5, 6\}$
27. $\bar{E} \cap \bar{L} = \{$Roll an odd number that is also at least a 3$\} = \{3, 5\}$ **29.** Yes   **31.** $\frac{3}{11}$ **33.** $\frac{5}{11}$ **35.** 0   **37.** $\frac{8}{11}$
39. Yes   **41.** $\frac{1}{9}$ **43.** $\frac{5}{9}$ **45.** 0   **47.** $\frac{8}{9}$

PROBLEM SET 6.2, PAGES 208–209

1. $\frac{3}{8}$ **3.** $\frac{2}{8} = \frac{1}{4}$ **5.** $\frac{6}{20} = \frac{3}{10}$ **7.** $\frac{13}{20}$ **9.** 0   **11.** $\frac{4}{52} = \frac{1}{13}$ **13.** $\frac{1}{52}$ **15.** $\frac{16}{52} = \frac{4}{13}$ **17.** .05   **19.** .19
21. Results may vary; yes   **23.** $\frac{4}{36} = \frac{1}{9}$ **25.** $\frac{6}{36} = \frac{1}{6}$ **27.** $\frac{4}{36} = \frac{1}{9}$ **29.** $\frac{2}{36} = \frac{1}{18}$ **31.** $\frac{7}{36}$ **33.** $\frac{18}{36} = \frac{1}{2}$
35. $\frac{8}{36} = \frac{2}{9}$ **37.** $\frac{1}{36}$ **39. a.** $\frac{1}{16}$ **b.** $\frac{2}{16} = \frac{1}{8}$ **c.** $\frac{3}{16}$

41.

	8	7	6	5	4	3	2	1
8	8,8	7,8	6,8	5,8	4,8	3,8	2,8	1,8
7	8,7	7,7	6,7	5,7	4,7	3,7	2,7	1,7
6	8,6	7,6	6,6	5,6	4,6	3,6	2,6	1,6
5	8,5	7,5	6,5	5,5	4,5	3,5	2,5	1,5
4	8,4	7,4	6,4	5,4	4,4	3,4	2,4	1,4
3	8,3	7,3	6,3	5,3	4,3	3,3	2,3	1,3
2	8,2	7,2	6,2	5,2	4,2	3,2	2,2	1,2
1	8,1	7,1	6,1	5,1	4,1	3,1	2,1	1,1

43. a. $\frac{9}{196}$ **b.** $\frac{1}{196}$ **c.** $\frac{1}{14}$ **d.** $\frac{1}{49}$

PROBLEM SET 6.3, PAGES 213–214

1. $\frac{10}{16} = \frac{5}{8}$ **3.** $\frac{4}{16} = \frac{1}{4}$ **5.** $\frac{12}{16} = \frac{3}{4}$ **7.** $\frac{6}{13}$ **9.** $\frac{3}{13}$ **11.** $\frac{10}{13}$ **13.** 1 to 3 **15.** $\frac{9}{10}$ **17.** 1 to 12 **19.** $\frac{1}{4}$
21. 0.008 **23.** $\frac{2}{17}$ **25.** $\frac{1}{4}$ **27.** $\frac{3}{8}$ **29.** $\frac{5}{16}$ **31.** $\frac{5}{16}$ **33.** 2 to 7 **35.** $\frac{3}{8} \approx .38$ **37.** $\frac{1}{20} \approx .05$
39. $\frac{1}{15} \approx .07$ **41.** .08 **43.** .31 **45.** .23 **47.** .01 **49.** .33 **51.** .14 **53.** .01 **55.** .21 **57.** .06
59. .00000154 **61.** .42256903

PROBLEM SET 6.4, PAGES 219–221

1. $\frac{3}{8}$ **3.** $\frac{1}{8}$ **5.** $\frac{3}{7}$ **7.** $\frac{7}{8}$ **9.** $\frac{1}{4}$ **11.** $\frac{1}{4}$ **13.** $\frac{2}{5}$ **15.** $\frac{1}{6}$ **17.** $\frac{2}{11}$ **19.** .36 **21.** .19 **23.** .53 **25.** .32
27. .59 **29.** $P(E|F) = .50, P(F|E) = .20$ **31.** $P(E|F) = \frac{1}{3}, P(F|E) = \frac{1}{2}$ **33.** $P(E|F) = .5, P(F|E) = .8$ **35.** 60%
37. $\frac{1}{2}$ **39.** Independent **41.** Independent **43.** Not independent **45.** Independent **47.** Independent
49. Not independent **51.** Not independent **53.** Not independent **55.** Answers vary

PROBLEM SET 6.5, PAGE 224

1. $\frac{1}{2}$ **3.** $\frac{5}{6}$ **5.** $\frac{1}{12}$ **7.** $\frac{2}{3}$ **9.** $\frac{4}{9}$ **11.** $\frac{11}{12}$ **13.** $\frac{1}{3}$ **15.** $\frac{5}{9}$ **17.** $\frac{1}{4}$ **19.** 0 **21.** $\frac{1}{2}$ **23. a.** $\frac{1}{3}$ **b.** $\frac{5}{12}$
c. $\frac{5}{9}$ **25. a.** $\frac{7}{12}$ **b.** $\frac{1}{6}$ **c.** $\frac{1}{18}$ **27.** $\frac{1}{12}$ **29.** $\frac{1}{6}$ **31.** $\frac{3}{4}$ **33.** $\frac{1}{2}$ **35.** $\frac{1}{3}$ **37.** $\frac{1}{2}$ **39.** $\frac{25}{64}$ **41.** $\frac{15}{32}$
43. $\frac{49}{64}$ **45.** $\frac{3}{28}$ **47.** $\frac{15}{56}$

PROBLEM SET 6.6, PAGES 230–231

1. .05 **3.** .04 **5.** .05 **7.** $\frac{1}{4}$ **9.** $\frac{4}{17}$ **11.** $\frac{1}{4}$ **13.** .04 **15.** .04 **17.** .05 **19.** $\frac{1}{4}$ **21.** .44 **23.** .59
25. $\frac{2}{3}$ **27. a.** .22 **b.** .09 **c.** .70 **29.** $\frac{5}{12}$

PROBLEM SET 6.7, PAGES 231–232

1. a. For a sample space S and event E, a real number $P(E)$ is associated, so that: (*i*) $0 \le P(E) \le 1$; (*ii*) $P(S) = 1$;
(*iii*) and, if E and F are mutually exclusive events, then $P(E \cup F) = P(E) + P(F)$ **b.** Their intersection is empty
c. One event does not affect the probability of the other   **3.** No. The probabilities of A, B, and C are $\frac{1}{4}, \frac{1}{4}$, and $\frac{1}{2}$, respectively
5. $\frac{18}{38} \approx .47$ **7. a.** $\frac{125}{375} \approx .333$ **b.** $\frac{125}{600} \approx .208$ **c.** $\frac{150}{625} = .24$ **d.** $\frac{250}{400} = .625$ **9. a.** .33   **b.** .27   **c.** .73

CHAPTER 7
PROBLEM SET 7.1, PAGES 238–239

1.

Inches	Tally	Frequency
63	||	2
64	||||	4
65	|||	3
66	|||	3
67	ⵘ	5
68	|||	3
69	||||	4
70	|||	3
71	||	2
72	|	1

3.

Number of spots	Tally	Frequency
2	|	1
3	|||	3
4	ⵘ	5
5	||	2
6	ⵘ ||||	9
7	ⵘ |||	8
8	ⵘ |	6
9	ⵘ |	6
10	||||	4
11	ⵘ	5
12	|	1

Let x = students' heights (rounded to the nearest inch), where x = 63, 64, ..., 72

Let x = sum of dice, where x = 2, 3, 4, ..., 11, 12

5.

Number of years	Frequency	Relative frequency
13	1	$\frac{1}{25} = .04$
14	2	$\frac{2}{25} = .08$
15	1	$\frac{1}{25} = .04$
16	3	$\frac{3}{25} = .12$
17	3	$\frac{3}{25} = .12$
18	0	$\frac{0}{25} = 0$
19	3	$\frac{3}{25} = .12$
20	3	$\frac{3}{25} = .12$
21	2	$\frac{2}{25} = .08$
22	4	$\frac{4}{25} = .16$
23	1	$\frac{1}{25} = .04$
24	1	$\frac{1}{25} = .04$
25	1	$\frac{1}{25} = .04$
	$\overline{25}$	$\overline{1}$

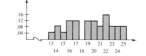

7.

Number waiting	Frequency	Relative frequency
2	20	$\frac{20}{50} = .40$
3	15	$\frac{15}{50} = .30$
4	7	$\frac{7}{50} = .14$
5	5	$\frac{5}{50} = .10$
6	2	$\frac{2}{50} = .04$
7	1	$\frac{1}{50} = .02$
	$\overline{50}$	$\overline{1}$

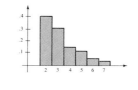

9.

Number of heads	Frequency	Relative frequency
3	18	$\frac{18}{150} = .12$
2	56	$\frac{56}{150} = .373$
1	59	$\frac{59}{150} = .393$
0	17	$\frac{17}{150} = .113$
	$\overline{150}$	$\overline{1}$

11.

Number of heads	Relative frequency
0	$\frac{1}{16} = .0625$
1	$\frac{4}{16} = .25$
2	$\frac{6}{16} = .375$
3	$\frac{4}{16} = .25$
4	$\frac{1}{16} = .0625$
	$\overline{1}$

13.

Number of acres	Relative frequency
0	$\frac{188}{221} = .851$
1	$\frac{32}{221} = .145$
2	$\frac{1}{221} = .005$
	$\overline{1}$

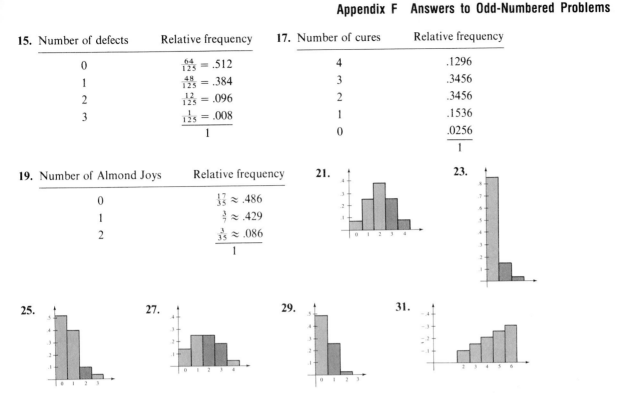

15.

Number of defects	Relative frequency
0	$\frac{64}{125} = .512$
1	$\frac{48}{125} = .384$
2	$\frac{12}{125} = .096$
3	$\frac{1}{125} = .008$
	1

17.

Number of cures	Relative frequency
4	.1296
3	.3456
2	.3456
1	.1536
0	.0256
	1

19.

Number of Almond Joys	Relative frequency
0	$\frac{17}{35} \approx .486$
1	$\frac{3}{7} \approx .429$
2	$\frac{3}{35} \approx .086$
	1

PROBLEM SET 7.2, PAGES 247–249

	Mean	Median	Mode	Range	Variance	Standard deviation
1.	3	3	None	4	2.5	1.58
3.	105	105	None	4	2.5	1.58
5.	6	7	7	8	6.7	2.58
7.	10	8	None	18	52.0	7.21
9.	91	95	95	17	50.5	7.11
11.	3	3	3	4	1.7	1.29

13. Answers vary   **15.** Mean = $15,800; median = $16,000; mode = $10,000
17. Mean = 68.09; median = 70; mode = 70   **19.** Mean = 56; median = 65; mode = 70; range = 90
21. Variance = $141\frac{2}{3}$; standard deviation = 11.90   **23.** Variance = $41,288,888.89; standard deviation = $6,425.64
25. Mean = .75; variance = .6875; standard deviation = .83   **27.** Mean = 3.4; variance = .84; standard deviation = .92
29. Mean = 2.85; variance = 2.3275; standard deviation = 1.53   **31.** Answers vary; all data are the same
33. Answers vary; first student 70, 70, 70, 70, 70, 70; second student 80, 70, 60, 90, 50, 70
35. Answers vary   **37.** Answers vary   **39.** 9.1596   **41.** 9.5736   **43.** 10.4198

PROBLEM SET 7.3, PAGES 254–255

1. $P(X = 0) = \binom{3}{0}(.2)^0(.8)^3 = .512$ **3.** $P(X = 2) = \binom{3}{2}(.2)^2(.8) = .096$ **5.** $1 - P(X = 0) = .488$

7. $P(X = 4) = \binom{4}{4}(.9)^4(.1)^0 = .6561$ **9.** $P(X = 0) = \binom{4}{0}(.9)^0(.1)^4 = .0001$ **11.** .948   **13.** .132

15. .1281   **17.** .016   **19.** .0002

	Mean	Variance	Standard deviation
21.	1.5	1.05	1.02
23.	7.8	2.73	1.65
25.	3.0	1.50	1.22
27.	0.7	0.63	0.79
29.	6.0	2.4	1.55

31. .4812   **33.** .5443   **35.** .484   **37.** 16
39. Mean = 1.2; standard deviation = .98
41. Mean = 30; standard deviation = 3.46   **43. a.** .109375
b. .1143   **45.** .3125   **47.** $n2^{1-n}$

PROBLEM SET 7.4, PAGE 260

1. .0228   **3.** .8413   **5.** .2709   **7.** .9712   **9.** .8413   **11.** .0455   **13.** .2548   **15.** .7926   **17.** 34 persons
19. 8 persons   **21.** 25   **23.** 1 student   **25.** .0228   **27.** .978   **29.** 40.5 inches   **31.** $P(X > 250) = .5$
33. $P(X < 220) = P(z < -1.2) = .1151$ **35.** $P(220 \leq X \leq 320) = .88$ **37.** .0228   **39.** $P(X > .43) = .0668$

PROBLEM SET 7.5, PAGE 265

1. Yes   **3.** No   **5.** Yes   **7.** Yes   **9.** No   **11.** $P(z < 1.29) = .9015$ **13.** $P(z > 2.84) = .002$ **15.** .2206
17. .567   **19.** .092   **21.** .0005   **23.** $P(z < .28) = .6103$ **25.** $P(z \geq -1.41) = .9207$ **27.** $P(z > -.57) = .7157$
29. $P(-1.41 < z < -.57) = .2050$ **31.** $P(-2.31 < z < -2.22) = .0028$ **33.** $P(-.61 < z < -.52) = .0306$
35. $P(z < 3.5) \approx .9998$ **37.** .5   **39.** $P(z > -1.17) = .8790$ **41.** $P(-8.16 < z < 3.5) = .9998$
43. $P(-2.39 < z < -2.27) = .0032$ **45.** $P(1.11 < z < 1.22) = .0223$
47. $np < 5$, $P(X > 6) = P(X = 7) + P(X = 8) + P(X = 9) + P(X = 10) = .0035$
49. $np < 5$, $1 - P(X = 0) - P(X = 1) - P(X = 2) = .068$ **51.** $P(X > 300) = P(z > 2.12) = .017$

PROBLEM SET 7.6, PAGES 271–273

1. Yes   **3.** No   **5.** Yes   **7.** No   **9.** No   **11.** No
13. $r = .995$, at 5%   **15.** $r = -.984$, at 1%   **17.** $r = .970$, at 1%

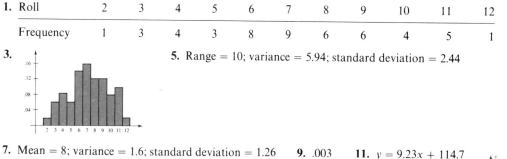

19. $y = 3.9x - 3$ **21.** $y = .56x + 41.6$ **23.** $y = -2.7143x + 30.0571$ **25.** $r = .842$; yes
27. $y = 2.4545x + 6.090$ **29.** $r = -.732$; not significant at 1% or 5%   **31.** $r = -.750$; not significant at 1% or 5%
33. $r = .924$; significant at 1%   **35.** $y = .00087x + 2.83518$

PROBLEM SET 7.7, PAGES 273–275

1. Roll	2	3	4	5	6	7	8	9	10	11	12
Frequency	1	3	4	3	8	9	6	6	4	5	1

3.

5. Range = 10; variance = 5.94; standard deviation = 2.44

7. Mean = 8; variance = 1.6; standard deviation = 1.26   **9.** .003   **11.** $y = 9.23x + 114.7$
13. D   **15.** A

PROBLEM SET 7.8, PAGES 275–277

1. a. $\{2\}$ **b.** $\{1, 3, 5, 7, 9\}$ **c.** $\{2, 4, 6, 8, 10\}$ **d.** $\{\ \}$ **e.** $\{1, 4, 6, 8, 9, 10\}$ **3.** 35   **5.** 32   **7. a.** $\frac{13 \cdot 12}{52 \cdot 51} \approx .0588$
b. $\frac{26 \cdot 13}{52 \cdot 51} \approx .127$ **c.** $\frac{13 \cdot 13}{52 \cdot 51} \approx .0637$ **d.** $\frac{13}{51} \approx .255$ **e.** 0   **9. a.** .15   **b.** Empirical   **11. a.** .2   **b.** .4
c. .08   **d.** .328

13.

Data	40	45	50	55	60	65	70	75	80	85	90	95
Probability	.07	0	0	.07	0	.13	.27	.07	.13	.07	.07	.13

15. Range $= 55$; variance $= 204.9$; standard deviation $= 14.3$ **17.** .6589   **19.** .1587   **21.** $y = -.16x + 41.6$
23. C   **25.** C   **27.** B

CHAPTER 8
PROBLEM SET 8.1, PAGES 284–286

1. a. No   **b.** Yes   **c.** No   **3. a.** No   **b.** Yes   **c.** No   **5. a.** Yes   **b.** No

7. $T^2 = \begin{bmatrix} \frac{3}{8} & \frac{5}{8} \\ \frac{5}{16} & \frac{11}{16} \end{bmatrix}$; $T^3 = \begin{bmatrix} \frac{11}{32} & \frac{21}{32} \\ \frac{21}{64} & \frac{43}{64} \end{bmatrix}$ **9.** $T^2 = \begin{bmatrix} .452 & .548 \\ .3151 & .6849 \end{bmatrix}$; $T^3 = \begin{bmatrix} .39724 & .60276 \\ .346587 & .653413 \end{bmatrix}$

11. $T^2 = \begin{bmatrix} .2075 & .605 & .1875 \\ .265 & .3675 & .3675 \\ .425 & .385 & .19 \end{bmatrix}$; $T^3 = \begin{bmatrix} .3524 & .4136 & .234 \\ .2603 & .4889 & .2509 \\ .2753 & .393 & .3318 \end{bmatrix}$ **13.** $\begin{bmatrix} 0 & 1 \\ .3 & .7 \\ .6 & .4 \end{bmatrix}$

15.

	Next Day	
	Increase	Decrease

Present $\begin{matrix} \text{Increase} \\ \text{Decrease} \end{matrix} \begin{bmatrix} .3 & .7 \\ .8 & .2 \end{bmatrix}$

17. a. No; the bottom row is not a probability vector
b. The probability that a person who rented a car from SFO will return it to SFO is .8   **c.** SJC

19. $T^2 = \begin{bmatrix} .69 & .17 & .15 \\ .33 & .53 & .15 \\ .46 & .29 & .41 \end{bmatrix}$ **21.** $T^2 = \begin{bmatrix} .22 & .52 & .26 \\ .13 & .70 & .17 \\ .13 & .34 & .53 \end{bmatrix}$ **23.** $\frac{3}{4}$ **25.** $\frac{39}{64}$ **27.** $\frac{9}{16}$ **29.** $\frac{7}{16}$ **31.** .99

33. .9890   **35.** .9891   **37.** .0109

PROBLEM SET 8.2, PAGES 290–292

1. Regular; approaches $\begin{bmatrix} \frac{1}{4} & \frac{3}{4} \\ \frac{1}{4} & \frac{3}{4} \end{bmatrix}$ **3.** Regular; approaches $\begin{bmatrix} \frac{2}{3} & \frac{1}{3} \\ \frac{2}{3} & \frac{1}{3} \end{bmatrix}$ **5.** Regular; approaches $\begin{bmatrix} \frac{4}{13} & \frac{9}{13} \\ \frac{4}{13} & \frac{9}{13} \end{bmatrix}$

7. Regular; approaches $\begin{bmatrix} \frac{5}{8} & \frac{3}{8} \\ \frac{5}{8} & \frac{3}{8} \end{bmatrix}$ **9.** Regular; approaches $\begin{bmatrix} \frac{2}{9} & \frac{4}{9} & \frac{1}{3} \\ \frac{2}{9} & \frac{4}{9} & \frac{1}{3} \\ \frac{2}{9} & \frac{4}{9} & \frac{1}{3} \end{bmatrix}$ **11.** Regular; approaches $\begin{bmatrix} \frac{1}{5} & \frac{1}{2} & \frac{3}{10} \\ \frac{1}{5} & \frac{1}{2} & \frac{3}{10} \\ \frac{1}{5} & \frac{1}{2} & \frac{3}{10} \end{bmatrix}$

13. a. No   **b.** Yes; $\begin{bmatrix} \frac{1}{2} & \frac{1}{2} \end{bmatrix}$ **c.** No   **15. a.** $0 \le a \le 1$ **b.** $\begin{bmatrix} \frac{1}{2} & \frac{1}{2} \end{bmatrix}$

17. $\begin{matrix} & C & P & D \\ C \\ P \\ D \end{matrix} \begin{bmatrix} 0 & 1 & 0 \\ \frac{2}{3} & 0 & \frac{1}{3} \\ \frac{2}{3} & \frac{1}{3} & 0 \end{bmatrix}$ approaches $\begin{matrix} & C & P & D \\ C \\ P \\ D \end{matrix} \begin{bmatrix} \frac{2}{5} & \frac{9}{20} & \frac{3}{20} \\ \frac{2}{5} & \frac{9}{20} & \frac{3}{20} \\ \frac{2}{5} & \frac{9}{20} & \frac{3}{20} \end{bmatrix}$

In the long run, she will buy canned food 40% of the time, packaged food 45% of the time, and dry food 15% of the time

19.
$$\begin{array}{c} \\ A \\ B \\ C \end{array}\begin{array}{ccc} A & B & C \\ \left[\begin{array}{ccc} .7 & .1 & .2 \\ .1 & .8 & .1 \\ .1 & .4 & .5 \end{array}\right] \end{array}$$

21. 25% ($\frac{1}{4}$) for Alpha, 54% ($\frac{13}{24}$) for Better Bean, 21% ($\frac{5}{24}$) for Carolyn's

23. 57% ($\frac{4}{7}$) Democrat, 12% ($\frac{6}{49}$) Republican, 31% ($\frac{15}{49}$) Independent   **25.** 22.5% Jersey, 27.5% Guernsey, 50% white-faced
27. The first company can expect 40% of the market and the second company can expect 60% of the market
29. $\frac{1}{3}$

PROBLEM SET 8.3, PAGES 296–297

1. Not absorbing   **3.** Not absorbing   **5.** Not absorbing   **7.** All three states are absorbing
9. States 1 and 2 are absorbing   **11.** State 3 is absorbing
13. $FP = [\frac{5}{4}][\frac{3}{10} \quad \frac{1}{2}] = [\frac{3}{8} \quad \frac{5}{8}]$; if the system begins in state 1, the probability that it will end up in absorbing state 2 is $\frac{3}{8}$, and the probability that it will end up in absorbing state 3 is $\frac{5}{8}$
15. $FP = [2][.15 \quad .35] = [.30 \quad .70]$; if the system begins in state 3, the probability that it will end up in absorbing state 1 is .30, and the probability that it will end up in absorbing state 2 is .70
17. $FP = \begin{bmatrix} 10 & 0 \\ \frac{20}{3} & \frac{10}{9} \end{bmatrix}\begin{bmatrix} \frac{1}{10} \\ \frac{3}{10} \end{bmatrix} = \begin{bmatrix} 1 \\ 1 \end{bmatrix}$; since there is only one absorbing state, the probability of ending in that state is 100% regardless of the beginning state

19. $FP = \begin{bmatrix} \frac{3}{2} & 0 \\ \frac{1}{2} & \frac{4}{3} \end{bmatrix}\begin{bmatrix} \frac{1}{3} & \frac{1}{3} \\ \frac{1}{4} & \frac{1}{4} \end{bmatrix} \approx \begin{bmatrix} \frac{1}{2} & \frac{1}{2} \\ \frac{1}{2} & \frac{1}{2} \end{bmatrix}$; if the system begins in state 2, the probability that it will end up in absorbing state 1 is $\frac{1}{2}$, and the probability that it will end up in absorbing state 4 is also $\frac{1}{2}$. If the system begins in state 3, the situation is identical.

21. $FP = \begin{bmatrix} \frac{20}{13} & \frac{6}{13} \\ \frac{10}{13} & \frac{16}{13} \end{bmatrix}\begin{bmatrix} .4 & .1 \\ .1 & .4 \end{bmatrix} = \begin{bmatrix} \frac{43}{65} & \frac{22}{65} \\ \frac{28}{65} & \frac{37}{65} \end{bmatrix}$; if the system is in state 2, there is a probability of $\frac{43}{65}$ that it will end up in absorbing state 1 and a probability of $\frac{22}{65}$ that it will end up in absorbing state 3. If the system begins in state 4, there are probabilities of $\frac{28}{65}$ and $\frac{37}{65}$ that it will end up in absorbing states 1 and 3, respectively

23. $FP = \begin{bmatrix} \frac{5}{3} & \frac{5}{3} & \frac{5}{6} \\ \frac{5}{9} & \frac{20}{9} & \frac{11}{18} \\ \frac{5}{9} & \frac{20}{9} & \frac{29}{18} \end{bmatrix}\begin{bmatrix} .1 & .1 \\ .2 & .2 \\ 0 & 0 \end{bmatrix} = \begin{array}{c} \\ 2 \\ 3 \\ 4 \end{array}\begin{array}{c} 1 \quad\; 5 \\ \begin{bmatrix} \frac{1}{2} & \frac{1}{2} \\ \frac{1}{2} & \frac{1}{2} \\ \frac{1}{2} & \frac{1}{2} \end{bmatrix}\end{array}$; if the system begins in state 2, 3, or 4, there is an equal chance of ending up in either absorbing state 1 or absorbing state 5

25. a.
$$\begin{array}{c} \\ 0 \\ 1 \\ 2 \end{array}\begin{array}{ccc} 0 & 1 & 2 \\ \left[\begin{array}{ccc} 1 & 0 & 0 \\ \frac{1}{2} & 0 & \frac{1}{2} \\ 0 & 0 & 1 \end{array}\right] \end{array}$$
    b. $FP = [1][\frac{1}{2} \quad \frac{1}{2}] = [\frac{1}{2} \quad \frac{1}{2}]$ **c.** 50%   **27.** $\frac{7}{8}$ **29.** $\frac{3}{8}$

PROBLEM SET 8.4, PAGE 298

1.

		Next		
		S	L	G
Present	S	.20	.75	.05
	L	.10	.60	.30
	G	0	.40	.60

3. $[\frac{5}{8} \quad \frac{3}{8}]$ **5. a.** 0 and 5 are absorbing states
b. The game does not favor either player   **c.** .2

CHAPTER 9
PROBLEM SET 9.1, PAGES 305–307

1. 2   **3.** 6   **5.** 2.05   **7.** 8¢   **9.** 2¢   **11.** $12   **13.** $318   **15.** $85   **17.** $-\$\frac{2}{38}$ or -5¢   **19.** -5¢
21. -5¢   **23.** $-\$\frac{3}{38}$ or -8¢   **25.** $-\$\frac{2}{38}$ or -5¢   **27.** No; the expectation is $-\$8,125$ **29.** $0
31. No; the expectation is -50¢   **33.** Two wells   **35.** The optimal number of cars is 12

PROBLEM SET 9.2, PAGES 314–316

1. Strictly determined; $[1 \quad 0]$ and $\begin{bmatrix} 0 \\ 1 \end{bmatrix}$; value is 0   3. $[\frac{1}{2} \quad \frac{1}{2}]$ and $\begin{bmatrix} \frac{1}{2} \\ \frac{1}{2} \end{bmatrix}$; value is 0

5. Strictly determined; $[1 \quad 0]$ and $\begin{bmatrix} 0 \\ 1 \end{bmatrix}$; value is -2 7. Strictly determined; $[1 \quad 0]$ and $\begin{bmatrix} 0 \\ 1 \end{bmatrix}$; value is -1

9. $[\frac{1}{3} \quad \frac{2}{3}]$ and $\begin{bmatrix} \frac{1}{6} \\ \frac{5}{6} \end{bmatrix}$; value is $-\frac{1}{6}$ 11. Strictly determined; $[1 \quad 0 \quad 0]$ and $\begin{bmatrix} 1 \\ 0 \\ 0 \end{bmatrix}$; value is 2

13. $[\frac{1}{6} \quad 0 \quad \frac{5}{6}]$ and $\begin{bmatrix} 0 \\ \frac{2}{3} \\ \frac{1}{3} \end{bmatrix}$; value is $-\frac{1}{3}$ 15. Strictly determined; $[0 \quad 1 \quad 0]$ and $\begin{bmatrix} 0 \\ 1 \\ 0 \end{bmatrix}$; value is -1

17. $[\frac{1}{4} \quad 0 \quad \frac{3}{4}]$ and $\begin{bmatrix} \frac{1}{4} \\ \frac{3}{4} \\ 0 \\ 0 \end{bmatrix}$; value is $-\frac{1}{4}$ 19. My Guess

| | Friend | | |
	10¢	25¢	50¢
10¢	10	-15	-40
25¢	-15	25	-25
50¢	-40	-25	50

; not strictly determined

21. Attacker

| | Defense | | |
	North	South	Both
North	0	1	1
South	1	0	1
Both	-1	-1	0

; not strictly determined   23. $[.2 \quad .8]$ and $\begin{bmatrix} .1 \\ .9 \end{bmatrix}$; value is 2.2

25. Holmes

| | Moriarty | |
	Canterbury	Dover
Canterbury	-100	0
Dover	50	-100

; $[.6 \quad .4]$ and $\begin{bmatrix} .4 \\ .6 \end{bmatrix}$ are the probability vectors for the strategies of Holmes and Moriarty, respectively. The value is -40

PROBLEM SET 9.3, PAGE 320

1. $[\frac{5}{6} \quad \frac{1}{6} \quad 0]$ and $\begin{bmatrix} \frac{1}{6} \\ \frac{5}{6} \end{bmatrix}$; value is $\frac{7}{6}$ 3. $[\frac{1}{3} \quad \frac{2}{3} \quad 0 \quad 0]$ and $\begin{bmatrix} \frac{2}{3} \\ \frac{1}{3} \end{bmatrix}$; value is $\frac{1}{3}$

5. $[\frac{1}{2} \quad \frac{1}{2}]$ and $\begin{bmatrix} \frac{5}{8} \\ 0 \\ \frac{3}{8} \end{bmatrix}$; value is $-\frac{1}{2}$ 7. $[\frac{2}{5} \quad 0 \quad \frac{3}{5}]$ and $\begin{bmatrix} 0 \\ \frac{2}{5} \\ \frac{3}{5} \end{bmatrix}$; value is $\frac{9}{5}$

9. Strictly determined; $[0 \quad 1 \quad 0]$ and $\begin{bmatrix} 0 \\ 0 \\ 1 \end{bmatrix}$; value is -1 11. $[0 \quad \frac{1}{6} \quad \frac{5}{6} \quad 0]$ and $\begin{bmatrix} 0 \\ \frac{1}{6} \\ 0 \\ \frac{5}{6} \end{bmatrix}$; value is $\frac{13}{6}$

13. Attacker

| | Defense | | |
	North	South	Both
North	0	1	1
South	1	0	1
Both	-1	-1	0

; $[\frac{1}{2} \quad \frac{1}{2} \quad 0]$ and $\begin{bmatrix} \frac{1}{2} \\ \frac{1}{2} \\ 0 \end{bmatrix}$; value is $\frac{1}{2}$

PROBLEM SET 9.4, PAGES 321–322

1. $[1,0]$ and $\begin{bmatrix} 1 \\ 0 \end{bmatrix}$, value is 3

3. Not strictly determined; $[.1 \quad .9]$ and $\begin{bmatrix} .8 \\ .2 \end{bmatrix}$, value is 2.2

5. Not strictly determined; $[\frac{3}{4} \quad 0 \quad \frac{1}{4}]$ and $\begin{bmatrix} \frac{1}{4} \\ \frac{3}{4} \\ 0 \end{bmatrix}$, value is $\frac{5}{4}$

7. 1   **9.** Optimal number is 24 cars   **11.** B

CHAPTER 10
PROBLEM SET 10.1, PAGES 332–333

1. 6, 16, 26, 36   **3.** 1, 4, 16, 64   **5.** 10, -50, 250, -1250 **7.** 4, 11, 25, 53   **9.** 1, 5, 13, 29   **11.** 3, 13, 63, 313
13. 32   **15.** 81   **17.** 10   **19.** 312   **21.** 143   **23.** $\frac{5}{8}$ **25.** $x_n = 5 + 25n$ **27.** $x_n = 4 - 2n$ **29.** $x_n = 4n$
31. $x_n = 3^n$ **33.** $x_n = 2^n$ **35.** $x_n = 3 - 3 \cdot 2^n$ **37.** $x_n = -\frac{9}{4} + \frac{17}{4} \cdot 5^n$ **39.** $x_n = -\frac{2}{3} + \frac{5}{3} \cdot 4^n$ **41.** 20
43. 49   **45.** 11   **47.** 80   **49.** $32,610.26   **51.** $x_8 \approx 188,470$ **53.** $x_{24} = 2^{24} = 16,777,216$

PROBLEM SET 10.2, PAGES 342–343

1. $400 simple interest; $469.33 compounded annually   **3.** $720 simple interest; $809.86 compounded annually
5. $12,000 simple interest; $43,231.47 compounded annually   **7.** $1,400 simple interest; $1,469.33 compounded annually
9. $2,720 simple interest; $2,809.86 compounded annually   **11.** $17,000 simple interest; $48,231.47 compounded annually
13. 9%; 5; 1.538624; $1,538.62; $538.62   **15.** 8%; 3; 1.259712; $629.86; $129.86   **17.** 2%; 12; 1.268242; $634.12; $134.12
19. 4.5%; 40; 5.816365; $29,081.83; $24,081.83   **21.** 5%; 40; 7.039989; $35,199.95; $30,199.95
23. 2%; 60; 3.281031; $13,124.12; $9,124.12   **25.** 4%; 5; 1.216653; $1,520.82; $270.82   **27.** $24,067.32
29. $19,898.24   **31.** $29,960.20   **33.** 6.17%   **35.** 12.75%   **37.** $16,889.10   **39.** $153,278.42   **41.** $5
43. $12   **45.** $2,691   **47.** $188   **49.** $168,187   **51.** $773,662   **53.** $3,655.96   **55.** $131,444.49
57. $2,218.89   **59.** Present value is $50,000 compared to $50,734.56; best to take $10,000 now and $45,000 in 1 year

PROBLEM SET 10.3, PAGES 346–347

1. $56,642   **3.** $59,498   **5.** $22,620   **7.** $61,173   **9.** $61,878   **11.** $23,299   **13.** $4,883   **15.** $220,498
17. $2,930   **19.** $72,433   **21.** $19,075   **23.** $413,755   **25.** $24,297   **27.** $1,374,583   **29.** Answers vary

PROBLEM SET 10.4, PAGES 351–352

1. $5,628.89   **3.** $5,655.87   **5.** $6,934.43   **7.** $4,313.07   **9.** $1,181.41   **11.** $5,250.32   **13.** $44.42
15. $272.19   **17.** $111.52   **19.** $133.78   **21.** $887.30   **23.** $4,802.66   **25.** $131,022   **27.** $1,780.28
29. The present value of the annuity is $25,906.15, so it is the most valuable

PROBLEM SET 10.5, PAGES 354–356

1. $1,193.20   **3.** $1,896.70   **5.** $4,485.45   **7.** $6,631.19   **9.** $2,076.83   **11.** $351.35   **13.** $5,426.39
15. $220,000 + $35,310 = $255,310   **17. a.** Present value   **b.** $56,255   **19. a.** Future value   **b.** $140,522
21. a. Present value   **b.** $131,444   **23. a.** Future value   **b.** $292   **25. a.** Present value   **b.** $165,135
27. a. Installment payments   **b.** $1,317   **29. a.** Ordinary annuity   **b.** $175,611

PROBLEM SET 10.6, PAGES 357–358

1. $1000(1.01)^{12(2.5)} = \$1,347.85$ **3.** $\dfrac{10,000}{(1.03)^{20}} = \$5,536.76$ **5.** $\dfrac{(10,000)(0.09)}{1.09^5 - 1} = \$1,670.92$

7. $10,000(1.0625)^5 = \$13,540.81$ **9.** $450\left(\dfrac{1.11^{26} - 1}{0.11} - 1\right) = \$57,149.45$ **11.** $\dfrac{1,000,000}{(1.12)^{37}} = \$15,098.48$

13. $\dfrac{5,000(0.115)}{(1.115)^{15} - 1} = \$139.62$ **15.** $1,000,000\left[\dfrac{1 - (1.14)^{-5}}{0.14}\right] = \$3,433,081$

17. $(150,000,000)(0.09)(1) + \dfrac{(150,000,000)(0.10)}{(1.10)^{50} - 1} = \$13,628,876$ **19.** C

APPENDIX A

1. a, b, and d are statements   **3.** a, b, and d are statements
5. $e \wedge d \wedge g$; where e = W. C. Fields is eating, d = W. C. Fields is drinking, g = W. C. Fields is having a good time
7. $\sim J \wedge \sim R$; where J = Jack will go tonight, R = Rosamond will go tomorrow
9. $(J \vee I) \wedge \sim P$; where J = The decision will depend on judgment, I = The decision will depend on intuition,
P = The decision will depend on who paid the most
11. $p \rightarrow \sim w$; p = We make a proper use of those means which the God of Nature has placed in our power, w = We are weak
13. $w \rightarrow n$; where w = It is work, n = It is noble
15. $(t \vee m) \rightarrow (\sim i \vee w)$; i = You must itemize deductions on Schedule A; w = You must complete the worksheet;
t = You have earned an income of \$2,300; m = You have earned an income of more than \$2,300

17.

p	q	$\sim p$	$\sim q$	$\sim p \wedge \sim q$
T	T	F	F	F
T	F	F	T	F
F	T	T	F	F
F	F	T	T	T

19.

r	s	$\sim r$	$\sim s$	$\sim r \vee \sim s$
T	T	F	F	F
T	F	F	T	T
F	T	T	F	T
F	F	T	T	T

21.

r	s	$r \wedge s$	$\sim s$	$(r \wedge s) \vee \sim s$
T	T	T	F	T
T	F	F	T	T
F	T	F	F	F
F	F	F	T	T

23.

p	q	$\sim p$	$\sim q$	$\sim p \vee \sim q$
T	T	F	F	F
T	F	F	T	T
F	T	T	F	T
F	F	T	T	T

25.

p	q	$\sim q$	$p \wedge \sim q$	$(p \wedge \sim q) \wedge p$
T	T	F	F	F
T	F	T	T	T
F	T	F	F	F
F	F	T	F	F

27.

p	q	$\sim q$	$p \vee q$	$p \wedge \sim q$	$(p \vee q) \vee (p \wedge \sim q)$
T	T	F	T	F	T
T	F	T	T	T	T
F	T	F	T	F	T
F	F	T	F	F	F

29.

p	q	$\sim p$	$\sim p \rightarrow q$	$p \rightarrow (\sim p \rightarrow q)$
T	T	F	T	T
T	F	F	T	T
F	T	T	T	T
F	F	T	F	T

31.

p	q	$\sim p$	$p \wedge q$	$\sim (p \wedge q)$	$\sim p \rightarrow \sim (p \wedge q)$
T	T	F	T	F	T
T	F	F	F	T	T
F	T	T	F	T	T
F	F	T	F	T	T

33.

p	q	$\sim q$	$p \rightarrow p$	$q \rightarrow \sim q$	$(p \rightarrow p) \rightarrow (q \rightarrow \sim q)$
T	T	F	T	F	F
T	F	T	T	T	T
F	T	F	T	F	F
F	F	T	T	T	T

35.

p	q	$\sim p$	$\sim q$	$p \to q$	$\sim q \to \sim p$	$(p \to q) \to (\sim q \to \sim p)$
T	T	F	F	T	T	T
T	F	F	T	F	F	T
F	T	T	F	T	T	T
F	F	T	T	T	T	T

37.

p	q	r	$\sim r$	$p \wedge q$	$(p \wedge q) \wedge \sim r$
T	T	T	F	T	F
T	T	F	T	T	T
T	F	T	F	F	F
T	F	F	T	F	F
F	T	T	F	F	F
F	T	F	T	F	F
F	F	T	F	F	F
F	F	F	T	F	F

39.

p	q	r	$\sim p$	$q \vee \sim p$	$p \wedge (q \vee \sim p)$	$[p \wedge (q \vee \sim p)] \vee r$
T	T	T	F	T	T	T
T	T	F	F	T	T	T
T	F	T	F	F	F	T
T	F	F	F	F	F	F
F	T	T	T	T	F	T
F	T	F	T	T	F	F
F	F	T	T	T	F	T
F	F	F	T	T	F	F

41. Direct   **43.** Indirect   **45.** Indirect   **47.** Indirect
49. If you can learn mathematics, then you understand human nature   **51.** All trebbles are expensive
53. If a nail is lost, then the kingdom is lost   **55.** $b \neq 0$ **57.** I will not eat that piece of pie
59. I will participate in student demonstrations   **61.** You obey the law   **63.** Babies cannot manage crocodiles
65. If they are my poultry, then they are not officers

APPENDIX B

Answers vary.

APPENDIX C

1. 4914   **3.** 18.96048   **5.** 5,517,840   **7.** 5937   **9.** 54,669.23077   **11.** 378,193,771   **13.** 0.77345
15. 17.61441762   **17.** 0.2644086835   **19.** 0.0078046506   **21.** 0.6764741715   **23.** 2.942851909   **25.** .454975922
27. 0.251485404   **29.** 2.259842248

Index